Q is for QUANTUM

An Encyclopedia of Particle Physics

JOHN GRIBBIN

edited by Mary Gribbin
illustrations by Jonathan Gribbin
timelines by Benjamin Gribbin

THE FREE PRESS

THE FREE PRESS
A Division of Simon & Schuster Inc.
1230 Avenue of the Americas
New York, NY 10020

Originally published in Great Britain in 1998 by Weidenfeld & Nicolson
Published by arrangement with Orion Publishing

THE FREE PRESS and colophon are trademarks
of Simon & Schuster Inc.

Manufactured in the United States of America

10 9 8 7 6 5 4 3 2 1

Library of Congress Cataloging-in-Publication Data

Gribbin, John R.
Q is for quantum : An Encyclopedia of particle physics /
John Gribbin ; edited by Mary Gribbin ; illustrations by
Jonathan Gribbin ; timelines by Benjamin Gribbin.
p.   cm.
Includes bibliographical references.
1. Particles (Nuclear physics)—Dictionaries.   2. Particles
(Nuclear physics)—Popular works.   I. Gribbin, Mary.   II. Gribbin, Jonathan.
III. Title.
QC793.2.G747   1999
539.7'03—dc21   98-9918
                                                                              CIP

ISBN 0-684-85578-X

Picture acknowledgements are given on page 453

CONTENTS

I think I can safely say that nobody understands quantum mechanics. … Do not keep saying to yourself, if you can possibly avoid it, 'But how can it be like that?' because you will go 'down the drain' into a blind alley from which nobody has yet escaped. Nobody knows how it can be like that.

Richard Feynman, The Character of Physical Law

INTRODUCTION

....

The quest for the quantum

This quick overview of a hundred years of scientific investigation of the microworld is intended to put the detail of the main section of this book in an historical perspective. All the technical terms are fully explained in the alphabetical section.

The quantum world is the world of the very small – the microworld. Although, as we shall see, quantum effects can be important for objects as large as molecules, the real quantum domain is in the subatomic world of particle physics. The first subatomic particle, the electron, was only identified, by J. J. Thomson, in 1897, exactly 100 years before this book, summing up our present understanding of the microworld, was completed. But it isn't just the neatness of this anniversary that makes this a good time to take stock of the quantum world; particle physicists have now developed an understanding of what things are made of, and how those things interact with one another, that is more complete and satisfying than at any time since Thomson's discovery changed the way people thought about the microworld. The standard model of particle physics, based upon the rules of quantum mechanics, tells us how the world is built up from the fundamental building blocks of quarks and leptons, held together by the exchange of particles called gluons and vector bosons.

But don't imagine that even the physicists believe that the standard model is the last word. After all, it doesn't include gravity. The structure of theoretical physics in the twentieth century was built on two great theories, the general theory of relativity (which describes gravity and the Universe at large) and quantum mechanics (which describes the microworld). Unifying those two great theories into one package, a theory of everything, is the Holy Grail that physicists seek as we enter the 21st century. Experiments that probe the accuracy of the standard model to greater and greater precision are being carried out using particle accelerators like those at CERN, in Geneva, and Fermilab, in Chicago. From time to time, hints that the standard theory is not the whole story emerge. This gives the opportunity for newspapers to run sensational headlines proclaiming that physics is in turmoil; in fact, these hints of something beyond the standard model are welcomed by the physicists, who are only too aware that their theory, beautiful though it is, is not the last word. Unfortunately, as yet none of those hints of what may lie beyond the standard model has stood up to further investigation. As of the spring of 1997, the standard model is still the best game in town.

But whatever lies beyond the standard model, it will still be based upon the rules of quantum physics. Just as the general theory of relativity includes the Newtonian version of gravity within itself as a special case, so that Newton's theory is still a useful and accurate description of how things work in many applications (such as calculating the trajectory of a space probe being sent to Jupiter), so any improved theory of the microworld must include the quantum theory within itself. Apples didn't start falling upwards when Albert Einstein came up with an improved theory of gravity; and no improved theory of physics will ever take away the weirdness of the quantum world.

By the standards of everyday common sense, the quantum world is very weird indeed. One of the key examples is the phenomenon of wave–particle duality. J. J. Thomson opened up the microworld to investigation when he found that the electron is a particle; three decades later, his son George proved that electrons are waves. Both of them were right (and they each won a Nobel Prize for their work). An electron is a particle, *and* it is a wave. Or rather, it is neither a particle nor a wave, but a quantum entity that will respond to one sort of experiment by behaving like a particle, and to another set of experiments by behaving like a wave. The same is true of light – it can behave either like a stream of particles (photons) or like a wave, depending on the circumstances. Indeed, it is, in principle, true of *everything*, although the duality does not show up with noticeable strength in the everyday world (which, of course, is why we do not regard the consequences of wave–particle duality as common sense).

All of this is related to the phenomenon of quantum uncertainty. A quantum entity, such as an electron or a photon, does not have a well-determined set of properties, in the way that a billiard ball rolling across the table has a precisely determined velocity and a precisely determined position at any instant. The photon and the electron (and other denizens of the microworld) do not know, and cannot know, both precisely where they are and precisely where they are going. It may seem an esoteric and bizarre idea, of no great practical consequence in the everyday world. But it is this quantum uncertainty that allows hydrogen nuclei to fuse together and generate heat inside the Sun, so without it we would not be here to wonder at such things (quantum uncertainty is also important in the process of radioactive decay, for substances such as uranium-235).

This highlights an important point about quantum physics. It is not just some exotic theory that academics in their ivory towers study as a kind of intellectual exercise, of no relevance to everyday life. You need quantum physics in order to calculate how to make an atom bomb, or a nuclear power station, that works properly – which is certainly relevant to the modern world. And you also need quantum physics in order to design much more domestic items of equipment, such as lasers. Not everybody immediately thinks of a laser as a piece of domestic equipment; but remember that a laser is at the heart of any CD player, reading the information stored on the disc itself; and the laser's close cousin, the maser, is used in amplifying faint signals, including those from communications satellites that feed TV into your home.

Where does the quantum physics come in? Because lasers operate on a principle called stimulated emission, a purely quantum process, whose statistical principles were first spelled out by Albert Einstein as long ago as 1916. If an atom has absorbed energy in some way, so that it is in what is called an excited state, it can be triggered into releasing a pulse of electromagnetic energy (a photon) at a precisely determined wavelength (a wavelength that is determined by the quantum rules) by giving it a suitable nudge. A suitable nudge happens when a photon with exactly the right wavelength (the same wavelength as the photon that the excited atom is primed to emit) passes by. So, in a process rather like the chain reaction of atomic fission that goes on in a nuclear bomb, if a whole array of atoms has been excited in the right way, a single photon passing through the array (perhaps in a ruby crystal) can trigger all of them to emit electromagnetic radiation (light) in a pulse in which all of the waves are marching precisely in step with one another. Because all of the waves go up together and go down together, this produces a powerful beam of very pure electromagnetic radiation (that is, a very pure colour).

Quantum physics is also important in the design and operation of anything which contains a semiconductor, including computer chips – not just the computer chips in your home computer, but the ones in your TV, hi-fi, washing machine and car. Semiconductors are materials with conducting properties that are intermediate between those of insulators (in which the electrons are tightly bound to their respective atomic nuclei) and conductors (in which some electrons are able to roam more or less freely through the material). In a semiconductor, some electrons are only just attached to their atoms, and can be made to hop from one atom to the next under the right circumstances. The way the hopping takes place, and the behaviour of electrons in general, depends on a certain set of quantum rules known as Fermi–Dirac statistics (the behaviour of photons, in lasers and elsewhere, depends on another set of quantum rules, Bose–Einstein statistics).

After semiconductors, it is logical to mention superconductors – materials in which electricity flows without any resistance at all. Superconductors are beginning to have practical applications (including in computing), and once again the reason why they conduct electricity the way they do is explained in terms of quantum physics – in this case, because under the right circumstances in some materials electrons stop obeying Fermi–Dirac statistics, and start obeying Bose–Einstein statistics, behaving like photons.

Electrons, of course, are found in the outer parts of atoms, and form the interface between different atoms in molecules. The behaviour of electrons in atoms and molecules is entirely described by quantum physics; and since the interactions between atoms and molecules are the raw material of chemistry, this means that chemistry is described by quantum physics. And not just the kind of schoolboy chemistry used to make impressive smells and explosive interactions. Life itself is based upon complex chemical interactions, most notably involving the archetypal molecule of life, DNA. At the very heart of the process of life lies the ability of a DNA molecule, the famous double-stranded helix, to 'unzip' itself and make two copies of the original double helix by building up a new partner for each strand of the original molecules, using each unzipped single molecule as a template. The links that are used in this process to hold the strands together most of the time, but allow them to unzip in this way when it is appropriate, are a kind of chemical bond, known as the hydrogen bond. In a hydrogen bond, a single proton (the nucleus of a hydrogen atom) is shared between two atoms (or between two molecules), forming a link between them. The way fundamental life processes operate can only be explained if allowance is made for quantum processes at work in hydrogen-bonded systems.

As well as the importance of quantum physics in providing an understanding of the chemistry of life, an understanding of quantum chemistry is an integral part of the recent successes that have been achieved in the field of genetic engineering. In order to make progress in taking genes apart, adding bits of new genetic material and putting them back together again, you have to understand how and why atoms join together in certain sequences but not in others, why certain chemical bonds have a certain strength, and why those bonds hold atoms and molecules a certain distance apart from one another. You might make some progress by trial and error, without understanding the quantum physics involved; but it would take an awful long time before you got anywhere (evolution, of course, does operate by a process of trial and error, and has got somewhere because it has been going on for an awful long time).

In fact, although there are other forces which operate deep within the atom (and which form the subject of much of this book), if you understand the behaviour of electrons and the behaviour of photons (light) then you understand everything that matters in the everyday world, except gravity and nuclear power stations. Apart from gravity, everything that is important in the home (including the electricity generated in nuclear power stations) can be described in terms of the way electrons interact with one another, which determines the way that atoms interact with one another, and the way they interact with electromagnetic radiation, including light.

We don't just mean that all of this can be described in general terms, in a qualitative, hand-waving fashion. It can be described quantitatively, to a staggering accuracy. The greatest triumph of theoretical quantum physics (indeed, of all physics) is the theory that describes light and matter in this way. It is called quantum electrodynamics (QED), and it was developed in its finished form in the 1940s, most notably by Richard Feynman. QED tells you about every possible interaction between light and matter (to a physicist, 'light' is used as shorthand for all electromagnetic radiation), and it does so to an accuracy of four parts in a hundred billion. It is the most accurate scientific theory ever developed, judged by the criterion of how closely the predictions of the theory agree with the results of experiments carried out in laboratories here on Earth.

Following the triumph of QED, it was used as the template for the construction of a similar theory of what goes on inside the protons and neutrons that make up the nuclei of atoms – a theory known as quantum chromodynamics, or QCD. Both QED and QCD are components of the standard model. J. J. Thomson could never have imagined what his discovery of the electron would lead to. But the first steps towards a complete theory of quantum physics, and the first hint of the existence of the entities known as quanta, appeared within three years of Thomson's discovery, in 1900. That first step towards quantum physics came, though, not from the investigation of electrons, but from the investigation of the other key component of QED, photons.

At the end of the 19th century, nobody thought of light in terms of photons. Many observations – including the famous double-slit experiment carried out by Thomas Young – had shown that light is a form of wave. The equations of electromagnetism, discovered by James Clerk Maxwell, also described light as a wave. But Max Planck discovered that certain features of the way in which light is emitted and absorbed could be explained only if the radiation was being parcelled out in lumps of certain sizes, called quanta. Planck's discovery was announced at a meeting of the Berlin Physical Society, in October 1900. But at that time nobody thought that what he had described implied that light only existed (or ever existed!) in the form of quanta; the assumption was that there was some property of atoms which meant that light could be emitted or absorbed only in lumps of a certain size, but that 'really' the light was a wave.

The first (and for a long time the only) person to take the idea of light quanta seriously was Einstein. But he was a junior patent office clerk at the time, with no formal academic connections, and hadn't yet even finished his PhD. In 1905 he published a paper in which he used the idea of quanta to explain another puzzling feature of the way light is absorbed, the photoelectric effect. In order to explain this phenomenon (the way electrons are knocked out of a metal surface by light), Einstein used the idea that light actually travels as a stream of little particles, what we would now call photons. The idea was anathema to most physicists, and even Einstein was cautious about promoting the idea – it was not until 1909 that he made the first reference in print to light as being made up of 'point-like quanta'. In spite of his caution, one physicist, Robert Millikan, was so annoyed by the suggestion that he spent the best part of ten years carrying out a series of superb experiments aimed at proving that Einstein's idea was wrong. He succeeded only in proving – as he graciously acknowledged – that Einstein had been right. It was after Millikan's experiments had established beyond doubt the reality of photons (which were not actually given that name until later) that Einstein received his Nobel Prize for this work (the 1921 prize, but actually awarded in 1922). Millikan received the Nobel Prize, partly for this work, in 1923.

While all this was going on, other physicists, led by Niels Bohr, had been making great strides by applying quantum ideas to an understanding of the structure of the atom. It was Bohr who came up with the image of an atom that is still basically the one we learn about when we first encounter the idea of atoms in school – a tiny central nucleus, around which electrons circle in a manner reminiscent of the way planets orbit around the Sun. Bohr's model, in the form in which it was developed by 1913, had one spectacular success: it could explain the way in which atoms produce bright and dark lines at precisely defined wavelengths in the rainbow spectrum of light. The difference in energy between any two electron orbits was precisely defined by the model, and an electron jumping from one orbit to the other would emit or absorb light at a very precise wavelength, corresponding to that energy difference. But Bohr's model introduced the bizarre idea that the electron did indeed 'jump', instantaneously, from one orbit to the other, without crossing the intervening space (this has become known as a 'quantum leap'). First it was in one orbit, then it was in the other, without ever crossing the gap.

Bohr's model of the atom also still used the idea of electrons as particles, like little billiard balls, and light as a wave. But by the time Einstein and Millikan received their Nobel Prizes, it was clear that there was more to light than this simple picture accounted for. As Einstein put it in 1924, 'there are therefore now two theories of light, both indispensable … without any logical connection'. The next big step, which led to the first full quantum theory, came when Louis de Broglie pointed out that there was also more to electrons than the simple picture encapsulated in the Bohr model accounted for.

De Broglie made the leap of imagination (obvious with hindsight, but a breakthrough at the time) of suggesting that if something that had traditionally been regarded as a wave (light) could also be treated as a particle, then maybe something that had traditionally been regarded as a particle (the electron) could also be treated as a wave. Of course, he did more than just speculate along these lines. He took the same kind of quantum calculations that had been pioneered by Planck and Einstein in their description of light and turned the equations around, plugging in the numbers appropriate for electrons. And he suggested that what actually 'travelled round' an electron orbit in an atom was not a little particle, but a standing wave, like the wave corresponding to a pure note on a plucked violin string.

De Broglie's idea was published in 1925. Although the idea of electrons behaving as waves was puzzling, this business of standing waves looked very attractive because it seemed to get rid of the strange quantum jumping. Now, it looked as if the transition of an electron from one energy level to another could be explained in terms of the vibration of the wave, changing from one harmonic (one note) to another. It was the way in which this idea seemed to restore a sense of normality to the quantum world that attracted Erwin Schrödinger, who worked out a complete mathematical description of the behaviour of electrons in atoms, based on the wave idea, by the end of 1926. He thought that his wave equation for the electron had done away with the need for what he called 'damned quantum jumping'. But he was wrong.

Also by 1926, using a completely different approach based entirely on the idea of electrons as particles, Werner Heisenberg and his colleagues had found another way to describe the behaviour of electrons in atoms, and elsewhere – another complete mathematical quantum theory. And as if that weren't enough, Paul Dirac had found yet another mathematical description of the quantum world. It soon turned out that all of these mathematical approaches were formally equivalent to one another, different views of the same quantum world (a bit like the choice between giving a date in Roman numerals or Arabic notation). It really didn't matter which set of equations you used, since they all described the same thing and gave the same answers. To Schrödinger's disgust, the 'damned quantum jumping' had not been eliminated after all; but, ironically, because most physicists are very familiar with how to manipulate wave equations, it was Schrödinger's variation on the theme, based on his equation for the wave function of an electron, that soon became the conventional way to do calculations in quantum mechanics.

This tradition was reinforced by the mounting evidence (including the experiments carried out by George Thomson in 1927) that electrons did indeed behave like waves (the ultimate proof of this came when electrons were persuaded to participate in a version of the double-slit experiment, and produced the classic diffraction effects seen with light under the equivalent circumstances). But none of this stopped electrons behaving like particles in all the experiments where they had always behaved like particles.

By the end of the 1920s, physicists had a choice of different mathematical descriptions of the microworld, all of which worked perfectly and gave the right answers (in terms of predicting the outcome of experiments), but all of which included bizarre features such as quantum jumping, wave–particle duality and uncertainty. Niels Bohr developed a way of picturing what was going on that was taught as the standard version of quantum physics for half a century (and is still taught in far too many places), but which if anything made the situation even more confusing. This 'Copenhagen interpretation' says that entities such as electrons do not exist when they are not being observed or measured in some way, but spread out as a cloud of probability, with a definite probability of being found in one place, another probability of being detected somewhere else, and so on. When you decide to measure the position of the electron, there is a 'collapse of the wave function', and it chooses (at random, in accordance with the rules of probability, the same rules that operate in a casino) one position to be in. But as soon as you stop looking at it, it dissolves into a new cloud of probability, described by a wave function spreading out from the site where you last saw it.

It was their disgust with this image of the world that led Einstein and Schrödinger, in particular, to fight a rearguard battle against the Copenhagen interpretation over the next twenty years, each of them independently (but with moral support from each other) attempting to prove its logical absurdity with the aid of thought experiments, notably the famous example of Schrödinger's hypothetical cat, a creature which, according to the strict rules of the Copenhagen interpretation, can be both dead and alive at the same time.

Although this debate (between Einstein and Schrödinger on one side, and Bohr on the other) was going on, most physicists ignored the weird philosophical implications of the Copenhagen interpretation, and just used the Schrödinger equation as a tool to do a job, working out how things like electrons behaved in the quantum world. Just as a car driver doesn't need to understand what goes on beneath the bonnet of the car in order to get from A to B, as long as quantum mechanics worked, you didn't have to understand it, even (as Linus Pauling showed) to get to grips with quantum chemistry.

The last thing most quantum physicists wanted was yet another mathematical description of the quantum world, and when Richard Feynman provided just that, in his PhD thesis in 1942, hardly anybody even noticed (most physicists at the time were, in any case, distracted by the Second World War). This has proved a great shame for subsequent generations of students, since Feynman's approach, using path integrals, is actually simpler conceptually than any of the other approaches, and certainly no more difficult to handle mathematically. It also has the great merit of dealing with classical physics (the old ideas of Newton) and quantum physics in one package; it is literally true that if physics were taught Feynman's way from the beginning, students would only ever have to learn the one approach to handle everything. As it is, although over the years the experts have come to accept that Feynman's approach is the best one to use in tackling real problems at the research level, the way almost all students get to path integrals is by learning classical physics first (in school), then quantum physics the hard way (usually in the form of Schrödinger's wave function, at undergraduate level) then, after completing at least one degree, being introduced to the simple way to do the job.

Don't just take our word for this being the simplest way to tackle physics – John Wheeler, Feynman's thesis supervisor, has said that the thesis marks the moment in the history of physics 'when quantum theory became simpler than classical theory'. Feynman's approach is not the standard way to teach quantum physics at undergraduate level (or classical physics in schools) for the same reason that the Betamax system is not the standard format for home video – because an inferior system got established in the market place first, and maintains its position as much through inertia as anything else.

Indeed, there is a deep flaw in the whole way in which science is taught, by recapitulating the work of the great scientists from Galileo to the present day, and it is no wonder that this approach bores the pants off kids in school. The right way to teach science is to start out with the exciting new ideas, things like quantum physics and black holes, building on the physical principles and not worrying too much too soon about the mathematical subtleties. Those children who don't want a career in science will at least go away with some idea of what the excitement is all about, and those who do want a career in science will be strongly motivated to learn the maths when it becomes necessary. We speak from experience – one of us (JG) got turned on to science in just this way, by reading books that were allegedly too advanced for him and went way beyond the school curriculum, but which gave a feel for the mystery and excitement of quantum physics and cosmology even where the equations were at that time unintelligible to him.

In Feynman's case, the path integral approach led him to quantum electrodynamics, and to the Feynman diagrams which have become an essential tool of all research in theoretical particle physics. But while these applications of quantum theory were providing the key to unlock an understanding of the microworld, even after the Second World War there were still a few theorists who worried about the fundamental philosophy of quantum mechanics, and what it was telling us about the nature of the Universe we live in.

For those who took the trouble to worry in this way, there was no getting away from the weirdness of the quantum world. Building from another thought experiment intended to prove the non-logical nature of quantum theory (the EPR experiment, dreamed up by Einstein and two of his colleagues), the work of David Bohm in the 1950s and John Bell in the 1960s led to the realization that it would actually be possible to carry out an experiment which would test the non-commonsensical aspects of quantum theory in a definitive manner.

What Einstein had correctly appreciated was that every version of quantum theory has built into it a breakdown of what is called 'local reality'. 'Local', in this sense, means that no communication of any kind travels faster than light. 'Reality' means that the world exists when you are not looking at it, and that electrons, for example, do not dissolve into clouds of probability, wave functions waiting to collapse, when you stop looking at them. Quantum physics (any and every formulation of quantum physics) says that you can't have both. It doesn't say which one you have to do without, but one of them you must do without. What became known as the Bell test provided a way to see whether local reality applies in the (for want of a better word) real world – specifically, in the microworld.

The appropriate experiments were carried out by several teams in the 1980s, most definitively by Alain Aspect and his colleagues in Paris, using photons. They found that the predictions of quantum theory are indeed borne out by experiment – the quantum world is not both local and real.

So today you have no choice of options, if you want to think of the world as being made up of real entities which exist all the time, even when you are not looking at them; there is no escape from the conclusion that the world is non-local, meaning that there are communications between quantum entities that operate not just faster than light, but actually instantaneously. Einstein called this 'spooky action at a distance'. The other option is to abandon both locality and reality, but most physicists prefer to cling on to one of the familiar features of the commonsense world, as long as that is allowed by the quantum rules.

Our own preference is for reality, even at the expense of locality; but that is just a personal preference, and you are quite free to choose the other option, the traditional Copenhagen interpretation involving both collapsing wave functions *and* spooky action at a distance, if that makes you happier. What you are *not* free to do, no matter how unhappy you are as a result, is to think that the microworld is both local and real.

The bottom line is that the microworld does not conform to the rules of common sense determined by our everyday experience. Why should it? We do not live in the microworld, and our everyday experience is severely limited to a middle range of scales (of both space and time) intermediate between the microworld and the cosmos. The important thing is not to worry about this. The greatest of all the quantum mechanics, Richard Feynman, gave a series of lectures at Cornell University on the theme *The Character of Physical Law* (published in book form by BBC Publications in 1965). In one of those lectures, he discussed the quantum mechanical view of nature, and in the introduction to that lecture he gave his audience a warning about the weirdness they were about to encounter. What he said then, more than 30 years ago, applies with equal force today:

> I think I can safely say that nobody understands quantum mechanics. So do not take the lecture too seriously, feeling that you really have to understand in terms of some model what I am going to describe, but just relax and enjoy it. I am going to tell you what nature behaves like. If you will simply admit that maybe she does behave like this, you will find her a delightful, entrancing thing. Do not keep saying to yourself, if you can possibly avoid it, 'But how can it be like that?' because you will go 'down the drain' into a blind alley from which nobody has yet escaped. Nobody knows how it can be like that.

That is the spirit in which we offer you our guide to the quantum world; take the advice of the master – relax and enjoy it. *Nobody knows how it can be like that.*

A few of the entries in the alphabetical section of this book overlap with entries in our earlier book, *Companion to the Cosmos*. In some cases, we have, in the spirit of quantum mechanics, tried to express the concepts in a complementary fashion on this occasion. But where we have found it impossible to improve on the form of words or analogy we used before, we have not made changes simply for the sake of making changes, although we have set the fragments of older material in their new context. See, for example, *arrow of time*.

Abelian group   A group of mathematical transformations which can be carried out in any order and still give the same end result. Simple multiplication is Abelian – 3 x 2 is the same as 2 x 3, or, in more general terms, *a* x *b* = *b* x *a*. This specific example, simple multiplication which forms an Abelian group, is said to follow a commutative law, or to commute. In the same way, in the everyday world addition is commutative, so that *a* + *b* = *b* + *a*.

But even in the everyday world, division does not obey the commutative law. *a*/*b* is not equal to *b*/*a*, and if, for example, you divide 4 by 2, the answer is 2, while if you divide 2 by 4, the answer is 0.5. Subtraction also does not commute; 4 – 2 is not the same as 2 – 4.

It is a key feature of the quantum world that many mathematical processes are non-Abelian, and in particular that multiplication is not necessarily commutative, which means that if *a* and *b* are quantum functions, not simple numbers, *a* x *b* may not be the same as *b* x *a*. Each component of a group (in this case, *a* and *b*) is called an element; if all the elements of the group commute, then the group is Abelian.

Named after the Norwegian mathematician Niels Henrik Abel (1802–1829).

absolute temperature   See *Kelvin scale*.

absolute zero   The lowest temperature that could ever be attained. At absolute zero, atoms and molecules would have the minimum amount of energy allowed by quantum theory. This is defined as 0 on the *Kelvin scale* of temperature; 0 K is –273.15 °C, and each unit on the Kelvin scale is the same size as one degree *Celsius*.

absorber theory   See *Wheeler–Feynman absorber theory*.

accelerator   A device which accelerates particles such as electrons and protons to very high speeds (close to the speed of light) using electric and magnetic fields. In modern accelerators, electrons can be accelerated to 99.999999986 per cent of the speed of light. The beams of fast-moving particles are then smashed into either stationary targets or beams of particles moving at similar speeds in the opposite direction. The way the particles bounce off the targets ('scatter') can be used to reveal details of the internal structure of the particles which make up the targets, rather as if those particles were being X-rayed. Also, when the fast-moving particles are brought to a halt, or dramatically slowed, in collisions, their energy of motion (kinetic energy) is converted into showers of new particles, in line with Albert Einstein's famous equation $E = mc^2$ (in this case, the important point being that $m = E/c^2$).

The showers of particles produced in this way can be studied to test theories of how the quantum world works. It is important to appreciate that in most cases there is no sense in which the particles in the showers were originally 'inside' the colliding particles and have been broken off, or knocked out, by the collisions. They have been created out of pure energy, and did not exist before the collision occurred. Indeed, the total mass of the particles in the shower can considerably exceed the rest mass of the particles involved in the collision.

In the most extreme experiments, beams of electrons collide with beams of positrons (antielectrons). Similarly, protons can be collided with antiprotons. When

Accelerator. Part of the particle accelerator ring at CERN.

a particle meets its antiparticle counterpart, as well as the kinetic energy from each particle the mass-energy is available to make new particles, as the particle/antiparticle pair annihilate one another entirely.

aces See *quarks*.

action A mathematical quantity which depends upon the mass, velocity and distance travelled by a particle. Action is also associated with the way energy is carried from one place to another by a wave, but it can be understood most simply by imagining the trajectory of a ball tossed in a high arc from one person to another.

One of the most fundamental laws of science is the law of conservation of energy. Energy cannot be created or destroyed, only converted from one form to another. The ball leaves the thrower's hand with a large kinetic energy, but as it climbs higher its speed slows down and the kinetic energy is reduced. But because the ball is higher above the ground (strictly speaking, because it is further from the centre of the Earth), it has gained gravitational potential energy. Leaving aside friction (which converts some of the energy of motion of the ball into heat energy as it passes through the air), the amount of gravitational energy it gains matches the amount of kinetic energy it has lost, for each point in its climb. At the top of its trajectory, the ball momentarily stops moving, so it has zero kinetic energy, but maximum gravitational energy for this particular trajectory. Then, as it falls towards the catcher, it gains kinetic energy at the expense of gravitational potential energy.

At any point along the trajectory, it is possible to calculate the kinetic energy and the potential energy of the ball. The total you get by adding the two is always the same. But if you subtract the potential energy from the kinetic energy, you get a

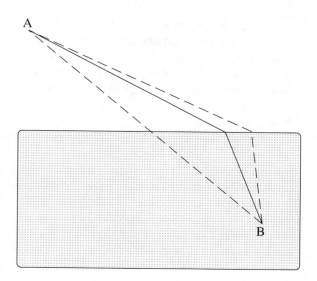

Action. The way light travels, to reach its destination by the quickest path possible, is an example of the principle of least action at work. The path which minimizes the time to go from A (in air) to B (in glass) is shown by the solid line; any other path takes longer.

different value of the difference at different points along the trajectory. If you add up this difference all along the trajectory, integrating the difference between the kinetic energy and the potential energy for the entire flight of the ball, the number you come up with is the action that corresponds to the flight of the ball. The action is a property not of a single point along the trajectory, but of the entire trajectory.

There is a value of the action for each possible trajectory of the ball. In a similar way, there is a value of the action corresponding to each trajectory that might be taken by, say, an electron moving in a magnetic field. The way we have described it here, you would calculate the action using Newton's laws of motion to describe the flight of the ball; but the process can be turned on its head, with the properties of the action used to determine the laws of motion. This works both for classical mechanics and for quantum mechanics, making the action one of the most important concepts in all of physics.

This is because objects following trajectories always follow the path of least action, in a way analogous to the way water runs downhill to the point of lowest energy available to it. There are many different curves that the ball could follow to get to the same end point, ranging from low, flat trajectories to highly curved flight paths in which it goes far above the destination before dropping on to it. Each curve is a parabola, one of the family of trajectories possible for a ball moving under the influence of the Earth's gravity. But if you know how long the flight of the ball takes, from the moment it leaves the thrower's hand to the moment it reaches its destination, that rules out all but one of the trajectories, specifying a unique path for the ball.

Given the time taken for the journey, the trajectory followed by the ball is always the one for which the *difference*, kinetic energy *minus* potential energy, added up all along the trajectory, is the *least*. This is the principle of least action, a property involving the whole path of the object.

Looking at the curved line on a blackboard representing the flight of the ball, you might think, for example, that you could make it take the same time for the journey by throwing it slightly more slowly, in a flatter arc, more nearly a straight line; or by throwing it faster along a longer trajectory, looping higher above the ground. But nature doesn't work that way. There is only one possible path between two points for a given amount of time taken for the flight. Nature 'chooses' the path with the least action – and this applies not just to the flight of a ball, but to any kind of trajectory, at any scale.

It's worth giving another example of the principle at work, this time in the guise of the principle of 'least time', because it is so important to science in general and to quantum physics in particular. This variation on the theme involves light. It happens that light travels slightly faster through air than it does through glass. Either in air or glass, light travels in straight lines – an example of the principle of least time because, since a straight line is the shortest distance between two points, that is the quickest way to get from A to B. But what if the journey from A to B starts out in air, and ends up inside a glass block? If the light still travelled in a single straight line, it would spend a relatively small amount of time moving swiftly through air, then a relatively long time moving slowly through glass. It turns out that there is a unique path which enables the light to take the least time on its journey, which involves travelling in a certain straight line up to the edge of the glass, then turning and travelling in a different straight line to its destination. The light seems to 'know' where it is going, apply the principle of least action, and 'choose' the optimum path for its journey.

In some ways, this is reminiscent of the way a quantum entity seems to 'know' about both holes in the *double-slit experiment* even though common sense says that it only goes through one hole; but remember that the principle of least action applies in the everyday world as well as in the quantum world. Richard Feynman used this to develop a version of mechanics, based on the principle of least action, which describes both classical and quantum mechanics in one package.

See also *sum over histories*.

WARNING! Unfortunately, physicists also use the word 'action' in a quite different way, as shorthand for the term 'interaction'. See *action at a distance*. This has nothing to do with the action described here.

action at a distance   The idea that interactions between objects, such as the gravitational interaction that holds the Earth in orbit around the Sun, operate without any intervening mechanism. The original version of the idea saw the interaction occurring instantaneously, regardless of the distance involved. Modern variations on the theme involve interactions that occur at a distance, but with a time delay related to the speed of light. See *transactional interpretation*, *Wheeler–Feynman absorber theory*.

Note that this use of the term 'action' as shorthand for 'interaction' has nothing to do with *action*.

adiabatic process   A process which occurs without heat entering or leaving a system. This usually means that the temperature of the system changes. For example, if a gas expands adiabatically, pushing a piston out of a cylinder, the gas cools because it has to do work to make the piston move. Contrast this with an *isothermal process*.

In particle physics, the term 'adiabatic' is used to describe interactions in which there is no input of energy.

ADONE   One of the first electron–positron colliders, built at Frascati, near Rome.

advanced wave   A wave that travels backwards in time, from the future, to arrive at its 'source'. See *transactional interpretation*, *Wheeler–Feynman absorber theory*.

Alhazen (Abu Ali al-Hassan ibn al-Haytham) (about 965–1038)   The greatest scientist of the Middle Ages, whose achievements were unsurpassed for more than 500 years, until the work of Galileo, Kepler and Newton. Usually referred to by the Europeanized version of his name.

Alhazen was born in Basra, now part of Iraq, in about 965. He later moved to Cairo, where he worked in the service of the Caliph al-Hakim, a reputedly mad tyrant. Having boasted that he could find a way to control the flooding of the Nile, Alhazen was sent south by the Caliph to make good his promise. When the expedition failed, in order to escape execution Alhazen himself had to pretend to be mad for several years, until the Caliph died in 1021. He then resumed his normal life.

Alhazen's greatest scientific contribution was his work on optics, contained in a series of seven books (what we would now call scientific papers) written on either side of the year 1000. This work was translated into Latin, the scientific language of the day, at the end of the 12th century, and influenced the thinking of, among others, Roger Bacon. It was published in Basle in 1572, more than 500 years after Alhazen's death, under the title *Opticae thesaurus* (*Treasury of Optics*).

The key insights in Alhazen's work included a logical argument that sight does not (as earlier thinkers had taught) work by the eye sending out rays to probe the world outside, but is entirely due to light, produced in a flame or by the Sun, bouncing off objects and entering the eyes from outside. His key analogy was with the way images are formed in a 'camera obscura' – a darkened room in which curtains are placed over the windows, with a pinhole in one curtain. Light from outside, passing through the pinhole, makes an image on the opposite wall of the world outside. This is indeed the way the eye (and a photographic camera) works. Alhazen realized that this means that light travels in straight lines, and thought of light as being made up of a stream of tiny particles that bounce off objects that they strike. This was the earliest introduction of the concept of what are now known as photons.

Alhazen measured both the reflection and the refraction of light, and tried to explain the occurrence of rainbows. He studied the Sun during an eclipse by using the camera obscura technique to cast its image on a wall, and he wrote scores of other 'books' on mathematics and scientific topics. He was the first scientific thinker to surpass the work of the ancient Greeks. He died in Cairo in 1038.

Alice matter   See *shadow matter*.

alpha decay   A process of radioactive decay in which the nucleus of an atom ejects an alpha particle. See *tunnel effect*.

alpha particle   The nucleus of an atom of helium-4, made up of two protons and two neutrons held tightly together by the strong nuclear force. This nucleus is unusually stable, and is held together so tightly that alpha particles, produced in alpha decay, do indeed behave in many ways like single particles. Such particles, each carrying two units of positive charge, were the first probes used to investigate the structure of atoms, by Ernest Rutherford and his colleagues, at the beginning of the 20th century.

alpha radiation   A stream of alpha particles, produced by alpha decay. The particles move at speeds of about 1,600 km per second, but can be stopped by a sheet of paper.

Alvarez, Luis Walter (1911–1988)   American physicist who received the Nobel Prize in 1968 for his work in high-energy particle physics, including the development of the hydrogen bubble chamber technique.

Alvarez was born on 13 June 1911 in San Francisco. His father was a physician, who later became a medical journalist. Alvarez studied at the University of Chicago, switching from chemistry to physics after starting his degree. He graduated in 1932, and stayed there to complete an MSc in 1934 and his PhD in 1936. He then moved to the Berkeley campus of the University of California, where he worked at the Lawrence Radiation Laboratory for the rest of his career, apart from wartime work on radar and on the Manhattan Project.

Back at Berkeley, Alvarez became a full professor in 1945, and played a key role in the development of the first practical linear accelerator, which (in 1947) could accelerate protons to energies of 32 MeV. In 1953, after a meeting with Donald Glaser (the inventor of the bubble chamber technique), Alvarez concentrated on developing a series of hydrogen bubble chambers, culminating in a large device, 72 inches in diameter, used to investigate the properties of particles at high energies. The data from these high-energy investigations provided key input for the theorists who developed the *eightfold way* concept and the idea of *quarks*.

Later, Alvarez became interested in less conventional ideas. He searched for hidden chambers in Egyptian pyramids using studies of cosmic rays to 'X-ray' the pyramids, and he gained worldwide fame in the 1980s for the suggestion, based on work carried out with his son Walter, that the 'death of the dinosaurs' was caused by the impact of a large meteorite with the Earth. He died in San Francisco on 1 September 1988.

amplitude   For a wave, the amplitude is half the height from the peak of the wave to the trough. In quantum mechanics, the amplitude of a process is a number that is related to the probability of the process occurring. If there are several alternative processes (for example, if there are several different ways an electron can get from A to B, then each has its own amplitude). The probability of the electron 'choosing' a particular route (in this example) is equal to the square of the amplitude. But the probability that the electron will choose any of the routes (the probability that it goes from A to B by any means, rather than going off to C or D instead) is given not by adding up the probabilities, but by adding up the amplitudes and squaring the total amplitude (see *sum over histories*).

The numbers which measure quantum mechanical amplitudes are so-called complex numbers, which involve the square root of –1. This affects the way in which the squares (the probabilities) are calculated (see *complex conjugation*), and has important implications for the *transactional interpretation* of quantum mechanics.

See *wave*.

a.m.u.   See *atomic mass unit*.

Anderson, Carl David (1905–1991)   American physicist who received the Nobel Prize in 1936 for the discovery of the positron – the first proof of the existence of antimatter.

Anderson, the son of Swedish immigrants, was born in New York on 3 September 1905. He was brought up in Los Angeles, and studied at the California Institute of Technology, where he graduated in 1927. He stayed on as a graduate student, working with Robert Millikan, and completed his PhD in 1930. Still at Caltech, where he stayed for the rest of his career, at Millikan's suggestion Anderson built a cloud chamber in order to study the tracks of electrons produced by cosmic rays. The detector soon (in 1932) revealed tracks produced by particles with the same mass as electrons but the opposite electric charge – positrons.

In the same year that he received the Nobel Prize, Anderson and his student Seth Neddermeyer announced the discovery of cosmic ray particles with mass larger than the mass of the electron but smaller than the mass of the proton. These particles became known as muons.

After wartime work on rockets, Anderson resumed his cosmic ray studies. He retired in 1976, and died on 11 January 1991.

Anderson, Philip Warren (1923–) American solid-state physicist, born in Indianapolis on 13 December 1923, who was awarded a share of the Nobel Prize in 1977 for his work on the behaviour of electrons in disordered (non-crystalline) solids. His work on semiconductors paved the way for the devices now used in computer memories.

Ångström, Anders Jonas (1814–1874) Swedish physicist who was one of the founders of the study of *spectroscopy*.

Born at Lödgö on 13 August 1814, Ångström was the son of a country chaplain. He studied at Uppsala, where he received his doctorate in 1839, and stayed to become first a lecturer and then (in 1858) professor of physics, a post he held until his death. Ångström's studies of spectra led him to conclude that gases emit and absorb light at characteristic wavelengths, and in 1861, building on his own work and that of Gustav Kirchoff, Ångström began a study of the solar spectrum which showed that hydrogen is present in the Sun. Because he worked quietly and did not seek publicity (and because he published his results in Swedish), the value of Ångström's work was only slowly recognized, but in 1870 he was elected a Fellow of the Royal Society, and after his death (at Uppsala on 21 June 1874) his name was given to a unit of wavelength used in spectroscopy.

ångström An obsolete unit of wavelength defined as one hundred millionth (10^{-8}) of a centimetre. Equal to one-tenth of a nanometre. Symbol Å.

angular momentum A property of rotating objects analogous to the *momentum* of an object moving in a straight line. The angular momentum of a spinning object depends on its mass, its size and the speed with which it is spinning. An object in orbit around another object (such as the Moon orbiting around the Earth) also has angular momentum, which depends on the mass of the object, the radius of its orbit and the speed with which it is moving. And the concept can be extended to any object moving in a curved trajectory.

In the quantum world, angular momentum is quantized, and can change only by amounts which are integer multiples of Planck's constant $\hbar$. One important implication of this is that if electrons are thought of as being in orbit around the central nucleus of an atom, they can only occupy distinct stable orbits which are separated by a whole number of angular momentum quanta. Similarly, the spin of an electron is quantized, and the electron can only exist with 'spin up' ($+\frac{1}{2}$) or 'spin down' ($-\frac{1}{2}$) measured against some external reference such as a magnetic field, not in an intermediate state.

annihilation   See *antimatter, pair production*.

anode   A positively charged electrode in, for example, a vacuum tube (electronic valve) or a battery (electric cell). Because electrons are negatively charged, they move towards an anode.

anomalous Zeeman effect   See *Zeeman effect*.

anthropic principle   The idea that the existence of life in the Universe (specifically, human life) can set constraints on the way the Universe is now, and how it got to be the way it is now.

The power of anthropic reasoning is best seen by an example. In order for us to exist, there has to be one star (the Sun), orbited at the appropriate distance by one planet (the Earth), made of the right mixture of chemical elements (particularly including carbon, nitrogen, oxygen and the primordial hydrogen left over from the Big Bang). At first sight, it may seem that the existence of the rest of the Universe, containing millions of galaxies scattered across billions of light years of space, is irrelevant to our existence.

But where did the elements of which we and the Earth are made come from? The Big Bang produced only hydrogen, helium and traces of a few light elements. Carbon and other heavy elements were manufactured inside stars (see *nucleosynthesis*) which had to run through their life cycles and explode, scattering the heavy elements across space to form clouds of material from which later generations of stars, including the Sun, and their attendant planets could form. This took billions of years. The evolution of life on a suitable planet to the point where intelligent beings could notice their surroundings and wonder about the size of the Universe took more billions of years. All the while, the Universe was expanding. After billions of years, it is inevitably billions of light years across. So the fact that we are here to ask questions about the size of the Universe means that the Universe must contain many stars, and must be billions of years old and billions of light years across.

This rests upon the assumption that there is nothing special about our place in the Universe, and nothing special about us – a proposition sometimes referred to as 'the principle of terrestrial mediocrity'.

Further reading: John Barrow and Frank Tipler, *The Anthropic Cosmological Principle*; John Gribbin, *Companion to the Cosmos*.

antimatter   A form of matter in which each particle has the opposite set of quantum properties (such as electric charge) to its counterpart in the everyday world. The classic example of a particle of antimatter (an antiparticle) is the antielectron, or positron, which has the same mass as an electron but a positive charge instead of a negative charge. The existence of antielectrons was predicted by Paul Dirac, at the end of the 1920s, when he found that the equation which represents a complete description of the electron in terms of both quantum mechanics and the special theory of relativity has two sets of solutions, one corresponding to negatively charged particles and one to positively charged particles. The exact meaning of this was not clear until 1932, when Carl Anderson discovered positrons from the traces they left in his cosmic ray detector.

Our visible Universe is almost entirely composed of matter, and very little antimatter has existed since the Big Bang in which the Universe was born. When an antiparticle meets its particle counterpart (for example, when a positron meets an electron), they annihilate, converting all of their rest mass into energy in line with Einstein's equation $E = mc^2$. Antiparticles can be made out of energy in the reverse of this process, but only if a particle counterpart for every antiparticle is produced as well. This happens naturally in high-energy processes involving cosmic rays, and also in high-energy experiments in accelerators on Earth. Because the world is over-whelmingly made of matter, however, any antiparticle produced in this way soon meets up with a particle counterpart and annihilates.

It is now clear that there are antimatter counterparts for all the matter particles – antiprotons, antineutrons, antineutrinos and so on. All of these are produced routinely in accelerator experiments. In September 1995, a team of researchers at CERN succeeded for the first time in making complete antiatoms (of hydrogen, the simplest element) in which a negatively charged antiproton is associated with a positively charged antielectron. Just nine atoms of antihydrogen were manufactured in this first experiment, and they each survived for only about 40 billionths of a second before colliding with ordinary matter and annihilating. The researchers hope that they will soon be able to trap antiatoms for long enough to study them (for example, by probing them with laser beams) and find out if, as present theories predict, the laws of physics work in exactly the same way for antimatter as they do for matter.

Further reading: Yuval Ne'eman and Yoram Kirsch, *The Particle Hunters*.

antiparallel vectors   Arrows that point in opposite directions. For example, the spin of an electron, measured against a background magnetic field, can point in only one of two directions, up or down. If two electrons have opposite spins, their spins can be said to be antiparallel. This allows pairs of electrons to share otherwise identical quantum 'orbits' in an atom.

antiparticles   See *antimatter*.

Arago, (Dominique) François Jean (1786–1853)   French physicist who made pioneering investigations into the nature of light, was instrumental in establishing the wave theory developed by Augustin Fresnel as a good description of how light travels, and paved the way for Leon Foucault's determination of the speed of light in the laboratory.

Arago was born in Estagel, near Perpignan, on 26 February 1786. He studied at the Ecole Polytechnique in Paris, then worked for the Bureau de Longitude, and was a member of an expedition to Spain in 1806 to measure an arc of the meridian. He returned to Paris in 1809 and became a professor at the Ecole Polytechnique. Arago carried out experiments in many branches of physics, including electricity and magnetism, made astronomical observations, and was chairman of the committee of the French Academy of Sciences which tested Fresnel's wave theory in 1817. In 1838 Arago proposed a practicable experiment to test the wave theory further by measuring the speed of light in air and in water. Partly because of the turmoil of the 1848 revolution (as a result of which he became a government minister and abolished slavery in the French colonies), he never completed the experiment before his eyesight failed, and he handed over his apparatus to Foucault.

He died in Paris on 2 October 1853, three years after Foucault (and, working independently, Armand Fizeau) had completed the air/water comparison and confirmed that light travels as a wave.

Aristotle (384–322 BC) Greek philosopher and scientist. The son of the court physician to the King of Macedon (the father of Philip II and grandfather of Alexander the Great), Aristotle was born at Stagira in 384 BC. He studied under Plato and became one of the greatest philosophers of his day; in 342 BC he was appointed by Philip II to be the tutor of Alexander. Among an enormous body of work in mathematics, astronomy, zoology and other sciences, Aristotle established the idea that everything in the material world is composed of four 'elements' – fire, earth, air and water. He died at Chalcis in 322 BC.

Arrhenius, Svante August (1859–1927) Swedish chemist who was the first person to explain that when a chemical compound dissolves in water it dissociates into electrically charged ions. Born at Uppsala on 19 February 1859, Arrhenius studied at the local university, graduating in 1878 and then beginning studies for a PhD before moving to Stockholm, where he submitted his thesis in 1884. That doctoral thesis contained the essence of his work on solutions, for which he received the Nobel Prize in 1903. After submitting his thesis, Arrhenius spent five years working at different universities around Europe, supported by a scholarship from the Swedish Academy of Sciences; he spent the rest of his career in Sweden, becoming director of the Nobel Institute of Physical Chemistry, in Stockholm, in 1905 and holding the post until shortly before he died. From the security of this post he broadened his research activities to cover what is now known as the greenhouse effect, and the idea that spores of life might cross space from one planet to another. He died on 2 October 1927.

arrow of time One of the greatest mysteries in science is the distinction between the past and the future. At a subatomic level, neither the old ideas of classical mechanics nor the modern theory of quantum mechanics distinguishes between the past and the future. In a typical interaction involving subatomic particles, two particles may come together and interact in some way to produce two different particles, which then separate. The laws of physics say that almost every such interaction can run equally well in reverse, with the 'final' two particles coming together and interacting to make the 'original' two particles. At this level, there is no way to distinguish the past from the future simply by looking at each pair of particles.

But at the macroscopic level of our human senses, the distinction between the past and the future is obvious. Things wear out; people get older. In the equivalent of the particle interaction, we can imagine a wine glass balanced precariously on the edge of a table, then falling to the floor and smashing. We never see smashed glasses reassembling themselves, even though each interaction involving the atoms of the wine glass as it smashes is, according to the known laws of physics, reversible. If we were shown two still photographs, one of the glass on the table and the other of the smashed glass on the floor, we would have no difficulty saying which one was taken first and which one was taken later in time. There is an inbuilt arrow of time, pointing from the past to the future, when we are dealing with complex systems which contain many particles.

It is, however, important to distinguish between an arrow which *points* into the future and one which *moves* into the future. The correct analogy is with the needle of a compass, which points to the north, but does not have to be moving north (or anywhere else) at all. If we had a movie of the glass falling off the table, instead of just

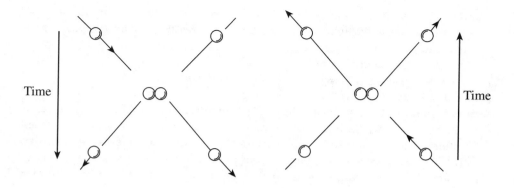

Arrow of time. At the level of colliding particles, the laws of physics make no distinction between the future and the past – the arrow of time can point either way and the laws are still the same.

two 'before and after' pictures, and if the individual frames of the film were cut up and mixed together, we would still be able to sort them out into the right order. The film does not have to be running through a projector for the distinction between the past and the future to be clear. There is still a distinction between the past and the future, which can be represented by an arrow pointing from the past into the future.

This distinction can be expressed mathematically. The science of thermodynamics is based on analysis of the way things change as we 'move' from the past into the future. The key insight is that the amount of disorder in the Universe is always increasing – glasses break, but do not re-assemble themselves. Physicists measure disorder in terms of a quantity called entropy; the most fundamental law of physics is that the entropy of a closed system always increases (the second law of thermodynamics).

You can seem to get round this law in an open system, which has an external source of energy. The second law seems to be violated on Earth, because living things grow and people can, for example, take a pile of bricks and turn it into a much more ordered structure, in the form of a house. But all of this depends on an input of energy, originally from the Sun. The decrease in entropy taking place on Earth is much less than the increase in entropy associated with the nuclear fusion reactions going on inside the Sun and the way it radiates heat out into space. The entropy of the whole Universe increases as time passes.

In the quantum world, however, the distinction between past and future is far from clearcut. Interactions described by Feynman diagrams work just as well 'backwards' in time as they do forwards in time. The only sense in which there is an arrow of time is that some states are more complex than others, and this complexity – a result of the outcomes of many different quantum interactions – can be regarded as lying in the future, while simplicity lies in the past.

Time does come into both Maxwell's equations (which describe the behaviour of electromagnetic radiation) and Schrödinger's equation (which describes the behaviour of quantum systems). But it does so in a completely symmetric way, so that there is no distinction between waves travelling from the past into the future and waves travelling from the future into the past. This may have important implications for understanding the nature of light (see *Wheeler–Feynman absorber theory*) and quantum phenomena (see *transactional interpretation*).

See also *CPT conservation*.

Further reading: John Barrow, *The World within the World*.

artificial radioactivity   See *induced radioactivity*.

Aspect experiment   Actually a series of experiments, carried out in Paris in the early 1980s by Alain Aspect and his colleagues, which established that what Albert Einstein called 'spooky action at a distance' really does operate in the quantum world. This was direct, experimental proof that the quantum world does not obey the same laws of common sense that apply in the everyday world of human experience.

The seed of what became the Aspect experiment was sown in 1935, when Einstein and two of his colleagues published a paper drawing attention to one of the seemingly paradoxical features of quantum mechanics (see *EPR experiment*). At the time, this was presented as a 'thought experiment', intended to demonstrate the absurdity of quantum physics by providing a logical contradiction; it was not intended that any such experiment should be carried out. But in 1952 David Bohm suggested a variation on the EPR theme (still as a thought experiment) that involved the behaviour of photons, and in 1964 John Bell showed how Bohm's variation on the EPR theme might, in principle, form the basis of a real experiment.

At the time, not even Bell thought that such an experiment was a practical possibility. But the experimenters almost immediately took up the challenge. Within twenty years, several groups had come close to making the appropriate measurements with the required precision; it is generally accepted that it was the results from Aspect's team, published in 1982, that finally established that Einstein (and common sense) was wrong, and that non-locality rules the quantum world. It is Bell's version of Bohm's variation on the EPR theme, as tested by Aspect and his colleagues, that we describe here.

The quantum property that is measured in these experiments is the polarization of a photon, which can be thought of as an arrow pointing either up or down. It is possible to stimulate an atom in such a way that it produces two photons simultaneously, heading off in different directions. Overall, the polarizations of the two photons must cancel – if one arrow is 'up' then the other arrow is 'down'. According to common sense, each photon starts out with a definite polarization, its partner with the opposite polarization, and they retain those properties as they fly through space. But according to the standard interpretation of quantum theory (see *Copenhagen interpretation*), any quantum entity which has such a choice of possibilities exists in a superposition of states, a mixture of both possibilities, until (in this case) its polarization is measured. Then, and only then, there is a 'collapse of the wave function', and it settles into one of the two states.

But the counterpart to the photon that is being measured must also have existed in a superposition of states until just that moment. Then, at precisely the moment that the measurement of photon A causes its wave function to collapse, the wave function of photon B (which could, in principle, by now be on the other side of the Universe) must collapse into the opposite state, without any measurement being carried out on it. This instantaneous response of photon B to what happens to photon A is Einstein's spooky action at a distance. Einstein also had grave doubts about the notion of the collapse of the wave function, but it is the action at a distance – or non-locality – that matters here.

The actual experiment is slightly more complicated than we have outlined, because the polarization is measured at an angle, which can be varied, across the up/down arrows. The probability of a photon with a certain polarization passing through a filter set at a certain angle depends on its own polarization and on the angle between its polarization and the filter. In a non-local world, changing the angle at which you choose to measure the polarization of photon A will alter the probability of photon B passing through a polarizing filter set at a *different* angle. And instead of just two photons being emitted, the experiment deals with beams of photons, sets of correlated pairs hurtling through the apparatus one after the other. The special feature of Bell's theorem is that it deals with the statistics of such an experiment, in terms of two numbers that each represent the sum of many sets of measurements that can be taken in this way. One of these numbers is always bigger than the other, if common sense prevails. This is called Bell's inequality. But the Aspect experiment (and many other experiments) shows that the first number is actually always smaller than the second number. In other words, Bell's inequality is violated and common sense does not prevail.

The power of the experiments is highlighted by the fact that they did not always give the answers the experimenters were looking for. The first good test of Bell's theorem was carried out by John Clauser, Michael Horne, Abner Shimony and Richard Holt at Berkeley, California, in the early 1970s (Stuart Freedman also worked with Clauser on a related experiment). The experiment was motivated by Clauser's desire to prove that the world is local – that there is no spooky action at a distance. In spite of this, his team found that Bell's inequality is violated. If an experimenter sets out to prove one thing and ends up proving the opposite, you can be pretty sure that the experiment is being run honestly, with no self-delusion going on (see **Millikan, Robert Andrews**).

There was one loophole in the Berkeley experiment. The experimental set-up was determined before the photons set out on their journey, and stayed the same throughout their brief flight from one side of the lab to the other. This allowed for the possibility that some influence from the polarization measuring filters might be leaking round to affect how the photons set out on their journey, letting them know in advance what kind of filter was waiting for them. The key feature of the Aspect experiment is that it closed this loophole, by switching between two different polarizing filters while the photon was in flight – after it had left the atom in which it was born. The improvement used a switch that changes the direction of a beam of light passing through it, directing it towards either one of two differently angled polarizing filters. Each flight took 20 nanoseconds (20 thousand-billionths of a second, 20×10^{-9} sec) for the photon to travel from its birthplace to the detector, but the switch could flip once every 10 nanoseconds. The direction of the switch was flipped at random, by a computer, while the photons in the Aspect experiment were in flight. And still Bell's inequality was violated.

The important feature of all these experiments is that they have directly detected non-locality. There is, in fact, no need to invoke the collapse of the wave function, or any other interpretation of quantum mechanics, or indeed to accept quantum theory at all. Although the motivation for the Aspect experiment came from quantum theory, Bell's theorem has broader implications and the combination of Bell's theorem and the experimental results reveals a fundamental truth about the Universe, that there are correlations which take place instantaneously, regardless of the separation between the objects involved (for one explanation of how this could happen, within the context of quantum theory, see ***transactional interpretation***).

There are, in fact, three fundamental assumptions underlying the derivation of the Bell inequality. The first is that there are real things in the Universe, which exist whether or not we observe them. The second is that it is legitimate to draw general logical conclusions from the outcome of consistent experiments or observations (in other words, if I notice that every time I push a pencil off my desk it falls downwards, I can infer that next time I push it off my desk it will not fall upwards). And the third is that no signal can travel faster than light. This combination of ideas is often referred to as 'local reality'. Because the Bell inequality is violated, at least one of these assumptions must be wrong.

If the second assumption is wrong, there is no point in trying to understand the world anyway, so if we want to understand the world we have to keep that one. If the first assumption is wrong, it is still possible to construct a logically self-consistent description of the way the world works, and some interpretations of quantum mechanics do just that (see ***participatory universe***). But if you want to believe in a real world that exists independently of your observations, and you want to believe that this world operates logically, then you have to accept that it also operates non-locally.

But there is no possibility of using this non-locality, the spooky action at a distance, to send useful information instantaneously from one place to another, faster than the speed of light. In effect, the non-locality would allow someone on the other side of the Universe from you to receive a set of random numbers, determined by the angle of your polarizing filter. If you change the angle, your friend will receive a different set of random numbers, but will have no way of knowing that they are different. Neither set of random numbers contains any useful information, so no useful information is conveyed!

Further reading: Nick Herbert, *Quantum Reality*.

associative law   A rule that a sequence of three or more mathematical operations gives the same answer regardless of which subcomponent of the calculation is done first. Addition is associative. If you want to add 12 + 6 + 3, you get the same answer if you work out (12 + 6) and then add 3, or if you work out (6 + 3) and then add 12. But division is not associative (12/6)/3 gives you 2/3, but 12/(6/3) is equal to 6. See also ***Abelian group***.

Aston, Francis William (1877–1945)   English physicist who invented the technique of mass spectroscopy, using it to show that many common elements come in different varieties (called isotopes), which have the same atomic number but different atomic masses.

Aston was born in Birmingham on 1 September 1877. He studied chemistry at Mason College (forerunner of Birmingham University), and after a short spell of research spent three years with a brewing firm, before returning to Mason College in 1903. He was interested in the behaviour of gases in so-called discharge tubes, in which an electric current is passed through a tenuous gas (neon tubes and strip lighting are discharge tubes). In 1909 he moved to the Cavendish Laboratory in Cambridge to continue this work with J.J. Thomson, the discoverer of the electron. They studied the way the positively charged 'rays' produced in discharge tubes (what we would now call ions) are deflected by electric and magnetic fields, and Aston eventually (after an interruption to his career caused by war work) developed this technique to measure the masses of the ions.

For this major contribution to the understanding of atoms (which to modern eyes seems to fall in the province of physics), Aston was awarded the Nobel Prize for Chemistry in 1922. The study of isotopes remained at the centre of Aston's career, and when he discovered that the masses of atoms are not quite exact multiples of the mass of a hydrogen atom (for example, each atom of the most common isotope of oxygen, oxygen-16, has a mass 15.9949 times the mass of a hydrogen atom), he realized that the 'missing' mass is locked up in the so-called binding energy of the atomic nucleus. Aston appreciated that this nuclear energy might be liberated by the fusion or fission of nuclei, and warned of the potential dangers.

Aston's other career interests included astronomy, and his skill in photography made him an invaluable member of several eclipse expeditions. He was also an accomplished musician. He died in Cambridge on 20 November 1945, having lived long enough to learn of the release of nuclear energy by fission in the first atomic bombs.

asymptotic freedom   Descriptive term for the way in which the *colour* force between quarks decreases, the closer the quarks are together. This seems peculiar at first sight, because we are used to thinking of forces like those of gravity and electricity, which get stronger when things are closer together. But we have all experienced a force which behaves like the colour force between quarks, when we have stretched an elastic band. The more you stretch the elastic band, the more effort you need to stretch it further – until, eventually, it breaks.

So you can think of the colour force between quarks (for example, the three quarks inside a proton) acting as if the quarks were joined to each other by elastic bands. When the quarks are close together, the elastic bands flop about, and they don't notice their confinement. But if one quark tries to move away from its companions, the elastic bands will stretch, tending to pull it back near them.

This picture also provides an image of why a single quark is never detected. Suppose that one of the quarks in a proton is given a strong push, perhaps by a collision with a fast-moving electron fired into the proton from outside. It will go much further away from its companions than usual, stretching the bonds trying to hold it in place. If enough energy is given to it by the incoming electron, the bonds will break. But 'enough energy', in this case, means enough to make a new quark, out of pure energy, on either side of the break, in line with Einstein's equation $E = mc^2$. So instead of a single quark escaping, the collision produces a pair of quarks (actually, a quark–antiquark pair, see *antimatter*) held together by one piece of elastic, while in place of the old quark a new quark stays behind in the proton, held to its two companions by the original piece of elastic.

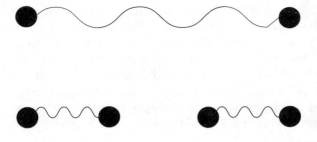

Asymptotic freedom. When a quark tries to escape from another quark, the stream of gluons holding the two together stretches until it breaks. It only breaks when there is enough energy to make new quarks on each side of the break. So an isolated quark is never seen.

Because the quark is effectively free at the moment it is struck by the electron, however, the way an electron scatters from a quark inside a proton is effectively the same as the way it scatters from an isolated particle. Studies of this kind of quasi-free scattering were important in the development of the quark model (see *parton model*).

The inability of the quark to get far from its companions under normal conditions is sometimes referred to as infrared slavery, a loose use of the term 'infrared' to mean 'long distance'.

atom The smallest component of an element that can take part in a chemical reaction. Atoms were originally thought of as the ultimate, indivisible building blocks of matter, but are now known to be made up of three components: electrons (which surround the atom and are responsible for its chemical properties) and protons and neutrons, which cluster together in a tiny central nucleus within the electron cloud. This internal structure of the atom is discussed under *atomic physics*.

One of the earliest known references to the concept of what would now be called atoms occurs in the writings of the Chinese philosopher Hui Shih, who lived in the 4th century BC. He introduced the idea of the 'small unit', the smallest possible entity that could exist in nature, with nothing inside itself. This is also a forerunner of a fundamental concept of quantum physics, that there is a limit to how much things can be divided. But these ideas did not catch on in the East.

In the West, the idea that matter is made up of fundamental, indivisible particles arose in Greek thought in the 5th century BC, giving us the word 'atom', from the Greek *atomos*. Greek atoms were not thought of as being necessarily very small; what mattered was that they were indivisible, so they could even have internal structure, provided that they could not be divided. This struck a chord with other cultures, including Islam, because some religions found the concept of atoms essential if God were to be able to take account of everything at the Last Judgement. The atomic idea was promoted by philosophers such as Democritus and Leucippus, but ironically the idea was preserved and passed on largely through the writings of Aristotle, who was an opponent of the theory.

The idea was revived in Europe by scientists such as Nicolaus Copernicus, and more strongly in the 17th century by researchers such as Christaan Huygens and Robert Boyle. In his *Opticks*, published in 1704 but written decades earlier, Isaac Newton wrote of matter being made of 'primitive Particles … incomparably harder than any porous Bodies compounded of them; even so very hard, as never to wear out or break in pieces'. In 1738 the Swiss mathematician Daniel Bernoulli described the behaviour of a gas in terms of the motion of many tiny particles, which bounced around, colliding with one another and with the walls of their container. And by the 19th century most physicists seemed to accept the idea of atoms as an established fact. Chemists, however, were much slower to recognize the existence of atoms as real entities, like little billiard balls, rather than as an abstract idea that helped in picturing how chemical reactions took place, but which might be no more than an heuristic device.

The first step towards the modern understanding of atoms was made by John Dalton at the beginning of the 19th century. It was Dalton who introduced the notion of an atom as the smallest unit of an element that can take part in a chemical reaction. This draws a distinction between atoms and molecules. Water, for example, is made up of hydrogen and oxygen, and water molecules can be broken apart by chemical means to produce hydrogen and oxygen. But chemical processes cannot break either hydrogen or oxygen atoms into anything simpler; the atoms of hydrogen and of oxygen are more fundamental, simpler entities than the molecules of water.

The next step in understanding atoms and molecules was made by Amedeo Avogadro in 1811. He realized that if you took a box of a certain size, and filled it with gas at a certain temperature and pressure, then whatever gas you chose to put into the box, there would always be the same total number of particles (atoms and/or molecules) bouncing around inside the box at that temperature and pressure. As far as the pressure on the walls of the box is concerned, it doesn't matter what the particles that hit it are made of, only how often they hit it, and how fast they are moving when they do so. The number of collisions depends on how many particles there are, and the speed of each collision depends on the pressure.

It was not until the 1850s that this idea began to have much impact on the way scientists thought about atoms – even Dalton did not accept it at first. But then people realized that if Avogadro was correct, his insight provided them with a means to calculate the sizes of atoms.

In these calculations, it is usual to work in terms of a box of gas at $0\,^{\circ}C$ and under 1 standard atmosphere of pressure. The size of the box is chosen to be just the right size to contain a mass of gas in grams equal numerically to the molecular weight of the gas. For example, for oxygen that would be 32 g of gas, because each molecule of oxygen weighs 32 times as much as one atom of the lightest element, hydrogen (each oxygen molecule contains two oxygen atoms; just as in the case of water molecules, oxygen molecules can be broken apart by chemical means, but the oxygen atoms cannot). Whatever the gas put into the box in this way, the number of molecules present will always be the same – a number now known as Avogadro's number, and equal (in round terms) to 6×10^{23} (a 6 followed by 23 zeros). This gives you some idea how small atoms are; in each cubic centimetre of air under the same conditions there would be 4.5×10^{19} molecules. But these numbers are based on modern experiments; even at the end of the 19th century, estimates of Avogadro's number were still uncertain. The person who put those estimates on a secure footing was Albert Einstein.

The first good 19th-century attempt at estimating the sizes of molecules was made by Thomas Young in 1816. His estimate was based on measurements of the surface tension of a liquid – the elasticity you can feel if you touch the surface gently. Surface tension is caused because molecules in the surface layer of the liquid are being tugged by molecules in the bulk of the liquid below them, and on either side, but not by the air above. Young guessed that the strength of surface tension must be related to the range of the forces between molecules, and used this range as a rough measure of the sizes of molecules themselves. He came up with an estimate that 'particles of water' must have a size in the range from 5 to 25 billionths of a centimetre, only about ten times bigger than the modern estimate.

Fifty years later, Johann Loschmidt used another technique, involving Avogadro's number. He assumed that in a liquid all the molecules are touching one another, but in a gas the molecules travel through empty space between collisions. Knowing the volume occupied by a certain amount of liquid, he knew how much of the space occupied by the same liquid when it had evaporated and filled a container must be just that – empty space. This gave him enough extra information to calculate that a typical molecule of air is a few millionths of a millimetre in size, and to estimate Avogadro's number as 0.5×10^{23}. Modern versions of the equivalent experiments tell us that each molecule travels, on average, 13 millionths of a metre between collisions (at 0°C), and an oxygen molecule at that temperature will be moving at just over 461 metres per second. So it undergoes more than 3.5 billion collisions every second.

Einstein's technique for measuring the sizes of molecules involved not gases but solutions, in which one kind of molecule (the solute) is spread through a liquid made up of another kind of molecule (the solvent). His solutions were sugar in water, and the calculations would apply very accurately to a cup of sweet tea. The experiments involve the way solute flows through a membrane placed between two solutions with different strengths. In his PhD thesis (in 1905), Einstein derived a value for Avogadro's number of 2.1×10^{23}, and estimated the sizes of molecules as a few ångströms (a few hundred millionths of a centimetre), in line with modern estimates. Although the ångström is now an obsolete unit, it is memorable that any individual atom is about 1 Å in diameter. In 1906, with better experimental data, Einstein improved his estimate for Avogadro's number to 4.15×10^{23}, and in 1911 he came up with a value of 6.11×10^{23}. By then, several researchers had found essentially the same number by a variety of techniques, and the reality of atoms was no longer in any doubt.

If all this sounds like the steady and inevitable progress of science, think again. Although the idea of a gas being made up of many tiny particles bouncing around and off one another made it possible to develop a mathematical theory of gases (known as statistical mechanics) in the 19th century, without direct proof of the existence of atoms even some physicists doubted the wisdom of this approach. In the 1890s one of the pioneers of this field, Ludwig Boltzmann, felt himself to be so much an individual struggling against a weight of establishment opinion (possibly true at the time in the German-speaking world, although elsewhere atoms were much more widely accepted) that in 1898 he published details of his calculations in the express hope 'that, when the theory of gases is again revived, not too much will have to be rediscovered'. This was *after* J. J. Thomson had discovered the electron! In 1906, ill and depressed, Boltzmann killed himself, unaware of Einstein's thesis work, which had been published just a few months earlier. Although Boltzmann was perhaps less misunderstood than he thought, his experience does show how the concept of atoms became fully established only in the 20th century, and how important the work of Einstein and his contemporaries in the first decade of that century was in establishing the concept. Almost immediately after submitting his thesis, in May 1905 Einstein sent for publication his classic paper on **Brownian motion**, the curious zigzag dance of pollen grains suspended in water, which is explained in terms of the buffeting that the grains receive from the molecules themselves, and made atoms and molecules physically real in the minds of all scientists.

Further reading: Steven Weinberg, *The Discovery of Subatomic Particles*.

atomic bomb   A device in which energy is released explosively by the fission of nuclei of unstable heavy elements such as uranium-235 or plutonium-239. The term is, in fact, a misnomer, and nuclear bomb would be more appropriate. See **nuclear weapons**.

atomic clock   General name for any of a variety of timekeeping devices which are based on regular vibrations associated with atoms. The first atomic clock was developed in 1948 by the US National Bureau of Standards, and was based on measurements of the vibrations of atoms of nitrogen oscillating back and forth in ammonia molecules, at a rate of 23,870 vibrations per second. It is also known as an ammonia clock.

The standard form of atomic clock today is based on caesium atoms. The **spectrum** of caesium includes a feature corresponding to radiation with a very precise frequency, 9,192,631,770 cycles per second. One second is now defined as the time it takes for that many oscillations of the radiation associated with this feature in the spectrum of caesium. This kind of atomic clock is also known as a caesium clock; it is accurate to one part in 10^{13} (one in 10,000 billion), or one second in 316,000 years.

Even more accurate clocks have been developed using radiation from hydrogen atoms. They are known as hydrogen maser clocks, and one of these instruments, at the US Naval Research Laboratory in Washington, DC, is estimated to be accurate to within one second in 1.7 million years. In principle, clocks of this kind could be made accurate to one second in 300 million years.

atomic energy   Energy released by the fission of nuclei of unstable heavy elements such as uranium-235 or plutonium-239. A better term is nuclear energy, or **nuclear power**. The energy involved in ordinary chemical reactions (reactions between atoms) could more accurately be described as atomic energy, but it is too late to change this now.

atomic mass   See *atomic mass unit*.

atomic mass unit (a.m.u.)   The unit in which masses of atoms are measured. 1 a.m.u. is defined as one-twelfth of the mass of a single atom of the isotope carbon-12. This is 1.66×10^{-27} kg, or roughly 931 MeV. Also known as 1 dalton.

atomic number   The number of protons in the nucleus of a single atom of a particular element. This is equal to the number of electrons associated with that atom, and therefore determines the chemical properties of the element. Denoted by the symbol Z. Also known as the proton number. See also *isotopes*, *mass number*.

atomic physics   The study of the internal structure of the atom. By and large, this means the way in which electrons are arranged around the nucleus of an atom, and the discovery of the way the numbers of protons and neutrons in the nucleus determine the properties of the atoms. Much of what is commonly referred to as 'atomic physics' (particularly the processes of fission and fusion) really fall within the province of *nuclear physics*.

By the 1890s, there was good evidence that atoms exist, but they were still regarded as the fundamental, indivisible building blocks of matter (see *atom*). Atomic physics began with the discovery, made in 1897 by J. J. Thomson at the Cavendish Laboratory in Cambridge, that the so-called cathode rays were in fact a stream of tiny charged particles broken off from atoms. These particles soon became known as electrons.

The discovery built from work carried out by William Crookes in the 1870s. He was studying the effects of electric currents passed through traces of gas in discharge tubes, similar to the tubes of a modern neon sign or strip lighting. The electricity flows between two electrically charged plates, one at each end of the tube – the positively charged anode and the negatively charged cathode. One effect of this is to produce a stream of positively charged particles, repelled from the anode. Crookes called these 'molecular rays', and showed that they travelled in straight lines unless deflected by a magnetic or electric field; these positively charged particles would now be called ions.

Like earlier researchers, Crookes also noted the existence of a stream of negatively charged particles, repelled from the cathode in a discharge tube. These cathode rays had first been discovered some twenty years earlier, by the glow they produced when they hit the wall of the glass tube. Crookes showed that cathode rays also travel in straight lines but are deflected by magnetic fields, suggesting that they are negatively charged particles; but he did not appreciate that they were not simply negatively charged counterparts to the positive molecular rays. Crookes also noticed that photographic plates kept near his discharge tubes became fogged, but he did not follow through on this discovery.

Interest in cathode rays received a boost in 1895, when Wilhelm Röntgen made the accidental discovery of *X-rays*, explaining (among other things) the fogging of Crookes' photographic plates. It turned out that X-rays were a previously unknown form of radiation produced in the glass wall of a discharge tube where it is struck by the beam of cathode rays. Thomson and his student Ernest Rutherford investigated the effects of X-rays on a gas. They found that gas that had been irradiated in this way could conduct electricity, and Thomson introduced the term 'ionized' to describe such an electrically conducting gas. Then Thomson turned his attention to the cathode rays that produced the X-rays.

A couple of years earlier, in 1894, the German physicist Philipp Lenard had shown that cathode rays could pass right through a piece of metal foil, so they could not (as Crookes had guessed) be electrically charged molecules. Lenard thought they must be a form of electromagnetic radiation, and the discovery of X-rays briefly encouraged this belief. But Thomson found that cathode rays produced in his discharge tubes travelled at only a fraction of the speed of light. He showed, in 1897, that they must be particles, much smaller than an atom, and measured the ratio of their electric charge (e) to their mass (m) by studying the way they moved in magnetic fields. He called these particles 'corpuscles', but the term was soon replaced by the name 'electrons', suggested by the Dutch physicist Hendrik Lorentz.

The discovery of electrons showed that atoms are not the smallest particles, and the presence of positively charged 'atoms' and negative electrons in discharge tubes clearly indicated that electrons could be knocked out of atoms. If atoms contained both positive and negative charge, removing some of the negative charge would leave a surfeit of positive charge behind.

After his studies of electrons, Thomson turned his attention to studies of positively charged ions, known at the end of the 19th century as 'canal rays'. He showed that they behaved exactly as if they were made up of streams of atoms carrying positive charge (that is, atoms from which negative electric charge had been removed). This work also led to the discovery, in the second decade of the 20th century, that the atoms of an element do not necessarily all have the same mass. By measuring the ratio e/m for ions which have the same amount of electric charge, Thomson found that, for example, neon comes in two varieties, particles (atoms) which have 20 times the mass of a hydrogen atom or particles which have 22 times the mass of a hydrogen atom. Such varieties are now known as isotopes.

Building from Thomson's discoveries, the British physicist Lord Kelvin put forward the first modern model of the atom in 1902. This was enthusiastically endorsed by Thomson, and became known as the Thomson model (since Lord Kelvin's name was originally William Thomson, although the two Thomsons were not related this is as good a name as any!). By the end of the 19th century several experiments had shown that all atoms are about the same size, roughly 1 ångström, or 0.1 of a nanometre, in diameter (see **atom**). The Thomson model imagined an atom to be a uniform sphere of positive charge, 1 Å across, with electrons embedded in it, like plums in a plum pudding. This was soon shown to be wrong.

In 1896 the French physicist Henri Becquerel, stimulated by the discovery of X-rays, had discovered another form of radiation, produced spontaneously by atoms of uranium. Rutherford, who had moved on to McGill University in Montreal in 1898, showed that there are actually two kinds of this atomic radiation, which he called alpha and beta rays (a third form of radiation discovered later was called gamma radiation). 'Beta rays' were soon shown to be fast-moving electrons, while alpha rays turned out to be particles with a mass four times that of a hydrogen atom and carrying two units of positive electric charge. They were the same as helium atoms from which two electrons had been removed (what are now known as helium nuclei).

In 1902 Rutherford and Frederick Soddy showed that when an atom emits either alpha radiation or beta radiation it is turned into an atom of a different element. Rutherford then set about using the alpha rays produced by natural radioactivity to probe the structure of other atoms.

By 1909 Rutherford was back in England, as professor of physics at the University of Manchester. There, two physicists working under his direction, Hans Geiger and Ernest Marsden, discovered that when a beam of alpha particles was fired at a thin sheet of metal foil most of the particles went straight through, but occasionally one bounced back from the foil.

The Thomson model could not explain this, because on that picture the foil would be uniformly made up of serried ranks of 'plum puddings', touching each other. There would be no hard centres for the alpha particles to bounce off. So Rutherford developed a new picture of the atom, the Rutherford model, in which a tiny central nucleus (much smaller than one ångström in diameter), which contains all the positive charge and almost all of the mass of the atom, is orbited by electrons in a way roughly analogous to the way the Sun is orbited by the planets. On this picture, most of any seemingly solid object, including a sheet of foil, is empty space. An alpha particle would easily brush through the electron clouds in the gaps between nuclei, but just occasionally an alpha particle would hit a nucleus more or less head on, and be deflected by a large angle.

The frequency (or scarcity!) of such high-angle deflections enabled Rutherford to work out the size of the nucleus. A nucleus is typically about 10^{-13} cm across, compared with the 10^{-8} cm diameter of the entire atom – just one hundred-thousandth of the size of the atom, equivalent to the size of a pinhead in the middle of the dome of St Paul's cathedral in London.

There was one obvious problem with the Rutherford model. A charged particle (such as an electron) travelling in an orbit would radiate electromagnetic energy and spiral rather rapidly into the nucleus. This problem was resolved with the development of the **Bohr model** of the atom, one of the first successes of quantum physics, but the picture of a tiny central nucleus surrounded by a cloud of even tinier electrons at a relatively large distance is still a good image of the atom.

In this context, the discovery of isotopes implies that all the atoms of an element have the same number of electrons, giving identical clouds which interact with the electron clouds of other atoms, and therefore they all have the same amount of positive electric charge in the nucleus, to balance the negative electric charge of the electrons. But they can have nuclei with different masses. In 1919 Rutherford discovered that sometimes when a fast-moving alpha particle strikes a nucleus of

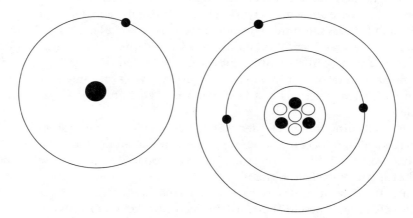

Atomic physics. The Bohr model represents hydrogen (left) as a single proton orbited by a single electron. In heavier atoms, the nucleus is made up of both protons and neutrons, with the appropriate number of electrons (one for each proton) in orbit around it (the atom represented on the right is lithium-7).

nitrogen, the nucleus is changed into one of oxygen, and a hydrogen nucleus is emitted. He had achieved the alchemists' dream of transmutation of the elements. The discovery implied that the liberated hydrogen nucleus must have been inside either the alpha particle or the nitrogen nucleus, and was knocked out in the collision. It was given the name 'proton'; a proton has one unit of positive charge, the same magnitude as the unit of negative charge carried by an electron.

Hydrogen is the simplest and lightest atom, so hydrogen nuclei (protons) seemed to be a fundamental building block of matter. But nuclei could not be made solely of protons, because the mass of a nucleus (in units of the mass of a proton) is not, in general, equal to its positive charge (in units of the charge on a proton). In the early 1920s, one possibility seemed to be that the nucleus contained a mixture of protons and electrons, so that the overall positive charge was reduced; but a more plausible explanation seemed to be that there must be a kind of electrically neutral particle, which had never been detected, which existed in nuclei and provided the extra mass. This was confirmed in 1932 when James Chadwick, working in Cambridge, identified the neutron and found that it had roughly the same mass as a proton.

The basic model of the atom was now complete. For a particular element, each nucleus contains a certain number of protons, with the same number of electrons in the cloud surrounding the nucleus. But there can be slightly different numbers of neutrons in the nucleus, explaining the occurrence of isotopes – atoms which have the same number of electrons (and therefore identical chemical properties) but different numbers of neutrons (and therefore different masses). Beta radiation is produced when an unstable nucleus emits an electron, gaining one unit of positive charge (in fact, because a neutron has turned into a proton; see *beta decay*) and turning into the nucleus of an atom of another element. Similarly, in alpha decay an unstable nucleus emits two protons and two neutrons, bound together as a helium nucleus, and changes into an atom of another element.

If you are wondering how positively charged protons can stick together in the nucleus, in spite of the electric repulsion between their charges, see *forces of nature*.

atomic pile   Early name for *nuclear reactor*.

atomic radius   See *atomic size*.

atomic size   The size of an atom is determined by the extent of the cloud of electrons orbiting its nucleus. Although the simplest atom, hydrogen, has only one electron in its 'cloud', while atoms of the heaviest elements each have more than 100 electrons, all atoms are much the same size, a few ångströms across. The largest atom, caesium, has a diameter of 5 Å: that is, 0.00000005 cm.

atomic spectroscopy   See *spectroscopy*.

atomic theory   See *atom*.

atomic time   Generally, time recorded by an *atomic clock*. Specifically, the standard time used today (the International Atomic Time Scale) based on atomic clock data and starting from zero at 0 h 0 min 0 sec GMT on 1 January 1958. It is this time scale that calibrates the time signals provided by radio stations.

atomic weight   See *relative atomic mass*.

atom splitting   A misnomer for nuclear fission. See *fission*. The first people to 'split the atom' were John Cockcroft and Ernest Walton, in 1932.

Auger effect A process in which a suitably excited atom (one with an excess of energy) can return to a lower energy state by emitting an electron, instead of emitting electromagnetic radiation. The emitted electron is called an Auger electron. The energy carried by the ejected electron is equivalent to the energy of a photon in the X-ray or gamma-ray part of the spectrum.

The effect is useful because the electrons are emitted with a precise energy (for a particular type of atom), and can therefore be used to calibrate detectors. The energy of the ejected electron also provides information about the spacing of energy levels within the cloud of electrons surrounding that particular kind of atom; investigation of the internal structure of the atom using measurements of the energies of Auger electrons is sometimes known as Auger spectroscopy.

Named after the French physicist Pierre Auger, born in 1899, who discovered the phenomenon in 1925; it had also been noticed two years earlier by Lise Meitner.

Auger electron See *Auger effect*.

Avogadro, Amedeo (1776–1856) Italian chemist and physicist who came from a noble family in Piedmont, gloried in the full name Lorenzo Romano Amedeo Carlo Avogadro di Quaregna e di Cerreto, and succeeded his father as Count of Quaregna in 1787.

Born in Turin on 9 August 1776, Avogadro came from a long line of lawyers, and studied and practised law before turning to science in 1800, initially alongside his legal work. He investigated electricity, gave up law in 1806 (when he was appointed as a demonstrator at the Academy in Turin) and became professor of Natural Philosophy at the Royal College in Vercelli in 1809. He had two spells as professor of mathematical physics at Turin, from 1820 to 1822, when the post was abolished for political reasons, and from 1834 (two years after it had been re-established) to 1850 (when he retired).

Avogadro showed (using the discoveries of Joseph Gay-Lussac) that the chemical formula for water is H_2O, not HO, and in 1811 he published the paper in which he set out the idea that equal volumes of gas at the same temperature and pressure contain equal numbers of molecules (Avogadro's law). The importance of this idea was not appreciated at the time, and it was only after Avogadro's death that his fellow countryman Stanislao Cannizaro took up the idea and (in 1860) demonstrated its importance. Even then, it was another twenty years before the idea and its implications were fully accepted. See *atom*.

Avogadro died in Turin on 9 July 1856.

Avogadro's Law See *Avogadro, Amedeo*. Sometimes known as Avogadro's hypothesis.

Avogadro's number The number of molecules (or atoms) of a substance contained in an amount of that substance with a mass in grams numerically equal to the molecular (or atomic) weight of the substance. The atomic weight of carbon is 12, so 12 g of carbon contain Avogadro's number of atoms. Although the atomic weight of hydrogen is 1, each molecule of hydrogen gas contains 2 atoms. So 2 g of hydrogen gas contain Avogadro's number of molecules. Avogadro's number (also known as the Avogadro constant) is 6.022×10^{23}. See *atom*.

axion Hypothetical subatomic particle required by some theories to explain details of the workings of *quantum chromodynamics* (QCD). If axions do exist, they each have a mass of around 10^{-5} eV, less than 10^{-12} of the mass of a proton, but there could be so many of them in the Universe, in the space between the stars and galaxies, that they contribute a large proportion of the overall mass of the Universe, in the form of dark matter.

The axion was postulated to explain why CP violation is not observed in inter-actions involving the strong force (see *CP conservation*), although it should be according to simple versions of QCD. It enables the theory to take account of the way the particle world distinguishes between left and right in some interactions, but not others. This is described in terms of axial symmetry, hence the name (although it is alleged that the name was also deliberately chosen to match that of a brand of washing powder, because the axion 'washes away' this particular problem).

Balmer, Johann Jakob (1825–1898) Swiss mathematician and schoolteacher who studied spectroscopy and found a simple formula which relates the frequencies of lines in the visible spectrum of hydrogen to one another. One of the early triumphs of the theory of atomic structure developed by Niels Bohr (and based on quantum principles) was that it explained Balmer's formula and this pattern of lines.

Born in Lausen, near Basle, on 1 May 1825, Balmer was the son of a successful farmer, who also served in the local government. He studied in Karlsruhe and briefly in Berlin before returning to Basle, where he worked as a schoolteacher while completing a PhD, awarded by the University of Basle in 1849. In 1850 he began work as a teacher of mathematics and Latin at a girls' school in Basle, where he stayed until he retired in 1890. Alongside this post, he also worked as a part-time lecturer in geometry (a *Privatdozent*) at Basle University from 1865 to 1890.

Although trained as a mathematician and working as a mathematics teacher, Balmer became interested in spectroscopy, which was very much at the cutting edge of scientific research at the time, and was encouraged by colleagues at Basle University to try to find an explanation for the way lines in a spectrum are distrib-uted – at first sight, in a random pattern. Although many other scientists had tried and failed to find some order in these patterns, in 1884, in his sixtieth year, Balmer succeeded in finding an equation that not only reproduced the relationship between the four bright lines known in the spectrum of hydrogen at that time, but also predicted the existence of a fifth line, at the edge of visibility, that was soon detected. The formula was published in 1885, the year Niels Bohr was born.

The five lines became known as the Balmer series. Balmer also predicted the existence of further series of hydrogen lines beyond the range of the visible spectrum, in the ultraviolet and in the infrared. These were later discovered, and named (after their discoverers) the Lyman series, the Paschen series, the Brackett series and the Pfund series. Balmer later found a similar explanation for the patterns of lines in the spectra of helium and of lithium, and published the discovery in 1897, when he was 72. Although Balmer's results were entirely empirical, based on the mathematical relationships alone with no attempt to find their underlying physical cause, this discovery that there was a mathematical order in the pattern of spectral lines was an important step towards developing an understanding of the spectrum in quantum terms.

Balmer died in Basle on 12 March 1898.

Balmer series A series of five lines in the visible spectrum of hydrogen. A formula developed in 1884 by Johann Balmer relates the frequencies of these lines to one another. The spacing between the lines decreases, so that the first line is in the red part of the spectrum, the next line in the green part of the spectrum, the third line in the violet, and the others also in the violet. Working in terms of wavelength λ (the common way to work in Balmer's day), rather than frequency, the formula is

$$1/\lambda = R(1/2^2 - 1/n^2)$$

where R is a number known as Rydberg's constant, and n is 3, 4, 5, 6 or 7 for the five lines studied by Balmer. Replacing the term $1/2^2$ by $1/1^2$ and allowing the values of n to start at 2 gives the Lyman series, while if the first term is $1/3^2$ and the ns start at 4, you get the Paschen series, and so on (see *Balmer, Johann Jakob*).

Bardeen, John (1908–1991) American physicist who was the first person to be awarded the Nobel Prize for Physics on two separate occasions.

Born in Madison, Wisconsin, on 23 May 1908, Bardeen obtained a BSc in electrical engineering from the University of Wisconsin in 1928 and an MSc in 1929; he then worked in industry before giving up his career in 1933 and enrolling as a research student under Eugene Wigner at Princeton in 1933. He was awarded his PhD in 1936 and worked on problems in solid-state physics at Princeton, Harvard and the University of Minnesota before carrying out war work at the Naval Ordnance Laboratory in Washington, DC, from 1941 to 1945. After the Second World War, Bardeen joined the Bell Laboratories, where (together with Walter Brattain and William Shockley) his work on semiconductors led to the development of the transistor and, in 1956, to his first Nobel Prize.

In 1951 Bardeen left Bell Labs to become professor of electrical engineering at the University of Illinois, where he stayed for the rest of his career. With Robert Schrieffer and Leon Cooper, he developed a theory of superconductivity, for which he received a second Nobel Prize in 1972. He died at Champaign, Illinois, on 30 January 1991.

bare charge The charge of a quantum entity, such as an electron, measured at very short range – specifically, the value of the charge corresponding to interactions at the vertices in a *Feynman diagram*. This can differ from the *dressed charge* of the entity, which it exhibits in long-range interactions, because of the way the entity is 'dressed' in a cloud of virtual particles.

bare mass The mass of a quantum entity, such as an electron, measured at very short range – specifically, the value of the mass corresponding to interactions at the vertices in a *Feynman diagram*. This can differ from the *dressed mass* of the entity, which it exhibits in long-range interactions, because of the way the entity is 'dressed' in a cloud of virtual particles.

barn A unit used in measuring *cross-sections* for particle interactions (as in 'you couldn't hit the side of a barn with a shotgun'). $1\,b = 10^{-28}$ square metres, about one-hundredth of the area of a cross-section through the nucleus of an atom.

baryon The name for any *fermion* that feels the influence of the strong interaction (see *forces of nature*). All baryons are therefore members of the hadron family. The most important baryons are the proton and neutron, which make up most of the mass of ordinary atoms. For this reason, everyday matter is often referred to as 'baryonic matter'.

baryonic matter   See *baryon*.

baryon number   A label used to denote which particles are baryons and which ones are not. Each baryon has a baryon number of 1, each antibaryon has a baryon number of –1. Every other particle has a baryon number of 0. The total baryon number is conserved in all particle reactions that have been carried out on Earth (so if you make a baryon, you have to make an antibaryon to balance it), but a slight asymmetry in the laws of physics allowed baryons to be created in the Big Bang. The same asymmetry implies that baryons will eventually decay into leptons and photons. See *grand unified theories*.

Basov, Nikolai Gennadiyevich (1922–)   Russian physicist who shared the Nobel Prize for Physics in 1964 for his work on the development of masers and lasers. After serving in the Red Army during the Second World War, Basov studied in Moscow, joined the Lebedev Institute in 1948 as a laboratory assistant, and worked his way up to become director of the Institute in 1973. He suggested the idea of using semiconductors to make lasers in 1958.

BCS theory   See *superconductivity*.

beauty   See *bottom*.

Becquerel, (Antoine) Henri (1852–1908)   French physicist who discovered radioactivity.

Henri Becquerel was the third member of a unique lineage of eminent French physicists. Born in Paris on 15 December 1852, he trained as an engineer and started teaching at the Ecole Polytechnique in 1875. But, following family tradition, he also began research in physics, initially without any formal appointment.

Henri's grandfather, Antoine (1788–1878), had investigated electric and luminescent phenomena, and had been so successful that in 1838 a chair of physics was set up for him at the French Museum of Natural History. Antoine's third son, Alexandre-Edmond Becquerel (1820–1891), helped his father with experiments and was drawn into the study of phosphorescent solids – crystals that glow in the dark. As early as 1858, he noted in a scientific paper that 'the bodies which produce the most brilliant effects are uranium composites'. When Antoine died in 1878, Edmond (as he was usually known) succeeded him as professor at the Museum.

Meanwhile, Henri had established himself as a physicist, receiving a PhD from the Faculty of Sciences of Paris in 1888, and in 1889 he was elected to the French Academy of Sciences. When Edmond died in 1891, Henri became the third person, and the third Becquerel, to become professor of physics at the Museum of Natural History. Alongside this work, he also served as chief engineer for the Department of Bridges and Highways in Paris. In due course, Henri's only son, Jean, succeeded him as professor. It was only in 1948, when Jean (who left no heir) retired that the professorship passed out of the Becquerel family, 110 years after the chair had been established. Out of all the Becquerel dynasty, though, it is Henri who is assured of scientific immortality, for a discovery he made on a grey March day in Paris in 1896.

X-rays had been discovered (by Wilhelm Röntgen) in 1895, and the discovery was announced on 1 January 1896. Becquerel was present at a meeting of the French Academy of Sciences on 20 January 1896, where he learned that the X-rays came from a bright spot where cathode rays struck the wall of a vacuum tube and made it fluoresce (see *atomic physics*). This suggested to Henri that phosphorescent objects might also emit X-rays, and he set out to test several samples, including some uranium salts that had been prepared fifteen years earlier, during work with his father.

The phosphorescent materials that Becquerel was studying had to be exposed to sunlight to make them glow. When they were kept away from sunlight, the glow faded away, and they had to be 'recharged'. His test simply consisted of wrapping a photographic plate in two sheets of thick black paper, and setting it out in the sunlight with a dish of phosphorescent crystals on top. When the plate was developed, it showed the outline of the phosphorescent material, and if a coin was placed between the dish and the photographic plate, the developed photograph showed an outline of the coin.

It looked as if the phosphorescence induced by sunlight was causing the salts to emit X-rays, just as the cathode rays could cause the glass of the vacuum tube to emit X-rays. At the end of February, Becquerel prepared another experiment, with a piece of copper in the shape of a cross between the dish of uranium salts and the photographic plate. But it was overcast for several days, and he left the whole set-up in a cupboard, waiting for the Sun. On Sunday, 1 March, tired of waiting, he developed the plate anyway, and found the outline of the cross on it. He seems to have been completely surprised; Jean, who was 18 at the time, recalled his father being 'stupefied' by the discovery. He had found that no outside agency was needed to make the uranium salts produce radiation that could penetrate sheets of paper and leave its trace on a photographic plate.

Although Becquerel carried out important investigations of the phenomenon, soon to be known as radioactivity, the baton was picked up and carried into the 20th century by Marie and Pierre Curie, who shared the Nobel Prize with Becquerel in 1903.

Becquerel died at Croisic, in Brittany, on 24 August 1908.

Further reading: Abraham Pais, *Inward Bound*.

becquerel   A unit of radioactivity, equal to one disintegration per second. Symbol Bq.

Bednorz, George (1950–)   German-born physicist who (jointly with Alex Müller) received the Nobel Prize for Physics in 1987 for his work on high-temperature super-conductivity.

Born in Germany on 16 May 1950, Bednorz graduated from the University of Münster in 1976, and obtained his PhD from the Swiss Federal Institute of Technology in 1982; his thesis supervisor was Alex Müller. He then joined the IBM Zurich Research Laboratory, but also lectures at the Institute of Technology and at the University of Zurich. The work for which he shared the Nobel Prize was carried out from 1983 onwards, and published in 1986, just a year before the award was made.

Bell, John Stuart (1928–1990)   Irish physicist who developed a practicable way to test some of the stranger predictions of quantum theory, using 'Bell's inequality.' Experiments carried out to measure Bell's inequality (see *Aspect experiment*) showed conclusively that the quantum world does not operate in accordance with everyday common sense, and that *local reality* is violated in the quantum world.

Bell was born in Belfast on 28 July 1928. His family could not afford to pay for him to study for a degree, and he started work at Queen's University in Belfast as a

John Bell (1928–1990).

technician. His abilities were quickly recognized by the academics he was working for, who arranged for him to be awarded scholarships in order to take a degree. He obtained his BSc in mathematical physics at Queen's before (in 1949) joining the staff at Harwell, where he was soon given leave of absence to study for a PhD at the University of Birmingham, under the supervision of Rudolf Peierls. His thesis provided a proof of **CPT conservation**. Bell stayed at Harwell until 1960, then moved on to work at CERN, where he spent the rest of his career. His early work at CERN concerned **CP violation**, but his major contribution to science came from his interest in the philosophical foundations of quantum mechanics. He expressed more clearly than any of his contemporaries the limitations of the different interpretations of quantum mechanics, and without ever committing himself to any one interpretation showed that the idea of **hidden variables** had been unreasonably neglected for far too long.

He died in Geneva on 1 October 1990.

Further reading: John Bell, *Speakable and Unspeakable in Quantum Mechanics*.

Bell's inequality A comparison between the results of two sets of measurements carried out on a pair of quantum entities that were once in contact but have since been separated. Developing from the ideas of David Bohm and from the *EPR experiment*, John Bell showed (in the mid-1960s) that it would be possible in principle to carry out such an experiment and make the appropriate measurements. By assuming common sense (*local reality*), one set of these numbers must always be larger than the other. This is Bell's inequality. When the appropriate experiments were carried out, they showed that Bell's inequality is violated, and that the quantum world does not obey common sense. See *Aspect experiment*. This violation of Bell's inequality means that quantum entities are connected by influences that take place instantaneously, what Albert Einstein referred to as 'spooky action at a distance'. See *transactional interpretation*.

beta decay   Process in which a neutron within the nucleus of an atom, or a free neutron, ejects an electron. The neutron is transformed into a proton, and a neutrino (strictly speaking, an electron antineutrino) is ejected along with the electron. See *neutrino*. Beta decay is mediated by the weak interaction (see *forces of nature*). In inverse beta decay, a proton absorbs an electron and becomes a neutron; this is the time-reversed version of beta decay.

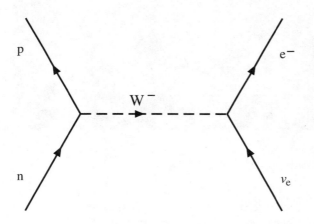

Beta decay. In beta decay, a neutron emits a W⁻ particle and becomes a proton. The W⁻ almost immediately converts itself into an electron and an electron anti-neutrino. This is exactly the same as if an incoming electron neutrino intercepts the W⁻ and is converted into an electron.

beta particle   Old name for *electron*, sometimes extended to include positrons.

beta radiation   A stream of electrons produced by *beta decay* of radioactive nuclei. The most energetic beta rays can travel through several metres of air, but are stopped by a thin sheet (a couple of millimetres thick) of aluminium.

betatron   Generic name for a type of particle accelerator which produces beams of high-energy electrons by accelerating them in a circle using magnetic fields.

Bethe, Hans Albrecht (1906–)   German-born American physicist who made major contributions to our understanding of how energy is generated inside the Sun and other stars, and to the development of quantum physics in the 1940s.

　　Bethe was born in Strassburg, Alsace-Lorraine, on 2 July 1906. At the time, this was in Germany; it is now Strasbourg, and part of France. He was the son of a university professor (in physiology) and studied at the universities of Frankfurt and Munich, receiving his PhD in physics from the latter in 1928. This was exactly at the time that quantum mechanics was being established by researchers such as Niels

Hans Bethe (1906-).

Bohr, Werner Heisenberg, Paul Dirac and Max Born; Bethe's thesis supervisor was Arnold Sommerfeld. Bethe worked in Germany on studies of the way charged particles lose energy as they travel through matter, which led to a technique for estimating the energies of particles by measuring their ranges in different materials. Between 1928 and 1933 he taught at the universities of Frankfurt, Stuttgart, Munich and Tübingen, and visited Ernest Rutherford in Cambridge and Enrico Fermi in Rome. But in 1933 he was dismissed from his post at Tübingen after Adolf Hitler came to power (Bethe's mother was Jewish) and left Germany, working briefly at the University of Bristol, in England, before moving on to the United States in 1935. He was professor of physics at Cornell University from 1937 (he had been assistant professor from 1935 to 1937) until his retirement in 1975, but he remained active in physics, and associated with Cornell, into his nineties.

In 1938 Bethe proposed a mechanism by which energy could be generated inside stars. The process, which became known as the carbon cycle, is a series of six nuclear reactions involving carbon, nitrogen and oxygen. The net effect is that protons (hydrogen nuclei) inside the star are combined to make helium nuclei, with energy being released. Bethe then (working with C. L. Critchfield) discovered a second process, the proton–proton chain, by which the same end could be achieved at a lower temperature. It turned out that the proton–proton chain is the process mainly responsible for the energy generated by the Sun and similar (or smaller) stars, while the carbon cycle dominates in more massive stars. In 1967 Bethe received the Nobel Prize for this work.

Between 1943 and 1946, after some war work on radar, Bethe was seconded from Cornell to be chief of the Theoretical Physics Division at the Los Alamos National Laboratory in New Mexico, working on the Manhattan Project. It was there that he met the young Richard Feynman, whom he persuaded to join him at Cornell after the war.

Back at Cornell, Bethe was the first person to apply the technique of **renormalization** to the puzzle of the **Lamb shift** in atomic spectra, pointing the way for the development of **quantum electrodynamics**. He had a major influence on physics after the Second World War not only through his own work, but through the clarity of his papers and the example he gave of the need to keep theoretical developments closely linked to experimental observations. Although he had nothing to do with the 'Alpher, Bethe, Gamow' paper which gave an early theory of the Big Bang (see **Gamow, George**), Bethe maintained his interest in astronomy with studies of the theory of supernova explosions and, in the 1980s, a major contribution to the attempt to find a theoretical understanding for the fact that the nuclear fusion processes going on inside the Sun produce fewer detectable neutrinos than simple theory predicts. He was also active in the campaign for a nuclear test ban treaty in the 1950s and early 1960s.

BeV Obsolete shorthand term for 1,000 million (that is, 1 billion) electron volts. See *GeV*.

Bevatron A particle accelerator (proton synchrotron) built at the Lawrence Berkeley Laboratory in California. It began operating in 1954, and could accelerate protons to an energy of 6 GeV.

Big Bang Popular term for the violent event in which the Universe as we know it was born. Observations of galaxies (systems like our Milky Way) show that they are receding from one another, and must have been closer together in the past. This matches a prediction of the general theory of relativity, that the space between the galaxies should be expanding. The implication is that the Universe was born out of a superhot, superdense fireball some 15 billion years ago; this has been confirmed by the discovery of a faint hiss of microwave radiation, filling the Universe, which is the afterglow of the fireball.

Conditions in the Big Bang were more extreme (in terms of density, temperature and pressure) than anything that has existed since or can be reproduced in particle accelerators on Earth. So the Big Bang provides a testing ground for quantum theories which attempt to describe the behaviour of matter and energy under such extreme conditions. Because different theories make different predictions about how the Universe would have emerged from the Big Bang, affecting how it would look today, cosmology (the study of the Universe on the largest scale) has become a test bed for particle physics (the study of the Universe on the smallest scale). For example, the way the Universe expands and the amount of helium in old stars showed that there could not be more than three families of neutrinos, *before* particle physicists were able to prove this in accelerator experiments. The marriage between particle physics and cosmology has been one of the most significant developments in physics in the last quarter of the 20th century.

Further reading: John Gribbin, *Companion to the Cosmos*.

billion 1,000 million, 10^9.

binding energy A measure of the strength with which a collection of particles is held together, especially for an atomic nucleus. The binding energy shows up as a difference in the mass of all the particles taken separately and, in this case, the mass of the nucleus itself. The total mass of, for example, two isolated protons and two isolated neutrons is more than the mass of a nucleus of helium-4, even though a helium-4 nucleus contains two protons and two neutrons. The difference in mass-energy (the binding energy) is released when nuclei form by fusion. This does not, however, mean that the individual nucleons inside the nucleus have different masses to the masses they have when they are free; the mass of the nucleus is an overall property of the system, and is less because the nucleus represents a more stable state than the equivalent number of free nucleons.

It is convenient to measure the binding energy of a nucleus by dividing this by the number of nucleons to give a figure for the 'binding energy per nucleon'. For nuclei up to those with a *mass number* of 50–60, this increases fairly rapidly, indicating that more massive nuclei are more stable (more tightly bound). It then decreases slowly, indicating that nuclei more massive than those of iron and nickel are slightly less tightly bound. This means that energy can be released either by making light nuclei fuse together to make heavier nuclei (up to iron and nickel) or by making heavy nuclei split to make lighter nuclei (down to iron and nickel).

The heavy nuclei themselves can be made from lighter nuclei only by putting energy in. This happens in the explosion of a star, called a supernova. Without such stellar explosions, there would be no elements heavier than iron in the Universe.

Binnig, Gerd Karl (1947–) German physicist who shared the Nobel Prize for Physics in 1986 for the development of the scanning tunnelling microscope.

Born in Frankfurt on 20 July 1947, Binnig (according to his own account) first learned how to be an effective member of a creative team by playing and composing with rock bands while a student in the 1960s. He obtained his PhD for work on super-conductivity from Goethe University in Frankfurt in 1978, then joined the IBM Zurich Research Laboratory, where he worked with Heinrich Rohrer on the new type of microscope.

Bjorken, James David (1934–) American physicist who played a major part in the development of the *parton model* of nucleons, which led to the establishment of the idea of *quarks* as a fundamental building block of matter.

Born on 22 June 1934, in Chicago, Illinois. Bjorken studied at MIT (BSc 1956) and then obtained his PhD from Stanford University in 1959, joining the faculty there the following year, and becoming a full professor at the Stanford Linear Accelerator Center in 1967. At that time, the SLAC machine was being used to investigate the structure of protons by firing electrons into them and observing the way the electrons were deflected (a process similar to the way Ernest Rutherford probed the structure of the atom by bombarding it with alpha particles; see *atomic physics*). Bjorken, a theorist, described these interactions using an abstract mathematical formalism known as current algebra (largely developed by Murray Gell-Mann); on a visit to SLAC, Richard Feynman explained these results in physical terms, in terms of the presence of particles (partons) within the proton. Parton theory was then developed further, alongside the experiments at SLAC, by Bjorken and his colleague

Emmanuel Paschos, and this combination of theory and experiment established the reality of quarks in the minds of most physicists by the end of the 1970s.

In 1979 Bjorken moved to Fermilab, where he became associate director of physics in 1984. He returned to SLAC in 1989.

black body   A hypothetical object which absorbs all electromagnetic radiation that falls on it. Such an object would, if heated, be a perfect radiator, producing *black body radiation*. Such a hot black body may no longer be black, because it will be radiating visible light. The radiation from many hot objects (including the Sun) is approximately the same as that of a black body at the appropriate temperature.

black body radiation   The radiation emitted by a hot *black body*. The best example of a black body in the laboratory is a container with a small hole in it, into which radiation shines and is trapped. If such a container is heated, radiation bounces around inside the cavity and eventually escapes from the hole as black body radiation (this is the origin of the older terminology 'cavity radiation').

The important point about black body radiation is that its properties depend only on its temperature. The curve representing the spectrum of the radiation is shaped like a smooth hill, with a peak at a frequency (colour) which depends on its temperature. The hotter the radiation, the higher the frequency at which the peak occurs. This can be seen by looking at what happens to an iron poker placed in a fire. When the poker is just warm, you can feel heat coming from it in the form of infrared radiation, but cannot see it. As the poker gets hotter, it glows first red, then orange and eventually white hot as the peak of the intensity of the black body radiation it emits shifts across the spectrum. Radiation from the Sun follows almost the black body curve, with a peak corresponding to a temperature of just under 6,000 K.

Until Max Planck developed the first quantum theory of radiation at the end of the 19th century, the shape of the black body curve was a mystery, since it could not be explained by the classical physics of Maxwell's equations of electromagnetism. The problem was that if electromagnetic waves are treated mathematically in the same way as waves on a violin string, and if the waves can be any size, classical theory predicts that when energy (heat) is put into any object and radiated as electromagnetic waves, the amount of energy radiated at each frequency is proportional to the frequency. So the higher the frequency, the more radiation there should be. A black body should emit huge amounts of energy in the highest frequency (shortest wavelength) part of the spectrum, in the ultraviolet and beyond. This was known as the 'ultraviolet catastrophe'.

After years grappling with the problem, in 1900 Plank found a way to avoid the ultraviolet catastrophe. It involved cutting radiation up, mathematically, into chunks, or quanta. At a particular frequency f, each quantum of radiation has an energy E given by the equation $E = hf$ where h is a constant of nature, now known as Planck's constant, determined by experiments (among other things, by studying black body radiation). It is easy to see how this removes the ultraviolet catastrophe. In any object, the energy is distributed among the atoms (and molecules) according to the temperature of the object. A few atoms have low energy, a few have high energy and a lot have a middling amount of energy. But what you mean by 'middling' increases as the temperature increases. Each atom can emit electromagnetic radiation. For very high frequencies (large values of f), the energy needed to emit one quantum of energy (E) is very large, and only a few of the atoms in the black body will have that much energy available, so only a few high-frequency quanta are radiated. At very low frequencies, it is easy for atoms to emit low-energy quanta, but they each have so little energy that even added together they do not amount to much. In between the two extremes, however, there are many atoms which each have enough energy to emit moderate-sized quanta of radiation, which add up to produce the peak in the black body curve. And the peak shifts to higher frequencies for hotter bodies, because in hotter bodies there are more individual atoms with greater amounts of energy. This was the first step towards a quantum theory of physics.

See also *photon*.

Blackett, Patrick Maynard Stuart (1897–1974)   British physicist who developed the *cloud chamber* and received the Nobel Prize as a result in 1948.

Born in Croydon, Surrey, on 18 November 1897, Blackett joined the Royal Navy in 1912 and served at the Battle of Jutland. He resigned from the Navy soon after the end of the First World War, and completed a physics degree at Cambridge in 1922, where he stayed (apart from a year at Göttingen) to carry out research with cloud chambers at the Cavendish Laboratory. He became professor of physics at Birkbeck College in London in 1933, using cloud chambers to study cosmic rays, and moved to the University of Manchester in 1937. From 1935 he was a member of the Air Defence Committee, and during the Second World War Blackett was a scientific adviser to the government; afterwards, he returned to Manchester, where he became interested in the Earth's magnetic field. In 1953 he returned to London as head of the physics department at Imperial College, and was once again a government adviser in the 1960s. He became a life peer (Baron Blackett of Chelsea) in 1969, and died in London on 13 July 1974.

black hole   A concentration of matter which has a gravitational field strong enough to curve spacetime completely round upon itself so that nothing can escape, not even light. This can happen either if a relatively modest amount of matter is squeezed to very high densities (for example, if the Earth were to be squeezed down to about the size of a pea), or if there is a very large concentration of relatively low-mass material (for example, a few million times the mass of our Sun in a sphere as big across as our Solar System, equivalent to about the same density as water).

The first suggestion that objects like this second kind of black hole might exist was made by John Michell, a Fellow of the Royal Society, in 1783. But the notion of 'dark stars' was forgotten in the 19th century and only revived in the context of Albert Einstein's general theory of relativity, when astronomers realized that there was another way to make black holes.

One of the first people to analyse the implications of Einstein's theory was Karl Schwarzschild, an astronomer serving on the eastern front in the First World War. The general theory of relativity explains the force of gravity as a result of the way spacetime is curved in the vicinity of matter. Schwarzschild calculated the exact mathematical description of the geometry of spacetime around a spherical mass, and sent his calculations to Einstein, who presented them to the Prussian Academy of Sciences early in 1916. The calculations showed that for *any* mass there is a critical radius, now called the Schwarzschild radius, which corresponds to such an extreme distortion of spacetime that, if the mass were to be squeezed inside the critical radius, space would close around the object and pinch it off from the rest of the Universe. It would, in effect, become a self-contained universe in its own right, from which nothing (not even light) could escape.

For the Sun, the Schwarzschild radius is 2.9 km; for the Earth, it is 0.88 cm. This does not mean that there is what we now call a black hole (the term was first used in this sense only in 1967, by John Wheeler) of the appropriate size at the centre of the Sun or of the Earth. There is nothing unusual about spacetime at this distance from the centre of the object. What Schwarzschild's calculations showed was that *if* the Sun could be squeezed into a ball less than 2.9 km across, or *if* the Earth could be squeezed into a ball only 0.88 cm across, they would be permanently cut off from the outside Universe in a black hole. Matter can still fall into such a black hole, but nothing can escape.

For several decades this was seen simply as a mathematical curiosity, because nobody thought that it would be possible for real, physical objects to collapse to the states of extreme density that would be required to make black holes. The Schwarzschild radius corresponding to any mass M is given by the formula $2GM/c^2$, where G is the constant of gravity and c is the speed of light.

In the 1930s, Subrahmanyan Chandrasekhar showed that even a white dwarf star, about as big as the Earth, could be stable only if it had a mass less than 1.4 times the mass of the Sun, and that any heavier dead star would collapse further. A few researchers considered the possibility that this could lead to the formation of neutron stars, typically with a radius only one seven-hundredth of that of a white dwarf, just a few kilometres across. But the idea was not widely accepted until the discovery of pulsars in the mid-1960s showed that neutron stars really did exist.

This led to a revival of interest in the theory of black holes, because neutron stars sit on the edge of becoming black holes. Although it is hard to imagine squeezing the Sun down to a radius of 2.9 km, neutron stars with about the same mass as the Sun and radii less than about 10 km were now known to exist, and it would be a relatively small step from there to a black hole.

Theoretical studies show that a black hole has just three properties that define it – its mass, its electric charge and its rotation (angular momentum). An uncharged, non-rotating black hole is described by the Schwarzschild solution to Einstein's equations; a charged, non-rotating black hole is described by the Reissner–Nordstrøm solution; an uncharged but rotating black hole is described by the Kerr solution; and a rotating, charged black hole is described by the Kerr–Newman solution. A black hole has no other properties, summed up by the phrase 'a black hole has no hair'. Real black holes are likely to be rotating and uncharged, so that the Kerr solution is the one of most interest.

Both black holes and neutron stars are now thought to be produced in the death throes of massive stars that explode as supernovae. If such an object happened to be in orbit around an ordinary star, it would strip matter from its companion to form an accretion disc of hot material funnelling into the black hole. The temperature in the accretion disc might rise so high that it would radiate X-rays, making the black hole detectable.

Since the early 1970s, several such objects have been found. But very many stars should, according to astrophysical theory, end their lives as neutron stars or black holes. Observers actually detect about the same number of good black hole candidates in binary systems as they do binary pulsars, and this suggests that the number of isolated stellar mass black holes must be the same as the number of isolated pulsars. This supposition is backed up by theoretical calculations.

There are about 500 active pulsars known in our Galaxy today. But theory tells us that a pulsar is only active for a short time, before it fades into undetectability. So there should be correspondingly more 'dead' pulsars (quiet neutron stars) around. Our Galaxy contains 100 billion bright stars, and has been around for thousands of million of years. The best estimate is that there are around 400 million dead pulsars in our Galaxy today, and even a conservative estimate would place the number of stellar mass black holes at a quarter of that figure – 100 million.

In 1994 observers using the Hubble Space Telescope discovered a disc of hot material, about 150,000 parsecs (some 500,000 light years) across, orbiting at speeds of about 2 million kilometres per hour (about 3×10^7 cm/sec, 0.1 per cent of the speed of light) around the central region of the galaxy M87, at a distance of about 15 million parsecs (3 million light years) from our Galaxy. A jet of hot gas, more than a kiloparsec long, is being shot out from the central 'engine' in M87. The orbital speeds in the accretion disc at the heart of M87 provide conclusive proof that it is held in the gravitational grip of a supermassive black hole, with a mass that may be as great as 3 billion times the mass of our Sun, and the jet is explained as an outpouring of energy from one of the polar regions of the accretion system.

A more speculative suggestion is that tiny black holes, known as wormholes, may form tunnels through the fabric of spacetime. A wormhole can be thought of as a shortcut through spacetime, a cosmic subway connecting two black holes. The 'other end' of a wormhole could be anywhere.

Solutions to the equations of the general theory of relativity that describe wormholes were actually found in 1916, shortly after the theory was developed, although they were not interpreted in this way at the time. Albert Einstein himself, working with Nathan Rosen at Princeton in the 1930s, discovered that the Schwarzschild solution actually represents a black hole as what they called a bridge (now known as an Einstein–Rosen bridge) between two regions of flat spacetime. Although these equations were studied as mathematical curiosities (notably by John Wheeler and his colleagues), before 1985 they were not regarded as real features of the Universe, because every example investigated mathematically opened up only very briefly, snapping shut again (according to the equations) before anything, even light, could traverse the tunnel.

Although the idea was loved by science fiction writers, it was generally accepted by scientists that there must be some law of nature preventing the existence of wormholes. But when relativists working at Caltech tried to prove this in the 1980s, they found that they could not. There is nothing in the general theory of relativity (the best theory of gravity and spacetime that we have, which has passed every test that it has been subjected to) which forbids the existence of wormholes. What's more, Kip Thorne and his colleagues found that there are, after all, solutions to Einstein's equations which allow for the existence of long-lived wormholes.

The 'mouth' of such a wormhole would look like the event horizon of a spherical black hole, but with one important difference. The event horizon is a one-way surface, and nothing can come out of it. But the surface of a wormhole mouth allows two-way traffic. Physicists are intrigued by the possibility that naturally occurring wormholes may exist on the scale of the *Planck length*, providing the basic foam-like structure of spacetime, weaving the fabric of spacetime itself (to mix the metaphor) out of wormhole strands.

If so, there are many curious possibilities. For example, such tiny (ultra-submicroscopic) wormholes may link far distant regions of the Universe, allowing information to leak through and ensuring that the laws of physics are the same here on Earth as they are in a distant quasar. In 1995 some calculations suggested that these tiny wormholes might be topologically equivalent to the *string* invoked by superstring theories to explain the structure of matter on the smallest scales (much, much smaller than protons or neutrons). If so, black holes may be the missing link required to complete the sought-for 'theory of everything'.

Further reading: John Gribbin, *In Search of the Edge of Time*; Kip Thorne, *Black Holes and Time Warps*.

Bloembergen, Nicolaas (1920–)   Dutch-born American physicist who received the Nobel Prize for Physics in 1981 for his development (independently of Arthur Schawlow) of the use of *lasers* in *spectroscopy*.

Born at Dordrecht, on 11 March 1920, Bloembergen studied at the University of Utrecht, where he obtained his bachelor's and master's degrees (in 1941 and 1943 respectively) before moving to the University of Leiden as a teaching assistant. He visited Harvard from 1946 to 1948, where he carried out the research that led to the award of his PhD, but the degree was actually conferred by the University of Leiden, in 1948. Bloembergen promptly returned to Harvard, where he became an associate professor in 1951 and spent the rest of his career, becoming an American citizen in 1958. He was appointed Gordon McKay Professor of Applied Physics at Harvard in 1957, Rumford Professor of Physics in 1974, and Gerhard Gade University Professor in 1980. His interest in masers and lasers began during his research at Harvard in the late 1940s, and formed the basis of his career. The work for which he received the Nobel Prize was largely carried out in the 1970s.

blue sky   The sky is blue because of the way light from the Sun is bounced around (scattered) by molecules in the air. Longer wavelengths of light, at the red end of the spectrum, are scattered less than shorter wavelengths of light, at the blue end of the spectrum. So sunsets and sunrises are red, seen by the leftover light reaching us more or less directly from the Sun, and the sky as a whole is blue, seen by the scattered sunlight. Although John *Tyndall* had realized in the 1860s that the blueness is caused by scattering, he had thought that the scattering was caused by small dust particles or droplets of liquid in the air (this is indeed the reason why sunsets and sunrises are so red). It was later suggested that molecules might be doing most of the scattering that makes the sky blue, but it was only in 1910 that Albert Einstein proved that this is the case, using the way that light is scattered by the atmosphere to provide an estimate of the size of molecules and of the value of *Avogadro's number.*

Bohm, David Joseph (1917–1992)   American-born physicist and philosopher of science who made major contributions to the interpretation of quantum mechanics.

Born in Wilkes-Barre, Pennsylvania, on 20 December 1917, Bohm became

David Bohm (1917–1992).

interested in science through reading science fiction at the age of eight, and then moved on to astronomy books before he began his formal training in physics. He graduated from Pennsylvania State College in 1939 and then worked under Robert Oppenheimer at the University of California, Berkeley. He followed Oppenheimer to Los Alamos, where he worked on the Manhattan Project, and received his PhD (from Berkeley) in 1943. In 1946 he became assistant professor at Princeton University. In 1951 he was fired from this position for his alleged Marxist leanings, in particular his refusal to implicate any of his colleagues as members of the Communist Party (this was at the time of the McCarthy witch-hunts against communism in the USA, and Bohm had been questioned by the Un-American Affairs Committee in 1949). In fact, Bohm was paid for the last year of his contract at Princeton only on condition that he did not actually enter the campus! He moved to Brazil, where he worked at the University of São Paulo, and then (in 1955) to the Technion in Israel, before finally (in 1957) settling in Britain. There he worked at the University of Bristol until 1961, when he became professor of theoretical physics at Birkbeck College, in London, where he stayed until he retired in 1983.

Bohm's early work involved the behaviour of plasmas (very hot gases in which electrons have been stripped from their atoms to leave ions) and the behaviour of electrons in metals, where they act very much as if they were in a plasma. After the Second World War he became intrigued by quantum theory and wrote a textbook, published in 1951, which is still regarded as one of the most accessible accounts of the standard Copenhagen interpretation. But in the process of setting out the standard theory clearly, Bohm became convinced that it was flawed, and he devoted most of the rest of his career to developing and promoting an alternative version of quantum theory, variously referred to as the pilot wave, undivided whole or hidden variables interpretation (much later, Bohm also used the term 'ontological interpretation' to refer to the same package of ideas).

One of the key ingredients in Bohm's interpretation involves non-locality, an instantaneous action at a distance operating between quantum entities. In the 1950s, and for decades thereafter, this was seen by most physicists as a major flaw with the interpretation and it was largely ignored. But in the 1980s, after the Aspect experiment had used Bell's inequality to show that the quantum world really is non-local, there was a revival of interest in this kind of interpretation. John Cramer's transactional interpretation of quantum mechanics, for example, is very much in the spirit of the pilot wave approach.

Bohm also worked on various philosophical problems linked with modern ideas in physics, and on the nature of consciousness.

Bohr, Aage Niels (1922–)   Danish physicist who received the Nobel Prize for Physics in 1975 for his work on the theory of the structure of the nucleus.

Aage Bohr was born in Copenhagen on 19 June 1922, the year that his father, Niels Bohr, was awarded the Nobel Prize for his work on the structure of the atom. Aage began studying physics at the University of Copenhagen, but in October 1943 the Bohr family secretly left Denmark for Sweden, and then on to Britain, fearing that Niels was about to be arrested by the occupying Germans. For the next two years Aage, technically employed by the Department of Science and Industrial Research in London, helped his father with his work as an adviser on the Manhattan Project. The family returned to Denmark at the end of the war, and Aage obtained his MSc in Copenhagen in 1946. In 1948 he visited the Institute for Advanced Study in Princeton, where he became interested in the way in which details of the spectrum of an atom might be interpreted to provide information about the properties of its nucleus. A year later, on a visit to Columbia University, he learned more about the possibilities of this approach from Leo Rainwater. Back in Copenhagen, he developed, with Ben Mottelson, a theoretical model of the nucleus in which the individual nucleons move around within the nucleus, while the nucleus as a whole changes its shape and orientation. This was the work for which Bohr, Mottelson and Rainwater shared the Nobel Prize. The usual analogy is with a swarm of bees, which behaves like a single object moving slowly through the air even though each individual bee buzzes about rapidly inside the swarm.

Aage Bohr completed his PhD in 1954, became professor of physics at the Institute for Theoretical Physics (now the Niels Bohr Institute) in 1956, and succeeded his father (who died in 1962) as director of the Institute in 1963, retiring from the post in 1970 but serving as director of the Nordic Institute for Theoretical Atomic Physics (NORDITA) from 1975 to 1981.

Bohr, Niels Hendrik David (1885–1962) Danish physicist who was awarded the Nobel Prize in 1922 for his theoretical model of the structure of the atom (see *Bohr model*), based on spectroscopy and the principles of quantum physics.

Bohr was born in Copenhagen on 7 October 1885. His father was professor of physiology at the University of Copenhagen, where Niels Bohr studied, obtaining his master's degree in 1909 and his PhD in 1911. He went to Cambridge to work with J. J. Thomson, but they failed to find a rapport, and Thomson expressed little interest in Bohr's work. So in 1912 he moved to Manchester, where Ernest Rutherford had just discovered the nucleus of the atom from analysis of his team's scattering experiments (see *atomic physics*). Bohr made the key breakthrough by postulating that the orbital momentum of an electron 'in orbit' around a nucleus is quantized, so that it can only 'jump' from one allowed orbit to another, and not spiral steadily into the nucleus. The energy associated with these quantum jumps explained the lines seen in the spectra of different elements, notably the *Balmer series* of hydrogen.

In 1912 Bohr returned to Copenhagen, where he developed these ideas as a lecturer at the university. The model (published in 1913) was an unashamed mixture of classical and quantum ideas, with the great merit that it explained spectroscopic observations. It was succeeded, in the 1920s, by a fully quantum-mechanical model, which Bohr also helped to create.

In 1914 Bohr was enticed back to Manchester by Rutherford to become a reader in theoretical physics, but he stayed only until 1916. The authorities in Denmark were so reluctant to lose him that they offered him an immediate professorship and the promise of his own institute. He returned to Copenhagen, and in 1918 the Institute for Theoretical Physics was established (largely funded by donations from the Carlsberg brewery), with Bohr as the first director of the Institute. He held the post until his death in 1962 (when he was succeeded by his son, Aage Bohr). The Institute is now known as the Niels Bohr Institute.

In Copenhagen, Bohr attracted most of the best theoretical physicists of the day for longer or shorter visits, providing a stimulus for the development of ideas in quantum mechanics – so much so that the standard interpretation of quantum mechanics, developed by Bohr and others at the end of the 1920s, and vigorously promoted and defended by Bohr, became known as the *Copenhagen interpretation*. To be fair, though, major contributions to this interpretation were made elsewhere, not least by Max Born and his colleagues in Göttingen. Bohr's forceful personality and prestige ensured that in spite of serious flaws the Copenhagen interpretation became *the* accepted 'explanation' of quantum mechanics, and it was only seriously challenged in the 1980s and 1990s.

Bohr became interested in the possibility of obtaining energy by the fission of uranium (although he did not invent the *liquid drop model* of fission often erroneously attributed to him), and discussed this with American researchers on a visit to the USA in 1939. When Copenhagen was overrun by the German army and occupied at the beginning of the Second World War he became concerned about the possibility of atomic weapons. In 1943, when it seemed likely that he was about to be arrested for his unrestrained patriotic views (his mother was Jewish), Bohr escaped with his family to Sweden and went on to Britain. Working for the British

Niels Bohr (1885–1962).

government, Niels and Aage Bohr went to the United States, where Niels Bohr contributed to the Manhattan Project.

After the war, Bohr was concerned about the dangers inherent in the existence of nuclear weapons and actively worked for their control, organizing the first Atoms for Peace conference in Geneva in 1955. He was one of the principal figures behind the foundation of CERN, the European particle research centre in Switzerland, and in the foundation of the Scandinavian research institute NORDITA. He was also active in promoting the peaceful uses of atomic energy, and served as the first chairman of the Danish Atomic Energy Commission. He died in Copenhagen on 18 November 1962, two days after chairing a meeting of the Danish Royal Academy of Sciences.

His brother, Harald Bohr (1887–1951) was a distinguished mathematician who became professor of mathematics at the University of Copenhagen.

Bohr model A theoretical model of the atom, the first model taking on board ideas from quantum theory, developed by Niels Bohr at Manchester and in Copenhagen in 1912 and 1913, and published in 1913. Although superseded by a fully quantum-mechanical treatment of the atom developed in the 1920s, the Bohr model still provides the most familiar image of the atom for most people, with electrons 'in orbit' around a tiny central nucleus, in a manner reminiscent of the way the planets orbit the Sun.

In 1911 Ernest Rutherford had discovered that an atom does indeed consist of a tiny central nucleus surrounded, somehow, by a cloud of electrons (see *atomic physics*). It was natural to guess that the structure might be like a planetary system. But there was one big problem with this idea. Any electrically charged particle that is accelerated, either by being made to slow down or speed up, or by being made to move in a curved path, radiates electromagnetic waves. Because electrons are electrically charged, they should radiate electromagnetic energy continuously as they orbit around the nucleus, losing energy steadily and falling into the centre of the atom. But Bohr took this idea of planetary electrons circling the nucleus and combined it with the idea that electromagnetic radiation can be absorbed or emitted only in chunks of certain sizes, called quanta. This idea came from Max Planck's explanation of the spectrum of *black body radiation*.

Bohr said that electrons could only occupy certain 'stable orbits' around the nucleus, each corresponding to a certain fixed amount of energy, a multiple of the basic quantum. But there were no in-between orbits, because they would correspond to fractional amounts of energy. Because circular orbits are involved, the basic quanta of energy are measured not in terms of Planck's original constant, h, but in terms of $h/2\pi$ (often written as $\hbar$). An electron might jump from one orbit to another – emitting a quantum of energy if it was moving closer to the nucleus, or absorbing a quantum of energy if the jump took it further out from the nucleus. But it could not spiral in steadily towards the nucleus.

So why didn't all the negatively charged electrons simply jump straight into the nucleus, attracted by its positive electric charge? Bohr added another ingredient to his model, arguing that each stable orbit around the nucleus in some sense has room for only a limited number of electrons. If the orbit was full up, then no matter how many electrons there might be in orbits with more energy, they could not give up that excess energy and jump down into the occupied orbit. Equally, it was simply forbidden for the electrons in the lowest-energy orbit to make the final jump into the nucleus itself. But if the lower orbit had room for it, an electron in a higher-energy orbit could jump down into it, radiating one quantum of energy (what we would now call a photon corresponding to a particular wavelength of light) as it did so; in

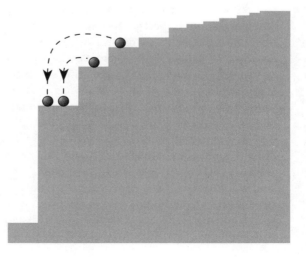

Bohr model. A schematic representation of the energy level 'staircase' in the Bohr model of the atom.

order to jump up from a lower-energy orbit to a higher one, an electron would have to absorb precisely one quantum with the right amount of energy. The image is rather as if each electron sits on one step of a staircase, and can only jump up or down by a whole number of steps, because there are no in-between steps to rest on – but the picture is complicated because the steps are not all the same height. Because every atom of a particular element was assumed to behave in the same way, with the same spacing of steps on the energy-level staircase available to its electrons, the model explained both the appearance of bright lines in the emission spectrum of an element, and the presence of dark lines at the same wavelengths in its absorption spectrum.

As all of this shows, Bohr's great genius was that he didn't worry about trying to develop a complete, self-consistent theory of the atomic world. He was quite prepared to take bits of quantum theory (energy quanta) and bits of classical theory (the idea of orbits) and to mix and match them as necessary to patch up something that worked after a fashion. And his model did work. With each jump from one orbit to another corresponding to a specific energy, and therefore to a specific frequency of light (in line with Planck's equation $E = hf$; see **black body radiation**), it explained the spectrum of light emitted by the simplest atom, hydrogen – in particular, the **Balmer series** of lines in the hydrogen spectrum.

Bohr described his model to the eighty-third annual meeting of the British Association for the Advancement of Science in September 1913. Many people liked the model (Sir James Jeans called it 'ingenious, suggestive and convincing'), while others (including J. J. Thomson) were not convinced. But as the model was developed further (with notable contributions from Arnold Sommerfeld in Germany), the doubters were persuaded, and in 1922 Bohr received the Nobel Prize for his work.

Progress was, though, painfully slow after 1913 (the outbreak of the First World War, restricting communication among scientists, didn't help, but caused much less disruption to normal scientific life than the Second World War would). By mixing classical ideas and quantum ideas in a hodgepodge, adding whatever seemed necessary to the mixture to patch it up and keep the model going, theorists seemed to take one step backwards for every two steps forward. Bohr's original model 'allowed' far more spectral lines than are actually seen, and arbitrary rules (like the one limiting the number of electrons in each orbit) had to be brought in to say that some transitions between different energy states in the atom were 'forbidden'. These properties were organized by assigning so-called quantum numbers to describe the state of the atom and make its behaviour match the observations, with no secure theoretical foundation to explain where the quantum numbers came from, or why certain transitions were forbidden.

Through it all, though, the model worked. It may have predicted too many spectral lines, but it also predicted some lines that had not previously been seen, but were now discovered in exactly the places in the spectrum that the model predicted. And when Albert Einstein finished his general theory of relativity in 1916 and turned his attention to the way atoms radiate energy, he found a convincing connection with known physics. By applying standard statistical techniques to work out the probability that an atom in a state corresponding to a particular set of quantum numbers would 'decay' into another state with lower energy and a different set of quantum numbers, he showed how to add up the effects of radiation from a large number of atoms in a hot object. The equation he derived for the combined radiation from a large number of atoms was exactly Planck's formula for black body radiation. It was this discovery of a link between the behaviour of individual atoms and black body radiation that finally ensured the acceptance of Bohr's model of the atom. But within five years of Bohr receiving his Nobel Prize, the whole idea of orbits and the last vestiges of classical mechanics were to be swept away from atomic theory as quantum mechanics came of age.

Further reading: John Gribbin, *In Search of Schrödinger's Cat*.

Bohr radius The minimum distance at which the electron can orbit the nucleus of a hydrogen atom in the Bohr model of the atom. The radius is determined in terms of the fundamental units of the mass and charge of the electron and Planck's constant; although the Bohr model has been superseded, this number, 5.292×10^{-11} m, is still a good indication of the size of an atom.

Bohr-Sommerfeld model See *Bohr model*. The key improvement made by Sommerfeld was to allow for the possibility of elliptical electron orbits.

Boltzmann, Ludwig (1844–1906) Austrian physicist who (among other things) made important contributions to the understanding of thermodynamics, and extended James Clerk Maxwell's theory of the distribution of velocities among gas molecules. All Boltzmann's work was based on a belief in the reality of atoms and molecules; although this was increasingly accepted elsewhere, in Vienna there was strong opposition to the atomic theory from the 'positivists', headed by Ernst Mach. Boltzmann suffered severe depression and, partly because of the belief that he was fighting a lone and losing battle against the positivists, killed himself in 1906.

Boltzmann was born in Vienna on 20 February 1844. He studied at the University of Vienna (where two of his tutors were Josef Stefan and Josef Loschmidt) and obtained his PhD in 1866. He held several posts in the German-speaking academic world, eventually returning to Vienna to succeed Stefan as professor of theoretical physics in 1894. In 1900 he moved to the University of Leipzig, but he returned to Vienna in 1902 to hold both Stefan's old chair and the chair of natural philosophy, from which Mach had just retired.

Boltzmann made many contributions to physics, including the study of gases and black body radiation (see *Stefan–Boltzmann Law*). His most important work was as a pioneer of the field of statistical mechanics, in which overall properties of matter (such as the pressure of a gas) are understood in terms of the combined properties of the component atoms and molecules, obeying simple laws of physics and the working of chance.

He committed suicide while on holiday at Duino, near Trieste, on 5 September 1906, unaware of Albert Einstein's work on Brownian motion, which convinced many remaining doubters of the reality of atoms.

bond See *chemical bond*.

Born, Max (1882–1970) German physicist who made major contributions to the development of quantum theory, and introduced the idea that the outcome of experiments or interactions involving quantum entities is not directly deterministic, but involves probability in an intimate way.

Born was born in Breslau, in Silesia (then part of Germany, now Wroclaw, in Poland), on 11 December 1882. His mother died when he was four, and he was largely brought up by her mother. Born's father was professor of anatomy at the university there, and Max Born was encouraged by him to study both arts and sciences at Breslau before studying at Heidelberg and in Zurich, where his main subject was mathematics. He then went on to Göttingen, where he was awarded a PhD in physics and astronomy in 1907. Just after this, in 1907–8, he spent six months in Cambridge, where he attended lectures given by J. J. Thomson, and then returned to Breslau, where he learned of Albert Einstein's special theory of relativity. At the end of 1908, Born moved back to Göttingen, where he began to work his way up the academic ladder and carry out research. His early work was on the theory of crystals.

In 1914 Born moved to be professor of physics in Berlin. He carried out war work as a technical expert in the artillery, but had time to strike up a friendship with Einstein, who was also working in Berlin at this time. In 1919 Born moved briefly to Frankfurt-am-Main, but in 1921 he returned to Göttingen University to become professor of physics and head of the physics department, which he developed into a centre of excellence in theoretical physics (especially quantum physics) second only to Niels Bohr's institute in Copenhagen – in the 1920s, Born had Werner Heisenberg, Pascual Jordan and Wolfgang Pauli working in the department. It was Born who coined the term 'quantum mechanics', in 1924. In 1925, when Heisenberg developed a new mathematical description of quantum physics, it was Born who

Max Born (1882–1970).

recognized its connection to matrix theory (see *matrix mechanics*) and worked with Heisenberg and Jordan on a scientific paper which provided the first complete, self-consistent version of quantum mechanics. A little later, Erwin Schrödinger came up with another version of quantum physics, based on treating quantum entities (such as electrons) as waves (see *wave mechanics*). Born showed that the waves in Schrödinger's version of quantum physics could be regarded as representing probability, and he became the leading proponent of the idea that the outcome of any interaction in the quantum world is determined, from the range of possible outcomes allowed by the laws of physics, by chance, in the strict mathematical sense. This led him into a correspondence with Einstein (who refused to accept that 'God plays dice') that lasted until Einstein's death.

The probabilistic interpretation, developed in Göttingen, is a key component of what became known, to Born's annoyance, as the *Copenhagen interpretation* of quantum mechanics. Born also felt distinctly left out when most of the other quantum pioneers were awarded Nobel Prizes in the late 1920s and early 1930s, an oversight that was only rectified (to the relief of many of his fellow physicists) when Born at last received the prize in 1954, when he was in his seventies. The citation specifically referred to his statistical interpretation of the wave function.

Born was a warm, friendly man who took unusual care over the well-being of his students at a time when German universities were still stiffly formal. He was also a pacifist and came from a Jewish family, so he was forced to leave Germany in 1933, when Hitler came to power. He moved first to Cambridge and then (in 1936) to Edinburgh University, where he succeeded Charles Darwin (grandson of 'the' Charles Darwin) as Tait Professor of Natural Philosophy, and became a British citizen in 1939. He stayed in Edinburgh until he retired in 1953, when he returned to Germany, to live in Bad Pyrmont, a spa town near Göttingen. An active opponent of the development of nuclear weapons, Born was one of the founders of the Pugwash movement. His book *Atomic Physics* became a standard university text. He died, at the age of 87, on 5 January 1970, in Göttingen.

Bose, Satyendra Nath (1894–1974)

Indian physicist who made one outstanding contribution to quantum theory, the development of what became known as *Bose-Einstein statistics*.

Born in Calcutta on 1 January 1894, Bose studied at Presidency College there, and became a lecturer in physics at the University of Calcutta, before moving to take up a lectureship at the University of Dacca when it was founded in 1921. In 1924 he found a way to derive Planck's equation for black body radiation using a statistical approach based entirely on the idea that light is made up of tiny particles (photons). This echoed the statistical mechanics approach of Ludwig Boltzmann to the behaviour of gases, but using a different statistical rule; it derives the black body relation entirely in quantum terms, without using the idea of electromagnetic radiation at all. Bose wrote a paper about his discovery and sent it to Albert Einstein, who immediately saw its significance, translated it into German and arranged for its publication in the prestigious *Zeitschrift für Physik*. Einstein developed the idea to apply to other kinds of particle, not just to a 'gas' of photons, which is why this approach is usually referred to as 'Bose–Einstein statistics'. Paul Dirac coined the name 'bosons' for particles which obey Bose–Einstein statistics (see also *Fermi–Dirac statistics*). It is no coincidence that the name 'photon' was coined for the particle of light only in 1926, after Bose had put the quantum theory of light on a secure mathematical footing.

Although Bose obtained leave to spend two years in Europe, where he visited Einstein and many of the other pioneers of quantum mechanics, he made no other major contribution to science. As he commented late in life, 'I was like a comet, a comet which came once and never returned again.' But the light shed by that comet changed the way physicists thought in the mid-1920s, at a crucial time in the development of quantum theory, and has affected the way they have thought ever since. Bose had a distinguished career in education, inspiring generations of young Indian scientists by his example and through his skill as a teacher. He died in Calcutta, on 4 February 1974.

Bose condensate = *Bose–Einstein condensate*.

Bose–Einstein condensate   A group of *bosons* which are all in the same quantum state, and behave like a single entity. In 1995 physicists at the Joint Institute of Laboratory Astrophysics (JILA), in Boulder, Colorado, succeeded in cooling about 2,000 atoms of rubidium gas to 170 billionths of a degree above absolute zero (that is, 170 nanokelvin), where they formed a Bose–Einstein condensate less than 100 micrometres across. The condensate lasted for about 15 seconds, and was cooled all the way down to 20 nanokelvin. If the technique can be extended to larger aggregates, it will make single 'quantum particles' visible.

Bose–Einstein statistics   The statistical rules which apply to the behaviour of a large number of quantum particles which each carry an integer amount of quantum *spin*. Such particles are called bosons; the archetypal example of a boson is the photon, the particle of light, which has spin 1. Any number of identical bosons can exist in the same quantum state, which is crucial in determining the statistical behaviour of large numbers of bosons.

boson   A particle which obeys Bose–Einstein statistics. All bosons have integer *spin* (1, 2 and so on). They are the particles associated with the transmission of forces (for example, the photon carries the electromagnetic force). Bosons are not conserved – for example, photons are created in billions every time you turn on a light, and disappear when they are absorbed by atoms.

Bothe, Walther Wilhelm Georg Franz (1891–1957)   German physicist who won a half-share in the Nobel Prize for Physics in 1954 for developing an improved particle detector technique (the coincidence counter, derived from the Geiger counter) which he used in his work on cosmic rays. Working with Hans Geiger, Bothe showed that the conservation laws (for energy and momentum) are precisely obeyed in particle interactions.

bottom   One of the six *flavours* that distinguish different types of *quark*. Previously referred to as 'beauty'.

Boyle, Robert (1627–1691)   English pioneer of chemistry, best remembered for 'Boyle's Law', which says that the volume of a given mass of gas at constant temperature is inversely proportional to the pressure on it (so if you double the pressure, you halve the volume, and so on).

Born on 25 January 1627 in Lismore Castle, Munster, in Ireland, Boyle was the fourteenth child of the Earl of Cork. He was educated at Eton, and was one of the founders of the Royal Society, which received its charter from Charles II in 1662. His best scientific work was carried out in Oxford, where he lived from 1654 until 1668. Among his many and varied contributions to science, the most important in the context of the present book was the way he made chemistry scientific, bringing it out from under the mystique of alchemy and teaching that the objective of chemistry was to find out what different substances are made of. He invented the use of the term 'analysis' for this kind of work.

In *The Sceptical Chymist*, published in 1661, Boyle rejected the Aristotelian idea of the four elements, and (following thinkers such as Pierre Gassendi) argued that all matter is made up from primary particles which join together in different ways, and whose behaviour and movement explains the observed properties of matter. This helped to spread the idea of atoms. He introduced the idea of elements and compounds in the same book.

Boyle died in London on 30 December 1691.

Bragg, Sir William Henry (1862–1942) and Bragg, Sir (William) Lawrence (1890–1971) Father-and-son team of English physicists who jointly won the Nobel Prize for Physics in 1915, for their studies of crystal structure using X-rays.

The elder Bragg (known as William) was born in Westward, in Cumberland (as it then was) on 2 July 1862. His father was a former seaman who had become a farmer. Bragg studied at Trinity College, Cambridge, graduating in mathematics in 1884. He

Sir William Bragg (1862–1942).

Sir Lawrence Bragg (1890–1971).

stayed in Cambridge until 1886, when he became professor of mathematics at the University of Adelaide, Australia. He began to experiment with the newly discovered X-rays, and partly as a result of this career change became professor of physics at Leeds, in England, in 1909. In 1912 Max von Laue and his colleagues in Munich discovered that X-rays could be diffracted by crystals (proving that X-rays behave like very-short-wavelength waves), and Bragg began work, now with his son, on using this discovery to probe the structure of crystals.

The younger Bragg (usually known as Lawrence) was born in Adelaide on 31 March 1890. He studied mathematics at Adelaide University and then (when his father moved to Leeds in 1909) at Trinity College in Cambridge, before switching to physics in 1910 at his father's instigation. He graduated in 1912 and in 1914 became a Fellow of Trinity College and a lecturer in the university. By then, as a young researcher at the Cavendish Laboratory, he had made his name with his first piece of work after graduating, a study of the diffraction patterns found by von Laue. This led him to discover Bragg's Law, which relates the spacing of bright and dark features in the diffraction pattern to the wavelength of the X-rays and the spacing of the atoms in the crystal.

William Bragg used this discovery in his experimental work on crystals, and father and son combined to write a book, *X Rays and Crystal Structure*, which appeared in 1915. By then, Lawrence was serving as a technical expert in the army in France, where he learned that he had become the youngest person (at the age of 25) to win a Nobel Prize. The Braggs are still the only father-and-son team to share this achievement for work carried out together. The X-ray crystallography technique was later instrumental in determining the structure of DNA.

In the same year, William Bragg moved to University College, London; in addition, he served as a government adviser during the war. He was knighted in 1920, and became director of the Royal Institution in 1923. He died in London on 12 March 1942.

Lawrence Bragg survived the war and in 1919 became professor of physics at Manchester University. He then served as director of the National Physical Laboratory in 1937 and 1938, as Ernest Rutherford's successor as Cavendish Professor in Cambridge from 1938 to 1953, and (following in his father's footsteps) as director of the Royal Institution from 1954 to 1966. He was knighted in 1941. Lawrence Bragg retired in 1966 and died in Ipswich in 1971.

Brattain, Walter Houser (1902–1987)   American physicist (actually born in China, but brought up on a cattle ranch in Washington state), who shared the Nobel Prize for Physics in 1956 with William Shockley and John Bardeen for their discovery of the transistor effect.

Born in Amoy, China, on 10 December 1902, Brattain studied at Whitman College (BSc 1924) and at the University of Oregon (MA 1926), before working for his PhD (awarded in 1928) at the University of Minnesota. After a year working for the National Bureau of Standards, Brattain joined Bell Laboratories in 1929 and stayed there throughout his career, except for wartime work on submarine detection. He retired in 1967 and went back to Whitman College, where he studied biological membranes for five years as adjunct professor and then in an honorary post. He died in Seattle, Washington, on 13 October 1987.

Brockhouse, Bertram Neville (1918–)   Canadian physicist who received the Nobel Prize for Physics in 1994 for studies of solids and liquids using neutron-scattering techniques.

Brockhouse was born in Lethbridge, Alberta, on 15 July 1918. After graduating from high school in 1935, Brockhouse worked as a laboratory assistant and as a self-employed radio repair man. During the Second World War, he served with the Royal Canadian Naval Volunteer Reserve, for some of the time as an electronics technician, and after the war he studied at the University of British Columbia (BSc 1947) and at the University of Toronto (PhD 1950). Brockhouse worked at the Chalk River Laboratories of the Atomic Energy Project of the National Research Council of Canada (later Atomic Energy of Canada Limited) from 1950 onwards. The work for which he shared the Nobel Prize with Clifford Shull (but which he carried out independently of Shull) was done between 1950 and 1962, and probed the structure of materials using *inelastic scattering* of slow neutrons. In 1962 Brockhouse became professor of physics at McMaster University, in Hamilton, Ontario. He retired in 1984.

Broglie, Louis-Victor Pierre Raymond, Prince de (1892–1987)   French physicist who received the Nobel Prize for Physics in 1929 for the proposal, put forward in his doctoral thesis in 1924, that all material 'particles', such as electrons, could also be described in terms of waves. This wave–particle duality lies at the heart of quantum physics.

De Broglie was born at Dieppe, on 15 August 1892. He came from an aristocratic family, and initially studied history at the Sorbonne, entering in 1909, being intended for a career in the diplomatic corps. But under the influence of his elder brother Maurice (seventeen years his senior), who became (very much against the wishes of his father) a pioneering researcher interested in X-ray spectroscopy, Louis began studying physics alongside history. Maurice had obtained his doctorate in 1908, and had been one of the scientific secretaries at the first Solvay Congress. Louis' study of physics was interrupted in 1913 by what should have been a short spell of compulsory military service, but was extended when war broke out. During the First World War, he served in the radio communications branch of the army, operating for a time from the Eiffel Tower. It was because of army service and his switch from history that de Broglie's scientific studies were delayed, so that his PhD thesis was only submitted (at the Sorbonne) in 1924, when he was in his thirties. But by then he had already published several important papers on the properties of electrons, atoms and X-rays, and he was one of the first physicists to accept fully the idea of light quanta, which he discussed in an article published in 1922. The thesis developed from Albert Einstein's work which showed that light (traditionally thought of as a wave) could also be explained in terms of particles (now known as photons) to propose that 'particles' could also behave as waves, with the wave and particle aspects of the quantum entity linked by the equation: wavelength x

Louis-Victor de Broglie (1892–1987).

momentum = h, where h is *Planck's constant*. De Broglie's supervisor, Paul Langevin, was nonplussed by this, and showed the thesis to Einstein. Einstein (who had just received Satyendra Bose's famous paper on light quanta) wrote back to assure Langevin that de Broglie's work was sound ('I believe', he said, 'that it involves more than a mere analogy'), and de Broglie duly received his PhD. It was in his second paper on the 'Bose gas' that Einstein made a reference to de Broglie's work which caught the attention of Erwin Schrödinger, and started him down the road that led to wave mechanics.

After completing his thesis, in 1924 de Broglie became a lecturer at the Sorbonne, taking up a joint post at the new Henri Poincaré Institute in 1928 and being appointed professor of theoretical physics, a post he held until 1962. Initially, de Broglie developed an interpretation of quantum mechanics in which the wave 'steered' a real, material particle; this was obscured by the Copenhagen Interpretation, which dominated thinking after 1927, but it bears many similarities to the 'hidden variables' interpretation put forward by David Bohm and others, and now once again taken seriously by many physicists.

Both Maurice and Louis de Broglie were involved with the peaceful development of atomic energy, and served on the French High Commission of Atomic Energy. Louis de Broglie inherited the title Duc from Maurice when his brother died in 1960; he already held the Austrian title Prinz, to which all male members of the family were entitled as a reward after an ancestor gave help to the Austrian side in the Seven Years' War. Louis de Broglie died at Louveciennes on 19 March 1987.

broken symmetry   Best explained by an example. We can think of a bar magnet as containing an enormous number of tiny magnets. When the bar is heated, the internal magnets jiggle and spin around and jostle against one another, so there is no overall magnetic field associated with the bar. It is, in magnetic terms, symmetric, with nothing to distinguish the two ends. But when the bar cools, the tiny magnets do not have enough energy to move around and they line up with one another, freezing into a state in which all their magnetic fields are working together. Now, the bar has an overall magnetism, with a north pole at one end and a south pole at the other. The symmetry has been broken.

Another way to think of broken symmetry is in terms of a ball balanced on a ridge. The situation is symmetric until the ball rolls off on one side of the ridge or the other, when the symmetry is broken.

In physics, there are high-energy symmetries associated with the *forces of nature*. At high energies, the electromagnetic force and the weak nuclear force are equally strong and described by one set of equations, as the electroweak force. At low energies, the symmetry is broken and the forces behave in different ways. The idea of symmetry, and symmetry breaking, is an important component of theories that attempt to describe all the forces of nature in one mathematical package, a theory of everything.

See *symmetry, spontaneous symmetry breaking*.

Brookhaven National Laboratory   American particle accelerator centre on Long Island, New York. One of the early machines at Brookhaven, built in 1953, was one of the first synchrocyclotrons capable of accelerating protons to energies above 1 GeV.

Brown, Robert (1773–1858) Scottish botanist who noticed, in 1827, that pollen grains suspended in water can be seen under the microscope to be in continuous erratic movement. He had no explanation for this Brownian motion, as it became known. In 1905 Albert Einstein proved that it is caused by molecules of water buffeting the pollen grains.

Brownian motion The erratic motion of any tiny particles (such as pollen grains) suspended in a fluid such as water (or in the air). Robert Brown discovered that the tiny grains move in an erratic zigzag fashion in 1827, and in the 1860s several physicists independently suggested (but could not prove) that this might be due to the buffeting they receive from the constant, but erratic, bombardment by the molecules of the fluid in which they are suspended.

When Albert Einstein came to the problem in 1905, he did so from the opposite direction, having convinced himself of the reality of molecules and trying to find evidence that they existed that would persuade others. In a sense, he predicted Brownian motion, starting from the molecular theory. He showed, in a paper published in 1905, that 'according to the molecular-kinetic theory of heat, bodies of microscopically visible size suspended in a liquid will perform movements of such magnitude that they can be easily observed in a microscope'. Each 'zig' or 'zag' of the particle, Einstein showed, was caused not by the impact of a single molecule, but by the combined statistical effect of many impacts (an idea put forward qualitatively, unknown to Einstein, by the Frenchman Louis Georges Gouy some time earlier). Each tiny grain feels a pressure from the bombardment it receives on all sides, and moves slightly in the direction in which a statistical fluctuation slightly reduces the pressure, or away from a direction in which the pressure temporarily increases. Einstein calculated the precise way in which the statistics worked, and predicted that the average distance of a particle from any chosen starting point increases as the square root of the time since it was first kicked away from the starting point. This 'random walk', as it is now known, means that, ignoring all the zigs and zags and measuring the straight-line distance from the starting point, the particle travels twice as far in 4 seconds as it does in 1 second, and takes 16 seconds to travel four times as far as it does in 1 second. This was a completely new prediction, made by Einstein and confirmed by observations. The random walk idea carries over into many other areas of science; and using Einstein's calculations other physicists used the rate at which suspended particles drifted away, in zigzag fashion, from their starting points to calculate the size of water molecules, and make an estimate of *Avogadro's number*.

bubble chamber A device which records the tracks of high-energy particles through a fluid by making a trail of bubbles along the line of flight of the particle. The principle is exactly the same as the way in which bubbles form in a bottle of fizzy drink when the top is opened – the bubbles grow because of a fall in pressure. But the efficiency of the bubble chamber depends on another trick as well, preparing the liquid in a 'superheated' state.

If a liquid under pressure is kept at a temperature close to the boiling point corresponding to that pressure, when the pressure is gently reduced the liquid will begin to boil. The lower the pressure, the lower the boiling point, as any mountaineer who has tried to make a good hot cup of tea will know. But if the pressure is reduced

Bubble chamber. One of the largest bubble chambers in the world, known as BEBC, operated at CERN. It was 3.7 metres across, and 6.3 million photographs of events inside the chamber were taken during its operational lifetime, which ended in 1984.

suddenly, the liquid will remain liquid (at least, for a time) even though its temperature is now higher than the boiling point corresponding to the lowered pressure. This is a superheated liquid. Such a state is unstable, and if a charged particle zips through the liquid, it will create a trail of bubbles along its line of flight, as the liquid begins to boil.

If nothing else were done, the trail would soon vanish in the froth of boiling liquid. But if the pressure is first lowered quickly and then almost immediately (after about a millisecond) restored to its higher value, the trail of bubbles grows big enough and remains long enough to be photographed, but the liquid as a whole does not boil. Then, the bubbles disappear and the whole process can be repeated. Don't be misled by the use of the term 'boil', though; a common working fluid in a bubble chamber would be liquid hydrogen, at a temperature far below the freezing point of water and a pressure several times that of the atmosphere at the surface of the Earth. The pressure is altered by the rapid movement of a large piston.

The bubble chamber was invented by Donald Glaser at the University of Michigan in the early 1950s, and developed by Luis Alvarez at Berkeley. Bubble chambers played a crucial role in the investigations of the particle world that led to the development of the standard model.

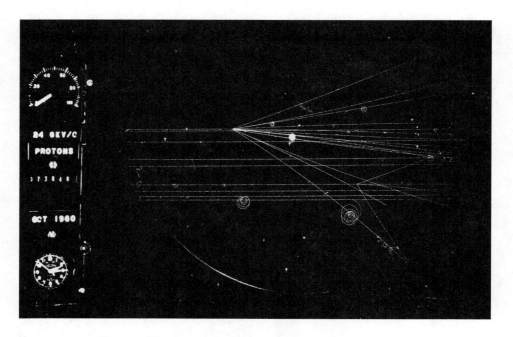

A photograph of the tracks inside a **bubble chamber**, showing particle interactions.

Bunsen, Robert Wilhelm (1811–1899)   German physicist and chemist best known for the burner that now bears his name, but which was actually invented by Michael Faraday and improved by Bunsen's assistant, Peter Desdega, who had the commercial acumen to market it under Bunsen's name. With the aid of the eponymous burner, Bunsen and Gustav Kirchoff invented the use of spectroscopy as a tool for chemical analysis at the end of the 1850s. They investigated the spectra of different elements, and used spectroscopy to discover new elements, opening up a whole new area of scientific investigation of atoms and molecules.

Cabibbo angle   A measure of the probability that one flavour of quark will change into another flavour when affected by the weak force. See *Cabibbo rotation, mixing angle.*

Cabibbo rotation   The transformation of one flavour of quark into another under the influence of the weak force. This is perceived as if the two kinds of quark were different aspects of a single particle, just as the head and the tail are separate facets of a single coin. When the underlying 'particle' is rotated, it changes its visible appearance so that, say, what previously looked like an up quark now looks like a down quark. See also *mixing angle, isotopic spin.*

caesium clock   See *atomic clock.*

Cameron, Alastair Graham Walter (1925–)   Canadian-born American physicist who made important contributions to the understanding of how elements are manufactured inside stars.

Born in Winnipeg, Cameron studied in Manitoba, where he obtained his BSc in 1947, and at the University of Saskatchewan, where he was awarded his PhD in 1952.

He worked as a research officer for Atomic Energy Canada Limited before emigrating to the United States in 1959. He became a naturalized US citizen in 1963 and has held posts at Caltech, the Goddard Institute for Space Studies, Yeshiva University, and Harvard University.

In the 1950s, Cameron developed a theoretical understanding of the way elements are formed inside stars, and he published this in 1957. Unluckily for him, Fred Hoyle and his colleagues had independently developed an even more comprehensive treatment of the problem, which was published in full in the same year, relegating Cameron to a footnote in history. See *nucleosynthesis*.

canal rays Obsolete name for the positively charged 'rays' produced (as well as the negatively charged cathode rays, or *electrons*) by a cathode ray tube; now known to be *ions*.

Cannizzaro, Stanislao (1826–1910) Italian chemist and political activist who was the first person to appreciate the difference between molecular weight and atomic weight, drawing on the work of Amedeo Avogadro to show that common gases such as hydrogen exist as molecules (in this case, H_2).

Born in Palermo, Sicily, on 13 July 1826, Cannizzaro studied in Palermo, Pisa and Turin. He worked in Pisa from 1845 to 1847, then returned to Sicily and took part in the failed Sicilian rebellion against the ruling Bourbons (at the time, his father, a local magistrate, was Chief of Police). He was sentenced to death after the 1848 'revolution', but fled to Paris, returning to Italy in 1851 to become professor of chemistry at Alessandria. In 1855 he moved to Genoa, where he carried out his most important scientific work. He then went back to Sicily and briefly served in Garibaldi's rebellion of 1860, became professor of chemistry at Palermo in 1861, and moved on to Rome in 1871, after the unification of Italy. There, he was appointed professor of chemistry at the university, and was also made a Senator in the new republic (he later became Vice President).

Cannizzaro was the first person to draw up a table of atomic and molecular weights based on the atomic weight of hydrogen as the fundamental unit of mass; this was an important step towards the *periodic table*. He also coined the name 'hydroxyl' for the OH radical. He died in Rome on 10 May 1910.

capture Any process in which an atom or a nucleus acquires an extra particle. For example, a nucleus may capture a neutron, changing into a different isotope of the same element; or an atom may capture an electron, becoming a negatively charged ion.

carbon The element with atomic number 6. The most common isotope of carbon has six protons and six neutrons in its nucleus, and its mass is defined as 12 atomic mass units, setting the standard against which all atomic and molecular weights are measured.

carbon dating Technique used to establish the age of samples of organic material. As well as stable isotopes of carbon, some of the carbon in carbon dioxide in the air is in the form of radioactive carbon-14, produced by the interaction of cosmic rays with nitrogen in the atmosphere. Along with other isotopes of carbon, the carbon-14 is taken up by plants when they absorb carbon dioxide during photosynthesis, and is incorporated in their tissues. From plants, the radioactive carbon-14 (sometimes known as radiocarbon) can get into the tissues of animals, when the plants are eaten.

When the living organism dies, no more carbon from the atmosphere can be incorporated into its remains. The proportion of carbon-14 in the remains (compared with the proportion of stable isotopes, in particular carbon-12) decreases as time passes, because the carbon-14 decays with a *half life* of 5,730 years. In principle, this makes measurements of the amount of carbon-14 remaining in the samples a good guide to ages up to about 45,000 years (eight half lives, by which time the amount of carbon-14 left in the sample is only 1/256 times the amount it started with). In practice, the situation is complicated because at different times in the past there has been more or less carbon-14 produced in the atmosphere, as the flux of cosmic rays has varied. However, for recent millennia the carbon-14 calendar can be calibrated by comparing radiocarbon dates for wood samples with the true dates of the wood samples, determined by counting tree rings. This comparison of tree-ring and carbon-14 data also supplies information about how the input of cosmic radiation to the atmosphere has varied.

carbon resonance   The existence of an energy level for an excited carbon-12 nucleus which allows carbon to form from reactions involving three nuclei of helium-4 inside stars. Without the existence of the carbon resonance, there would be no heavy elements in the Universe and we would not exist.

The main process by which elements are built up inside stars is the addition of helium-4 nuclei (alpha particles) to existing nuclei (see *nucleosynthesis*). But there is a bottleneck at the first link in the chain because beryllium-8, the nucleus made by putting two alpha particles together, is unstable, and lives for only 10^{-19} sec. The impact of a third alpha particle with this unstable nucleus during its brief lifetime might be expected to break it apart completely. But in the 1950s Fred Hoyle realized that the three alpha particles could get together to make carbon-12 if the carbon nucleus itself had an excited energy state corresponding to the combined energy of three alpha particles. If so, instead of the double collision blowing everything apart the three alpha particles would stick together, settling, as it were, on a high-energy 'step' of the ladder of energy levels belonging to the carbon-12 nucleus (analogous to the energy levels of the *Bohr model*). The newly formed nucleus could then radiate its excess energy away and settle into the ground state.

The combined energy of a beryllium-8 nucleus plus an alpha particle is 7.3667 MeV. Hoyle badgered experimenters at Caltech to measure the energy levels of the carbon-12 nucleus, searching for a level in this range, and they found an excited state at 7.6549 MeV. The extra 0.3 MeV is just the sort of energy that would be carried into the collision by the kinetic energy of the third alpha particle, sliding the triplet smoothly into place on the energy step. Hoyle's insight provided one of the most powerful examples of logical inference in the whole of science; the head of the group who carried out the experiments suggested by Hoyle, Willy Fowler, later received a Nobel Prize for this work, but, ridiculously, Hoyle was snubbed by the Nobel committee because of his unconventional views in other areas of science.

Further reading: John Gribbin, *In the Beginning*.

Casimir, Hendrik Brugt Gerhardt (1909–)   Dutch physicist who, in a wide-ranging career, made particularly important contributions to the theory of superconductivity (developing an early version of the 'two-fluid' model), and predicted the *Casimir effect*.

Born at The Hague on 15 July 1909, Casimir studied at Copenhagen and at Leiden University, obtaining his PhD in 1931. Apart from a brief spell in Zurich, he stayed at Leiden until 1942, when he moved to the Philips Research Laboratories at Eindhoven because of the difficulties faced by Dutch universities during the German occupation. He stayed there for the rest of his career, retiring in 1972 and becoming emeritus professor at Leiden.

Casimir effect   A quantum force which pulls two parallel metal plates, placed a short distance apart, towards one another.

Quantum uncertainty allows energy to appear spontaneously from nothing, as long as it disappears again swiftly. Since matter is a form of energy, this allows particles to appear briefly out of nothing at all, so that 'empty space' has to be thought of as seething with 'virtual particles' (strictly speaking, particle–antiparticle pairs) which exist only fleetingly before disappearing again.

The easiest particles to make are photons, since they have no rest mass. In the 1940s, the Dutch physicist Hendrik Casimir suggested an experiment to measure the influence of these virtual photons. If two metal plates are placed close together, face to face, the virtual photons of the quantum vacuum in the gap between the plates will bounce between the highly reflective surfaces of the plates. The effect of this is not to push the plates apart, but to pull them together, for the following reason.

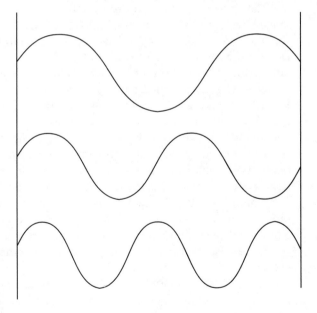

Casimir effect. Standing waves bouncing between two conducting plates.

Think of the photons in terms of waves of light, and make an analogy with the vibrations of a plucked guitar string. Because the string is fixed at each end, it can vibrate only at certain frequencies – certain wavelengths. In the same way, only photons of light with certain wavelengths will fit neatly into the gap between the two metal plates, 'fixed' at each end. This means that all other possible wavelengths are excluded from the gap, so there are *fewer* photons in each cubic centimetre of vacuum between the plates than there are in the vacuum outside. So, in effect, there is an excess pressure from outside pushing the plates together.

The resulting force is very small, but it has been measured (for plates separated by gaps of a few nanometers), showing that Casimir was right and proving that quantum fluctuations of the vacuum are a real phenomenon.

cathode A negatively charged electrode in, for example, a vacuum tube (electronic valve) or a battery (electric cell). Because electrons are negatively charged, they move away from a cathode. This is the origin of their old name, cathode rays.

cathode rays Old name for electrons, which were first discovered as a stream of 'radiation' emitted by a cathode. See *atomic physics*.

'cat in the box' thought experiment See *Schrödinger's cat*.

Cat paradox See *Schrödinger's cat*.

Cavendish, Henry (1731–1810) English scientist who made pioneering investigations in chemistry and used a torsion balance experiment, devised by John Michell, to make the first accurate measurements of the mean density of the Earth and the strength of the gravitational constant. He also carried out pioneering work on electricity, but much of his work was not published in his lifetime, and only became widely known when Cavendish's papers were edited and published by James Clerk Maxwell in 1879.

Cavendish could afford not to publish his results because he did not have to make a living out of science. Born on 10 October 1731 at Nice, in France, Cavendish was the son of Lord Charles Cavendish, and grandson of both the 2nd Duke of Devonshire (on his father's side) and the Duke of Kent (on his mother's side). His father, himself a Fellow of the Royal Society, was administrator of the British Museum. Henry Cavendish studied at Cambridge University from 1749 to 1753, but left without taking a degree (not particularly unusual in those days), and studied in Paris for a year before settling in London. He lived off his private fortune, and devoted his time to the study of science. Apart from his scientific contacts, he was reclusive and published little, although he used some of his money to found a library, open to the public, located well away from his home. He was once described as 'the richest of the learned, and the most learned of the rich'.

Among his unpublished discoveries, Cavendish anticipated Ohm's Law and much of the work of Michael Faraday and Charles Coulomb. He also showed that gases could be weighed, and that air is a mixture of gases, not a pure substance.

Cavendish died on 28 February 1810, and left more than a million pounds in his will. The famous Cavendish Laboratory in Cambridge, named after Henry Cavendish, was founded in 1871 with funds provided by the 7th Duke of Devonshire, a relative of Cavendish and himself a talented mathematician.

Cavendish Laboratory A centre of excellence in Cambridge, England, where many of the pioneering investigations of the subatomic world were carried out. Established in 1871 and named after *Henry Cavendish*.

cavity radiation See *black body radiation*.

Celsius, Anders (1701–1744) Swedish astronomer best remembered for the temperature scale that now bears his name.

Born at Uppsala on 27 November 1701, Celsius became professor of astronomy there in 1730, a post he held until his death, on 25 April 1744. But although based in

Uppsala, Celsius travelled and worked in France, Germany and Italy. He presented his idea for a temperature scale to the Swedish Academy of Sciences in 1742.

Celsius scale   Temperature scale originally devised by Anders Celsius in 1742. He suggested a scale in which the boiling point of water was defined as 0 degrees and the freezing point as 100 degrees. Shortly after his death, colleagues at Uppsala University revised the scale by switching the numbers associated with these two fixed points. At one time this was known as the centigrade scale, but the name was changed in 1948 to avoid confusion with the grade, a unit of angular measure.

centigrade scale   Old name for the *Celsius scale.*

Cerenkov, Pavel Alekseyevic (1904–1990)   Russian physicist who received the Nobel Prize for Physics in 1958 for the discovery and interpretation of the phenomenon now known as *Cerenkov radiation.* Born on 25 July 1904 (15 July, Old Style) in Novaya Chigla, where his parents were peasants, Cerenkov had to work for a living while attending school. He did not enter college until he was twenty, but then studied at Voronezh State University, where he graduated in 1928. In 1934 he discovered the radiation effect now named after him, when he was a research student at the Institute of Physics of the Academy of Sciences in Moscow. He received his doctorate in 1940, and stayed at the Institute throughout his career, contributing to the development of the synchrotron facility there and becoming an outstanding teacher; he became a full professor the year after receiving the Nobel Prize, and died on 6 January 1990.

Cerenkov counter (Cerenkov detector)   A particle detector which monitors the *Cerenkov radiation* produced by particles moving rapidly through a transparent liquid (often water, but also organic liquids) or solid (such as fused quartz) to count the number of particles. Such detectors can also be used to monitor the flux of neutrinos. Although neutrinos do not themselves produce Cerenkov radiation, they may interact with electrons in the fluid, boosting the recoil electrons to a high enough speed to make this radiation.

Cerenkov radiation   Radiation, in the form of bluish light, produced by charged particles moving through a transparent medium at a speed greater than the speed of light in the medium. This is the optical equivalent of a sonic boom. The light is emitted in a cone around the direction the particle is travelling, and the angle of the cone depends on the speed of the particle. The radiation was discovered in 1934 by the Russian physicist Pavel Cerenkov.

CERN   European particle physics research centre, founded in 1954. Originally named Conseil Européen pour la Recherche Nucléaire, from which this acronym comes, but later renamed (with a sex change) Organisation Européenne pour la Recherche Nucléaire, although still known as CERN, not as OERN. The laboratory straddles the Swiss–French border, near Geneva, and is the home of the Large Electron Positron collider (*LEP*). (See illustration on p.72.)

CESR = Cornell Electron Synchrotron Ring, an accelerator at Cornell University.

Chadwick, James (1891–1974)   British physicist who was awarded the Nobel Prize in 1935 for his discovery of the neutron.

Born at Bollington, near Macclesfield, on 20 October 1891, Chadwick attended Manchester Grammar School and studied at the University of Manchester, graduating in 1911. He stayed on to work with Ernest Rutherford on radioactivity,

CERN. The tunnel of the Large Electron-Positron collider (LEP) at CERN.

and was awarded an MSc in 1913 before moving to Berlin to work with Hans *Geiger*. This unfortunately timed move led to him being interned from 1914 to 1918, in a camp near Spandau. After the First World War he took up a post in Cambridge, and in 1923 he became assistant director of the Cavendish Laboratory, under Rutherford. In 1935 Chadwick became professor of physics at the University of Liverpool, where he directed the construction of the first cyclotron in Britain, and in the Second World War he was the head of the British delegation sent to the United States to work on the development of the atomic bomb. He was knighted in 1945, left Liverpool in 1948 to become master of Gonville and Caius College in Cambridge, and retired in 1958. He died in Cambridge on 24 July 1974.

Chadwick's most significant contribution to physics was carried out in 1932, when he followed up the discovery by Walther *Bothe* and Irène *Joliot-Curie* and her husband of artificial radioactivity. They had found that when light elements such as beryllium are bombarded with alpha particles, some form of uncharged (and therefore difficult to detect) radiation is produced from the light target element, and this in turn causes protons (which are easy to detect) to be ejected from paraffin. At first, it was thought that this artificial radioactivity was a form of intense gamma radiation. Chadwick explained that the alpha radiation knocks neutrons out of the nuclei of beryllium (or whatever the target might be), and that the neutrons in turn knock protons (hydrogen nuclei) out of the paraffin molecules, which are rich in hydrogen. In further experiments using boron as the target, Chadwick determined the mass of the neutron, slightly greater than the mass of the proton.

chain reaction   A nuclear fission reaction that maintains itself because neutrons produced by the fission (splitting) of some unstable atomic nuclei (such as those of uranium-235) strike other nuclei nearby, causing them to split and eject more neutrons. In a runaway chain reaction, each atom that splits triggers the fission of at least two more nuclei, producing a nuclear explosion. In a controlled chain reaction, control material (a moderator) added to the fissile material absorbs some of the neutrons harmlessly, so that on average each atom that splits triggers just one more fission. Energy is still released, but in a controlled fashion. This is the principle used in all nuclear power stations to date.

    If each fission triggers exactly one more fission, the chain reaction is said to be critical. If each fission triggers more than one other fission, it is supercritical. If each fission, on average, fails to trigger even one other fission, the reaction is subcritical.

Chamberlain, Owen (1920–)   American physicist who shared the Nobel Prize with Emilio Segre in 1959 for their discovery of the antiproton, only the second antiparticle (after the positron) to be identified. Born in San Francisco on 19 July 1920, Chamberlain graduated from Dartmouth College in 1941 and joined the Manhattan Project, where he first worked with Segre. After the Second World War, he studied for his PhD (awarded in 1949) at the University of Chicago, where he was supervised by Enrico Fermi. Most of his career was spent at the University of California at Berkeley, where the antiproton work was carried out, using the **Bevatron**, in 1955.

chaos   Unpredictable behaviour occurring in response to precisely deterministic laws. An essential feature of a chaotic system is that its behaviour is non-linear, so that a small change in the initial conditions of an experiment (or of a situation in the real world) may have a very large influence on the outcome of the experiment. This means that what are called 'predictive errors' resulting from the imprecision of our knowledge about the initial conditions of the system get bigger as time passes, until, beyond a certain point, we cannot predict how the situation will develop at all.

    In a typical non-chaotic (that is, linear) situation, errors still accumulate, but they do so roughly in proportion to the time that has elapsed, so predictions can be made more or less reliable. The fundamental feature of chaotic (that is, non-linear) systems is that if two identical systems are given slightly different nudges (or if the same system starts out from the same place twice, in two very slightly different ways), the differences between them diverge exponentially fast. Instead of the error (or difference) increasing by the same amount in each second that passes, in one second the divergence may be as large as in all of the preceding seconds since the experiment began, and so on for each second that passes. Small errors quickly grow to overwhelm the power of any predictive calculation.

The best example of this is the behaviour of a pendulum that is mounted so that it is free to swing in any direction – it could simply be a ball suspended from a piece of string attached to a pivot. If the pivot is jiggled in and out by a motor in a smooth, regular fashion, the ball will start to swing about. After a while, it may settle down into a steady pattern, with the ball tracing out a roughly elliptical path, going round and round exactly the same loop, with the same period as the rhythm of the motion of the pivot. But if the period of this pivotal motion happens to be close to some critical value, when the frequency of the driving rhythm is altered slightly the movement of the ball may become chaotic, with the pendulum first swinging one way and then another, doing a few clockwise turns and then a few anticlockwise turns, in a completely random manner. The same laws of physics (in this case, essentially Newton's laws) apply to the system; but where it was previously behaving in a deterministic fashion, it is now behaving chaotically.

The difference between the two situations can be pictured with the aid of a couple of simple diagrams. If we drop a lump of lead on to a lawn, we can be sure that within a very small range of errors (caused by the wind, perhaps) the lead will end up exactly below the place it was released. This is represented on the first diagram by two horizontal lines. There is a one-to-one correspondence between the points on the two lines, and shifting the starting point P by a small amount (to P') only shifts the end point Q by a small amount (to Q'). A small error in our knowledge of the initial state implies only a small error in our prediction of the final state.

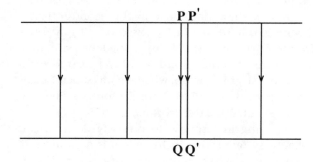

Chaos 1. If you drop a ball vertically onto a flat surface, a small change in where you drop it from produces a small change in where it lands.

But a chaotic system is better represented by the second diagram, in which the starting points are on a circle, and the end points are on a horizontal line. There is still a one-to-one correspondence between the points on the circle and the points on the line. But the lines connecting the two sets of points fan out, so that very slight changes in the starting point can produce huge changes in the end point. This symbolizes chaos, where the system is extremely sensitive to the initial conditions.

This is not just because of human inability to measure things precisely enough to work out what happens. It is impossible for nature herself to 'know' the state of the Universe to the required precision. To take just one simple example, known to the Ancient Greeks, the very concept of labelling the points on a straight line is an approximation to the truth. To label every point on a line (no matter how short the line is) you would need an infinite number of numbers. In the interval from 0 to 1,

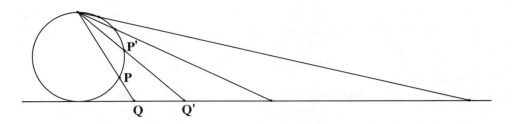

Chaos 2. If the direction in which the ball moves is defined as explained in the text, a small change in the starting conditions can make a much bigger difference to where the ball ends up.

many of the points can be expressed in terms of the familiar fractions, such as ½ or ¼ (or by less familiar fractions, such as 137/731). But some of the points can only be represented by 'irrational' numbers, numbers which cannot be expressed as the ratio of two integers in a fraction, and which require an infinite number of digits after the decimal point in order to be described fully. Exact predictability requires that the position of the pendulum bob (in our simple example) or, say, a planet in its orbit around the Sun is determined precisely. But this infinite precision in determining position (just one of the factors to be taken account of in the real world) is impossible to achieve. Specifying just one point on a line precisely can require an infinitely large computer. The Universe itself cannot 'know' its own workings with absolute precision, and therefore cannot 'predict' what will happen next in every detail. Some things really are random (over and above the uncertainty inherent in quantum mechanics). Happily for us, though, very many systems are more or less linear and behave in a reasonably predictable way – the orbits of the planets around the Sun, for example, although strictly speaking subject to chaos are stable on timescales of billions of years.

In some complex systems, even whether or not the system is chaotic is not predictable. Weather forecasters, for example, try to predict the weather two or three days ahead by running computer simulations (models) starting out from the known conditions today. They find that sometimes if they run the same model from the same starting conditions several times, they get more or less the same forecast for a couple of days ahead. For that particular configuration of weather, the system is, to an extent, deterministic. On other occasions, they find that running the same simulation from the same starting conditions several times gives wildly different forecasts. Under those starting conditions, the weather system is chaotic, and it is literally impossible to predict the weather a couple of days ahead.

Because nature is not, in fact, deterministic anyway, thanks to quantum uncertainty, you might guess that quantum uncertainty and chaos would combine to make the Universe even more unpredictable. Curiously, though, computer simulations of systems operating on the atomic and molecular scale suggest that quantum effects seem to damp down chaos, in some way which is still poorly understood. Some systems which are chaotic when described mathematically at the level of classical (Newtonian) mechanics do not behave in a chaotic fashion when quantum effects are introduced into the calculations. On the large scale, though, these effects are not important, and chaos proceeds as we have outlined it here. The important point is that because of this the only 'computer' that can simulate in every detail the behaviour of the entire Universe is the Universe itself. This means that the future can never be predicted in all its details, and that the future of the Universe is not irredeemably fixed.

Further reading: Paul Davies and John Gribbin, *The Matter Myth*.

charge A measure of the strength with which some elementary particles interact with one another. The most familiar example is electric charge, which comes in two varieties, positive and negative. Two particles which have the same kind of electric charge will repel one another, while two particles with opposite electric charge attract one another. The strength of the force, in either case, depends on how much charge each of the particles carries. The basic unit of electric charge is the (negative) charge carried by one electron, which is 1.602×10^{-19} coulombs.

In order to explain the interactions between quarks in **quantum chromodynamics**, it is necessary to assign quarks a different kind of charge, called colour charge, which is a measure of the strength with which they interact with one another through the strong force. Quarks also, as it happens, carry electric charge, but this operates independently of colour charge. Colour charge comes in three varieties, not two, which are labelled by giving them names usually associated with colours (red, green and blue). It has nothing to do with electric charge, and nothing to do with colour in the everyday use of the term (the names might just as well be tom, dick and harry). But, as with electric charge, it is a fundamental feature of the theory that colour charge is conserved – the total amount of each kind of charge in the world stays the same. Colour charge is the source of the strong interaction between particles, just as electric charge is the source of the electromagnetic interaction.

charge conjugation The (imaginary) process of swapping particles for antiparticles (and antiparticles for particles) in a particle interaction. The process of reversing the property of elementary particles that determines the difference between a particle and its antiparticle counterpart is usually denoted by the symbol C. In the case of electrically charged particles, the equivalent antiparticle carries the opposite electric charge, which is the origin of the term 'charge conjugation'. But electrically neutral particles, such as the neutron and the neutrino, also have antiparticle counterparts. If you apply the 'operation' C on a system of interacting particles, this means turning every particle into its antiparticle equivalent, and vice versa. So an interaction in which an electron collides with a neutron would become an interaction in which a positron collides with an antineutron.

The interaction you obtain by applying the C operation to a possible particle interaction is also allowed by the laws of physics, and in most cases follows exactly the same laws as its counterpart. But because of a tiny asymmetry in the workings of the Universe, processes involving the weak interaction are not quite invariant under C. So, for example, although beta decay (when a neutron decays to produce a proton, an electron and an antineutrino) does have its counterpart (in which an antineutron decays into an antiproton, a positron and a neutrino), this decay is not quite the same as the interaction described by applying the C operation to beta decay itself, and takes place at a different rate. So the antimatter world is not quite the same as the matter world, even allowing for charge conjugation. This fundamental difference between matter and antimatter may explain why matter, rather than antimatter, was left over from the Big Bang to make the stars and galaxies of the visible Universe, and ourselves. See **CP conservation**, **CPT conservation**.

charged current   An interaction in which charge is carried by the **boson** that mediates the interaction. For example, the beta decay of a neutron can be thought of as being mediated by a charged boson, the W⁻, which carries one unit of negative charge away from the neutron, leaving it with one unit of positive charge as it is converted into a proton. The negative charge carried by the W⁻ is transferred to an electron created by the interaction (in effect, by the transmutation of a neutrino), so that overall charge is conserved. This is an example of the weak interaction at work, but not all weak interactions involve charged currents – see **neutral current**.

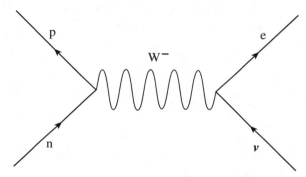

Charged current. Unlike the photon which mediates the electromagnetic force, the carriers of the weak interaction (in this case, the W⁻) can carry electric charge. This is an example of a charged current interaction.

charm   One of the six **flavours** that distinguish different types of **quark**.

charmonium   General name for any system in which a charmed quark and an anticharmed quark are held together by the strong interaction, in a sense 'in orbit' around one another. The basic charmonium state, which has the lowest energy level, corresponds to the **J/psi particle**. See **positronium**.

Charpak, Georges (1924–)   Polish-born French physicist who was awarded the Nobel Prize in 1992 for his work on developing the kind of detector known as a multiwire proportional chamber (used, among other things, to discover the **J/psi particle**). Born in Dabrovica, Poland on 1 August 1924, Charpak moved with his family to Paris when he was seven. He was a member of the Resistance during the Second World War, in 1943 he was imprisoned by the Vichy government and then, in 1944, he was sent to Dachau concentration camp. The camp was liberated in 1945 and Charpak returned to Paris, where he became a French citizen in 1946. He studied in France and obtained his PhD from the Collège de France in 1955. He worked at the French National Centre for Scientific Research, then, from 1959 onwards, at CERN, seeking ways to pick out very rare particle events from a mass of data. See **wire chamber**.

chemical bond   A link which holds atoms together to make molecules, essentially as a result of quantum effects. This can be thought of most simply in terms of the **Bohr model**, in which electrons occupy 'shells' with different energy levels around the central nucleus of an atom. The shells can be thought of as like layered onion skins, each containing a certain number of electrons.

The most stable (lowest-energy) configuration for each shell occurs when it contains the maximum number of electrons allowed by the quantum rules – two for the innermost shell, eight for most other shells. Chemical binding results from an attempt by the atoms to reach this desirable state.

One way in which this can be achieved is by sharing electrons between two atoms. For example, each hydrogen atom has one electron, and would 'like' to have two, to fill its only occupied shell. Each carbon atom has six electrons, two in the (full) innermost shell, and four in the (half-empty) outer shell. If four hydrogen atoms surround a carbon atom in the right way, the four hydrogen atoms each get a part-share in one of the outer four electrons of the carbon atom, and the carbon atom gets a part-share in each of the four electrons associated with the hydrogen atoms, so that all of the atoms in the resulting molecule (methane, CH_4) have the illusion of a full outer shell. This is a covalent bond.

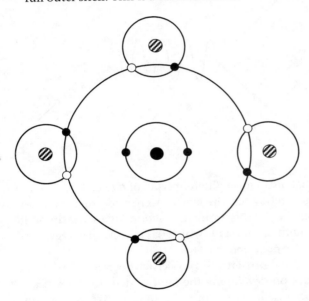

Chemical bond 1. In molecules such as methane, the link between atoms can be regarded as being formed by sharing pairs of electrons.

Another type of bond keeps sodium and chlorine bound together in common salt. Each sodium atom has eleven electrons: two in a full innermost shell, eight in another full shell, and one on its own outside. If it could lose the outermost electron, it would be left with an energy efficient full shell as its outermost face to the world. Chlorine atoms, on the other hand, each have seventeen electrons: two in a full innermost shell, eight in the next (full) shell, and seven in the outermost occupied shell. They need one additional electron to fill the outermost shell. In effect, when sodium and chlorine combine, each sodium atom gives up one electron to a chlorine atom, so both have obtained the nirvana of a full outermost occupied shell. But this leaves each sodium atom with one unit of positive charge (as a positive ion) and each chlorine atom with one unit of negative charge (as a negative ion). So the charged ions are held together by electric forces, arranged in a crystal lattice. This is an ionic bond.

Chemical bond 2. In molecules such as sodium chloride, the link between atoms can be regarded as being formed when one electron (denoted here by the open circle) transfers completely from one atom to the other.

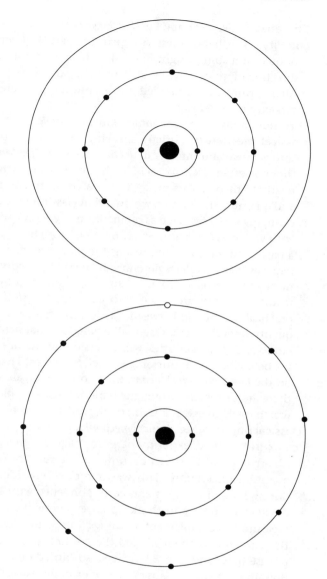

In practice, all chemical bonds are a mixture of these two simple forms, and the situation is further complicated because the electrons are not, of course, little charged particles in orbit around the nucleus, but are better regarded as spread out clouds of charge, whose shape and orientation are determined by quantum probability. The first full description of the chemical bond in those terms was made by Linus *Pauling* in the 1920s. See *quantum chemistry*.

chemistry   The branch of physics that deals with the way in which atoms combine to form molecules, and the way molecules interact with each other. See *quantum chemistry*.

Cherenkov   See *Cerenkov, Pavel Alekseyevic.*

chirality   The 'handedness' of a quantum particle. Chirality is related to the property of quantum spin. A right-handed particle has its spin oriented along the particle's direction of motion, while a left-handed particle has its spin oriented the other way. All neutrinos are left-handed, and all antineutrinos are right-handed. Other particles can exist in either state.

chromodynamics   See *quantum chromodynamics.*

classical mechanics   Strictly speaking, the laws of physics that describe the way things move and interact in accordance with *Newton's laws of motion*. The term is, though, often used to include the modifications to Newton's laws required by relativity theory. This makes the clearcut distinction between classical mechanics and quantum mechanics, which is that classical mechanics deals with continuously changing variables (such as momentum or energy), while quantum mechanics deals with variables (such as momentum or energy) that can change only by multiples of a basic unit, or quantum. Just as the most accurate version of classical mechanics requires that the effects described by relativity theory are taken into account, so the most accurate version of quantum mechanics requires that the effects described by relativity theory are taken into account, in *relativistic quantum mechanics*. The practical distinction between classical mechanics and quantum mechanics is that quantum mechanics is essential to describe what happens on the scale of molecules, atoms and smaller entities, while classical mechanics is a good tool to use to describe the behaviour of anything big enough to be seen. This rule of thumb, which depends on the fact that *Planck's constant* is so small, breaks down under conditions where there are extremely strong gravitational fields, in black holes or in the Big Bang in which the Universe was born, but it works well in the rest of the Universe.

classical physics   See *classical mechanics.*

'clock in the box' experiment   An imaginary *thought experiment* dreamed up by Albert Einstein to try to circumvent the rules of quantum physics, in particular quantum uncertainty. This was one of several ideas that Einstein came up with during long discussions and correspondence with Niels Bohr. The game was that Einstein would try to think up an experiment in which it would be possible to measure two complementary things (in the quantum sense; see *complementarity*) at the same time – the mass and the position of a particle, or its precise energy at a precise time, and so on. Einstein also carried on a long correspondence with Max Born in this vein, but this particular example was debated with Bohr.

Einstein asked his colleague to imagine a box which has a hole in one wall that can be opened and closed by a shutter worked by a clock inside the box. Apart from the clock and the shutter mechanism, the box is filled with radiation. At some precise, predetermined time, the clock automatically opens the shutter long enough to allow just one photon to escape before the shutter is closed again. The box is weighed before the photon escapes, and again afterwards. Because mass and energy are equivalent, it will weigh slightly less the second time, and the difference in the two weights tells us the energy of the photon that escaped. So we know, in principle (the beauty of thought experiments is that they never have to be put into practice!), the exact energy of the photon and the exact time that it passed through the hole, violating the uncertainty principle.

Or do we? Bohr found the flaw in Einstein's argument, as he always did in these debates, by looking at the practicalities. How, he asked, do you weigh the box? You might attach it to a spring, in a gravitational field. Before the photon escapes, you note the position of a pointer, firmly attached to the box, against a scale. Afterwards, you add weights to the box to bring the pointer back to the same place. The amount of weight you add will equal the mass-energy of the photon that escaped.

But all of this measurement process is itself afflicted by quantum uncertainty! The position of the pointer against the scale can be determined only within the limits set by Heisenberg's uncertainty principle, and when the box moves as the spring takes up (or releases) the strain, its momentum changes. There is an intrinsic uncertainty in the momentum of the box associated with the uncertainty in the position of the pointer. Bohr showed that it would be impossible, in principle, ever to beat the limits set by the uncertainty relation.

The important point underlying all of this debate is that the results of all experiments have to be interpreted in terms of the language of classical physics, the language of everyday reality. Even if what you are describing is 'only' a thought experiment, it is no good saying 'imagine measuring the position of a photon precisely' without specifying exactly how the measurement is going to be made. You don't actually have to carry out the experiment, but just thinking through, in practical fashion, how you would set about it always shows that you cannot get round the quantum rules. So the clock in the box experiment, first debated at the end of the 1920s, has gone down in scientific history for exactly the opposite reason from the one Einstein intended – it shows that you cannot beat quantum uncertainty, no matter how hard you try.

Further reading: Abraham Pais, *Subtle is the Lord*.

cloud chamber   The first type of detector to show the tracks of elementary particles, using the same principle as the way in which 'vapour trails' form behind high-flying aircraft. The chambers developed from the work of Charles Wilson at the Cavendish Laboratory in the 1890s. He was interested in creating artificial mist, in order to investigate its effect on light, and did so by building a desktop-sized apparatus in which a glass chamber full of moist air was connected to a piston which could be suddenly moved outward, lowering the pressure and causing mist (or cloud) to form in the chamber. The mist droplets grow on tiny particles of dust in the air (cloud condensation nuclei). But, to his surprise, Wilson found that even when all the dust had been removed from the chamber, when the piston was rapidly moved out over a large distance a very thin mist still formed in the chamber. He surmised that the droplets were condensing around electrically charged particles (ions), and proved this, early in 1896, by operating the cloud chamber alongside a source of X-rays (X-rays had only been discovered in the preceding year) and seeing it fill up with condensation as the X-rays passing through it ionized the atoms in the air inside the chamber.

Wilson did not develop the idea further until 1910, when he fired alpha and beta radiation through a cloud chamber and, for the first time, saw tracks of individual particles as thread-like clouds. The technique was taken up and developed by one of Wilson's colleagues at the Cavendish, Patrick Blackett, who devised an automatic chamber that cycled once every 10–15 seconds, while recording the tracks produced

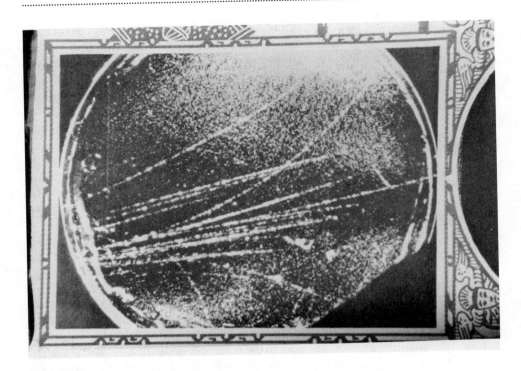

Cloud chamber. The cloud chamber photograph which revealed the existence of positrons in 1933. Particles enter the chamber from the left; the tracks curving to the right as they move (downward here) are due to positively charged electrons.

on cine film. This opened the way to the study of cosmic rays. Cloud chambers became an essential tool of physics (it was a cloud chamber photograph, for example, that first revealed the existence of the positron) and grew much larger, many being several metres across. They are still used, although superseded for many purposes by other detectors, including bubble chambers and wire chambers.

Cockcroft, John Douglas (1897–1967) English physicist who, with Ernest Walton, carried out the first experiment in which one element was deliberately transmuted into another artificially (popularly referred to as the first 'splitting of the atom').

Cockcroft was born at Todmorden, in Yorkshire, on 27 May 1897. He started an undergraduate course in mathematics at Manchester University in 1914, but left in 1915 to join the army, serving for three years as a signaller in the Royal Field Artillery and being involved in many of the great battles on the Western Front. At the end of the First World War he joined the engineering firm Metropolitan Vickers as an apprentice, and was sent by them to study electrical engineering at Manchester College of Technology, where he obtained a master's degree in 1922. He then switched to Cambridge University, where he took a degree in mathematics, graduating in 1924, and joined Ernest Rutherford's group at the Cavendish Laboratory.

The work which made Cockcroft famous (and won him the Nobel Prize for Physics in 1951) was carried out with Walton using an early form of particle acceler-

John Cockcroft (1897–1967).

ator to bombard a lithium target with a beam of protons (hydrogen nuclei). It was based on George Gamow's suggestion (made to Cockcroft on a visit by Gamow to Cambridge in 1928) that in these circumstances the protons could 'tunnel' into the target nuclei at much lower energies than had previously been thought. In 1932 this experiment produced traces of helium (in the form of alpha particles), from a reaction in which a lithium-7 nucleus first captures a proton, becoming an unstable nucleus of beryllium-8, and then almost immediately 'splits' into two nuclei of helium-4. Cockcroft helped in the design of new equipment for the Cavendish, including its first cyclotron, and was Jacksonian Professor of Natural Philosophy at Cambridge from 1939 to 1946.

He worked on radar during the Second World War, and also directed the construction of the first nuclear reactor in Canada. He was the first director of the Atomic Energy Research Establishment at Harwell, from 1946 until 1959, and then became master of Churchill College, in Cambridge. He was knighted in 1948, and died in Cambridge on 18 September 1967.

Cockcroft–Walton machine   The first particle accelerator, built by John Cockcroft and Ernest Walton at the Cavendish Laboratory in Cambridge in the early 1930s. The machine was a linear accelerator which boosted the energy of protons by accelerating them across an electric potential difference of 300 kilovolts (down a vertical tube a few metres tall); it was funded by a grant of just £1,000, and was soon boosted to

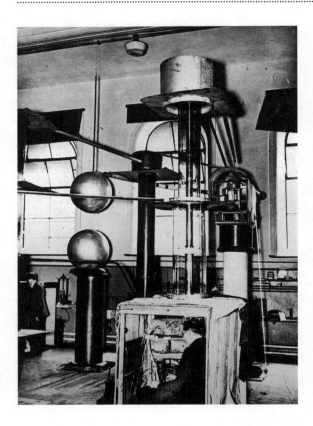

Cockcroft-Walton machine. The equipment with which John Cockcroft and Ernest Walton (inside the box housing the detector apparatus) first 'split the atom'.

provide an accelerating potential of 800 kV. It was this machine that was used in the first experiments in which one element (lithium) was artificially transformed into another element (helium).

cold fusion   Any form of nuclear fusion that takes place more or less at room temperature, instead of at the high temperatures (many millions of degrees) at the heart of a star or in a nuclear bomb. There was a flurry of excitement in the 1980s about claims that a form of cold fusion could occur under certain conditions involving electrolysis of deuterium oxide, but these claims have not stood up to closer investigation. A more promising, but still limited, possibility is that atoms similar to those of deuterium (heavy hydrogen) could be produced by replacing the electron in a deuterium atom by a muon, a particle exactly like an electron but 207 times heavier. The resulting 'atom' would be much smaller than an ordinary deuterium atom, because the muon would be much closer to the nucleus, which itself consists of a proton and a neutron. This muonic deuterium atom would therefore be able to get close enough to another deuterium atom for fusion to occur, with the two nuclei combining to make one nucleus of helium-4. The muon itself would be liberated to replace the electron in another atom, allowing fusion to continue step by step, with the muon acting as a catalyst. Unfortunately, the lifetime of the muon is so short (just over 2 microseconds) that it can only catalyse a few reactions before it decays into an electron, a muon neutrino and an electron antineutrino.

collapse of the wave function   See *Copenhagen interpretation.*

collider   See *storage ring.*

colour   A property possessed by *quarks* and *gluons*, which determines the strength of their interactions with one another through the strong force. Colour is the strong force equivalent of electric charge. See *charge.*

colour charge   See *charge, quantum chromodynamics.*

colour force = *glue force.*

commutative law   See *Abelian group.*

compactification   See *Kaluza–Klein theory.*

complementarity   The quantum nature of certain pairs of variables which prevents both of them having precise values at the same time. The classic example of a pair of complementary properties is position and momentum (essentially equivalent to velocity). It is impossible for a quantum object to have a precise position and a precise momentum at the same time.

This is not just a result of the imperfection of our human ability to make measurements. It is a key feature of the quantum world that there is always uncertainty in the value of at least one of the complementary properties, so that, in this case, the quantum entity itself cannot 'know' both exactly where it is and exactly where it is going. The amount of this uncertainty is given by Heisenberg's uncertainty relation.

In a similar way, it is impossible to define precisely the energy of a particle (or a quantum system) at a precisely determined time (see *'clock in the box' experiment*).

Complementarity is closely related to the fact that a quantum entity can be described either as a particle or as a wave, which are regarded as two complementary properties of the quantum entity. Position is a 'particle-like' property, and when an electron, say, makes a spot of light on a screen you know exactly where it is, but not how it got there. But a wave is a spread-out thing, with no well-defined position but a very well-defined direction of motion, so when the electron is travelling its position is uncertain even though it 'knows' where it is going.

Complementary properties such as position and momentum are also known as *conjugate variables.*

complementary properties   See *complementarity.*

complex conjugate   The number you get by reversing the sign in front of the complex part of a *complex number*, so that, for example, the complex number $(x + iy)$ is transformed into $(x - iy)$.

complex conjugation   The way in which the squares of *complex numbers* are calculated. For ordinary numbers (known as 'real' numbers), the square is just the number multiplied by itself: so, for example, the square of x is $x \times x$. But the square of a complex number is the number multiplied by its *complex conjugate*, so that, for example, the square of $(x + iy)$ is $(x + iy) \times (x - iy)$. This has important implications for an understanding of the behaviour of waves in the quantum world, because those waves can be described in terms of a complex variable with the form $(x + it)$, where the imaginary part of the number describes the way the wave changes as time passes. This is known as the wave function. It applies to electromagnetic waves, or to the probability waves that are an essential ingredient of, for example, the *Copenhagen interpretation* of quantum mechanics. Since actual probabilities are calculated by taking the square of the probability wave function, this means that all calculations of quantum probabilities automatically take account of waves that are travelling backward in time, as well as waves that are travelling forwards in time. This is the basis of John Cramer's *transactional interpretation* of quantum mechanics. See also *Wheeler–Feynman absorber theory*.

complex number   A number which is made up of two components, an ordinary number (like 1, 2, 3; but not necessarily an integer) and an imaginary number, in which an ordinary number is multiplied by the square root of –1, denoted by *i*. So a complex number has the basic form ($x + iy$), where x and y are ordinary numbers. x is the 'real part' of the complex number, and iy is the 'imaginary part'.

Compton, Arthur Holly (1892–1962)   American physicist who received the Nobel Prize for Physics in 1927 for his discovery of the *Compton effect*.

Compton was born in Wooster, Ohio, on 10 September 1892, and studied at the College of Wooster (where his father was dean and professor of philosophy), graduating in 1913. He moved on to Princeton University, where he received his master's degree in 1914 and PhD in 1916. After a brief spell at the University of Minnesota and two years working for the Westinghouse Corporation, Compton visited the Cavendish Laboratory in Cambridge in 1919–20, and returned to the USA to become head of the Physics Department at Washington University, St Louis. In 1923 he settled at the University of Chicago, where he was involved in the development of the first nuclear reactor, but he returned to St Louis in 1945, and remained associated with Washington University there for the rest of his career. He died at Berkeley, California, on 15 March 1962.

Compton effect   An increase in the wavelength of X-rays (or gamma rays) when they are scattered by electrons. The effect was discovered by Arthur Compton in the early 1920s. If the X-rays behaved as classical waves, it would be impossible to explain this effect, because there is no reason why the scattering should alter the wavelength. But Compton realized that it could be explained by treating the X-rays as a stream of particles, now known as photons. When a high-energy photon scatters off an electron (like a fast-moving billiard ball striking a stationary ball), it loses energy, which is taken up by the electron. The loss of energy is equivalent to an increase in wavelength (or a decrease in frequency, *f*) in line with Planck's equation $E = hf$, where *h* is Planck's constant.

This explanation, put forward by Compton and Peter Debye and confirmed in a series of careful experiments, was clinching proof of the dual wave–particle nature of electromagnetic radiation, and instrumental in encouraging Louis de Broglie to develop the idea of electrons as having the same kind of dual wave–particle nature. Before Compton's work, the idea that light really could be described in terms of particles was very far from being fully accepted, but after 1923 it could no longer be doubted, although it was only in 1926 that the term 'photon' was introduced as the name for the particle of light, by Gilbert Lewis, at Berkeley in California.

computer model   See *models*.

configuration space   The conceptual arena in which quantum waves interact. An ordinary wave on a string (such as a plucked guitar string) is moving in one dimension, up and down the string. But the mathematical description of the way quantum waves interact is equivalent to movement in a space which possesses three dimensions for each quantum entity involved in the interaction. So even dealing with the description of a hydrogen atom (a single electron interacting with a single proton) is equivalent to describing a wave moving in six dimensions. Fortunately, it is not necessary to be able to picture what is going on in configuration space in order to manipulate the equations.

confinement   Shorthand term for the way in which quarks are held together by their colour charge, so that they only exist in triplets (inside hadrons) or as quark–antiquark pairs (inside mesons), and are never seen in isolation.

conjugate variables   Pairs of quantum properties which are related to one another by *complementarity* and *uncertainty*. The most important such pairs are position/momentum and energy/time. Also known as conjugate pairs.

conservation laws   Laws which state that the total amount of some quantity does not change during an interaction. For example, electric charge obeys a conservation law, so when a neutron decays to form a proton and an electron (plus an antineutrino), it starts out with zero charge and ends up with one unit of positive charge and one unit of negative charge, which precisely cancel each other out. Leaving aside any other considerations, a neutron could not decay to produce a proton, an electron and a positron, because the decay products would have an overall charge of +1, and this would not balance with the zero charge the neutron started with. See also *invariance*.

consistent histories interpretation   An interpretation of quantum mechanics based on the idea that out of all the possible ways in which an observed experimental result might have been caused, only a few make sense in terms of the rules of quantum physics. The histories have to be consistent with the rules of quantum physics. In a simple experiment, a particle may interact with another particle in such a way that it is scattered in one of two directions, and makes a flash on one of two detector screens. The paths by which the particle gets from the scattering object to the detectors are the histories. This is conceptually different from the idea that a wave function spreads out in all directions from the scattering object, and only collapses in one detector at the moment when the particle is detected. But the usual probabilities come into the calculation, which turns out to be mathematically equivalent to the Copenhagen interpretation. In particular, the consistent histories approach is no help with understanding the *double-slit experiment* – it is not possible to regard the path of a single particle through either slit as a consistent history, and in this case the consistent history is still that a single particle goes through both slits and interferes with itself.

See *quantum interpretations*.

Cooper, Leon Niels (1930–)   American physicist who won a share of the Nobel Prize for Physics in 1972 for his work with John Bardeen and Robert Schrieffer on the theory of superconductivity. Born in New York City on 28 February 1930, Cooper attended Bronx High School and Columbia University, where he was awarded his PhD in 1954. After a year at the Institute for Advanced Study in Princeton, he moved to the University of Illinois, where he worked with Bardeen and Schrieffer, and then on to Ohio State University and, in 1958, to Brown University, on Rhode Island.

Cooper pairs   See *superconductivity*.

Copenhagen interpretation   The standard 'explanation' of what goes on in the quantum world, which held sway from the 1930s to the 1980s, and is still taught in most textbooks and many university courses. This is by no means the only way to interpret quantum mechanics (see *quantum interpretations*), and it achieved its position of pre-eminence largely through the forceful personality of Niels Bohr. Because Bohr worked in Copenhagen, that is how the interpretation got its name; major parts of the interpretation, especially those relating to quantum probability, were, however, developed by Max Born and his colleagues in Göttingen, and the mathematical formalism describing the collapse of the wave function was developed later, by John von Neumann in Princeton.

The Copenhagen interpretation bears the classic stamp of Bohr's pragmatic approach to physics, welding together different ideas to make a workable package without any real underlying theory to explain everything in a coherent fashion. Indeed, the interpretation never has been set down in a definitive, coherent fashion. The nearest anyone ever came to that was when Bohr presented the package (at that time, without a name) to a conference held in Tomo, Italy, in September 1927. Although the Copenhagen interpretation is unsatisfactory in many ways, that conference marked the completion of a consistent theory of quantum mechanics in a form where it could be used by any competent physicist to solve problems involving atoms and molecules, with no great need to think about the fundamentals but a simple willingness to follow the instructions in the quantum recipe book and turn out the required answers. For half a century, very few people worried about what it all meant; but any attempt to spell out what the Copenhagen interpretation is immediately highlights its weirdness.

Bohr stressed the importance of experiments in our attempts to probe the quantum world. All we really know is what we measure with our instruments, and the answers we get depend on the questions we ask. The questions we ask are inevitably coloured by our everyday experiences, so that we may design an experiment to measure, say, the momentum of an electron. When we do, we get an answer which we interpret as the momentum of the electron. A different experiment may 'measure' the wavelength of the electron; but all we really know is that, if we probe the system in a certain way, we get certain answers – readings on our meters and dials, or flashes of light on a detector screen. The experiments are essentially part of the world of classical physics, even though we know that classical physics does not work as a description of the quantum world. And Bohr also stressed that the only way we can measure anything is by disturbing the quantum world, probing it with our instruments. So, according to the Copenhagen interpretation, it is meaningless to ask what atoms and other quantum entities are doing when we are not looking at them. All we can do, as Born explained, is to calculate the probability (*never* a certainty) that a particular experiment will come up with a particular result.

Heisenberg's uncertainty relation, which says that, for example, a quantum entity does not have a precise momentum and a precise position at the same time, is also an essential ingredient of the Copenhagen interpretation, as is the idea of complementarity, the way in which quantum entities have attributes of both particle and wave. Schrödinger's wave equation, originally intended as the mathematical description of a real physical wave describing entities such as electrons, is reinterpreted as the mathematical description of the probability that the electron (or whatever) is in a particular state. All possible wave functions (all the possible probabilities) are intermingled into what is called a superposition of states, until a measurement is made. The act of measurement forces the quantum entity into one state (chosen in accordance with the statistical rules of chance), giving us a unique answer to the question posed by the experiment. But as soon as the measurement has been made, the quantum entity begins to dissolve again into a superposition of states.

The absurdity of this interpretation is forcefully demonstrated by the famous example of *Schrödinger's cat*; it is also highlighted by the experiment with two holes (see *double-slit experiment*).

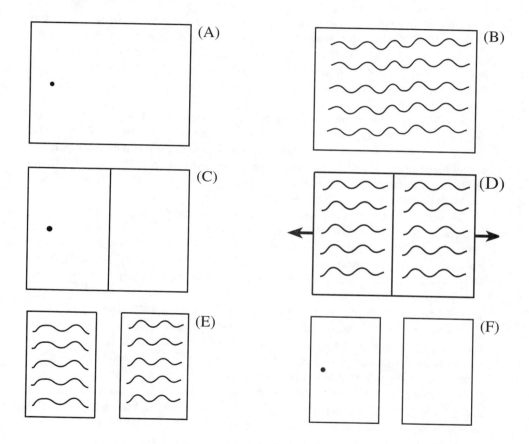

Copenhagen interpretation. Common sense tells us that an electron in a box has a definite location even if we don't know where it is (A). The Copenhagen interpretation says that the electron exists as a wave filling the box, and could be anywhere inside (B). At the moment we look for the electron, the wave function collapses at a certain location (A). Common sense says that if we slide a partition into the box without looking, the electron must be in one half of the box (C). The Copenhagen interpretation says that as long as we don't look, the electron wave still occupies both halves of the box (D), and it only collapses on one side of the barrier when we look inside (C). As long as we don't look, even if we move the two halves of the box far apart the wave still fills both boxes (E). Even if the boxes are light years apart, it is only when we look into either one that the electron wave function collapses, instantaneously, and the electron 'decides' which box it is in (F).

According to the Copenhagen interpretation, an electron which leaves an electron gun on one side of the experiment immediately dissolves into a series of overlapping probability waves, a superposition of states. The probability of finding the electron at a particular place (assuming you are looking for it) is determined by the square of the value of the appropriate wave function. The waves pass through both holes in the experiment, and interfere with one another to create a new superposition of states, which now carries information about the experiment – the arrangement of the two holes. So there is a new set of probabilities regarding the likelihood of finding the electron in a particular place. On the far side of the experiment, the detector screen in effect makes a measurement of the quantum system. At that point, the wave function describing the superposition of states collapses, and the electron becomes a real particle at a definite point on the screen. But it doesn't stay in that well-defined state, and begins to dissolve away again as a new superposition (giving yet another set of probabilities) as soon as the measurement has been made.

As Heinz Pagels (at the time, president of the New York Academy of Sciences) expressed it in his book *The Cosmic Code*, according to the Copenhagen interpretation 'there is no meaning to the objective existence of an electron at some point in space, for example at one of the two holes, independent of actual observation. The electron seems to spring into existence as a real object only when we observe it' and 'reality is in part created by the observer'.

It is this package of ideas – uncertainty, complementarity, probability and the collapse of the wave function when the system is disturbed by an observer – that is referred to as the Copenhagen interpretation. When Erwin Schrödinger saw the way his wave function had been hijacked and turned into this semi-mystical probability wave, he commented, 'I don't like it, and I wish I'd never had anything to do with it.'

See *quantum interpretations*.

correspondence principle   An idea developed by Niels Bohr in the early days of quantum mechanics, which holds that because the classical laws of physics work very well when describing systems much larger than individual atoms (so-called macroscopic systems), the laws of quantum mechanics, which work very well when describing small-scale systems (so-called microscopic systems) must become indistinguishable from the laws of classical mechanics when applied on a large enough scale. It may seem like a statement of the blindingly obvious, but in the early days this principle was a great help to Bohr when he was trying to develop a model of the atom by a combination of guesswork and the application of the quantum rules, because it enabled him to eliminate guesses which did not produce small-scale behaviour which tended towards the correct large-scale behaviour.

cosmic rays   Energetic particles from space (primary cosmic rays), including electrons and protons, some of which interact with the nuclei of atoms in the atmosphere of the Earth to produce showers of secondary cosmic rays, dominated by mesons, which penetrate to the surface of the Earth. Some primary cosmic rays also reach the surface. Before the development of particle accelerators, cosmic rays provided physicists with their only source of high-energy particles to study (the positron was found in this way), and in the heyday of cosmic ray studies, in the 1920s and 1930s, pioneers such as Robert Millikan (who coined the term 'cosmic rays')

Tracks in a photographic emulsion which show a **cosmic ray** (in this case, a magnesium nucleus) entering from the top left and colliding with a bromine nucleus in the emulsion, producing a spray of particles (many of them protons) as the two nuclei are broken apart in the collision.

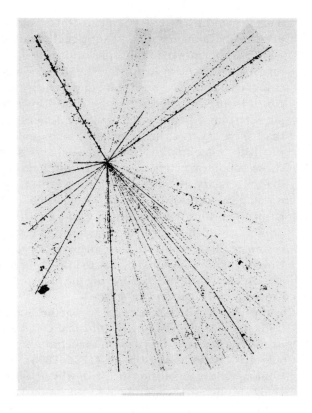

would trek up mountains to set up their detectors as close as possible to the top of the atmosphere, or launch them on unmanned balloons to even higher altitudes.

The first clue that cosmic rays existed came in 1910, when Theodor Wulf used an early detector to measure the natural background radiation on top of the Eiffel Tower, and found more than at ground level. In 1911 and 1912 the Austrian Victor Hess made manned balloon flights which showed that the intensity of the radiation is five times greater at an altitude of 5,000 m than it is at sea level, and the subject literally took off in the 1920s with the development of high-altitude studies using unmanned balloons and improved detectors such as the Geiger counter.

cosmic string Hypothetical entities left over from the Big Bang in which the Universe was born. It is argued that when the Universe cooled from a superhot state the various forces of nature froze out from an original superforce, in a so-called phase transition. This is like the way a ferromagnetic material, such as iron, has no magnetism when it is at a high temperature (above the so-called Curie point), but becomes magnetic as a result of a sudden phase transition at the Curie point when it cools. The iron does not settle down into one uniform magnetic orientation, but is divided into different magnetic domains. Each domain has its own magnetic orientation (its own north and south poles), but neighbouring domains can have quite different magnetic orientation. The domains are separated by sharply defined walls, known as defects. Cosmic string could be produced in the very early stages of the Big

Bang, when the Universe was about 10^{-35} sec old, as a topologically similar kind of defect to the domain walls in ferromagnetic material.

The simplest way to think of cosmic string is as a tiny (in terms of its diameter) tube filled with energy in the state the Universe was in when it was 10^{-35} sec old. Such strings could not have open ends, or the energy would leak out. So they either form closed loops, or stretch across the entire Universe. Even though the string has a diameter only 10^{-14} times that of an atomic nucleus, the energy it contains is so great that a piece of string a metre long (in a loop) could weigh as much as the Earth, while a loop of cosmic string a few hundred light years across could provide a 'seed' with a strong enough gravitational pull to hold a cloud of gas in the early Universe together to form a galaxy. There is, alas, no direct evidence that cosmic string exists.

cosmological constant   A force of repulsion (a kind of antigravity) that arises naturally in some theories of the very early Universe involving the break-up of a primordial superforce into the four forces of nature known today. The effect of this cosmological constant is to cause an extremely rapid (exponential) expansion of the Universe (known as inflation) in the first split-second of its existence, before the constant fades away to zero, leaving the more sedate expansion that we see today.

cosmology   See *quantum cosmology*.

Cosmotron   A particle accelerator built at the Brookhaven National Laboratory in the USA, which began working in 1952 and could accelerate protons to an energy of 3 GeV. The first operational proton synchrotron.

Coulomb, Charles Augustin de (1736–1806)   French physicist who discovered Coulomb's Law (published in 1785), which says that the force between two small charged spheres is proportional to the product of the two charges divided by the square of the distance between them – an inverse square law similar to Newton's law of gravity.

Born in Angoulême on 14 June 1736, Coulomb was the son of a wealthy man who lost his money through financial speculation. The family moved to Paris in Coulomb's youth, then he followed his father to Montpellier after an argument with his mother. After service as an army engineer in Martinique, he returned to France in 1779 and held a series of public offices before the revolution. Forced to give up his work and leave Paris, he then moved to Blois, but returned in 1795 and again held public office under Napoleon. He carried out experiments in mechanics, as well as in electricity and magnetism, duplicating many of the results of Henry Cavendish, but (unlike Cavendish) actually publishing them. Coulomb died in Paris on 23 August 1806.

coulomb   The amount of electricity carried by a current of 1 ampere in 1 second.

Coulomb scattering   The scattering of charged particles off one another as a result of the electric force between them – for example, when an alpha particle is scattered from an atomic nucleus. The two particles do not physically touch, but the alpha particle is deflected entirely by the electrostatic repulsion.

Coulomb's Law   See *Coulomb, Charles Augustin de*.

coupling constant   A measure of the strength of the interaction between two particles produced by one of the forces of nature. The most familiar example is the constant of gravity, which determines how strongly two masses of a certain size attract one another (although, as it happens, gravity is the weakest of the four forces of nature, and has by far the smallest coupling constant). The name comes from the idea of a particle being 'coupled' to the field associated with a particular force, like a wagon being coupled to a railway engine.

covalent bond   See *chemical bond*.

Cowan, Clyde Lorrain, Jr (1919–1974)   American physicist who worked with Frederick Reines on the experiment that first detected neutrinos, in 1956. Cowan studied at the University of Missouri and at Washington University, where he received his PhD in 1949. He later worked at Los Alamos (where he was based when the neutrino experiment was carried out), at George Washington University and at the Catholic University of America.

CP conservation   Although neither the property known as *charge conjugation* nor the property known as *parity* is precisely conserved in all particle interactions, for some time (in the late 1950s and early 1960s) physicists hoped that the way in which C and P are violated would always precisely cancel each other out, so that a combination of the two, CP, would always be conserved in particle interactions. But in 1964 it was discovered that the decay of a particle known as the neutral kaon (K°), to produce particles known as pions, does not conserve CP. This decay involves the weak interaction, so CP is not precisely conserved under the weak interaction. In particular, there is a tiny tendency for the decay of kaons to increase the number of positrons in the Universe, since these decays produce marginally more positrons (on average) than they do electrons.

Most theories of the weak interaction treat CP as being conserved, with a small perturbation producing occasional violations of this rule; the most important philosophical implication of CP violation is that it means that there is a natural 'handedness' to nature, and that therefore we could communicate what we mean by our concepts of 'left' and 'right' to scientifically sophisticated aliens by describing the way kaons decay. See *CPT conservation*.

CPT conservation   Although a particle interaction need not look precisely the same if it is imagined with a reversal of the property known as *charge conjugation*, or if it is imagined with the property known as *parity* reversed, or even if it is imagined running backwards in time (see *time reversal symmetry*), it is an absolute rule of nature that if you were to do the equivalent of making a film of a particle interaction, then turning all particles into antiparticles (and vice versa, in C reversal), swap left for right (P reversal) *and* run the film backwards (T reversal), what you would see would also be an entirely valid particle interaction. In other words, the combined CPT symmetry is conserved. This is true even though CP can together be violated; where this happens, the way in which T is violated precisely cancels out the way in which the combination CP is violated.

In practice, all of these violations are very weak, and become important only under extreme conditions, such as those in the Big Bang. Most 'everyday' particle interactions (such as beta decay, or the other processes described in this book using Feynman diagrams) conserve C, P and T separately, as well as together. See *CP conservation*.

CP violation   See *CP conservation*.

Cramer, John Gleason (1934–)   American physicist, based at the University of Washington, Seattle, who developed the *transactional interpretation* of quantum mechanics in the 1980s.

Born in Houston, Texas, on 24 October 1934, Cramer studied at Rice University, Texas, where he obtained his first degree in physics in 1957, his master's degree in 1959 and his PhD in 1961. He worked at Indiana University from 1961 to 1964, then moved to the University of Washington, Seattle, where he spent the rest of his career, being appointed a full professor in 1974. His main research work has been in experimental physics, including the behaviour of ultra-heavy ions and experiments involving pions and kaons. He is also interested in high-energy astrophysics. But the most intriguing of his interests, and the one most relevant to the present book, concerns the interpretation of quantum mechanics. He has also written some excellent science fiction stories based on real physics.

critical mass   The minimum mass of a particular radioactive material required to produce a self-sustaining *chain reaction*. The critical mass for a sphere of uranium-235 is 16 kg, corresponding to a ball with a diameter of about 12 cm.

critical reaction   See *chain reaction*.

Cronin, James Watson (1931–)   American physicist who shared the Nobel Prize for Physics in 1980 with Val Fitch for their discovery that the decay of neutral kaons violates *CP conservation*.

Cronin was born in Chicago on 29 September 1931 and studied at Southern Methodist University, where he received his BSc in 1951, and at the University of Chicago (MSc 1953, PhD 1955). He spent three years at Brookhaven National Laboratory, worked at Princeton University from 1958 to 1971, and returned to Chicago in 1971, dividing his time between the university and Fermilab. It was while at Princeton that he carried out his most important work, with Fitch. Although the tiny asymmetry in the laws of physics that they discovered is still not understood, it may explain why there is more matter (that is, some!) than there is antimatter (essentially none!) in the Universe. See *Sakharov, Andrei Dimitrievich*.

Crookes, William (1832–1919)   British physicist and chemist best known for his experiments with high-voltage discharge tubes ('Crookes tubes').

Born in London on 17 June 1832, Crookes was the eldest of sixteen children. He studied at the then-new Royal College of Chemistry at the end of the 1840s, and worked there in the early 1850s. He briefly held posts at the Radcliffe Observatory in Oxford (1854–5) and as lecturer in chemistry at Chester College of Science (1855–6), but then he inherited enough money from his father to become financially independent, and returned to London where all of his important work was carried out in his own private laboratory (although he did also work as editor of the *Journal of the London Photographic Society*). In 1859, he founded the weekly *Chemical News*, which he edited until 1906. Crookes made pioneering studies in spectroscopy, and discovered the element thallium as a result, in 1861.

In 1875 Crookes invented his famous 'radiometer', a small four-bladed paddle wheel mounted in a glass vessel in a vacuum. Each blade of the paddle is black on one side (so it absorbs light) and shiny on the other (so it reflects light). When light shines on it, the paddle-wheel spins rapidly. James Clerk Maxwell explained that this is because the black side of each paddle gets hotter than the shiny side, as it absorbs more energy from the light, so that the few molecules of air left in the vessel bounce off the dark side of the paddle more vigorously, giving it a push (in line with Isaac Newton's adage 'action and reaction are equal and opposite').

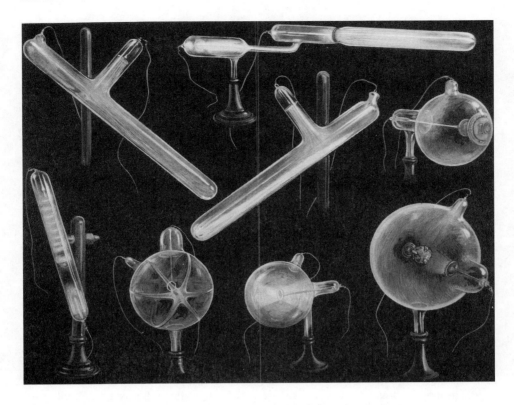

Crookes tubes. A variety of late-19th-century fluorescent tubes, like the ones developed by William Crookes, in operation.

It was also in the 1870s that Crookes developed the discharge tubes which bear his name, which operate rather like modern neon tubes. Observations made with these tubes led to the discovery of electrons (see *cathode rays*) and X-rays.

He was knighted in 1897 and died in London on 4 April 1919.

Crookes tube   See *Crookes, William.*

cross-section   A measure of the probability that a particular particle interaction will take place. You can imagine the cross-section as a disc surrounding the 'target' particle; the bigger the cross section, the greater the chance that the interaction will occur. Cross-sections are measured in units of area, such as square metres or *barns*. For an interaction between two hadrons, a typical cross-section would be a few millibarns ($1\,\text{mb} = 10^{-27}\,\text{cm}^2$).

crystal ball   A spherical particle detector at SLAC which uses sodium iodide crystals to detect the products of particle interactions.

Curie, Irène   See *Joliot-Curie, Frédéric and Irène.*

Curie, Marie (1867–1934) and Curie, Pierre (1859–1906) French husband-and-wife team who shared the Nobel Prize for Physics in 1903 (along with Henry Becquerel) for their pioneering investigations of radioactivity. Marie Curie (who had been born in Poland) also received the Nobel Prize for Chemistry in 1911 for her discovery of radium and polonium; their daughter Irène received the Nobel Prize for Chemistry in 1935, jointly with her husband Frédéric Joliot (see **Joliot-Curie, Frédéric and Irène**).

Pierre Curie was born in Paris on 15 May 1859. The son of a physician, he was educated at home before going to the Sorbonne at the age of sixteen. He was awarded a master's degree in 1877, and taught at the Ecole de Physique et Chimie in Paris for 22 years. His early work involved electricity (where, together with his brother Jacques, he discovered the piezoelectric effect) and magnetism, where he discovered the way in which the magnetic properties of a substance change dramatically at a critical temperature, now known as the Curie point. Pierre Curie obtained his PhD in 1895, for his study of magnetism. Apart from his well-known work on radioactivity, he also helped to develop the idea of symmetry as a tool for physicists.

In 1894 he met Marya Sklodowska, and they were married on 25 July 1895. They worked together on their investigation of radioactivity, and although it is sometimes suggested that she concentrated on the chemical side of the work and he concentrated on the physics, there is no way to disentangle their contributions.

After the Curies received the Nobel Prize, a chair in physics was created for Pierre at the Sorbonne in 1904. But he was plagued by ill health (possibly related to his work with radioactive materials) and was killed by a wagon when he slipped while crossing the Rue Dauphine in Paris on 19 April 1906. The unit of radioactivity was named after him in 1910 (see **curie**); the original definition was written by Marie Curie.

Marya Sklodowska was born in Warsaw on 7 November 1867, when Poland was part of the Russian Empire. Her parents were teachers, but they lost their jobs soon after Marya was born, and made a living by taking in boarders (her father had also lost money through unwise investments). Although Marya did well at school, there were no funds to pay for her to continue her education (and no scientific education available for women in Poland), so she worked for eight years as a governess, sending money to her sister Bronya in Paris to help pay for Bronya's medical studies. When her sister qualified and married in 1891, Marya went to join her in Paris and studied at the Sorbonne, coming first in the physics examinations in 1893 and second in mathematics in 1894, the year she met Pierre Curie.

With the discovery of radioactivity (which was actually given this name by Marie) in 1896 following hot on the heels of the discovery of X-rays in 1895, the Curies began to investigate radioactive materials in 1896, working under very difficult conditions, literally in a shed. Marie Curie, as she was now known, received no pay for her work until 1904, when she was officially appointed as her husband's assistant (when he became a professor at the Sorbonne), although in fact they were equal partners in their work. They identified a new element, which Marie called polonium, extracted from uranium ore, and a second new element, named radium, in pitchblende. In 1903 Marie became the first woman to be awarded a doctorate in France.

After Pierre's death, Marie's life revolved for several years around her work and the upbringing of her two daughters, Irène and Eva. The physics chair at the Sorbonne created for Pierre was passed to her, and she became the first woman to teach there. In 1910 she published a definitive book on radioactivity, and in 1914 she became head of the radioactivity laboratory of the Radium Institute, which had just been founded in Paris. During the First World War, with the help of private donations

Marie (1867–1934) and **Pierre** (1859–1906) **Curie.**

and assisted by Irène, she equipped ambulances with portable X-ray equipment, and became head of the radiological services of the Red Cross. After the war, she travelled widely as a spokeswoman for science, and was instrumental in establishing a stock of radium in Paris which was an important source of radioactive material for research purposes before the development of particle accelerators. Some of this material was used by the Joliot-Curies in their pioneering work which led to the discovery of artificial radioactivity.

In later life Marie Curie suffered from ill health related to her long exposure to radioactive materials (her notebooks are highly radioactive and now regarded as too dangerous to handle for long), and died (of leukaemia) in a sanatorium in the French Alps on 4 July 1934. The element curium (atomic number 96) was named after her.

curie   A unit which defines the quantity of a radioactive substance: one curie (Ci) is the amount of a radioactive material which undergoes 3.7×10^{10} disintegrations per second. This seemingly curious definition arises because it is equal to the radioactivity of 1 g of radium-226. Now obsolete; the preferred unit is the ***becquerel*** ($1 \, Ci = 3.7 \times 10^{10} \, Bq$).

current   In electricity, the amount of charge that flows across one unit of area (say, a square centimetre) in one unit of time (say, a second).

In particle physics, the concept of current is extended to describe the flow of bosons that mediate an interaction. See ***charged current, neutral current***.

cyclotron   An early type of particle accelerator, in which the electrically charged particles are accelerated around in a spiral path by a combination of magnetic and electric fields. The maximum energy that can be achieved in a cyclotron is limited by the way the mass of the particles being accelerated increases as a result of relativistic effects as the particles move faster; the cyclotron uses regularly spaced pulses to 'kick' the particles to higher energy, and depends on the particles taking the same time to complete each circuit, but as they get heavier they actually take longer to complete each circuit. For protons, the limit is about 25 MeV, when the particles are moving at about 20 per cent of the speed of light, and the increase in mass is about 2 per cent. The first cyclotron, accelerating protons to an energy of just 80 keV, was built by Ernest Lawrence at Berkeley and began operating in 1930.

Dalton, John (1766–1844)   English chemist who pioneered the use of atomic theory to explain chemical reactions, arguing that the atoms of different chemical elements can be distinguished by differences in their weights.

Dalton's exact date of birth is not known, but it was early in September 1766, and certainly at Eaglesfield, in Cumberland. He was the son of a poor weaver, and was brought up as a Quaker. The local schoolmaster, another Quaker, was Elihu Robinson, who was an excellent teacher and had a keen interest in science, especially meteorology. When he was twelve, Dalton began teaching as an assistant in the local school; he also worked on the land, and three years later he joined his older brother as an assistant at a Quaker school in Kendal, run by their cousin. When the cousin retired, Dalton became principal in 1785, and he stayed until 1793, when he became a tutor at the Manchester Academy (it would, incidentally, have been impossible for Dalton to attend either Oxford or Cambridge, since at the time they were only open to members of the Church of England). Both Dalton and his brother were colour blind, and in 1794 he wrote up a description of the condition, which became widely known as 'Daltonism'. In 1799 he left the Academy in order to have more time for research, and earned his living by giving private tuition until 1833, when he was awarded a government pension of £150 a year, increased to £300 per year in 1836.

Dalton was a leading member of the Manchester Literary and Philosophical Society, which provided him with rooms at a house on George Street for his teaching and research from 1799 onwards. Alongside his chemical investigations, he kept a daily weather record which has proved invaluable to historical meteorologists; indeed, it may have been his interest in weather that led him to the atomic theory, as he became interested in explaining why the mixture of gases that makes up the atmosphere stays as a mixture, instead of separating into layers of different gases. Dalton showed that in a mixture of gases, each constituent exerts the same pressure that it would if it were the only gas present, but filled the same volume (Dalton's Law). He published the first table of atomic weights in 1803, and developed the atomic theory in his book *New System of Chemical Philosophy*, the first volume of which was published in 1808. Dalton put forward the idea that atoms of different elements have different weights, and concluded that atoms could be neither created nor destroyed, with chemical reactions representing a rearrangement of the atoms. This was a fundamental step forward which paved the way for a great deal of new work in chemistry and physics in the 19th century.

Dalton was widely honoured in his lifetime, both in Britain and abroad. He died in Manchester on 27 July 1844, and 40,000 people filed past his coffin, on display in Manchester Town Hall, to pay tribute.

dalton   See *atomic mass unit.*

Dalton's Law   See *Dalton, John.*

dark matter   Astronomers know, from the way that galaxies move, that there is a great deal more dark matter in the Universe than there is bright stuff in the form of visible stars and galaxies. This dark matter is revealed by its gravitational influence on the bright stuff. Some of this dark matter may be made of baryons, the same sort of stuff that stars and galaxies (and people) are made of. But many cosmologists believe that the best way to explain how the Universe was born (a theory known as inflation) requires that there should be enough matter in the Universe to make it gravitationally 'closed', so that the present expansion will eventually be brought to a halt. If that is the case, there must be much more dark matter, and the amount of helium produced in the Big Bang and seen in old stars sets a limit on how much of this can be baryonic. At least ten times, and possibly 100 times, more matter must exist if the Universe is closed. Some of this non-baryonic matter, but not all, could be explained if neutrinos have a very small mass. At least two-thirds, however, must be in a form never detected on Earth, as particles of cold dark matter (also known as weakly interacting massive particles, or WIMPs). This is exciting for particle physicists, since their favoured grand unified theories predict the existence of just the right kind of particles, given names like axions, photinos and zinos (see *super-symmetry*), which would have been produced in large quantities in the Big Bang if these theories were correct. This marriage between particle physics and cosmology is one of the most important developments in science since the beginning of the 1980s.

Further reading: John Gribbin and Martin Rees, *The Stuff of the Universe.*

Davisson, Clinton Joseph (1881–1958)   American physicist who shared the Nobel Prize for Physics in 1937 with George Thomson for their independent confirmation of the wave nature of the electron.

Davisson was born on 22 October 1881, at Bloomington, Illinois. He studied part time at the University of Chicago (BSc 1908), where Robert Millikan was one of his teachers, supporting himself by working as an instructor at Princeton University, where he obtained his PhD in 1911. Davisson then spent six years at the Carnegie Institute of Technology in Pittsburgh, taking time off in the summer of 1913 to visit the Cavendish Laboratory. He carried out war work with the Western Electric Company in New York, and stayed with the company (which became Bell Telephone) after the war. When Davisson retired from Bell in 1946, he spent eight years as a visiting professor at the University of Virginia, Charlottesville.

In the 1920s, Davisson was studying the way electrons reflect from metal surfaces. In 1926 he attended a meeting of the British Association for the Advancement of Science, and learned of Louis de Broglie's suggestion that electrons might behave as waves. Davisson realized that some of the observations he had carried out with his assistant Lester Germer could be explained in this way, and Davisson and Germer carried out further experiments to confirm this in January 1927. Davisson died in Charlottesville on 1 February 1958.

Davy, Humphrey (1778–1829)   English chemist who discovered the *elements* sodium and potassium, and whose name is immortalized through his design of a safety lamp for use in coal mines.

Davy was born on 17 December 1778, at Penzance, in Cornwall. He came from a moderately wealthy family and received the beginnings of a good classical education. But when his father died in 1794 Davy had to give up his studies and became an apprentice to a Penzance surgeon-apothecary. He became interested in chemistry in 1797, when he read Antoine Lavoisier's *Traité élémentaire*, and in 1798 began work at the Medical Pneumatic Institution in Bristol, investigating the possible medical uses of gases such as nitrous oxide ('laughing gas').

In 1800 Davy began research on electrochemistry, and a year later he moved to the new Royal Institution in London, where he worked for the rest of his career. He was a hugely popular lecturer, and made the Royal Institution a financial success (and himself famous) through popularizing science. Davy also made important contributions to the development of chemistry, but was a late convert to John Dalton's atomic theory, and, with hindsight, was not as great a scientist as might be indicated by his contemporary fame and honours, which included a knighthood in 1812. Davy's most important contribution to science was probably his decision to appoint Michael Faraday as his assistant, in 1813. Davy became ill in 1827, and moved abroad in the hope of improving his health; he died of a heart attack in Geneva, on 29 May 1829.

de Broglie   See *Broglie, Louis-Victor Pierre Raymond, Prince de.*

de Broglie wavelength   The wavelength associated with all quantum entities, including those (such as electrons) that we are used to thinking of as particles. The wavelength is related to the momentum of the quantum entity by the equation: wavelength x momentum = h, where h is Planck's constant. This wave–particle duality is intimately bound up with quantum *uncertainty*.

Debye, Peter Joseph Willem (1884–1966)   Dutch-born chemist (originally Petrus Josephus Wilhelmus Debije) who became a US citizen in 1946. He was awarded the Nobel Prize for Chemistry in 1936 for his work on the structure of molecules.

Debye was born in Maastricht on 24 March 1884. He studied at the Technische Hochschule in Aachen (just over the border in Germany) and qualified as an electrical engineer in 1905. He then worked with Arnold Sommerfeld (who had been one of his teachers at Aachen) at the University of Munich, and was awarded his PhD in 1910. Over the next decade and a half, Debye held a succession of high-level posts, ending up in Leipzig in 1927. Along the way, he worked on specific heat, X-ray diffraction and the idea that molecules have permanent electric dipoles (the electrical equivalent of a bar magnet). In 1934 he went to Berlin to take charge of the Kaiser Wilhelm Institute of Physics (later renamed the Max Planck Institute), but ran into difficulties with the Nazi authorities because he was Dutch. He had the good sense to be visiting Cornell University in 1940 when Germany invaded the Netherlands, and stayed there as professor of chemistry until he retired in 1952. Debye remained active in science until his death, at Ithaca, New York, on 2 November 1966.

decay   The process whereby an unstable particle or nucleus spontaneously transforms itself into other particles or nuclei, with a release of energy. These processes have characteristic timescales (see *half life*) associated with the particular interactions involved. For an example of the decay of a nucleus (radioactive decay), see *alpha decay*; for an example of particle decay, see *beta decay* (although beta decay can take place inside a nucleus, the relevant point is that it involves the transformation of a single neutron into a proton, an electron and an antineutrino).

decoherence   In quantum mechanics, all states are thought of as being composed of a mixture of other states, a superposition (the classic example is the 'dead and alive' cat in Erwin Schrödinger's thought experiment). Alternatively, we can think of the quantum world in terms of a *sum over histories*, in which all possible outcomes of experiments occur (the cat is dead in one world and alive in the world next door). Decoherence is the process which 'untangles' the quantum states and produces a single version of reality at the macroscopic level (we see only a dead cat or a live cat, not a mixture of both). It is decoherence that makes it possible to assign probabilities to the outcome of an experiment (we know that the cat will be either dead or alive, with a fifty-fifty chance), even though we cannot say which outcome will result.

It is argued that decoherence results from the inevitable averaging over all the interactions that might influence a quantum entity – for example, all of the photons from the Sun that scatter off the Moon produce an average effect which both contributes to the decoherence of the orbit of the Moon and produces an image of the Moon which we see with our eyes. The classical behaviour of an object like the Moon (its obedience to the laws of *classical mechanics*) results from our ignorance about the details of all the possible quantum histories that are being averaged over, in this case, every time a photon bounces off the Moon.

On this picture, it is the simplicity of idealized quantum states which makes them behave in ways which run counter to everyday common sense. A photon faced with a choice of going through one or the other hole in the experiment with two holes exists in a superposition of states, passing through both holes (or following both possible histories) and interfering with itself because the system is free from involvement with the outside world. When the outside world gets involved – when we look to see which hole the photon goes through – the histories disentangle and we find that it does indeed go through only one hole.

The obvious flaw with this argument is that nobody knows where to draw the line. A single photon passing through the experiment with two holes is in a superposition of states; a cat is not. But at what point between the two extremes does a quantum system have enough contact with the outside world to start behaving like a classical object? In spite of this difficulty, a combination of the idea of decoherence and many histories is taken seriously by many researchers today, and is the subject of intense debate among quantum theorists and philosophers of science.

See *quantum interpretations*.

decomposition   The description of a quantum system in terms of a superposition of states. For example, Erwin Schrödinger's famous 'dead and alive' cat can be 'decomposed' into two superposed states – a dead cat and a live cat. See *decoherence*.

deep inelastic scattering   *Inelastic scattering* that takes place at very high energies, usually destroying the target particle in the process. See *partons*.

degeneracy pressure   See *degenerate gas*.

degenerate gas   A gas in which the density is so high that the particles no longer obey the laws of classical mechanics and their behaviour is constrained by quantum statistics. The electrons that conduct electricity in a metal form a degenerate gas obeying **Fermi–Dirac statistics**. Because no two fermions can occupy the same state, they are held apart and the pressure in such a gas is greater than the classical pressure. This is important in white dwarf stars and neutron stars, where the degeneracy pressure, as it is known, is responsible for supporting the stars and stopping them collapsing under their own weight to form black holes.

degenerate level   An energy level which is shared by two or more different quantum states of a system.

degenerate states   Two (or more) quantum states that occupy the same energy level.

degrees of freedom   The number of different parameters needed to specify completely the state of a particle or system. For example, the state of a single particle is described by its velocity and its position. But in order to specify its position in three-dimensional space, you need three parameters, and in order to specify its velocity in three-dimensions you need another three parameters. So a single particle has six degrees of freedom, not two. A cat weighing 1 kg would contain about 10^{26} atoms, so even if we ignore the particles the atoms are made of (the electrons, protons and neutrons) and treat each atom as a fundamental particle, you would need 6×10^{26} numbers to specify the state of the cat. Some physicists argue that it is the large number of degrees of freedom associated with everyday ('macroscopic') objects that makes them obey the laws of classical mechanics, rather than the laws of quantum mechanics. See **decoherence**.

delayed choice experiment   Originally a thought experiment dreamed up by John Wheeler to demonstrate the strangeness of the quantum world, but one which has been turned into a practical reality by experimenters. In this variation on the theme of the experiment with two holes, Wheeler started from the proven fact that if photons are fired through the experiment one at a time, they still build up an interference pattern on the other side, as if they had gone through both holes at once and interfered with themselves. But if the experiment is set up so that a detector monitors which hole each photon goes through, each photon is indeed observed to be going through only one hole, and there is no interference pattern.

Wheeler pointed out that it would be possible to set up a detector not at the holes themselves, but intermediate between the two holes and the ultimate detector screen, looking to see which route a particular photon was taking after it had passed the two holes but before it arrived at the screen. Quantum theory says that if we choose to turn this new detector off and not look at the photons, they will form an interference pattern. But if we look at the photons to see which hole they went through, even if we look *after* they have gone through the hole, there will be no interference pattern. The 'delayed choice' comes into the story because we can make the decision whether or not to look at the photon (or the decision can be made at random by a fast computer) *after* the photon has already passed through the hole(s). The decision we make, according to quantum theory, seems to affect how the photon behaved at the time it was passing through the hole(s), a tiny fraction of a second in the past.

Two independent experiments, one carried out at the University of Maryland and the other at the University of Munich, confirmed in the mid-1980s that this really is what happens. The behaviour of the photons in these experiments was seen to be affected by the experimental set-up, even though the set-up was changed while the photons were in flight; that implies that the photons have some precognition about how that set-up will be changed before they set out on their journey.

The timescales involved are tiny – only a few billionths of a second. But as Wheeler also pointed out, you can imagine a similar kind of experiment on a literally cosmic scale. It would use light from a single distant object (a quasar) which has reached us by two different routes, having been bent around a massive galaxy in the line of sight in the process known as gravitational lensing. Several such gravitational lenses are known. In principle, it would be possible to combine the light from the two quasar images to make an interference pattern, proving that it had travelled across the Universe like a wave following both possible routes. Or you could monitor individual photons, checking to see by which path they had reached us, in which case no interference pattern would form. Since the quasar might be 10 billion light years away, it would seem that our choice about which measurement to make must have affected the way the light set out on its journey 10 billion years ago – 5 billion years before our Solar System even existed. If that version of the delayed choice experiment is ever carried out, it will provide the most dramatic proof that the quantum world is influenced by connections which operate, from our everyday point of view, backwards in time.

de Maupertuis   See *Maupertuis, Pierre Louis Moreau de.*

Democritus of Abdera (about 460–370 BC)   Greek philosopher best known for his development of the atomic theory, which he got from Leucippus. Democritus suggested that the world is made up of only vacuum and atoms – an infinite number of tiny, hard, indestructible particles which combine with one another in different ways to produce the variety of everything in the world, both living and non-living.

density   The mass of a substance in each unit of volume. Standard units are kilograms per cubic metre; often measured in grams per cubic centimetre.

Descartes, René du Perron (1596–1650)   French mathematician and philosopher who developed the Cartesian system (named after him) of representing mathematical (algebraic) relationships in terms of geometry, including simple graphs showing the form of expressions such as the equation $y = x^2$ as curves (in this case, a parabola).

Born on 31 March 1596 at La Haye, Touraine, Descartes studied at the Jesuit College at La Flèche and went on to the University of Poitiers, where he graduated in law (although much of his time was spent studying science and mathematics) in 1616. He served as a military engineer in order to be able to see Europe at someone else's expense, but was far from being a conventional soldier and liked to spend as much of his time as possible lying in bed thinking. He returned to France in 1622, sold his inherited estate near Poitou to finance further travels, and eventually settled in the Netherlands in 1628. In 1649, when Descartes was in his fifties, he was invited to attend the court in Stockholm to teach Queen Christina philosophy. It seemed like another excellent opportunity to see the world at someone else's expense. Unfortunately, when he arrived in Stockholm and it was too late to back out of the deal, Descartes learned that the Queen required him to give the tuition at 5 a.m. each day. The shock to Descartes' comfort-loving system of such early rising in the depths of the Scandinavian winter quickly proved too much; he caught a chill, and died on 11 February 1650.

DESY   Acronym for Deutsches Elektronen SYnchrotron, a particle accelerator

René Descartes (1596–1650).

laboratory located in a suburb of Hamburg. Together with SLAC, DESY was for many years one of the leading electron–positron accelerators, and helped (in the 1970s) to establish the existence of quarks.

determinant   See *Jacobian*.

determinism   The philosophy that all actions (including all human actions) are pre-determined by preceding events, so that free will is an illusion. Strict determinism is one way (perhaps the only way) to avoid the implication from quantum mechanics and experiments such as the ***delayed choice experiment*** that quantum communications occur instantaneously across any distance, or even travel backwards in time.

deuterium   The isotope of hydrogen which contains one proton and one neutron in its nucleus. Also known as heavy hydrogen, occasionally as hydrogen-2. Symbol D.

deuteron   The nucleus of a deuterium atom.

diamagnetism   See *permeability*.

dielectric constant   See *permittivity*.

diffraction   The way in which waves bend around corners, or spread out after passing through a hole that is small compared with the wavelength of the waves themselves. It is easy to see diffraction at work in the ripples on a pond, or in waves marching steadily in from the sea, when they meet an obstruction. Diffraction of sound waves also explains why we can hear round corners.

We cannot see round corners because light waves have much shorter wavelengths than sound waves, and are only bent around very sharp corners, such as the

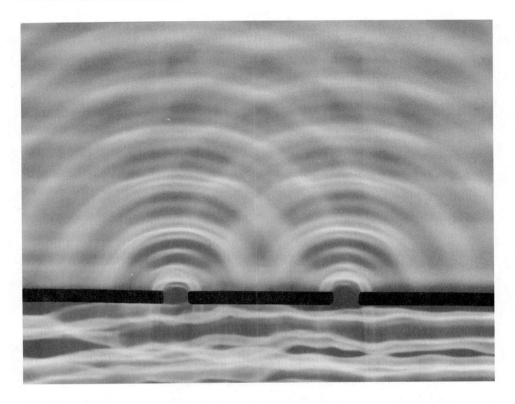

Diffraction. Water waves passing through two apertures in a barrier demonstrate both diffraction (as they bend to form semi-circular waves spreading out from each hole) and interference (as the two sets of semi-circular waves interact to make a more complicated pattern). See **double slit experiment.**

edge of a razor blade, or when passing through very narrow slits or holes. But when experiments were carried out early in the 19th century to search for interference patterns produced by the diffraction of light in such circumstances, the appropriate patterns were found and taken as proof that light is indeed a wave (see **Fresnel, Augustin Jean**). In fact, you can see this for yourself. If you hold your hand up in front of a light source, with the palm towards you, and close the fingers together until you are looking at the light through a tiny crack between two fingers, you will see a dark line (if you are lucky, several dark lines) in the gap between the two fingers. The dark line(s) is produced by interference resulting from diffraction.

Diffraction patterns are also produced, for example, by the interaction of X-rays with the regularly spaced atoms in a crystal, and the resulting diffraction patterns can be analysed to provide information about the structure of the crystal. One important application of this technique was in determining the structure of DNA.

diffraction grating   A set of regularly spaced slits which diffract light (or other waves) passing through them. A standard form of diffraction grating is a glass plate scored with parallel lines (as many as 1,000 lines per millimetre) by a diamond. Light passing through the grating is diffracted by an amount which depends on its wavelength, so white light is split into a rainbow spectrum, and a diffraction grating can be used in spectroscopy to analyse the components of light from a source.

Diffraction also occurs when light is reflected from a series of tiny, regularly spaced ridges; an ordinary compact disc works as a diffraction grating in this way, which is why it reflects a rainbow pattern of light.

diffraction pattern   See *diffraction*.

dipole   Although the term 'dipole' is widely used (for example, to describe a certain type of radio antenna), in particle physics it usually refers to a distribution of either magnetism or electricity so that it acts as if it has two opposite poles (or charges) a certain distance apart. An ordinary bar magnet is the most familiar kind of dipole, behaving as if it were made of a north and a south magnetic pole separated by a short distance. The distribution of electric charge around a molecule may leave it with an excess of electrons at one end, so that it behaves as if the charge were concentrated in a positive lump at one end and a negative lump at the other end. The dipole moment is equal to the charge on one end of the dipole multiplied by the distance between the two charges. Even individual particles which are neutral overall can have electric or magnetic dipole moments because of their internal structure.

dipole moment   See *dipole*.

Dirac, Paul Adrien Maurice (1902–1984)   English physicist and pioneer of quantum theory, who received the Nobel Prize for Physics in 1933, jointly with Erwin Schrödinger.

Dirac was born on 8 August 1902, in Bristol. His father, who came from Switzerland, taught French at the Merchant Venturers' Technical College in the city, where Paul was educated (Dirac's mother was English). He went on to Bristol University, where he took a degree in electrical engineering, graduating in 1921 but staying on to complete a mathematics degree in 1923. He then went to Cambridge to work for his PhD under the supervision of Ralph Fowler. It was only when he arrived in Cambridge that Dirac learned anything about quantum theory.

In July 1925, Werner Heisenberg gave a talk in Cambridge, where Dirac was in the audience. Heisenberg did not discuss his new ideas about quantum physics in that talk, but he mentioned them privately to Fowler, and followed up by sending Fowler an advance copy of his first paper on the matrix mechanics approach to quantum theory, in the middle of August. Fowler showed the paper (which had not yet been published) to Dirac, and Dirac jumped off from this, using his mathematical background to develop his own version of quantum theory (known as operator theory or quantum algebra), which was published in the *Proceedings of the Royal Society* in December 1925.

After he was awarded his PhD in 1926, Dirac visited Niels Bohr's institute in Copenhagen, where he showed that both Heisenberg's matrix mechanics and Erwin Schrödinger's wave mechanics were special cases of his own operator theory, and were therefore exactly equivalent to one another. In the same year, he married Margit Wigner, the sister of Eugene Wigner. He also visited the University of Göttingen before returning to England.

In 1927 Dirac introduced the idea of *second quantization* to quantum physics, pointing the way for the development of quantum field theory. In the same year, he became a Fellow of St John's College and a university lecturer in Cambridge, but also

Paul Dirac (1902–1984).

taught in the United States and visited both Japan and Siberia over the next few years. What is widely regarded as his greatest contribution to physics came in the following year, when he found an equation which incorporates both quantum physics and the requirements of the special theory of relativity to give a complete description of the electron. One of the most remarkable features of this equation was that it had two sets of solutions, corresponding to positive energy electrons and negative energy electrons; the 'negative energy electrons' are now called positrons. Dirac had predicted the existence of antimatter, although even Dirac was not entirely clear what the equations meant until the positron was discovered by Carl Anderson in 1932. It is no coincidence that Dirac received his Nobel Prize a year later. It was also in 1932 that Dirac was appointed Lucasian Professor in Cambridge.

Dirac also worked out the statistical rules which describe the behaviour of large numbers of particles which have half-integer spin, such as electrons. Because the same statistical rules were worked out independently by Enrico Fermi, they are now known as Fermi–Dirac statistics, and the particles whose behaviour they describe are called fermions.

In 1930 Dirac published a book, *The Principles of Quantum Mechanics*, which was the first systematic treatment of the subject and became the bible for generations of physicists (for example, it was an inspiration to the young Richard Feynman), and, revised down the years, is still in use. Dirac made no major contributions to physics after that (although in an obscure paper published in 1950 he did suggest that fundamental particles might have string-like properties), and was extremely unhappy about the process of *renormalization* which became an integral part of quantum field theory (see *quantum electrodynamics*). At a lecture in New Zealand in 1975, he commented:

> I must say that I am very dissatisfied with the situation, because this so-called 'good theory' does involve neglecting infinities which appear in the equations, neglecting them in an arbitrary way. This is just not sensible mathematics. Sensible mathematics involves neglecting a quantity when it turns out to be small – not neglecting it just because it is infinitely great and you do not want it!

Dirac was a great traveller, and worked as a visiting lecturer in several American universities (and at the Institute of Advanced Study in Princeton) during his time as Lucasian Professor. When he retired from Cambridge in 1969, he became a research professor at Florida State University, in Tallahassee, where he died on 20 October 1984. He is generally regarded as the only English theorist who can rank with Isaac Newton in the pantheon of physics.

disintegration   Any process in which an atomic nucleus breaks apart, either through spontaneous fission or decay, or because it has been struck by another particle.

Doppler, Christian Johann (1803–1853)   Austrian physicist who predicted what is now known as the *Doppler effect* in 1842, when he was a professor at the State Technical Academy in Prague. The prediction was tested and confirmed in 1845 in Holland, using a steam locomotive to haul an open carriage carrying several trumpeters. He also predicted that the effect would be observed with light.

Born in Salzburg on 29 November 1803, Doppler studied at the Vienna Polytechnic and worked as a private tutor before (in 1835) becoming a teacher at a school in Prague. In 1841 Doppler was appointed professor of mathematics at the Technical Academy in Prague, and in 1847 he became a professor at the Mining Academy in Chemnitz, in Germany. He moved back to Austria as professor of geometry at the Vienna Technical University in 1849, and became professor of experimental physics at the University of Vienna in 1850. He died, as a result of lung disease, in Venice, on 17 March 1853.

Doppler effect   A change in the frequency of *light*, or in the pitch of a sound, caused by the motion of the object emitting the light or making the sound.

The Doppler effect, predicted by Christian Doppler in 1842, is familiar in everyday life through its effect on sound waves. When a vehicle such as an ambulance is moving rapidly towards you with its siren wailing, the note you hear is higher than when the same vehicle has passed you and is moving away at high speed. This is because when the vehicle is moving towards you sound waves are squashed together by its motion (making a higher frequency), while when it is moving away the sound waves are stretched out (making a lower frequency). The abrupt change in pitch of the siren as the ambulance passes you is called a 'down Doppler'.

The Doppler effect is important in astronomy because it operates in exactly the same way for light, and for other *electromagnetic radiation*. The light waves from a star (or other object) that is moving towards you are squeezed together, producing a blueshift in its *spectrum*, and the light waves from a star that is moving away from you are stretched to longer wavelengths, producing a redshift.

DORIS. Part of DORIS opened up for maintenance, showing some of the powerful magnets used to focus the beams of electrons and positrons. The three magnets in the foreground are half of a set of six magnets which encircle the particle beam pipe (the blocked-off white circle at the top of the photograph) when the machine is running. Behind these three magnets, we can see half of a set of four magnets which does a similar job. There are many sets of each kind of magnet in DORIS.

The famous cosmological redshift of galaxies, associated with the expansion of the Universe, is not, however, due to the Doppler effect, because it is caused by space itself stretching, not by galaxies moving through space.

DORIS   Acronym for DOuble RIng Storage facility, built at the DESY laboratory in the mid-1970s. DORIS was built with two rings one on top of the other, so that two beams of electrons, or one beam of electrons and one of positrons, could be accelerated at the same time, one in each ring, and then made to collide with one another. It operated at an energy of 7 GeV and helped to establish the existence of the tau particle.

double-slit experiment   The experiment which, in the words of Richard Feynman, encapsulates 'the central mystery' of quantum mechanics. It is 'a phenomenon which is impossible, *absolutely* impossible, to explain in any classical way, and which has in it the heart of quantum mechanics. In reality, it contains the *only* mystery … the basic peculiarities of all quantum mechanics' (*Lectures on Physics*, Vol. III).

Most people first encounter the experiment with two holes, as Feynman used to refer to it, in the classic demonstration of the wave nature of light. Early in the 19th century, Thomas Young used such an experiment to demonstrate the wave nature of light, and it is still sometimes referred to as 'Young's double-slit experiment'. In this version of the experiment, light is shone on to a screen (maybe a simple sheet of

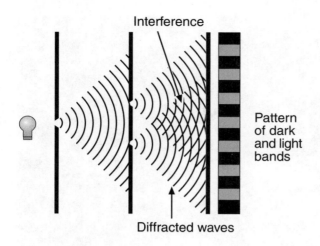

Interference

Diffracted waves

Pattern of dark and light bands

Double-slit experiment 1. When a beam of pure light passes through the experiment with two holes, the diffracted waves interfere to produce a characteristic pattern of light and shade.

cardboard) which has a small hole (or a narrow slit) cut in it. After passing through this hole, the light arrives at a second screen which has two holes in it. Light spreading out from the two holes in the second screen finally falls on a blank wall or screen, where it makes a pattern of light and shade.

The pattern of light and shade made in this way is called an interference pattern, and it is explained as a result of light spreading out from the two holes (as a result of diffraction) in a series of overlapping waves. In some places, the two sets of waves march in step and add together to make a bright patch of light; in other places, the two sets of waves move out of step and cancel one another, leaving darkness. You can see exactly the same kind of interference pattern in the ripples on a pond if you drop two stones in to it at the same time (or if you wiggle a couple of fingers around in your bath water).

This is not the way a stream of particles hurtling through the experiment would behave. If you stood behind a wall in which there were two holes, and hurled rocks through the holes, you would end up with two piles of rocks, one behind each hole.

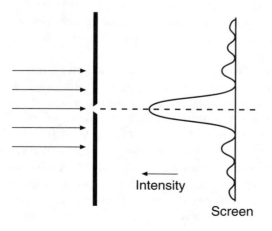

Intensity

Screen

Double-slit experiment 2. If you threw rocks through a hole in a wall, you would expect them to pile up with most rocks right behind the hole. If electrons are particles, this is the kind of pattern you would get building up on a detector screen if you fired them through a small hole one at a time.

There would be no interference. So if light were a stream of particles, you would expect to find just two bright patches on the detector screen, one behind each hole, not an interference pattern. One of the most distinctive features of the interference pattern is that the brightest patch of light is not behind either of the two holes, but exactly halfway between them, behind the obscuring screen. There is a very low intensity of light just either side of this peak, and then a succession of lesser peaks, separated by lows, repeating alternately as we move out along the detector screen.

The brightness of a patch of light is measured in terms of intensity, which is equal to the square of the amplitude of the wave at that point; in the interference pattern, for two waves whose amplitudes are represented by H and J, the intensity at any point is not given by $(H^2 + J^2)$, but by $(H + J)^2$, which works out as $(H^2 + J^2 + 2HJ)$. It is the extra term, $2HJ$, which is the contribution due to interference from the two waves, and, making allowance for the fact that the Hs and Js can be negative or positive, it precisely explains the highs and lows in the interference pattern. If both waves have the same amplitude, as they have in the classic Young's slit experiment, the brightest patch of light in the interference pattern is four times brighter than either wave would produce alone (because $H = J$, and $HJ = H^2 = J^2$), and at the other extreme the term $-2HJ$ (which is equal to $-2H^2$ or to $-2J^2$) exactly cancels out the other two terms to leave utter blackness. It all works beautifully as long as we are dealing with waves.

Curiously, though, exactly the equivalent of Young's experiment can be carried out using a beam of electrons, which we are used to thinking of as particles, and it produces exactly the same results. Electrons fired through a double-slit experiment produce an interference pattern on the detector screen (in this case, a screen rather like a TV screen, where the arrival of each electron makes a single point of light). Even this is not yet the 'central mystery', but notice what we have just said. Electrons are fired through the experiment and make an interference pattern, so they must be travelling as waves. But the arrival of each electron at one particular place on the

Double-slit experiment 3. If you fire an electron beam through the experiment with two holes, you get an interference pattern, as if the electrons were waves. The brightest part of the pattern is midway between the two holes. You do not get the pattern you would expect by adding up the two patterns corresponding to particles going through each of the two holes independently, which would give you two bright peaks, one behind each hole.

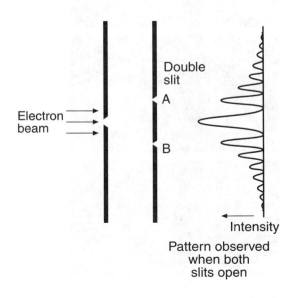

Electron beam

Double slit

A

B

Intensity

Pattern observed when both slits open

screen makes a single spot of light. So they are arriving as particles! Quantum entities travel like waves but arrive as particles. And although light is 'proved' by Young's experiment to be a wave, there is other evidence that light also can be regarded as a stream of particles (see **photon**).

This would be worrying enough if we were only dealing with beams of electrons (or photons) containing large numbers of particles. In that case, we might hope to explain the interference pattern as some sort of statistical effect. But the central

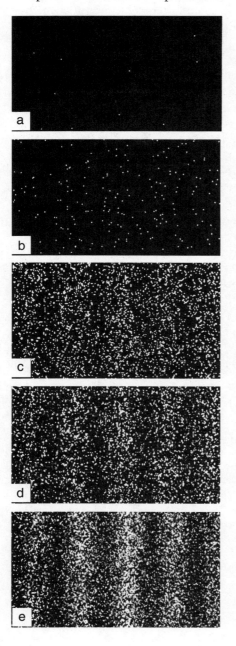

When a Japanese team at the Hitachi Research Laboratories, headed by Akira Tonomura, fired electrons one at a time through the experiment with two holes, the classic interference pattern associated with waves gradually built up on a detector screen (like a TV screen) on the other side of the experiment. Each electron made a single spot on the screen, proving that it was a particle; but the interference pattern proves that electrons are waves.

(a) The pattern after 10 electrons had arrived.

(b) The pattern after 100 electrons had passed through the experiment.

(c) The pattern after 3000 electrons had arrived.

(d) The pattern for 20,000 electrons.

(e) The pattern for 70,000 electrons.

Remember that each electron passed through the experiment on its own; the screen 'remembers' each spot of light as the pattern builds up.

mystery of quantum mechanics is revealed in all its glory when single quantum entities (either photons or electrons) are fired *one at a time* through the experiment, and the pattern they make on the detector screen is allowed to build up gradually.

We stress that this really has been done, both for photons and for electrons, with the pattern they make being allowed to build up on a TV-type screen (or on photographic film) as the spots made by each arriving photon or electron accumulate. Now, *single* particles are travelling one at a time through the experiment, and each makes a single spot on the screen. You might think that each particle must go through only one or the other of the two holes. But as more and more spots build up on the screen, the pattern that emerges is the classic interference pattern for waves passing through both holes at once. The quantum entities not only seem to be able to pass through both holes at once, but to have an awareness of past and future, so that each can 'choose' to make its own contribution to the interference pattern, in just the right place to build the pattern up, without destroying it.

There's more. If you think this is fishy, and set up a detector to tell you which hole each particle is going through, all of this mysterious behaviour disappears. Now, you do indeed see each particle (photon or electron) going through just one hole, and you get two blobs of light on the detector screen, without interference. The quantum entities seem to know when you are watching them, and adjust their behaviour accordingly (again, we emphasize that this version of the experiment really has been carried out). Each single quantum entity seems to know about the whole experimental set-up, including when and where the observer is choosing to monitor it, and about the past and future of the experiment (see also **delayed choice experiment**). And this doesn't just apply to electrons and photons, although they are relatively easy to work with and have been studied most intensively. It applies to all quantum 'particles', and similar experiments (with similar results) have been carried out with neutrons, protons and even whole atoms.

Hold on to these ideas, and remember them when you are reading about the other mysteries of the quantum world. As Feynman summed the situation up, in his book *The Character of Physical Law*, 'any other situation in quantum mechanics, it turns out, can always be explained by saying, "You remember the case of the experiment with two holes? It's the same thing."' See also **probability wave**, **Copenhagen interpretation, transactional interpretation.**

doublet   A closely spaced pair of lines in a spectrum – for example, the two lines that together make up the bright yellow sodium D line. They are explained by electron spin, which allows two electrons with opposite spin states (up or down) to share almost the same energy level in an atom.

down   One of the six *flavours* that distinguish different types of *quark.*

dressed charge   The charge of a quantum entity, such as an electron, measured at long range. This can differ from the *bare charge* of the entity, which it exhibits in short-range interactions, because of the way the entity is 'dressed' in a cloud of virtual particles.

dressed mass   The effective mass of a quantum entity, such as an electron, measured at long range. This can differ from the *bare mass* of the entity, which it exhibits in very short-range interactions, because of the way the entity is 'dressed' in a cloud of virtual particles.

Drift chamber. A drift chamber, dubbed WA1, at CERN. This particular example is 20 metres long and weighs 1500 tonnes.

drift chamber   The electronic equivalent of a *bubble chamber*, in which parallel wires are strung out in a chamber full of gas, across which an electric field is applied. When a charged particle passes through the gas, it produces an ionized trail which drifts in the electric field and triggers the detectors linked to the wires. Developed in the 1970s by Georges Charpak at CERN.

Dubna   Location of a major particle physics research laboratory (JINR) in Russia, 100 km north of Moscow. Home of a 10 GeV proton synchrotron accelerator.

Dyson, Freeman John (1923–)   British-born American physicist, best known for his clarification of the theory of quantum electrodynamics in the late 1940s, when he showed that the theories of Richard Feynman, Julian Schwinger and Sin-Itiro Tomonaga were equivalent to one another.

   Dyson was born on 15 December 1923 in Crowthorne. His father later became the director of the Royal College of Music. He studied at Winchester College and moved on, in 1941, to Cambridge, where he made original contributions to mathematical theory while an undergraduate. In 1943 Dyson left Cambridge and carried out wartime work at the headquarters of RAF Bomber Command, where he realized that the way the bombing effort was being carried out was a needless waste of airmen's lives, but was unable to persuade his superiors to change their approach to the war. After the war, he worked for a time at Imperial College in London as a demonstrator (a very junior teaching post) in mathematics, while writing a thesis which gained him a Fellowship at Trinity College, Cambridge, which he took up in 1946. It was at Trinity that he switched from mathematics to theoretical physics.

In 1947 Dyson went to the United States (initially to Cornell, where he met Feynman, then to Princeton) to carry out research, but he never bothered to complete a PhD. In spite of this, after a brief spell back in England and another at Princeton he became a professor at Cornell in 1951 (the year he became an American citizen) and a staff member at the Institute for Advanced Study in Princeton in 1953, and stayed there for the rest of his career. He has made intriguing contributions to many areas of physics, including speculations about space travel and cosmology, without ever making the major contribution that might have been expected from such a gifted mathematician. He has, however, written some superb books about his life in science.

Further reading: Silvan Schweber, *QED and the Men Who Made It*.

Eddington, Arthur Stanley (1882–1944)   British physicist who pioneered the investigation of stellar structure, and became the first astrophysicist. He was also a leading popularizer of the general theory of relativity, and tried unsuccessfully to unify relativity theory and quantum mechanics.

Born at Kendal, in what was then Westmorland, on 28 December 1882, Eddington was a Quaker (his father was headmaster of the school at which John Dalton had taught). He graduated from Owens College in Manchester in 1902, and went on to Cambridge, graduating with high honours in 1905. He moved to the Royal Greenwich Observatory in 1905, and returned to Cambridge in 1913, first as Plumian Professor of Astronomy and Experimental Philosophy (succeeding George Darwin, son of Charles Darwin) and from 1914 as director of the university observatories. Eddington led the expedition in 1919 which proved the accuracy of the general theory of relativity by measuring the bending of starlight by the Sun, observed during an eclipse, and wrote a definitive book, *The Mathematical Theory of Relativity*, published in 1923. This was followed by *The Internal Constitution of the Stars*, in 1926. Eddington was one of the first people to appreciate that the energy of the stars (including the Sun) must come from the conversion of matter into energy, in line with Einstein's equation $E = mc^2$. His later work centred on the unsuccessful attempt to construct a unified theory. He was knighted in 1930, and died in Cambridge on 22 November 1944.

effective mass   = *dressed mass*.

Ehrenfest, Paul (1880–1933)   Austrian physicist who was one of the first people to appreciate the significance of Max Planck's suggestion that energy is quantized. Although not a major contributor to the development of quantum theory, he was an inspired teacher who encouraged the development of many young researchers, including Enrico Fermi.

eigenfunction   = *eigenstate*.

eigenstate   A 'pure' quantum state, described by a unique *state vector* (or wave function). Most of the time, most quantum systems are in a *superposition of states*.

eigenvalue   The specific value of a quantum property associated with a particular *eigenstate*. For example, an electron in a particular energy level in an atom may be described by an eigenfunction, and the energy associated with that level is an eigenvalue of that particular eigenfunction. There will be other eigenvalues (for example, the spin state of the electron) associated with the same eigenstate.

eightfold way   Classification scheme for elementary particles, proposed independently by Murray Gell-Mann and by Yuval Ne'eman in 1961. The scheme grouped the particles known at the time according to their properties, in a manner reminiscent of the way Dmitri Mendeleyev had grouped the chemical elements according to their properties almost a hundred years earlier. As with Mendeleyev's periodic table of the elements, the eightfold way classification showed up gaps in the pattern, leading to the prediction that there must be particles, as yet undiscovered in 1961, with properties that would fit them into these gaps. It was one of the great triumphs of the classification system that particles (in particular, the omega minus) were duly found with the predicted properties.

The eightfold way got its name because the groups of particles with associated properties naturally associated in octets; it was deliberately chosen from 'the noble eightfold way' of the Buddhist religion (to Ne'eman's disappointment, he was unable to make the particles fit a pattern based on the six-pointed star of David). The patterns that make up the eightfold way were later explained in terms of the arrangements of quarks within the particles that form these patterns – just as the patterns in Mendeleyev's periodic table were later explained in terms of the electrons, protons and neutrons that make up the atoms in that table.

Further reading: Yuval Ne'eman and Yoram Kirsh, *The Particle Hunters*.

Einstein, Albert (1879–1955)   German-born physicist best known for his two theories of relativity, but who also made major contributions to the development of quantum theory (for which he received the Nobel Prize for Physics in 1922).

Einstein was born in Ulm on 14 March 1879. His father had his own business involving the manufacture of electrical equipment, but lack of success in business forced the family to move repeatedly. Einstein attended school in Munich, but before he had completed his education there the family moved to Italy, leaving him behind. Einstein left the school of his own volition and spent a year wandering around Italy before taking the entrance examination for the Eidgenössosche Technische Hochschule, a first-class technical university in Zurich. He failed the examination, and had to spend a year in a Swiss school catching up before being allowed to enter the ETH in 1896.

At the ETH, Einstein did the minimum amount of work necessary (one of his teachers, Hermann Minkowski, described him as 'a lazy dog') and spent most of his time in more enjoyable pursuits, including getting his girlfriend (later his first wife, Mileva) pregnant. The baby was adopted, and is thought to have died in infancy. Einstein passed his final examinations only with the help of his friend Marcel Grossman, who lent Albert lecture notes (from the lectures Einstein had not bothered to attend but Grossman had), which he swotted up on. But the reputation for laziness and Einstein's disrespect for authority meant that he was unable to get an academic job after graduating in 1900, and he ended up as a technical officer in the Swiss Patent Office in Berne. He became a Swiss citizen in 1901.

While working as a patent officer, Einstein completed a PhD thesis and the special theory of relativity (among other things) in his spare time (and during time at his desk when he should have been studying patent applications). In 1905 he submitted his thesis and published (just *before* the thesis was accepted, when he was still 'Herr Einstein', not yet 'Herr Doktor Einstein') not only the special theory of

Albert Einstein (1879–1955).

relativity but two other key papers, one on Brownian motion (which proved the reality of atoms) and one on the photoelectric effect (which suggested that light should be regarded as a stream of particles). It was for this last work that he received the Nobel Prize (actually the 1921 prize, held over until 1922).

In spite of this astonishing outpouring of ideas, it was not until 1909 that Einstein took up his first academic post, at the University of Zurich. He moved on to Prague, then back to Zurich, and then, in 1914, to the Kaiser Wilhelm Institute (now the Max Planck Institute) in Berlin, where he completed the general theory of relativity, published in 1915.

After the First World War, Einstein travelled widely, and made important contributions to the development of quantum theory, including ***Bose–Einstein statistics*** and the basics of stimulated emission of radiation from atoms, later used in the development of lasers, and gave the nod of approval which led to the rapid acceptance of Louis de Broglie's ideas about matter waves. But he never came to terms with the ***Copenhagen interpretation*** of quantum theory, arguing that 'God does not play dice' and that there must be some underlying, deterministic 'clockwork' running the Universe and giving the appearance of probability at work in quantum systems.

After Adolf Hitler came to power in Germany, Einstein (a Jew) took up a post at the Institute for Advanced Study in Princeton, where he stayed from 1933 until he died, on 18 April 1955, having become a US citizen in 1940. His later years were spent in a fruitless attempt to find a unified theory combining gravity and quantum mechanics (in particular, electromagnetism). The element einsteinium (atomic number 99) was named after him.

Further reading: Michael White and John Gribbin, *Einstein: A life in science*.

Einstein–Podolsky–Rosen experiment   See *EPR experiment*.

elastic collision   A collision in which there is no change in the kinetic energy of the particles involved – the total kinetic energy of the particles going towards the collision is the same as that of the particles heading away from the collision. In the everyday world (for example, when two billiard balls collide), some of the kinetic energy is converted into heat, so strictly speaking the collision is inelastic. In some quantum-scattering events (for example, when a proton approaches an atomic nucleus and is repelled because both the proton and the nucleus have positive charge), the collision can be essentially elastic. But in others, kinetic energy may be converted into 'new' particles, making the scattering highly inelastic.

elastic scattering   See *elastic collision*.

electric charge   A property possessed by some elementary particles, including electrons, protons and quarks, which determines the strength of their interaction with one another through the electromagnetic force. See *charge*. The basic unit of electric charge is the (negative) charge on the electron, 1.602×10^{-19} coulombs.

electric constant   The electrical equivalent of the constant of gravity. The force between two charges Q and q separated by a distance r in a vacuum is given by the equation:

$$F = Qq/r^2 4\pi \varepsilon$$

where ε is the electric constant, formerly known as the permittivity of free space. It has a value of 8.854×10^{-12} farads per metre, in the SI system of units, and 1 in any sensible system of units.

The same number also appears in the measurement of the way an electric field (for example, between two charged metal plates) is affected by the presence of a substance (a dielectric), in the same way that the *magnetic constant* is related to *permeability*. This should logically be 1, since free space has less effect on an electric field than anything else has. The problem is removed by dividing all permittivities by the electric constant for the system of units you are working in, to give relative permittivities, in which the value of the electric constant is normalized to 1. See *permittivity*.

electric dipole moment   A hypothetical electric counterpart to the *magnetic moment*. An electric dipole moment arises if the centre of concentration of the charge on an electron (or anything else) is displaced from the centre of concentration of the mass of the electron (or the something else), so that there will be a twisting effect if the electron (or whatever) moves through an electric field. The *standard model* predicts that the electron's electric dipole moment is zero; experimenters at the University of Sussex, England, are trying to measure the dipole moment for electrons to increasingly high accuracy in order to test the standard model. So far, the experiments show that the dipole moment is less than 10^{-27} e cm (where e is the charge on the electron). This corresponds to an accuracy the same as measuring the position of the centre of the Solar System to within 1 per cent of the thickness of a human hair.

electric field   The *field* associated with *electric charge*.

electricity   One component of the *electromagnetic force*.

electrode   A conductor that absorbs (*anode*) or emits (*cathode*) electrons in an electronic device.

electromagnetic force (electromagnetic interaction)   See *forces of nature*.

electromagnetic radiation   Radiation, including light, which is produced by interacting electric fields and magnetic fields moving together through space (or a suitably transparent medium) at the speed of light.

In the 19th century, Michael Faraday found that moving electricity creates magnetism, and that moving magnetism creates electricity. This is the basis of the electric motor and the electric generator. James Clerk Maxwell developed this idea in terms of field theory to explain how electromagnetic waves could move. The simplest analogy is to think of a rope, with one end tied to a post and the other in your hand. By shaking the rope, you can send waves rippling along it. You could make these vertical waves, moving up and down, or horizontal waves, moving left and right. Now imagine both sets of ripples, vertical and horizontal, moving along the rope at the same time. Suppose that the vertical ripples correspond to an electric wave, while the horizontal ripples correspond to a magnetic wave. At any point along the rope, the strength of the electric field is constantly changing as the ripples pass through. This changing electric field produces a changing magnetic field. So at every point along the rope the magnetic field is constantly changing as the ripples pass through. This changing magnetic field produces a changing electric field. The two changing fields march in step, each produced by the other, driven by the energy released from the source (a moving electric charge, equivalent to your hand shaking the rope, in, say, a light bulb or a TV antenna).

Maxwell found that such electromagnetic waves must move at a certain speed, which is the speed of light, thereby proving that light travels as an electromagnetic wave. He also predicted that longer-wavelength electromagnetic radiation, now known as radio waves, must exist.

But although Maxwell's equations seem to provide a complete picture of the way light travels, the behaviour of light and other electromagnetic radiation can also be explained in terms of tiny particles, called photons. This wave–particle duality lies at the heart of quantum physics.

electromagnetism   See *forces of nature*.

electron   One of the *elementary particles*, a *lepton* with a rest mass of $9.1093897 \times 10^{-31}$ kg ($= 0.5110034$ MeV), an electric charge of $-1.60218925 \times 10^{-19}$ coulombs and a spin of $\frac{1}{2}$, which obeys *Fermi–Dirac statistics*. See also *atomic physics*.

electron diffraction   The diffraction of a beam of electrons – for example, by the atoms in a crystal. The fact that electrons can be diffracted shows that they behave as waves, although in many experiments they behave as particles. This wave–particle duality lies at the heart of quantum physics. See *Broglie, Louis-Victor Pierre Raymond, Prince de*.

electron gun   An electrical device in which a series of electrodes are used to produce a narrow beam of fast-moving electrons. Electrons produced in this way have been used, for example, in the equivalent of the *double-slit experiment* for light. The beam of electrons that paints the picture on your TV screen is produced from an electron gun within the TV tube.

Electron microscope. Image of a cat flea, obtained using a scanning electron microscope.

electron microscope An electrical device in which a beam of electrons is used instead of a beam of light to produce an image of a small object. The principle is similar to that of an optical microscope, but the electron beams are focused by magnetic fields instead of by lenses. Because the wavelength of an electron is much smaller than the wavelength of light, an electron microscope can pick out finer details than can be seen using an optical microscope; the images are displayed on a screen like a TV screen. Electrons accelerated to an energy of 100 keV have a wavelength of 0.004 nanometres and can produce images of features as small as 0.2 nm. The electron microscope works only because electrons do indeed behave as waves.

electron neutrino See *neutrino*.

electron shell One of the energy levels allowed for individual electrons in an atom on the *Bohr model*. You can think of shells as like onion skins wrapped around one another on the outside of the atom. The image is still useful, especially for an elementary understanding of chemistry, where all that matters is the number of electrons in the outermost shell of an atom. But it has largely been superseded by the concept of an *orbital*, in which an electron in an atom is envisaged as spread out over a volume which may be non-spherical.

electron volt A measure of energy, introduced in 1912, equal to the energy gained by a single electron when it is accelerated across an electric potential difference of 1 volt. $1 \, eV = 1.602 \times 10^{-19}$ joules. Because the unit is so small, it is more commonly encountered as keV (thousand electron volts), MeV (million electron volts), or GeV (billion electron volts). Since Albert Einstein showed that mass and energy are interchangeable, dividing electron volts by the square of the speed of light gives a mass. The masses of elementary particles are usually given in terms of electron volts, but with the c^2 term usually ignored (equivalent to defining the speed of light as 1). Thus the mass of a proton is often written as 938.2796 MeV (0.9382796 GeV), although strictly speaking it should be 938.2796 MeV/c^2 (0.9382796 GeV/c^2). The mass of an average-sized person would be 4×10^{31} MeV (which is another way of saying that the average person is made up of about 4×10^{28} baryons).

If you dropped this book through a height of about 5 cm, it would be accelerated by gravity to a kinetic energy of about 1 billion billion electron volts. A 100 watt light bulb burns energy at a rate of 6.24×10^{20} eV per second. It takes only 13.6 eV to knock the electron right out of an atom of hydrogen, but the energies of particles produced in radioactive decay are typically several MeV; this gives an accurate indication of the difference in energies associated with chemical and nuclear processes.

electroscope   A simple instrument for detecting electric charge and electrically charged particles. In the most common design, two pieces of gold leaf (chosen because gold is an excellent conductor) hang side by side from an insulated metal support in a draught-proof case. If the support is given an electric charge, the leaves separate because of their mutual repulsion. In this charged state, the leaves will stay apart until something neutralizes their charge. A stream of charged particles passing through the air in the electroscope will ionize the molecules in the air, and either electrons knocked out of the molecules or the ionized molecules themselves will be attracted to the charge on the gold leaf, neutralizing it and allowing the leaves to close. So the electroscope was one of the earliest detectors used to record the existence of streams of charged particles, also known as ionizing radiation.

electrovalency   = *valency*.

electroweak interaction   A combination of electromagnetism and the weak interaction in one mathematical description. Formally, this is a ***non-Abelian gauge theory*** with the ***symmetry group*** SU(2) x U(1); the U(1) part corresponds to electromagnetism, and the SU(2) part to the weak interaction. See ***forces of nature***.

electroweak theory   Description of the electromagnetic force and the weak nuclear interaction in one mathematical package. See ***forces of nature***.

elementary particle physics   The study of *elementary particles* and their interactions. Not necessarily easy particle physics!

elementary particles   Strictly speaking, only the quarks and leptons (the only material particles not known to be made up from other particles) and the force-carrying bosons qualify as elementary particles. But for historical reasons, and because individual quarks themselves cannot exist in isolation, the baryons (which are each composed of three quarks) such as the proton and the neutron are often regarded as elementary particles, and the mesons (which are each composed of two quarks) may also be regarded as elementary particles for some purposes.

There is a pleasing symmetry between quarks and leptons at the truly fundamental level. There are six varieties of quark (up, down, strange, charm, top and bottom) and their antiparticle counterparts. They (and the particles which they combine to form, the hadrons) feel the strong force (and can be involved in weak interactions). There are six types of lepton (electron, muon, tau, electron neutrino, muon neutrino and tau neutrino) and their antiparticle counterparts. They feel the weak force (but are never involved in strong interactions).

The bosons which carry the forces of nature are the gluons (carriers of the strong force), intermediate vector bosons (carriers of the weak force), photons (carriers of electromagnetic forces) and gravitons (carriers of gravity).

elements An element is a substance that cannot be broken down into simpler components by chemical means. This means that it only contains one kind of atoms, which all have the same number of electrons, and therefore the same chemical properties. All the atoms of the same element also have the same number of protons in their nuclei (equal to the number of electrons surrounding the nucleus), and therefore the same atomic number; but different isotopes of the same element have different numbers of neutrons in their nuclei, and therefore have different masses (different atomic weights). Most elements have at least two isotopes; fluorine has only one, and tin has ten.

There are 92 elements (those with atomic numbers 1 to 92) known to occur naturally on Earth; 81 of these are stable, and the others are radioactive, transforming themselves spontaneously (but in some cases quite slowly) into other elements. Apart from hydrogen, helium and a tiny trace of very light elements such as lithium (which were created in the Big Bang in which the Universe was born), all of these elements have been manufactured inside stars by nuclear fusion, and scattered across space to form the raw material of systems like our Solar System in stellar explosions known as supernovae. Everything on Earth, including the atoms in your body, has been cooked inside stars in this way. More than 20 heavier elements (the number keeps going up), with atomic numbers greater than 92, have been created artificially in particle accelerators. All are extremely unstable and decay rapidly into lighter elements.

The idea that the complexity of the world might be made up of relatively few 'elements' arranged in different ways goes back to the Ancient Greeks and philosophers such as Thales, Democritus and Empedocles, who suggested that all substances are composed of fire, earth, air and water. It was only in the second half of the 17th century, however, that Robert Boyle took the first steps towards the modern understanding of elements and the way in which different elements combine to form compound substances (for example, water is a combination of the elements hydrogen and oxygen, with two atoms of hydrogen joined to each atom of oxygen to form a molecule of water). In 1789 Antoine Lavoisier published the first list of elements based on Boyle's definition, but he included substances now known to be compounds (such as silica) and retained a flavour of the old Greek ideas by including caloric (heat) as an element. It was John Dalton who then made the fundamental breakthrough by realizing that the atoms of different elements are distinguished by their weights.

In the middle of the 19th century, Dmitri Mendeleyev realized that when the chemical elements are arranged in order of increasing atomic weight, there is a repeating pattern in their properties. This became known as the Periodic Law, a discovery also made independently by Lothar Meyer (1830–95), a German contem-

The Periodic Table.

Period	1 Group Ia	2 IIa	3 IIIb	4 IVb	5 Vb	6 VIb	7 VIIb	8 VIII	9 VIII	10	11 Ib	12 IIb	13 IIIa	14 IVa	15 Va	16 VIa	17 VIIa	18 O
1	1 H																	2 He
2	3 Li	4 Be											5 B	6 C	7 N	8 O	9 F	10 Ne
3	11 Na	12 Mg											13 Al	14 Si	15 P	16 S	17 Cl	18 Ar
4	19 K	20 Ca	21 Sc	22 Ti	23 V	24 Cr	25 Mn	26 Fe	27 Co	28 Ni	29 Cu	30 Zn	31 Ga	32 Ge	33 As	34 Se	35 Br	36 Kr
5	37 Rb	38 Sr	39 Y	40 Zr	41 Nb	42 Mo	43 Tc	44 Ru	45 Rh	46 Pd	47 Ag	48 Cd	49 In	50 Sn	51 Sb	52 Te	53 I	54 Xe
6	55 Cs	56 Ba	57 La *	72 Hf	73 Ta	74 W	75 Re	76 Os	77 Ir	78 Pt	79 Au	80 Hg	81 Tl	82 Pb	83 Bi	84 Po	85 At	86 Rn
7	87 Fr	88 Ra	89 Ac **	104 Unq	105 Unp	106 Unh	107 Uns	108 Uno	109 Une	110 Uun								

6 *	58 Ce	59 Pr	60 Nd	61 Pm	62 Sm	63 Eu	64 Gd	65 Tb	66 Dy	67 Ho	68 Er	69 Tm	70 Yb	71 Lu
7 **	90 Th	91 Pa	92 U	93 Np	94 Pu	95 Am	96 Cm	97 Bk	98 Cf	99 Es	100 Fm	101 Md	102 No	103 Lr

porary of Mendeleyev, and to some extent presaged by the work of John Newlands in Britain. In 1869 Mendeleyev published the first periodic table, in which the elements were grouped according to their chemical properties. The key features of Mendeleyev's table were that some of the elements had to be rearranged slightly so that they were not exactly in order of increasing atomic weight (we now know that this is because of the occurrence of different isotopes, so in modern versions of the periodic table the elements are arranged by atomic number, not atomic weight) and that there were gaps, corresponding to unknown elements with well-defined chemical properties and specified atomic weights. This led to a search for the missing elements, which were duly found, confirming that the periodic table contained deep information about the nature of atoms and elements. The table shown here is a modern version of the periodic table with all of the elements up to atomic number 110. The best handy source of information about the individual elements is John Emsley's book *The Elements*.

A full explanation of the relationships between elements in the periodic table had to wait until Niels Bohr developed the first quantum-mechanical understanding of the atom (see ***Bohr model***). On this picture, the electrons in atoms with successively higher atomic numbers are added in shells around their nuclei. Chemical properties are determined almost entirely by the number of electrons in the outer occupied shell, regardless of how many filled shells there are beneath. And if the outermost shell is fully occupied, the atoms of that particular element are chemically inert, and reluctant to react with anything.

The simplest element, hydrogen, has a lone electron in its only occupied shell, and is highly reactive. The next element, helium, has two electrons forming a filled (or 'closed') shell, and is chemically inert. Lithium has three electrons, two in the filled innermost shell and one on its own outside. So it behaves very much like hydrogen, and is highly reactive. That second shell has room for eight electrons, so the elements with atomic numbers 3 (lithium) to 10 (neon) all have different properties. Because the eighth electron in the second shell fills it up, however, neon itself is inert, and behaves in a similar way to helium. The pattern then repeats, from the active sodium (atomic number 11) through to the inert argon (atomic number 18).

As we continue up the ladder to greater atomic numbers, complications set in because after element number 20 (calcium) it becomes more efficient, in quantum terms, to slide the next ten electrons in at a deeper level, closer to the nucleus than the outermost occupied shell. This gives a family of elements with very similar chemical properties, called transition elements, from scandium (atomic number 21) to zinc (atomic number 30), before the outermost occupied shell continues to fill up. But this pattern is itself repeated as we go to still greater atomic numbers, with other groups of transition elements formed by the belated filling in of inner electron shells.

Vertical columns in the periodic table are known as groups; horizontal rows in the table are called periods. All of the atoms in a group have the same number of electrons in the outermost occupied shell, but as you go down a group the atoms are slightly larger because there are more filled shells underneath the outermost occupied shell. Going across a period, the atoms are all essentially the same size, but with different numbers of electrons in the outermost occupied shell (except in the case of transition elements, which, for each period, all have the same number of electrons in the outermost occupied shell). Moving down a group, there is an increase in the metallic character of the elements, linked to the increased size of the atoms, which reduces the grip of the nucleus on the outermost electrons. Going across a period from left to right, there is a change of character from more metallic (atoms with fewer electrons in the outermost occupied shell) to less metallic. So non-metallic elements are concentrated in the top right of the periodic table, with metallic elements everywhere else (most elements are, in fact, metals).

The details of the way electrons behave in atoms have since been explained in slightly different terms (see *quantum chemistry*), and this is important for a full understanding of the way atoms combine to form molecules; but this simple shell model is all that you need to see why the elements are different from one another, and why the properties of the elements repeat throughout the periodic table.

elements, origin of   See *nucleosynthesis*.

Empedocles (about 494–434 BC)   Greek philosopher who was one of the first to suggest that the world is made up of four elements, fire, earth, air and water.

energy   A measure of the ability of a body to 'do work'. A moving object possesses kinetic energy, and one way to make this do work might be in a vehicle that is rolling along a flat road and meets an incline. If we ignore friction and imagine the vehicle to be rolling smoothly with the engine turned off and the gears disengaged, the kinetic energy will decrease as it does work lifting the vehicle up the hill, so the vehicle slows down. In fact, in this case the kinetic energy has been turned into gravitational potential energy, and could be made to do work by allowing the vehicle to roll back down the hill, picking up speed as it goes. If you stop the vehicle in a more conventional way, by applying the brakes, the kinetic energy is turned into heat energy in the brakes. You could imagine making this do work, boiling a small amount of water to make a cup of tea. Heat is, in fact, a measure of the kinetic energy of individual atoms and molecules in a substance; heat also exists in the form of electromagnetic radiation. There is also electrical potential energy associated with atoms and molecules, and chemical processes are driven by differences in this electrical potential energy. Left to their own devices, systems always seek out the lowest energy state accessible to them (cars roll downhill, not up).

Mass itself is a form of energy, and can be made to 'do work' in the explosion of a nuclear bomb, or the production of heat which is used to drive turbines and make electricity in a nuclear power station, or in nuclear fusion reactions at the heart of a star. With mass included in the definition of energy, energy can be neither created nor destroyed, but only converted from one form into another. Ultimately, this means that all energy will end up as radiant heat. See *arrow of time, entropy*.

energy levels   The allowed states of a quantum system corresponding to different amounts of stored energy. In an atom, for example, an electron has a well-defined amount of energy corresponding to its place in the structure of the atom – the *orbital* that it occupies (strictly speaking, it is the atom that has a certain amount of energy corresponding to the arrangement of all the electrons in their orbitals). You can think of this as like the electron sitting on a specific step on a staircase. If precisely the right amount of energy is available (a quantum of energy provided by a photon of the right wavelength), the electron can absorb this energy and jump to another well-defined energy level, corresponding to a higher step on the staircase. It can then fall back to its initial position, radiating the appropriate quantum of energy as it does so.

Other quantum systems, including molecules and nuclei, also have well-defined energy levels. It is a key feature of the quantum world that a system passes directly from one energy level to another, with no in-between state (this is the famous quantum leap). Unlike a real ball moving up or down a staircase, the atom is simply on one energy level at one instant, and then on the other energy level, instantaneously. See *Bohr model.*

ensemble interpretation An interpretation of quantum mechanics originally developed by Albert Einstein in the hope of removing some (or all!) of the mystery from quantum theory. The basic idea is that each quantum entity (such as an electron or a photon) has precise quantum properties (such as position and momentum), and the quantum wave function is related to the probability of getting a particular experimental result when one member (or many members) of the ensemble is somehow selected by experiment. There are many difficulties with the idea, but the killer blow was struck when individual quantum entities such as photons were observed behaving in experiments in line with the quantum wave function description. The ensemble interpretation is now only of historical interest.
See *quantum interpretations.*

entropy A measure of the amount of disorder in the Universe, or of the availability of the energy in a system to do work. As energy is degraded into heat, it is less able to do work, and the amount of disorder in the Universe increases (see *arrow of time*). This corresponds to an increase in entropy. In a closed system, entropy never decreases, so the Universe as a whole is slowly dying. In an open system (for example, a growing flower), entropy can decrease and order can increase, but only at the expense of a decrease in order and an increase in entropy somewhere else (in this case, in the Sun, which is supplying the energy that the plant feeds off).

Epicurus (about 342–271 BC) Greek philosopher who, following the example of Leucippus, taught that the Universe is made up of innumerable indestructible atoms, which differ from one another only in size, shape and position, moving in an infinite void. The idea was not widely accepted until the revival of atomism in the 17th century.

EPR experiment (EPR paradox) A thought experiment originally put forward in 1935 by Albert Einstein, Boris Podolsky and Nathan Rosen to demonstrate (as they thought) the logical impossibility of quantum mechanics. The basic idea was adapted by David Bohm in the 1950s and refined by John Bell in the 1960s to become a practicable experiment (see *Aspect experiment*) that was actually carried out in the 1980s, establishing that nature really does behave in a non-commonsensical way.

In the original version of the thought experiment – sometimes referred to as the 'EPR paradox', although it is not really a paradox – the EPR team imagined a pair of particles that interact with one another and then separate, flying far apart and not interacting with anything else at all until the experimenter decides to look at one of them. At the time the particles interact, it is possible to measure the total momentum of the system, and this cannot change if they do not interact with anything else. So if the experimenter chooses, much later, to measure the momentum of one particle, it is possible to calculate the momentum of the other particle, far away, by subtracting the measured momentum from the total. We know that quantum physics requires that by measuring the momentum of the first particle we destroy any information about its position, because of the *uncertainty principle*. But the EPR team suggested that quantum uncertainty could be circumvented by measuring the momentum of the first particle and the position of the second particle, while calculating the momentum of the second particle in the way we have outlined. The only alternative would be that by measuring the momentum of the first particle we destroy information about the position of the second particle (or prevent such information ever existing), instantaneously, no matter how far apart they are.

Einstein referred to this as a 'spooky action at a distance', arguing that it was both logically absurd and impossible for any communication to travel faster than light, so quantum theory must be flawed. But the experiments show that this kind of *non-locality* is indeed a feature of the quantum world, and that measurements made on one particle of such a quantum pair really do affect its counterpart, instantaneously, no matter how far away that counterpart may be. The one straw of comfort for Einstein, had he lived to see those experiments, is that although the correlation occurs instantaneously, there is no way to send useful information faster than light using such quantum connections.

equivalence principle   See *relativity theory*.

Esaki, Leo (1925–)   Japanese physicist who shared the Nobel Prize for Physics in 1973 with Ivar Giaever and Brian Josephson for the discovery of tunnelling in semiconductors. This work was carried out while he was working for Sony in Japan. Esaki moved to the United States in 1960, where he worked at the IBM Thomas J. Watson Research Center in Yorktown Heights, New York, but has retained his Japanese citizenship.

Euler, Leonhard (1707–1783)   Swiss mathematician (born at Basle on 15 April 1707) who was extraordinarily prolific. Among his many contributions, he developed the idea of the principle of least action and the calculus of variations, pointing the way for the later work of Joseph Lagrange, which was developed by Richard Feynman into a key tool in the path integral approach to quantum mechanics. After studying at the University of Basle, graduating (with a master's degree) at the age of sixteen, Euler moved to St Petersburg in 1727, becoming professor of physics at the new Academy of Sciences (founded by Catherine II) there in 1730, and professor of physics in 1733. He moved to Berlin in 1741 (at the invitation of Frederick the Great), but returned to St Petersburg (at the invitation of Catherine the Great) in 1766. Among his many lasting achievements, Euler introduced mathematical notations, such as π, e and i, which have become standard. He died at St Petersburg on 18 September 1783.

event horizon   See *black hole*.

Everett, Hugh (1930–1982)   American physicist who developed the *many worlds interpretation* of quantum mechanics while a graduate student at Princeton in the 1950s, and published it in 1957. (His PhD supervisor was John Wheeler, who also acted in this capacity for Richard Feynman.)

Born in Washington, DC, on 11 November 1930, Everett graduated from the Catholic University of America in 1953, and received his master's degree from Princeton in 1955, and his PhD, also from Princeton, in 1957. His thesis was titled 'On the Foundations of Quantum Mechanics', and presented the many worlds interpretation. But Everett did not stay in academic research, and spent most of his career working for the Pentagon, where he made many contributions to computer science and the application of games theory to strategic planning (we are talking war games here, not chess or Happy Families). In 1973 he founded the DBS Corporation, an information-science and data management organization, where he was chairman until his death, of a heart attack, on 19 July 1982 in McLean, Virginia.

exchange force   A force produced by the exchange of particles – for example, the strong force that operates through the exchange of *gluons* between *quarks*. See *forces of nature*.

excited state   When a quantum system, such as an atom or molecule, has more energy than it would have if it were in its *ground state*, and therefore occupies a higher *energy level*. Left to its own devices, a system in an excited state will radiate energy and return to the ground state.

exclusion principle   The principle, developed by Wolfgang Pauli in 1925, which holds that no two *fermions* can occupy the same quantum state (they cannot have the same set of *quantum numbers* as each other). It is this quantum exclusion that requires the electrons in an atom to occupy different *energy levels* instead of them all congregating in the lowest energy level. Without quantum exclusion, there would be no chemistry.

experiment with two holes   See *double-slit experiment*.

Faraday, Michael (1791–1867)   English chemist and physicist who was largely responsible for introducing the concepts of *fields* and *lines of force* into physics.

Faraday was born in Newington, Surrey, on 22 September 1791. At that time, this was still a village; it has now been absorbed into the urban sprawl of London. His father was a blacksmith, and Faraday received only a basic education in the 'three Rs' (reading, (w)riting and (a)rithmetic). When he was thirteen, Faraday began working as an errand boy for a bookbinder and bookseller who had a shop just off Baker Street, in London; the following year he was apprenticed to the bookbinder to learn the trade. There, he came across a copy of the *Encyclopaedia Britannica* that had been brought in for rebinding, read the article on electricity, and was entranced.

Although longing to become a scientist, Faraday seemed stuck in his working-class niche, although in 1810 he did join the City Philosophical Society, attending regular lectures on scientific topics, joining in experimental work and keeping careful notes which he bound up into book form. His employer used to show the bound volumes of Faraday's notes to customers in the shop, and one of those customers was so impressed that he arranged for the bookbinder's apprentice to attend lectures being given by Sir Humphrey Davy at the Royal Institution. This fired Faraday's enthusiasm for science even more and, nearing the end of his apprentice-ship (in 1812), he decided that he would try for a career in science. He wrote up Davy's lectures and bound them into a book, and sought, unsuccessfully, for any kind of employment in science. When Davy was temporarily blinded in an accident, Faraday was taken on to act as his temporary helper; but when Davy's sight returned, he was back to square one.

But then Faraday got his opportunity. Davy's regular assistant got into a fight and was sacked by the Royal Institution. The job was offered to Faraday, who started work at the RI on 1 March 1813, at the age of 21.

Michael Faraday (1791–1867).

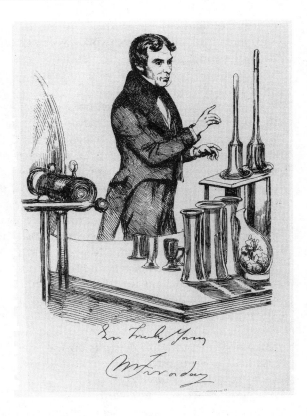

Davy was not a good employer. He treated Faraday like a servant, and was jealous and dismissive when Faraday began to make original contributions to science. But all this rolled off Faraday's back as he settled into his work. He began to publish scientific papers in 1816, was elected a Fellow of the Royal Society in 1824 (in spite of opposition from Davy, who was then President of the Society), and in 1825 became director of the laboratory at the Royal Institution; the following year he started the hugely successful series of popular Christmas Lectures on science, which continue to this day. Davy became increasingly bitter as Faraday's fame and success grew, but after Davy died, in 1829, Faraday and the Royal Institution were almost synonymous. He twice turned down the offer of the presidency of the Royal Society, and once refused a knighthood, saying that 'I have always felt that there is something degrading in offering rewards for intellectual exertion, and that societies or academies, or even kings and emperors, should mingle in the matter does not remove the degradation.' But this did not stop him accepting a house at Hampton Court, provided by Prince Albert, to live in when he retired in 1861. He died there on 25 August 1867, a month before his 77th birthday.

In Faraday's chemical work, he espoused an early form of atomic theory and made many important, but not ground-breaking, contributions. His greatest achievements, which made an impact that still echoes through physics, came through his studies of electricity and magnetism, which led him to develop the concepts of lines of force and fields. Along the way, he invented the electric motor and generator. Unusually, this work was carried out after 1830, when Faraday was in his forties and fifties – most scientists have finished their best work in their twenties or thirties, and even Faraday's late start cannot account for his continuing originality at a relatively great age.

The idea of fields was taken up and developed by James Clerk Maxwell in his theory of electromagnetism, but even here Faraday had pointed the way in 1846 with his suggestion that light is 'a high species of vibration in the lines of force which are known to connect particles'. In his own words, the idea was an attempt 'to dismiss the aether [the hypothetical fluid through which light waves were thought to move] but not the vibrations'. In some ways, this idea was 60 years ahead of its time, since the final dismissal of the ether only came with Albert Einstein's special theory of relativity, in 1905. Faraday is widely regarded as the greatest experimental physicist there has ever been; Einstein said that there had been two great couples in physical science: Galileo and Newton, and Faraday and Maxwell. In the 20th century, we might add Einstein and Feynman to that list.

Faraday effect   The way in which the plane of polarization of electromagnetic radiation (including light) is rotated when it passes through certain substances (including quartz and water) while exposed to a strong magnetic field. The effect is also seen in radio waves passing through a plasma when a magnetic field is present. The amount of rotation depends on the strength of the magnetic field and the distance the waves travel through the medium.

Faraday rotation   = *Faraday effect*.

fast breeder reactor   = *fast reactor*.

fast neutrons   Term used to denote high-speed neutrons in two ways. Sometimes referring to neutrons with kinetic energy greater than 0.1 MeV (a number chosen completely arbitrarily), but more commonly used to denote neutrons which have enough energy to trigger nuclear fission in uranium-238, which means more than 1.5 MeV.

fast reactor   A type of *nuclear reactor* enriched with plutonium-239 or uranium-235 that produces *fast neutrons* which encourage the fission of uranium-238. Because some of the neutrons produced by the radioactive processes are captured by other atoms of uranium-238 to make plutonium-239, such a reactor can produce more plutonium than is needed to start the process off. So it is sometimes called a fast breeder reactor. In terms of generating power, there is no great advantage in a fast reactor; its value (such as it is) lies in producing plutonium for use in nuclear weapons.

feedback   Any process in which the output from a system directly affects its performance. The most common example of positive feedback occurs when a microphone connected through an amplifier to a loudspeaker is placed close to that speaker, so that the sound from the speaker is picked up by the microphone, is amplified and comes out of the speaker, going round and round the circuit and getting louder and louder in a howl of noise. This is a runaway process which continues until something in the system gives. The other form of feedback is negative feedback. This occurs when the output from the system reduces the strength of the process responsible for this output. A simple example is the safety valve on an old-fashioned steam engine. When the engine runs faster, the valve opens, releasing steam pressure so that the engine runs more slowly and the valve shuts. This is a stabilizing process which keeps the system operating steadily within certain limits. Both positive and negative feedback occur in natural systems.

Fermat, Pierre de (1601–1675) French lawyer who indulged in mathematics as a hobby and made many important contributions which were preserved only in his letters to various scientists, not in formal publications. In the present context, his most important insight was Fermat's Principle, that the path followed by a light ray is the one which takes the least time to get from one place to another. This is now understood as an example of the principle of least *action*. Interestingly, Fermat's Principle works (as an explanation of refraction) only if light travels more slowly in a denser medium; this was eventually to be recognized as proof that light travels as a wave.

Fermat was born in Beaumont de Lomagne, on 20 August 1601; he came from a Basque family. Educated locally in his youth, Fermat moved to Bordeaux in his twenties and attended the University of Toulouse, and possibly the University of Bordeaux (little is known about his early life). He gained his law degree at the age of 30 and practised in Toulouse (where he also held government office), although he lived nearby in Castres, a small provincial town. Some time in the 1630s he added the 'de' to his name, but there is no evidence that this was more than an affectation. In 1652 he survived an attack of plague, from which he nearly died and after which he spent most of his time studying mathematics rather than practising law. He lived to a respectable old age, dying in Castres on 12 January 1675.

Fermat's Principle See *Fermat, Pierre de.*

Fermi, Enrico (1901–1954) Italian physicist, born in Rome on 29 September 1901, who contributed to the development of quantum theory and received the Nobel Prize for Physics in 1938, but is best remembered as head of the team that built the first nuclear reactor, then known as an atomic pile.

Fermi came from an affluent background, his father rising from work as a railway official to a senior position in the civil service. Enrico studied in Pisa, where he received his PhD in 1922, and worked with Max Born at Göttingen and Paul Ehrenfest in Leiden before becoming a lecturer in mathematical physics at the University of Florence in 1924; there, he worked out the properties of a gas made up of particles which obey the *exclusion principle*. Paul Dirac made the same calculation independently; the equations are now known as *Fermi–Dirac statistics*, and particles which obey Fermi–Dirac statistics are called *fermions*. In 1927 Fermi moved to Rome University, becoming the first professor of theoretical physics in Italy (in fact, though, Fermi was unusual among the physicists of his generation in being both a superb theorist and an expert experimenter, and he made important contributions in both branches of physics over the next few years). He wrote the first textbook of modern (that is, quantum) physics to be published in Italy, *Introduzione alla Fisica Atomica*, which appeared in 1928. In 1933 Fermi took up the suggestion, made by Wolfgang Pauli, that a 'new' particle was needed to explain details of beta decay, and he was responsible for giving that particle its name – the *neutrino*. He also suggested that the process of beta decay was associated with a new kind of force, the weak force. His paper introducing these concepts was rejected by *Nature* as being too speculative, and was published in an Italian journal.

Enrico Fermi (1901–1954).

Fermi also studied radioactivity, following up the discovery of artificial radioactivity made by Irène and Frédéric Joliot-Curie in 1934, and was involved (with Emilio Segre) in work which showed how neutrons could be slowed down by passing through water or paraffin, providing a means by which the rate of a nuclear reaction could be controlled. He used the *slow neutrons* produced in this way to irradiate just about anything he could lay his hands on, creating a large number of previously unknown radioactive isotopes (but he did not realize that in some of these experiments he was actually causing nuclei of uranium-238 to fission). It was this work that led to the award of his Nobel Prize, and possibly saved his life.

Fermi, whose wife was Jewish, was unhappy with political developments in Italy in the 1930s, and when he was awarded the Nobel Prize in 1938, he used the opportunity to leave the country (with his wife and two children) for Stockholm for the prize ceremony. He moved on straight from there to the United States, where he worked first at Columbia University and then in Chicago, where the first working atomic pile was constructed in the university squash court. It began operating on 2 December 1942 – the first controlled, self-sustaining nuclear reactor on Earth, capable of being switched on and off as the experimenters wished (using blocks of graphite, rather than paraffin, as the *moderator* to slow neutrinos down). In 1943 Fermi went to Los Alamos to work directly on the Manhattan Project (the design and construction of the first atomic bombs). He stayed there until the end of the Second World War, when he returned to Chicago, became a US citizen and carried out more research on radioactivity. He also served on the general advisory committee of the Atomic Energy Commission, and was involved in the design of the first synchrocyclotron at the University of Chicago. This still left him time to fight military censorship of nuclear research, and to stand up in the defence of Robert Oppenheimer when Oppenheimer became a victim of the communist witchhunts of the postwar period. A warm character and an inspiring teacher, Fermi died of cancer (possibly related to his work with radioactive materials) in Chicago on 28 November 1954; a year later, the element fermium (atomic number 100) was discovered and named after him. The US National Accelerator Laboratory, Fermilab, near Chicago, was also named in his honour.

fermi   Unit of length, named after Enrico Fermi. Although now obsolete, it is equal to 1 femtometre, and the abbreviation for that unit could just as well stand for fermi. $1\,\text{fm} = 10^{-15}\,\text{m}$.

Fermi constant   A number which expresses the strength of the weak interaction (see *forces of nature*). Confusingly given the symbol g, but not to be confused with the constant of gravity. $g = 294\,\text{GeV}^{-2}$.

Fermi–Dirac statistics   The statistical rules which apply to the behaviour of a large number of quantum particles which each carry a half-integer amount of quantum *spin*. Such particles are called fermions; the archetypal example of a fermion is the electron, which has spin of ½. No two identical fermions can exist in the same quantum state, which is crucial in determining the statistical behaviour of large numbers of fermions.

Fermilab   US National Accelerator Laboratory at Batavia, near Chicago, opened in 1972 and named in honour of Enrico Fermi. The main machine at Fermilab is the Tevatron, two ring-shaped accelerators one on top of the other, which can accelerate protons to a kinetic energy of 20 teraelectronvolts (1 TeV is equal to 1,000 GeV). This was the machine that found the evidence for the top quark in the mid-1990s.

Fermi National Accelerator Laboratory   = *Fermilab*.

fermion   A particle which obeys Fermi–Dirac statistics. All fermions have half-integer *spin* (½, ⅜ and so on). They are the particles that make up what we usually think of as the material world (for example, the electron and the proton). Fermions are conserved – the total number of each kind of fermion stays the same, provided that in any interaction an antiparticle is counted as 'minus one' particles.

ferromagnetism   The kind of magnetism associated with an ordinary bar magnet. A bar of ferromagnetic material can be thought of as made up of an enormous number of tiny internal magnets (corresponding to groups of individual atoms). When the ferromagnetic material is hot, these tiny internal magnets spin around and jostle one another at random, so there is no overall magnetic field. In this state, the material is said to be paramagnetic. But when the bar cools, at a certain temperature (known as the Curie temperature, or Curie point) the internal magnets quite suddenly line up with one another so that there is an overall magnetization, with each of their tiny magnetic fields adding up to produce the familiar overall magnetism of a bar magnet.

At high temperature (above the Curie point), the lowest energy state available to the system corresponds to zero overall magnetization; at low temperature (below the Curie point), the lowest energy state available to the system is when all the internal magnets are lined up. If there is an external magnetic field (such as the Earth's magnetic field), they will tend to line up along the existing lines of force. This switch from a uniform state in which there is no overall magnetism to one in which there is a bar magnet with distinct north and south poles is an example of symmetry breaking.

Feynman, Richard Phillips (1918–1988)   The greatest physicist of his generation, ranking with Isaac Newton and Albert Einstein, Feynman reformulated quantum mechanics to put it on a secure logical foundation in which classical mechanics is naturally incorporated (in a manner reminiscent of the way Newtonian gravitational theory is incorporated within the general theory of relativity). He also made major contributions to the theory of superfluidity in liquid helium, and to studies of both the strong and weak forces (see *forces of nature*), and provided stimulating insights into the way to approach a quantum theory of gravity. He developed the path integral approach to quantum physics (using *Feynman diagrams*), from which he derived the clearest and most complete version of *quantum electrodynamics* (QED), which stands alongside the general theory of relativity as one of the two most successful and well-established theories in physics.

Feynman was also an inspiring teacher and popularizer of science, and is known to many people who know little or nothing about his work as the human face of physics in the second half of the 20th century. As this icon of physics, his reputation is awesome; for once, though, the real achievements of the icon are even more impressive than the reputation.

Feynman was born in New York City on 11 May 1918. His father was a moderately successful businessman, a maker of uniforms, and although not rich the family survived the depression years without hardship. Feynman's unusual abilities

Richard Feynman (1918–1988).

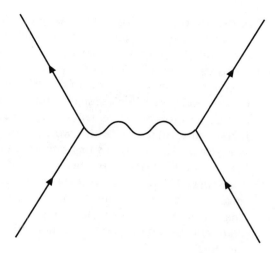

Feynman. The classic Feynman diagram, representing two particles (straight lines) that interact with one another by the exchange of a third particle (wavy line). The two particles might be electrons, for example, interacting by the exchange of a photon.

as a mathematician were recognized at school, where one of his teachers, Abram Bader, introduced him to the principle of least action, which was to become a guiding influence on much of Feynman's work.

He studied at MIT, where he made the move from mathematics to physics, graduating in 1939. Feynman then moved to Princeton, where he was nearing completion of his PhD (under the supervision of John Wheeler) when America was drawn into the Second World War by the Japanese attack on Pearl Harbor. Although he began working on the Manhattan Project before finishing his PhD, he found time to complete his thesis and receive his doctorate in June 1942, before moving to Los Alamos early in 1943 to work on the development of the atomic bomb. Feynman's abilities were quickly appreciated by Hans Bethe, the head of the Theory Division at Los Alamos, who made Feynman the youngest (by far) group leader on the project. Alongside his intense work at Los Alamos, Feynman suffered personal tragedy as his wife (and former childhood sweetheart) Arline lay dying of TB; he let off steam by indulging in outrageous pranks and practical jokes, including cracking the safes of his colleagues that held the secrets of the atomic bomb.

In 1946, his war work completed, Feynman moved to Cornell University (Bethe's base) to become a professor of theoretical physics. There, he completed his theory of QED, the work for which he received the Nobel Prize in 1965. In 1950 he moved to Caltech, staying there for the rest of his career, apart from visits on sabbatical or leave of absence to other research centres around the world. In the 1950s, Feynman developed the theory of superfluidity, and discovered a fundamental law describing the behaviour of the weak force – both achievements at least the equal of many which have been rewarded with the Nobel Prize. After a disastrous and short-lived second marriage, which ended in divorce, in 1960 he married Gweneth Howarth, from Yorkshire; they stayed together happily for the rest of his life.

In the early 1960s, Feynman gave the series of lectures that became the famous *Feynman Lectures on Physics*, a set of books that influenced the teaching of physics around the world. Like all Feynman's books, these were essentially transcripts of his spoken words, conveying a lively feel for his personality and speaking directly to the reader. He then developed his theory of **partons** to describe what happens when electrons scatter from protons in deep inelastic collisions. This was a major input to the development of the theory of quarks, gluons and the strong interaction in the 1970s. Alongside all this work, almost as a hobby, Feynman investigated the theory of gravity and laid the basics for the development of a quantum theory of gravitation. Nobody else has made significant contributions to our understanding of each of the four forces of nature. This is partly because scarcely anyone else has produced high-quality, original scientific work over as long a span as Feynman – almost 40 years. He was still producing first-rate, original scientific work in the 1970s, long after his own fiftieth birthday. He then produced a series of popular books, notably *QED: The Strange Theory of Light and Matter* and *Surely You're Joking, Mr Feynman!*, which introduced a wider public to his physics and himself. He also served on the Challenger inquiry, investigating the causes of the explosion of the Space Shuttle in 1986, and became known even more widely through the TV broadcasts of the commission's work.

By then, Feynman was seriously ill with cancer. He died in Los Angeles on 15 February 1988, attended by his wife and sister Joan (herself a successful scientist), and mourned by thousands of people who had never even met him.

Further reading: John and Mary Gribbin, *Richard Feynman: A life in science.*

Feynman diagram   A representation of the way particles interact with one another by the exchange of bosons. Although they look like simple graph-like representations of the way particles and bosons move through spacetime, the structure of the diagrams obeys specific rules, and in particular each vertex (each point where two or more lines meet) is in effect a shorthand notation for a series of (often complicated) equations. In Feynman's own words, 'the diagrams were intended to represent physical processes and the mathematical expressions used to describe them' (Jagdish Mehra, *The Beat of a Different Drum*, p. 290). Even the physical processes represented by the diagrams are not as simple as they seem at first sight. Each diagram actually represents an average over all the possible ways in which the interaction depicted can occur – a sum over histories, or **path integral**. It is easier, though, to manipulate the diagrams and translate them into mathematical terms after the manipulation than it is to manipulate the equations directly.

This is rather like the way in which a curve on an ordinary graph can represent a mathematical equation. For example, a parabola can be represented by the equation $y = x^2$. If you want to know the value of y that corresponds to a certain value of x (or vice versa), you can either read it off the graph, or work out the calculation. An expert can, in the same sort of way, 'read off' the mathematical expression corresponding to the input of a Feynman diagram, and the equivalent mathematical expression for the output, without having to calculate all the complicated intermediate mathematical steps (and, like a graph, the diagrams can be read both ways, to tell you the output if the input is known, or the input if the output is known).

So the diagrams can be used both as a simple pictorial model for non-specialists to get an image of how the particle world works (the way they are used in this book) and as a refined mathematical tool for the experts to use in working out details such as the rates at which interactions occur, and the probability that particles will interact in certain ways.

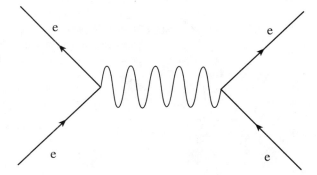

Feynman diagram. If we want to be specific about the interaction represented by a Feynman diagram, we can label the particles involved.

field   The concept, developed by Michael Faraday, that interactions such as gravity, electricity and magnetism extend their influence through space without any need for a medium – the ether – to 'carry' the force. This idea developed from Faraday's initial concept of lines of force reaching outwards from all magnetic, electrically charged or gravitating bodies. Instead of thinking of atoms as tiny lumps of solid, impenetrable matter, he suggested that they should be regarded solely as centres of the concentration of these forces.

The image of lines of force is familiar from the simple experiment in which a bar magnet is placed on a table, a sheet of paper is placed on top of the bar magnet, and iron filings are sprinkled on the paper. When the paper is tapped gently, the iron filings arrange themselves in lines, linking the north and south poles of the magnet. These are the lines of force. Together, they make up the magnetic field. A hypothetical lone north pole placed in the magnetic field would drift along a line of force until it touched the south pole of the magnet; a tiny compass needle placed in the magnetic field will align itself along a line of force.

Faraday introduced these ideas to Victorian England in two lectures given at the Royal Institution, one in 1844 and the other in 1846. In the first lecture, in a classic example of a thought experiment, Faraday asked his audience to imagine the Sun alone in space. What would happen if, by magic, the Earth were suddenly placed at its appropriate distance from the Sun? How would the Earth be aware of the Sun's

Field 1. Faraday introduced the idea of lines of force which reach out across space and act like stretched elastic bands to pull opposite magnetic poles together.

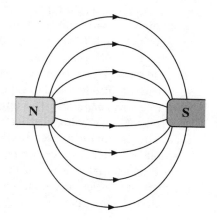

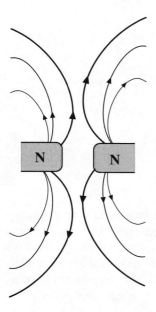

Field 2. But when two like poles are pushed together, the lines of force squash up in the gap like a squeezed block of rubber, and keep the poles apart.

existence? Faraday argued that even when the Earth was not there, the Sun's gravitational influence extended through space, in the form of lines of force. As soon as the Earth was put in place, it would feel the influence of the Sun. The response of the Earth to the Sun's gravitational influence is a direct response to the existence of the field at the locality of the Earth; to the Earth, all that matters are the lines of force.

In the second lecture, Faraday suggested that light could be explained in terms of vibrations of the electric lines of force. Although James Clerk Maxwell's equations of electromagnetism were developed in the 1860s and showed that light and other electromagnetic radiation can be explained as a wave phenomenon associated with varying electric and magnetic fields, Faraday's dismissal of the ether did not find favour with other scientists until the Michelson–Morley experiment failed (in the 1880s) to find any evidence that the Earth moves through the ether, and Albert Einstein's special theory of relativity combined Maxwell's equations with the equations of motion to provide (in the first decade of the 20th century) a system of mechanics – relativistic mechanics – in which there was no need for an ether. Einstein's general theory of relativity, completed in the second decade of the 20th century, is a fully worked out field theory of gravity, in which the field takes the tangible form of distortions in the fabric of spacetime.

Fields are also invoked to explain the behaviour of the strong and weak interactions (see ***forces of nature***). The simplest way to think of a field is as a set of numbers which label every point in spacetime with the strength of the appropriate force at that point. But in most cases this is indeed a *set* of numbers, not a single number. For example, in order to describe the value of the electromagnetic field at any point in three-dimensional space, you need a set of eight numbers – one to describe the intensity of the electric component and three to specify its direction, plus a similar set to describe the intensity and direction of the magnetic component.

This is a powerful and highly respectable model which has been used to make enormous progress in understanding the quantum world. Like all our models, though, it should not be regarded as anything more than a model, and there are alternative ways to describe the way forces operate between particles, in particular, through the exchange of bosons. See also *action at a distance*.

See also *quantum field theory, scalar field, vector field*.

Further reading: John Gribbin, *Schrödinger's Kittens*.

field theory   See *field*.

fifth force   Hypothetical force invoked in the 1980s to explain seemingly anomalous results from experiments to measure the strength of the force of gravity (see *forces of nature*) with very high precision. Further experiments showed that there is no fifth force, and that the seemingly anomalous results could all be explained by local variations in the Earth's gravitational field caused by differences in the density of near-surface rocks.

film chamber   Particle detector which uses a stack of photographic film to record the tracks of the particles. Named by analogy with *bubble chamber*.

fine structure   Details in the spectrum of light from atoms, which show up as closely spaced spectral lines. They are explained in terms of the existence of energy levels that have almost exactly the same energy as one another, but are split by small effects associated with the spin of the electron. Molecular spectra show an equivalent fine structure associated with the rotation and vibration of the molecules. Hyperfine structure is an even smaller-scale effect seen in atomic spectra, caused by the influence of the nucleus (the fact that the nucleus is not at a mathematical point) on the allowed energy levels of the electrons in the atom.

See *Bohr model*.

fine structure constant   A number which gets its name because it appears in calculations of the *fine structure* of atomic spectra, but which has a deeper and universal significance. It is a pure number (that is, it is dimensionless) which is formed out of a combination of four basic constants of physics: the charge on an electron, the speed of light in a vacuum, Planck's constant, and a constant known as the permittivity of free space or the *electric constant* (this can be defined as 1; with this choice of units the fine structure constant becomes equal to $e^2/\hbar c$). Using old-fashioned cgs units, the combination, usually denoted by the Greek letter alpha, is approximately equal to 1/137 (closely equal to 1/137.036), and this number can be regarded as a measure of the strength of the electromagnetic interaction (see *forces of nature*). The equivalent calculations of the strengths of the other forces of nature give a value of 1 for the strength of the strong interaction, 10^{-13} for the strength of the weak interaction and 10^{-38} for the strength of gravity.

first quantization   See *second quantization*.

fission   The process whereby the nucleus of a heavy atom splits into two or more parts, releasing energy and two or three free neutrons as it does so.

Fission releases energy because the most efficient way to store mass-energy in a nucleus is in the form of nuclei of iron and related elements (see *binding energy*). There is an essentially continuous gradient of energy storage efficiency from the heaviest elements down to iron, so that any process in which heavy nuclei can split to form lighter nuclei (down to iron) is favoured energetically. If nuclei of heavier elements can be split, forming lighter nuclei, this will release energy. In spite of this, many of these heavier nuclei are stable, once formed inside stars, even though it requires an input of energy (from a supernova) to make them.

Unstable heavy nuclei, such as those of uranium-235, may fission spontaneously. Fission can also be triggered if a fast-moving neutron strikes an unstable nucleus. Since fission itself releases neutrons from the nuclei that are splitting, if enough of a radioactive material such as uranium-235 is put together in one place, the spontaneous fission of one nucleus will trigger the fission of two or more nearby nuclei, each of which triggers the fission of at least two more nuclei, and so on in a cascade called a chain reaction. This is the process that releases energy in a so-called atomic bomb (really a nuclear bomb) and (in a controlled, slow process) in nuclear reactors used to generate electricity. In a bomb, the chain reaction runs away explosively because each fission triggers the fission of several more nuclei. In a nuclear reactor, the rate at which the reaction goes on is controlled by inserting material that absorbs some of the neutrons into the pile of uranium (or other radioactive material), so that on average each fission leads to the fission of just one more nucleus.

The fission of every nucleus in 1 kg of uranium-235 would yield 20,000 megawatt hours of energy (literally enough energy to keep a 20 megawatt power station running for a thousand years), the same as would be released by burning 3 million tonnes of coal.

Fitch, Val Lodgson (1923–) American physicist who shared the Nobel Prize for Physics in 1980 with James Cronin for their discovery that the decay of neutral kaons violates *CP conservation*.

Born in Merriman, Nebraska, on 10 March 1923 and raised on a cattle ranch, Fitch graduated from McGill University in 1948 and obtained a PhD from Columbia University in 1954. He then worked at Princeton University for the rest of his career, becoming chairman of the physics department, and serving from 1970 to 1973 on the US President's Science Advisory Committee. The discovery for which he shared the Nobel Prize with Cronin was made when they were working together at Princeton in the early 1960s. It shows that there are physical processes which can distinguish between the forward and backward directions of time in individual particle reactions, suggesting a link between the quantum world and the arrow of time defined by the outburst of the Universe from the Big Bang.

fixed target machine A kind of particle accelerator in which a beam of particles is fired at a fixed target – as opposed to a collider in which two beams of particles are smashed into each other.

Fizeau, Armand Hippolyte Louis (1819–1896) French physicist who worked with Leon Foucault in taking the first detailed photographs of the Sun, and who made the first reasonably accurate terrestrial measurements of the speed of light.

Fizeau came from a wealthy family. He was born in Paris on 23 September 1819 (four days after Foucault), and like Foucault first studied medicine, before turning to optics. After their joint work on astronomical photography, Fizeau devised an experiment to measure the speed of light by sending a beam of light through the gap in a rotating toothed wheel (like the gap in the battlements of a castle), along an 8 km path between the hilltops of Suresnes and Montmartre, off a mirror and back through another gap in the toothed wheel. As the speed of the rotating wheel was changed, sometimes the returning light was blocked because a tooth in the wheel had move into the gap, and sometimes the light could get through because the wheel had moved so much that the next gap was available for the passage of the light. Knowing the speed of the rotating wheel and the distance between the mirrors, in 1849 Fizeau measured the speed of light to an accuracy within 5 per cent of the best modern determination. At the time, this was the best ground-based measurement of the speed of light, but not as accurate as astronomical measurements. The following year, independently of Foucault, he showed that light travels more slowly through water than through air.

Fizeau was also the first person to investigate the Doppler effect for light. He died in Paris on 18 September 1896.

flavour   The property that distinguishes quarks from one another; quarks come in six flavours, known as up, down, strange, charm, top and bottom. The term is sometimes also extended to describe different 'flavours' of lepton; in that case, the varieties are electron, electron neutrino, muon, muon neutrino, tau and tau neutrino.

fluorescence   Short-lived emission of light from a substance after it has been stimulated by some outside influence other than a rise in temperature. The influence could, for example, be the effect of **X-rays**, or electrons striking the coating of a TV screen. The definition of 'short lived' is arbitrarily set as any **luminescence** lasting less than 10 nanoseconds (10^{-8} sec) after the stimulation has stopped.

forces of nature   There are only four forces that are known to operate between elementary particles. Two of these, gravity and electromagnetism, are familiar in the everyday world because they have a long range. The other two, the strong and weak nuclear interactions (sometimes referred to just as the strong force and the weak force), have a range roughly comparable to the size of an atomic nucleus (this is not a coincidence; it is because the forces have that range that nuclei have that size), and were only discovered when physicists began probing the structure of matter on that scale. The forces of nature are also known as the fundamental forces, or as the fundamental interactions.

Gravity (although it is, in fact, the weakest of the four forces) was the first one to be investigated scientifically, and the first to be described by a satisfactory scientific theory, Isaac Newton's theory of gravity. This is because although gravity is weak, it is additive. Every speck of matter that you put into a lump of stuff contributes something to the overall gravity of the lump. Gravity also has a very long range – it falls off as 1 over the square of the distance from the lump, and in principle that means it extends for ever, to infinity. So an object like the Sun, with a mass of 1.9891×10^{30} kg, exerts a powerful gravitational influence on its surroundings, sufficient to hold the planets in elliptical orbits around the Sun and thereby provide Newton with evidence of the inverse square law of gravity at work.

Electromagnetism is much stronger than gravity, but neither electricity nor magnetism adds up in the way that gravity does. Both electricity and magnetism come in two varieties – positive and negative charge, and north and south poles. These varieties tend to cancel each other out, reducing the overall influence. Electricity and magnetism used to be thought of as two separate forces. They were described in terms of two separate sets of equations, each, like gravity, obeying an

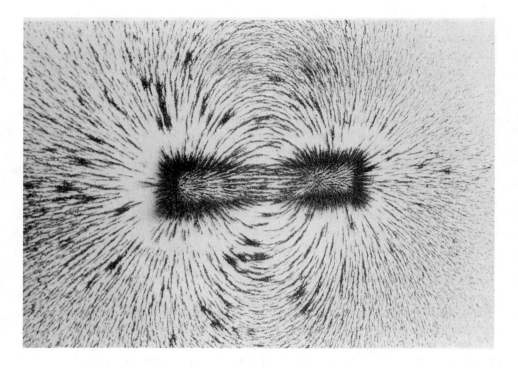

Forces of nature. We can actually see one of the forces of nature, magnetism, at work, using a simple experiment in which a bar magnet is placed underneath a sheet of stiff paper and iron filings are lightly sprinkled on the top. When the paper is tapped gently, the filings arrange themselves along the *lines of force* linking the two magnetic poles. See *Faraday, Michael.*

inverse square law, but with the added complication, in each case, of the two varieties. But in the second half of the 19th century, James Clerk Maxwell discovered a unified set of equations that describe both electricity and magnetism (complete with the inverse square law) in one package. This was the first successful attempt at unification – Maxwell's equations represent a unified theory of electricity and magnetism, showing that they are different facets of a single force, electromagnetism. Just as the discovery of the inverse square law of gravity provided the inspiration for physicists to develop inverse square law descriptions of electricity and magnetism, so the unification of electricity and magnetism has provided the inspiration for physicists to attempt to achieve further unification of the forces of nature, with the goal of finding one set of equations that will describe all of the forces as different facets of a single superforce.

Although he had no way of knowing it, when Newton watched an apple fall from a tree, he was also witnessing a demonstration of the relative strengths of gravity and electromagnetism. Electromagnetic forces come into their own on the small scale (the size of molecules, atoms and below) where one variety of electric charge, in particular, can dominate over the other in a small volume. It is electromagnetic forces that hold electrons and nuclei together in atoms, and which hold atoms together to make molecules. So the stalk of an apple is held together by elec-tromagnetic forces, operating between neighbouring atoms and molecules. When the stalk breaks and the apple falls from the tree, it is because the gravitational pull of the entire Earth, containing 5.976×10^{24} kg of matter, has just barely succeeded in breaking the electromagnetic bond between a few molecules in the stalk.

If electromagnetism is so strong, though, why doesn't the concentration of positive electric charge in the nucleus of an atom blow the nucleus apart? Because there is an even stronger force, the strong nuclear force, which overwhelms the electric repulsion on the scale of an atomic nucleus, and holds the nucleons (both protons and neutrons feel the strong force) together. This force does not obey an inverse square law. The force which acts to hold neutrons and protons together in nuclei is actually a vestige of a deeper force, operating within nucleons. This is the true strong interaction, which actually operates directly between quarks, through the exchange of gluons. It has a limited range because gluons have mass (unlike the photons which carry the electromagnetic force, or the gravitons which carry gravity). The range of force carriers like gluons is inversely proportional to their mass, so the more mass they have, the shorter is their range. The force leaks out of individual nucleons to influence the particles next door, but cannot reach outside the nucleus (in a roughly similar way, the overall electric charge on a proton or a neutron is actually the sum of the charges on its constituent quarks; but in that case, provided there is an overall charge, its influence tails off only as the inverse square of distance). The strong force is about 100 times stronger than the electromagnetic force, and as you might expect from this the heaviest stable nuclei have just under 100 protons in each nucleus (plutonium has 94). Add any more, and the overall electrical repulsion (which *does* add up if all the charges are the same) will overwhelm the strong force, which only operates between next-door nucleons.

The strong force has one particularly distinctive property. Up to the limit of its range, it is actually *stronger* for quarks that are further apart. This is why quarks are confined within hadrons and are never detected in isolation. See **confinement**.

The weak force is even less like the everyday concept of a force based on our experience of gravity and electromagnetism, which is why it is preferable in some ways to refer to fundamental interactions, rather than forces of nature. It operates, like the other forces, by the exchange of messenger particles (in this case, the ***intermediate vector bosons***), and, like the gluons, these messenger particles have mass, so the range of the weak force is limited. But instead of just pushing particles apart (or pulling them together), the exchange of these particular messengers changes the character of the particles that swap them. The weak interaction affects leptons and also influences hadrons, in particular through the process known as **beta decay**. In beta decay, a neutron emits an electron and an antineutrino, converting itself into a proton. In terms of the weak interaction, this is described as an interaction between an incoming neutrino (the same as an outgoing antineutrino) and a down quark inside the neutron. The down quark exchanges a W^- particle (one of the intermediate vector bosons) with the neutrino, turning itself into an up quark and the neutrino into an electron. The bosons involved have such a short range (because of their large mass) that these weak interactions take place essentially at a point.

One of the greatest triumphs of theoretical physics in the second half of the 20th

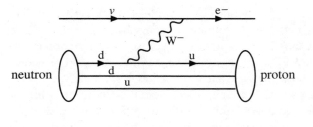

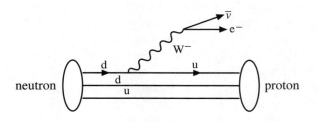

Forces of nature. The process of beta decay, by which a neutron converts itself into a proton, can be understood in terms of the weak interaction operating on the quarks within the neutron. A down quark changes into an up quark by emitting a W particle. The W particle very quickly decays into an electron and an anti-neutrino (bottom), which is equivalent to it meeting a neutrino and converting the neutrino into an electron (top).

century was the discovery, made independently by Abdus Salam and by Steven Weinberg in 1967, of a way to describe the weak interaction and the electromagnetic interaction in one mathematical formalism, as a single force, the electroweak interaction (the theory was extended and generalized by Sheldon Glashow in 1970). The theory requires three intermediate vector bosons to explain weak interactions (they are called W^+, W^- and Z^0), analogous to photons but with mass (and in two cases, electric charge). It also predicts the masses of these bosons, and they were duly observed in experiments at CERN in the early 1980s, with exactly the predicted masses, confirming the accuracy of electroweak theory. The combination of electromagnetism with the weak interaction is not quite as perfect as the way in which electricity and magnetism are combined in one package to make electromagnetism, but it has become the standard model of particle physics, the archetype used by physicists in their attempts to achieve further unification by incorporating the strong force into the package. So perhaps we should really think in terms of three forces of nature – gravity, the electroweak force and the strong force – not four.

The relative strengths of the four forces can be expressed numerically by setting the strength of the strong force to be 1. The precise numbers that you get for the relative strengths depend on exactly how you choose to measure them, but in terms of the relationships between the fundamental constants involved, in round numbers the strength of the electromagnetic force is 1/137 (or about 10^{-2}), the strength of the weak force is 10^{-13} and the strength of gravity is 10^{-38}.

See also *fifth force*, *grand unified theories*.

Further reading: Frank Close, *The Cosmic Onion*; G. D. Coughlan and J. E. Dodd, *The Ideas of Particle Physics*.

Foucault, (Jean Bernard) Leon (1819–1868) French physicist who made the first laboratory measurements of the speed of light accurate to within 1 per cent, invented the gyroscope and devised the pendulum method which bears his name for demonstrating the rotation of the Earth.

Foucault was born in Paris, on 19 September 1819. He was a sickly child and was educated at home before starting to study medicine. He very soon shifted into science, and from 1844 he supported himself by writing scientific textbooks and popular articles about science for a newspaper, the *Journal de Débats* – he was arguably the first professional science writer. This continued for eleven years while he carried out experiments in physics at home, but in 1855 Foucault became a professional scientist when he took up a post as physicist at the Paris Observatory.

In the 1840s, Foucault worked with Armand Fizeau on the development of photography for scientific uses, and they obtained the first detailed photographs of the surface of the Sun in 1845. It was through this interest in astronomical photography that Foucault developed his famous pendulum. In order to take clear photographs of the Sun, he needed a clockwork mechanism which would turn the camera slowly to keep the Sun centred in the field of view, as it 'moves across the sky' because of the rotation of the Earth. The mechanism Foucault devised was driven by a pendulum, and he noticed that the pendulum continued to swing in the same plane while the Earth rotated underneath it. After making an impressive demonstration of the effect by hanging a pendulum 67 metres long, with a 28 kg ball, from the dome of the Panthéon in Paris in 1851, Foucault extended the idea to spinning objects, which always maintain the same orientation in space, and invented the gyroscope in 1852. Both his pendulum and the gyroscope are examples of inertia at work (see **Mach's Principle**).

Alongside his interest in inertia, Foucault worked on measuring the speed of light. He used a system in which a beam of light was bounced off a rotating mirror, on to a stationary mirror and back to the rotating mirror (the technique had been suggested by Dominique Arago, and Foucault's first experiments used equipment that he took over from Arago after Arago's sight failed). Because the rotating mirror has moved slightly by the time the light gets back to it, the beam of light is deflected

Foucault pendulum. A 19th-century engraving illustrating Jean Foucault's demonstration of the rotation of the Earth in 1851, in Paris, with a long, heavy pendulum.

by a small amount. The amount of deflection indicates how far the mirror has rotated while the light was in flight, and since the speed of rotation of the mirror is known, this reveals the speed of light. Using this technique, in 1850 Foucault showed that the speed of light is greater in air than it is in water (as it must be if light travels as a wave) and by 1862 he had obtained a measurement of the speed of light, 298,005 km/sec, within 1 per cent of the best modern determination. This improved on slightly less accurate measurements made at the end of the 1840s by Fizeau, using a different technique.

At the Paris Observatory, Foucault made important contributions to improving the mirrors and lenses used in telescopes, and invented accurate regulators for driving machines at constant speed. Although they were devised for use with telescope motors to keep the telescopes tracking accurately across the sky, the regulators also found uses in industry. Foucault died in Paris on 11 February 1868, a victim of brain disease, in his 49th year.

Fowler, William Alfred (Willy) (1911–1995) American physicist who received the Nobel Prize in 1983 for his work with Fred Hoyle on the theory of how elements are manufactured inside stars – stellar nucleosynthesis.

Born in Pittsburgh on 9 August 1911, Fowler studied at Ohio State University and at Caltech, where he obtained his PhD in 1936. Apart from war work on various projects, Fowler spent his career at Caltech. His special skill lay in carrying out accurate laboratory determinations of the rate at which nuclear reactions, especially those important for energy generation inside stars, take place. After his work on stellar nucleosynthesis (in which Geoffrey and Margaret Burbidge were also involved), he worked with Hoyle and Robert Wagoner on the theory of how light elements are made in the very early Universe – Big Bang nucleosynthesis. He died in Pasadena on 14 March 1995.

Franck, James (1882–1964) German-born American physicist who shared the Nobel Prize for Physics with Gustav Hertz in 1925 for their experiment which showed that atoms can accept energy only in discrete units – quanta.

Born in Hamburg on 26 August 1882, Franck went to Heidelberg University in 1901 to read law and economics, but there met Max Born (his exact contemporary) who encouraged Franck to switch to science. He moved through geology and chemistry into physics, completing a PhD at the University of Berlin in 1906. His work with Hertz was carried out in 1914 when they were both junior researchers (the equivalent of modern-day 'postdocs') in Berlin. They showed that when mercury atoms are bombarded with electrons they always absorb precisely 4.9 eV of energy, leaving the mercury atom in an excited state, from which it returns to the ground state by emitting a photon with a wavelength of 2,537 ångströms. This was the first experimental confirmation of Niels Bohr's theory of energy levels in atoms.

Franck was awarded the Iron Cross twice during the early stages of the First World War, and worked from 1916 to 1920 at the Kaiser Wilhelm Institute of Physical Chemistry in Berlin. He then became professor of experimental physics at Göttingen University, where his old friend Born was professor of theoretical physics – Born had made it a condition of accepting his own post that a job be found for Franck as well. With Born, Franck was instrumental in making Göttingen a centre of excellence during the years in which quantum mechanics was developed. When the Nazis came to power, Franck was one of the few Jews allowed to keep their academic posts (because of his distinguished war record); but when he was ordered to dismiss other Jews from their jobs he refused and moved to Copenhagen and then on to the United States, where he became professor of physics at Johns Hopkins University in 1935. In 1938 he moved to the University of Chicago, as professor of physical chemistry, and studied the chemistry of photosynthesis.

During the Second World War, Franck worked on metallurgical studies related to the Manhattan Project, and became a US citizen in 1943. He headed a group of scientists that became known as the Franck Committee, which urged (unsuccessfully) that the atomic bomb should be demonstrated over an unpopulated part of Japan (or the sea), not dropped on a city. He retired in 1949 and died in Göttingen on 21 May 1964, while on a visit to friends.

Frank, Ilya Mikhailovitch (1908–1990)   Russian physicist who received the Nobel Prize in 1958 (shared with Pavel Cerenkov and Igor Tamm) for his investigations of the Cerenkov effect. Born on 23 October 1908 (10 October, Old Style) in St Petersburg; died 22 June 1990.

Fraunhofer, Josef von (1787–1826)   German physicist who developed the spectroscope and studied the dark lines in the spectrum of the Sun which now bear his name.

Fraunhofer was born in Straubing, Bavaria, on 6 March 1787. His father died in 1798, and he worked as an apprentice to a mirror-maker before joining the optical workshop of a Munich instrument company in 1806. He was so skilful that his work became an invaluable asset to the firm, and he became a partner in 1809 and a director of the company in 1811. Fraunhofer worked on the development of improved lenses, and developed the prism spectrometer into a refined scientific instrument, with which he studied the spectrum of the Sun between 1814 and 1817, discovering the Fraunhofer lines. He also studied diffraction of light, inventing both the transmission diffraction grating and the reflection diffraction grating in the early 1820s.

In 1823 he became director of the Physics Museum of the Bavarian Academy of Sciences. He died of tuberculosis at the age of 39, in Munich on 7 June 1826.

Fraunhofer diffraction   The kind of diffraction that occurs when a series of parallel wavefronts passes a diffracting object. Such parallel wavefronts would come from a light souce at an infinite distance away, but in practice the parallel beam of light is produced by an arrangement of lenses. After the light has passed the diffracting object, it is focused by another lens, so that the diffraction effects are seen in the focal plane of this second lens. The diffraction itself occurs when the light is travelling as plane parallel waves. This kind of diffraction was first studied by Josef Fraunhofer, early in the 19th century.

Fraunhofer lines   See *Fraunhofer, Josef von.*

frequency   The number of oscillations of a wave in 1 second. Usually given the symbol f or v; measured in hertz (1 Hz = 1 cycle per second). See also ***amplitude.***

Fresnel, Augustin Jean (1788–1827)   French physicist who played a major part in establishing the wave nature of light.

Born on 10 May 1788, in Broglie, Normandy, Fresnel was kept away from the turmoil of the French Revolution because his family moved to a quiet region near

Caen to avoid the troubles. There, he was educated by his parents (his father was an architect) until the age of twelve, when he entered the Ecole Centrale in Caen. In 1804 he moved on to the Ecole Polytechnique in Paris, starting an engineering course, and in 1806 switched to the Ecole des Ponts et Chaussées, where he completed a three-year course and became a civil engineer, working for the highways division of the government, in 1809. He worked on several road projects in France, developing an interest in optics as a hobby, and he was unaware of the work by Christiaan Huygens and Thomas Young on the wave theory of light. When Napoleon was defeated and sent to exile in Elba, Fresnel actively supported the Royalist cause, and when Napoleon returned to power briefly in 1815, Fresnel either resigned his post in protest or was fired – accounts differ. Either way, he was placed under house arrest at home in Normandy, where he had the leisure to develop his ideas on optics into a proper theory of light. When Napoleon was finally defeated and exiled once more (this time for good), Fresnel went back to his engineering work, and optics became a hobby once again. Even as a part-time interest, however, it still took a lot of his attention, and in 1820 he developed a type of lens, now known as a Fresnel lens, which found widespread use in concentrating the beams of light from lighthouses.

Fresnel developed his wave theory of light entirely on his own. The key ingredient was that he envisaged light as transverse waves, not as longitudinal waves (see *electromagnetic radiation*). At the time, there was considerable debate in scientific circles about whether light was made up of waves or particles, but Fresnel was not part of the scientific establishment and not involved in the debate. In 1817

Augustin Fresnel (1788–1827).

the French Academy of Sciences offered a prize for anyone who could provide the best experimental study of diffraction, and a theory to explain what was going on. It received only two entries. One was from a crackpot, whose name is not recorded in the annals of the Academy. The other was from Fresnel. His comprehensive (135-page) explanation received the prize eventually, but only after opposition from the judges, who largely supported the Newtonian idea that light is made up of particles (the corpuscular hypothesis).

One of the judges, Simeon Poisson, used Fresnel's theory to predict that if light is a wave then there should be a bright spot in the centre of the shadow cast by a circular object, caused by diffraction. This, to the judges, seemed absurd. But, being a good scientist, the chairman of the judges, François Arago, arranged for an experiment to be carried out to test the prediction. The spot, now known as the Poisson spot, was duly found, and both Fresnel and the wave theory were vindicated.

Fresnel's fame was assured, and even as an amateur he carried out further studies of light (the most important being his study of polarization in terms of the transverse wave idea) both on his own and with Arago, suggested the experiment to test the wave theory further by measuring the speed of light in water and air (see *Foucault, Leon* and *Fizeau, Armand Hippolyte Louis*) and was elected to the French Academy of Sciences in 1823.

He died, of tuberculosis, at Ville D'Avray, near Paris, on 14 July 1827; he was just 39.

Further reading: John Gribbin, *Schrödinger's Kittens*.

Fresnel diffraction The kind of diffraction produced when the source of light or the screen on which the interference pattern is produced (or both) is relatively close to the object which is causing the diffraction (strictly speaking, if the distances involved are less than infinity). This is the classic set-up with a light source, diffracting object and screen laid out on a laboratory bench. Under these circumstances, the light waves cannot be regarded as parallel plane waves (see *Fraunhofer diffraction*) and account has to be taken of the curvature of the wavefronts. This kind of diffraction was first studied by Augustin Fresnel, early in the 19th century.

Fresnel lens A lens made up of a large number of concentric circular pieces, each curved on one side (so that the pieces fit together to make a smooth curve like the front of a conventional convex lens) and stepped on the other, so that the lens is overall much thinner than an equivalent conventional convex lens. Such a lens is light and strong, but has relatively poor optical quality. Originally designed by Augustin Fresnel for use in lighthouses, Fresnel lenses are widely used today in, for example, spotlights and camera viewfinders.

Friedman, Jerome Isaac (1928–) American physicist who received a share of the Nobel Prize in 1990 for his contribution to the experimental work which confirmed the existence of particles within neutrons and protons – the quarks.

Friedman was born in Chicago on 28 March 1928, studied at the University of Chicago, where he received his PhD in 1956, and also worked there for a while before moving on to Stanford University and then, in 1960, to MIT. At the end of the 1960s, although still based at MIT, he worked at SLAC (with Henry Kendall and Richard Taylor) on experiments which probed the structure of nucleons (protons and neutrons) using beams of high-energy electrons. Together with the theoretical interpretation of these experiments provided by James Bjorken and Richard Feynman, they showed that the structure within the nucleons is caused by the presence of point-like particles. In the 1970s, further experiments confirmed that the particles could be identified with quarks. See *partons*.

Frisch, Otto Robert (1904–1979)   Austrian physicist who worked with Lise Meitner (who was his aunt) on the theory of fission. Frisch was born in Vienna on 1 October 1904. He studied at the University of Vienna and received his doctorate in 1926; he spent a year working for an inventor and three years at the German National Physical Laboratory, before moving to the University of Hamburg in 1930. Three years later, when the Nazis came to power, Frisch (who was Jewish) had to leave Germany, and he went to Birkbeck College in London. A year later, he moved on to Niels Bohr's Institute in Copenhagen. It was while visiting Lise Meitner in Sweden at Christmas in 1938 that he learned of Otto Hahn's discoveries concerning the behaviour of uranium, and Frisch and Meitner together came up with an explanation of Hahn's results. In 1939 Frisch wisely returned to England, where he worked at the University of Birmingham with Rudolf Peierls; they were probably the first people to appreciate the possibility of triggering an explosive chain reaction using uranium-235, and alerted the British government to the possibility of developing a nuclear bomb. Frisch also worked at the University of Liverpool and became a British citizen before joining the Manhattan Project in Los Alamos in 1943. After the Second World War, Frisch spent two years as head of the nuclear physics division at Harwell, before becoming Jacksonian Professor in Cambridge, where he stayed. He retired in 1972 and died in Cambridge on 22 September 1979.

fundamental constants   The parameters which have the same value everywhere in the Universe, and whose sizes determine the way the world works. The key constants are: the speed of light in a vacuum (c), the charge on an electron (e), Planck's constant (h), the constant of gravity (G), the electric constant (ε), and the magnetic constant (μ).

fundamental equation of quantum mechanics   The equation which specifies the way in which quantum properties such as position and momentum do not commute (see *Abelian group*). If the position of a quantum entity is specified by a parameter q and its momentum is specified by a parameter p, then the fundamental equation becomes:

$$pq - qp = \hbar/i$$

where $\hbar$ is Planck's constant divided by 2π and i is the square root of -1. In fact, the symbols p and q in this equation represent matrices, not simple numbers. The importance of this equation to an understanding of quantum processes was first appreciated by Werner Heisenberg, Max Born and Pascual Jordan, in the context of *matrix mechanics*, and was developed fully by Paul Dirac. The ideas of *complementarity* and *uncertainty* are implicit in this equation. The *Schrödinger equation*, which is also sometimes referred to as the fundamental equation of quantum mechanics, is a special case of this equation, referring specifically to waves.

fundamental forces   See *forces of nature*.

fundamental interactions   See *forces of nature*.

fusion   The process whereby light nuclei fuse together to make one heavier nucleus, releasing energy as they do so.

Fusion releases energy because the most efficient way to store mass-energy in a nucleus is in the form of nuclei of iron and related elements (see *binding energy*). There is an essentially continuous gradient of energy storage efficiency from the lightest elements up to iron, so that any process in which light nuclei can fuse to form heavier nuclei (up to iron) is favoured energetically. Although nuclei of lighter elements can exist in stable forms, if the nuclei are squeezed tightly enough together (for example, by the extreme pressures that exist inside stars), they will overcome the repulsion caused by the positive electric *charge* on each nucleus and combine with one another, releasing energy as they do so.

Inside stars, fusion releases energy through two main processes, known as the carbon cycle and the proton–proton reaction. The extreme conditions required to force nuclei to fuse make it very difficult to encourage fusion under controlled conditions on Earth, but this is the process which provides the energy in a so-called hydrogen bomb, where the fusion is triggered by the explosion of an atomic bomb. There are hopes of achieving controlled nuclear fusion in reactors that will provide electricity.

The fusion of a pair of deuterium nuclei releases about 2 per cent as much energy as the fission of a single nucleus of uranium-235. Because each uranium-235 nucleus weighs about 100 times as much as a nucleus of deuterium, however, there are 100 times as many deuterium nuclei in a kilogram of deuterium as there are uranium-235 nuclei in a kilogram of uranium-235. So if all those deuterium nuclei fused in pairs, they would generate roughly the same amount of energy as the complete fission of a kilogram of uranium-235, the equivalent of burning 3 million tonnes of coal.

fusion bomb = *hydrogen bomb.*

gal Unit of acceleration, equal to 1 cm per second per second, named after Galileo.

Galilei, Galileo See *Galileo.*

Galileo (Galileo Galilei) (1564–1642) Italian physicist and astronomer who laid the foundations of the scientific method of working out the laws of nature by carrying out experiments and using observations to test ideas about the way things behave. This was an important step forward from the old idea of trying to work out the laws of nature by pure reason, without ever carrying out experiments to test hypotheses.

Galileo was born in Pisa on 15 February 1564. His father, Vincenzio Galilei, was a mathematician and a successful musician, who was able to provide Galileo with a private tutor until he was eleven, when the family moved to Florence. From then until 1581 he studied at a monastery, then went back to Pisa, officially to study medicine. But Galileo was more interested in mathematics than in medicine, and also began to develop an interest in physics. It was while he was a student in Pisa, probably in 1583, that he noticed that provided the length of a pendulum is fixed, the time it takes for the pendulum to complete one swing is always the same, whether it swings through a large arc or a small one. He made this discovery in the cathedral in Pisa, when, bored by the service, he was watching the swinging chandeliers, which he timed using his pulse.

This regular swing of a pendulum became the basis for timekeeping in pendulum clocks. Galileo designed such a clock, but as far as we know never built one. His son did build a clock to Galileo's design, but the idea was only properly developed in practical form by Christiaan Huygens, in the 1650s.

Galileo (1564–1642).

Galileo left Pisa in 1585, without completing his degree, and went home to Florence, where he studied the works of the Ancient Greek scientists (including Euclid and Archimedes). He became so proficient that in 1589 he became professor of mathematics at the University of Pisa – the same university he had not bothered to graduate from four years earlier.

It was while he was a professor in Pisa that Galileo refuted the idea that objects with different weights fall at different speeds, although there is, alas, no evidence that he tested the idea by dropping a cannon ball and a musket ball simultaneously from the leaning tower.

In 1592 Galileo became professor of mathematics at Padua, where he carried out his most important work, investigating the behaviour of falling bodies and studying acceleration by rolling balls down inclined planes and timing them with pendulums. While in Padua, Galileo lived with a Venetian girl, Marina Gamba, and although they never married the couple had a son and two daughters.

In 1609 Galileo learned about the invention of the telescope (actually a reinvention by Dutch experimenters of a device previously invented by Leonard Digges in England), built his own telescopes and used them to study the heavens. He observed mountains on the Moon, discovered four of the moons of Jupiter and saw thousands of stars invisible to the naked eye. His book about all this, *The Starry Messenger*, was published in 1610 and proved a sensation. The same year, on the back of his resulting fame, Galileo gave up his university post to become an independent thinker, under the patronage of the Grand Duke of Tuscany.

In this capacity, back in Florence (but without Marina, who stayed behind in

Padua and soon married someone else), he carried out experiments in hydrostatics, and spoke out in support of the Copernican idea that the Earth moves round the Sun. Although the Catholic Church regarded this idea as heresy, and forbade Galileo to discuss it further, he was allowed to continue his other work. When Urban VIII became Pope in 1623, Galileo got permission to write a book ostensibly setting out the balanced arguments for and against the Copernican idea. The book, *Dialogue Concerning the Two Chief World Systems*, was published in 1632, but it was so clearly biased in favour of Copernicus that it was promptly banned by Rome (ensuring that it was eagerly read wherever an illicit copy could be found), and Galileo was tried for heresy and condemned in 1633, when he was 69. He was sentenced to life imprisonment, but was allowed to serve his sentence under house arrest at his villa near Florence, where he stayed and worked until he died on 8 January 1642 – the year Isaac Newton was born. The sentence was formally revoked by Pope John Paul II on 31 October 1992. Galileo's work paved the way for Newton's discovery of the law of gravity and the laws of motion.

gamma radiation   Electromagnetic radiation with wavelengths in the range from 10^{-10} to 10^{-14} m, corresponding to energies of 10,000 electron volts to 10 million electron volts per photon. Similar to *X-rays*, but with higher energy.

gamma ray   Electromagnetic radiation with wavelength in the range from 10^{-14} to 10^{-10} of a metre, corresponding to photons with energies in the range from 10 MeV to 10 keV.

Gamow, George (1904–1968)   Russian-born American physicist who made major contributions to quantum physics and cosmology, and also investigated the nature of the genetic code.

Born on 4 March 1904 in Odessa (now part of the Ukraine), Gamow studied at Leningrad University, completing his PhD in 1928, and visited Göttingen, Cambridge and Copenhagen in the late 1920s, just at the time that quantum mechanics was becoming established. He came up with a key insight in 1928, when he pointed out that *alpha decay* could be explained in terms of *tunnelling* (the American Edward Condon came up with the same idea independently). This was the first application of quantum mechanics to the understanding of nuclei, and pointed the way for the 'atom splitting' experiments of John Cockcroft and Ernest Walton. While working in Copenhagen, at the end of 1928 Gamow also invented the *liquid drop model* of the nucleus (often incorrectly credited to Niels Bohr), and about the same time he worked with Robert Atkinson and Fritz Houtermans on calculations of the rate at which nuclear reactions take place inside stars (which depend crucially on the tunnel effect).

Back in the USSR, Gamow worked as master of research at the Academy of Sciences in Leningrad from 1931 to 1933, and also became professor of physics at Leningrad University in 1931, but he did not fit into the bureaucratic Stalinist regime of the time. Allowed out of the Soviet Union in 1933 to attend an international scientific meeting in Brussels, he kept on going, settling at the George Washington University in Washington, DC, from 1934 until 1956, when he moved to the University of Colorado, in Boulder, where he stayed for the rest of his life. In 1936 he worked with Edward Teller on an early version of the theory of beta decay.

In the middle to late 1940s, Gamow was the leading proponent of the Big Bang

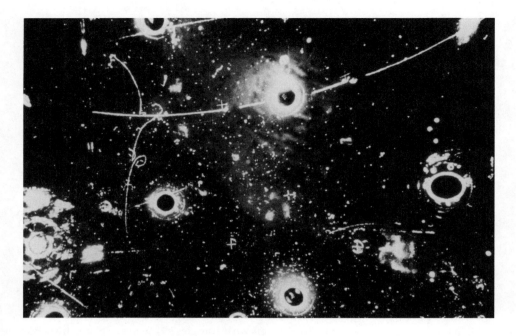

Inside **Gargamelle.** These seemingly insignificant bubble tracks actually provided the first direct evidence for *neutral currents.*

model of the Universe, and instrumental in getting the idea established. In the 1950s, he pointed out that the sequence of four bases along a molecule of DNA in a gene could carry a coded message, written in a four-letter alphabet. Alongside his scientific work Gamow wrote many entertaining popular science books, most notably the 'Mr Tompkins' series, which is still in print. He died in Boulder on 20 August 1968.

Gargamelle A large *bubble chamber* at CERN, used (among other things) in the discovery of neutral current interactions mediated by the *Z particle*. Probably the last great bubble chamber, Gargamelle was completed in 1971; bubble chambers have since been superseded by electronic detectors.

Gassendi, Pierre (1592–1655) French physicist and philosopher who actually carried out the experiment of dropping a ball from the top of the mast of a moving ship and showing that it landed at the foot of the mast, demonstrating that motion is relative.

Born in Champtercier on 22 January 1592, Gassendi studied theology, obtaining a doctorate from Avignon in 1616 and being ordained a year later, when he was also appointed professor of philosophy at Aix. In 1624 he moved to Digne, as provost of the cathedral there, and in 1645 he became professor of mathematics at the Royal College in Paris. He resigned because of ill health in 1648 and died in Paris on 21 October 1655.

Although more of a philosopher and theorist than an experimental scientist, Gassendi's place in history is secured by the famous ship experiment. In 1640, two years before the birth of Isaac Newton, Gassendi borrowed a galley (the fastest, smoothest means of transport available in those days) from the French Navy and, with the rowers working flat out across the calm Mediterranean Sea, used it to carry out his tests. Contrary to the 'common sense' of the day, the balls he dropped from the masthead were not left behind by the ship's motion, but, as Gassendi expected, fell at the foot of the mast, straight down in the frame of reference of the moving ship.

Gassendi also espoused the atomic theory, and believed that light is a stream of particles; he gave the aurora borealis its name.

gauge boson   Any particle that mediates particle interactions ('carries a force') in gauge theory – the photon for electromagnetism, the W and Z particles for the weak interaction, the gluons for the strong interaction, and the graviton for gravity. See also *intermediate vector boson*.

gauge field   See *gauge theory*.

gauge group   See *group theory*.

gauge symmetry   A concept used in field theory to describe a field for which the equations describing the field do not change when some operation is applied to all particles everywhere in space (it is also possible to have local symmetry, where the operation is applied in some particular region).

The term 'gauge' simply means 'measure', and the point is that fields with gauge symmetry can be regauged (or remeasured) from different baselines without affecting their properties.

The classic example is gravity. Imagine a ball sitting on a step on a staircase. It has a certain amount of gravitational potential energy. If the ball moves down to another step on the staircase, it loses a specific amount of gravitational energy, which depends only on the strength of the gravitational field and the difference in height of the two steps. You can measure the gravitational potential energy from anywhere you like. It is usual to measure from either the surface of the Earth or the centre of the Earth as your baseline, but you could choose any of the steps, or any point anywhere in the Universe as the zero for your measurements. The point is that the *difference* in energy between the two steps will always be the same, no matter how you regauge your baseline. So gravity is a gauge theory.

gauge theory   A field theory which has the property of *gauge symmetry*. Gravity and electromagnetism are gauge theories, and the requirement of gauge symmetry was one of the key inputs used in the development of the theory of the weak interaction and the theory of quantum chromodynamics in terms of quantum fields. The situation is more complicated in these quantum field theories than in the simple example of gravity (see *gauge symmetry*), but can be pictured with the help of an analogy developed by Heinz Pagels in his book *The Cosmic Code*.

Pagels asks us to imagine an infinite sheet of paper painted a uniform shade of grey. It is completely uniform and there is no way to tell where you are on the sheet of paper – it is globally invariant. The same is true whatever the exact shade of grey of the paint, an example of gauge symmetry ('regauging' the colour makes no difference). Now imagine a similar sheet of paper painted in different shades of grey. The symmetry is broken and it is easy to tell different regions of the sheet from one another. But the symmetry – the global invariance – can be restored if we lay over the multishaded paper a sheet of clear plastic which has been painted in shades which exactly balance the pattern in the paper – dark where the paper is light, light where the paper is dark. The combined effect will be to produce a uniform shade of grey, with the global invariance restored.

The sheet of multishaded paper represents a visible quantum field. The sheet of plastic painted in complementary shading represents a gauge field, which restores the symmetry. The gauge field is sometimes called a Yang–Mills field after the two researchers who developed this approach to quantum field theory in the 1950s.

The key point is that a completely globally invariant field is undetectable because it is the same everywhere. Fields express themselves, as it were, only when the symmetry is broken and there are differences from place to place. It was this idea of **broken symmetry** in gauge theory that led Steven Weinberg and Abdus Salam (working independently of one another) to develop the basis of the electroweak theory in 1967, pointing the way for all subsequent attempts to develop a **grand unified theory**.

Gauss, Karl Friedrich (1777–1855)   German physicist, mathematician and astronomer who made many contributions to science, including pioneering work in non-Euclidean geometry (later important in the development of the general theory of relativity).

Born in Brunswick on 30 April 1777, Gauss was the son of a gardener. His precocious talent for mathematics was recognized by a teacher who drew the attention of the Duke of Brunswick to the boy genius; the Duke paid for Gauss's further education, which took him to the Collegium Carolinum in Brunswick and then to the University of Göttingen from 1795 to 1798. The Duke was mortally wounded fighting agaist Napoleon's army at Jena in 1806, and Gauss then became director of the Göttingen Observatory, a post he held until he died, in Göttingen, on 23 February 1855. Although his greatest achievements were mathematical, Gauss also contributed to astronomy and physics, notably to the theory of electricity and magnetism.

gauss   A unit of magnetic flux density, symbol G. 1 G is equal to 10^{-4} tesla. Named after Karl Gauss.

Gay-Lussac, Joseph Louis (1778–1850)   French chemist who made an important contribution to the understanding of the behaviour of gases.

Born at Saint-Leonard on 6 December 1778, Gay-Lussac was the son of a judge (who was later imprisoned during the French revolution). He studied at the Ecole Polytechnique in Paris, graduated in 1800 and became an assistant to the chemist Claude-Louis Berthollet (1748–1822), a friend of Napoleon. After travelling with the explorer and natural historian Alexander von Humboldt (1769–1859), and also carrying out hot-air balloon ascents to see if the composition of the atmosphere changes with altitude, Gay-Lussac became professor of physics at the Sorbonne, in 1808, and (in addition) professor of chemistry at the Ecole Polytechnique, in 1810. He later held several government posts, including Chief Assayer to the Mint from 1829 onwards, and served in the French parliament. In 1832 he was appointed professor of chemistry at the Museum of Natural History. He died in Paris on 9 May 1850.

Gay-Lussac's most important scientific work was his discovery of the law that gases combine with one another in simple whole-number proportions by volume, published in 1808 and known as Gay-Lussac's Law. This was a key step towards developing a proper atomic theory of matter which incorporated the existence of molecules, and pointed the way for the key work of Amedeo Avogadro.

gedanken experiment   See *thought experiment.*

Geiger, Hans Wilhelm (1882–1945)   German physicist who invented the eponymous counter, for measuring radioactivity.

Born at Neustadt on 30 September 1882, Geiger studied at the University of Munich and at the University of Erlangen, where he was awarded his PhD in physics in 1906. He then went to Manchester, where he worked with Ernest Rutherford and Ernest Marsden on the experiments which revealed the nuclear structure of the atom.

In 1912 Geiger returned to Germany to become head of the Radioactivity Laboratories of the Physikalische Technische Reichsanstalt (the German physical laboratory) in Berlin. He served in the artillery in the First World War, then returned to this post. He moved on to the University of Kiel, as professor of physics, in 1925; to the University of Tübingen in 1929; and back to Berlin in 1936 as head of the physics department at the University of Charlottenburg. He became ill during the Second World War, lost his home and possessions in the latter stages of the war, and died in Potsdam on 24 September 1945. His famous counter was developed from a device he used in 1908 to count alpha particles during his work in Manchester; with Rutherford, he used this instrument to establish that an alpha particle carries two units of positive charge. There were many refinements over the years, until the final form of the detector was developed by Geiger and Walther Müller in 1928; it is often known as the Geiger–Müller counter.

Geiger counter (Geiger–Müller counter)   An instrument for measuring electrically charged particles ('ionizing radiation' such as alpha rays and beta rays). The basis of the detector is a cylindrical cathode wrapped round a wire anode, with the space between them filled with a mixture of methane and inert argon gas at a pressure of a few tenths of an atmosphere. The potential difference between the cathode and the anode is kept at about 1,000 volts. A charged particle passing through the tube triggers an electric discharge, which is often used to make a 'click' on a loudspeaker. The device can also detect energetic photons (gamma rays), if they carry enough energy to knock an electron out of an atom in the gas.

Geiger–Müller tube   The tube in a Geiger counter.

Gell-Mann, Murray (1929–)   American physicist who won the Nobel Prize for Physics in 1969 for his work on the classification of fundamental particles, and was one of the people who introduced the idea of quarks (which he named).

Born on 15 September 1929 in New York City, Gell-Mann was a child prodigy who entered Yale University when he was only fifteen years old, graduated in 1948 and received his PhD (from MIT) at the age of 22. After a year at the Institute for Advanced Study, in Princeton, in 1952 he moved to the University of Chicago, where he worked with Enrico Fermi, and in 1955 he became associate professor of physics at Caltech, where he was promoted to full professor in 1956 and stayed for the rest of his career. Intensely competitive and used to being best at everything, Gell-Mann felt slightly in the shadow of Richard Feynman at Caltech. One of Gell-Mann's former students (who asked to remain anonymous) told us that 'Murray was clever, but you always felt that if you weren't so lazy and worked really hard you could be as good as him; nobody ever felt that way about Dick.'

In 1953 Gell-Mann and the Japanese physicist Kazuhiko Nishijima independently hit upon the idea of explaining some of the properties of fundamental particles (many of which were being discovered for the first time in particle accelerator experiments in the 1950s) by assigning them a property dubbed 'strangeness' (simply because these particles had strangely long lifetimes, compared with otherwise similar particles, although still measured in tiny fractions of a second). This led Gell-Mann and Yuval Ne'eman (again, working entirely independently), at the beginning of the 1960s, to come up with a classification scheme for particles which was called the eightfold way. This system arranged particles according to their properties in much the same way that Dmitri Mendeleyev had arranged the elements according to their chemical properties.

In 1962, for the third time Gell-Mann had a bright idea at the same time as someone else, when he and George Zweig independently realized that many of the properties of the particles arranged in the eightfold way pattern could be explained if particles such as protons and neutrons, once thought to be indivisible, were actually composed of smaller units – just as the properties of the chemical elements in Mendeleyev's table can be explained if atoms, once thought to be indivisible, are explained in terms of composite structures made of protons, neutrons and electrons. For example, particles which exhibit strangeness contain a strange quark.

Zweig actually called these subparticles 'aces'; Gell-Mann called them 'quarks'. Zweig was a student, and Gell-Mann a professor with a forceful personality. Gell-Mann's name stuck. Even in 1969, however, there was so much doubt about the quark model that it was specifically not included in the citation for Gell-Mann's Nobel Prize. Gell-Mann also developed a theory of the weak interaction in the late 1950s, at the same time that Feynman (completely independently; he had been away in Brazil) came up with a similar idea. The two (mostly friendly) rivals were encouraged by their colleagues at Caltech to publish their work jointly, the only major piece of physics they did together. Curiously, for all his ability Gell-Mann never made an important discovery that wasn't made by somebody else, entirely independently, at about the same time.

general theory of relativity   See *relativity theory*.

generation   Term used to describe the three different sets of leptons and quarks. The first generation consists of the electron and its neutrino (leptons) and the up and down quarks. These are all the particles you need to construct everyday matter. The second generation consists of the muon (a heavy electron) and its neutrino, and the charmed and strange quarks. These have heavier masses than their first-generation counterparts, but otherwise seem to duplicate their properties. The third generation is made up of the tau particle and its neutrino plus the top and bottom quarks – another variation on the same theme at still higher masses. Nobody knows why nature repeats itself in this way.

Gerlach, Walther (1889–1979)   German physicist best known for his work with Otto Stern on the experiment which revealed the existence of electron spin (the Stern–Gerlach experiment).

Born at Biebrich am Rhein on 1 August 1889, Gerlach studied at the University of Tübingen, where he became a *Privatdozent* after completing his PhD in 1916. He worked briefly in similar junior posts at Göttingen and Frankfurt (where he carried out the famous collaboration with Stern) before returning to Tübingen as professor of physics in 1925. He became professor of physics at the University of Munich in 1929, and retired in 1957. He died in Munich on 10 August 1979.

Germer, Lester Halbert (1896–1971) American physicist who worked with Clinton Davisson on the experiment that proved electrons are waves, for which Davisson received a share of the Nobel Prize. Born in Chicago on 10 October 1896, Germer was a research student at Columbia University, supervised by Davisson, when this work was carried out. He died at Gardiner, New York, on 3 October 1971. See *Davisson, Clinton Joseph*.

GeV 1 billion *electron volts*.

ghost fields See *quantum gravity*.

Giaever, Ivar (1929–) Norwegian-born American physicist who shared the Nobel Prize for Physics in 1973 with Leo Esaki and Brian Josephson for his work on the tunnel effect. Born at Bergen on 5 April 1929, Giaever studied electrical engineering at the Norwegian Institute of Technology, and (after army service in 1952 and 1953) worked at the Norwegian Patent Office before emigrating to Canada in 1954, where he worked for the Canadian General Electric Company. In 1956 he transferred to General Electric's research headquarters in Schenectady, New York, where he worked on the tunnel effect in superconductors. He was awarded a PhD by the New York Rensselaer Polytechnical Institute in 1964, the same year that he became a US citizen.

Glaser, Donald Arthur (1926–) American physicist who invented the *bubble chamber*.

Glaser was born in Cleveland, Ohio, on 21 September 1926, the son of Russian immigrants. He studied at the Case Institute of Technology in Cleveland, graduated in 1946 and went on to Caltech, where he was awarded his PhD (for research involving cosmic rays) in 1949. He worked at the University of Michigan from 1949 to 1959, when he moved to the Berkeley campus of the University of California. He stayed at Berkeley, but switched his main area of interest from physics to molecular biology in 1964.

Glaser built his first bubble chamber, just a few centimetres across, in 1952, while he was at the University of Michigan. He was awarded the Nobel Prize for this work in 1960.

Glashow, Sheldon Lee (1932–) American physicist who shared the Nobel Prize for Physics in 1979 with Abdus Salam and Steven Weinberg for their contributions to the development of the electroweak theory, pointing the way towards the *standard model* of particle physics.

Born on 5 December 1932, in New York City, Glashow attended the same class at school (Bronx High School) as Weinberg. He graduated from Cornell University in 1954, and obtained his PhD from Harvard (where his supervisor was Julian Schwinger) in 1959. At the suggestion of Schwinger, Glashow attempted to develop a unified theory of the electromagnetic and weak forces, but although he found the right mathematical description for such a theory, he never completed it. The same idea was arrived at independently (and completed) by Abdus Salam and Steven Weinberg, but in its initial form 'worked' only for leptons.

Glashow visited the Niels Bohr Institute and CERN after completing his PhD, and worked from 1961 to 1967 at the Berkeley campus of the University of California, before returning to Harvard as professor of physics. In 1971 he was involved in work which extended the electroweak theory to cope with all elementary particles and led to the prediction of the fourth known quark, charm; he went on to contribute to the theory of quantum chromodynamics.

Glashow–Weinberg–Salam theory   = *electroweak theory*.

global invariance   See *gauge theory*.

global symmetry   See *gauge symmetry*.

glueball   A neutral meson that consists of a ball of *gluons* (sometimes referred to as gluonium). Because gluons have colour, like quarks they feel the strong force (even though they are the carriers of the force) and are confined by it so that a free 'coloured' gluon can never be seen. The force is always with them.

During 1995 there were claims that two separate particles manufactured in accelerator experiments might be glueballs. One has a mass of about 1,500 MeV (1.5 times the mass of a proton), the other a mass of about 1,700 MeV; both masses are in the range predicted by theory for glueballs, but the jury is still out. To give you some idea how difficult it is to test these ideas, the 1,500 MeV candidate particle lives for only about as long as it takes light to cross a single atomic nucleus, before it decays into other particles.

glue force   Name for the strong force between quarks (and between gluons themselves), used to distinguish this from the older use of the term 'strong force' to describe the force between baryons, which itself results from a leakage of the glue force from the quarks in one baryon across to the quarks in the baryon next door. Also known as the 'colour force'.

gluino   The supersymmetric counterpart to a gluon. See *SUSY particles*.

gluonium   See *glueball*.

gluons   The carriers of the glue force that holds *quarks* together. The equivalent in *quantum chromodynamics* of the *photon* in *quantum electrodynamics*. The best direct evidence for the existence of gluons comes from 'three-jet events', first observed in experiments at the PETRA collider in Hamburg. In these events, an electron and a positron collide and annihilate one another, producing a quark and an antiquark out of the energy of the two original particles. Almost immediately, either the quark or the antiquark radiates a gluon. But before any of these three particles has gone more than 10^{-15} m, they each turn into a shower of other particles, producing three *jets* radiating out from the site of the positron–electron collision.

Goeppert-Mayer, Marya (1906–1972)   One of only two women to have been awarded the Nobel Prize for Physics. Marya Goeppert-Mayer was a German-born American physicist who received that honour in 1963 (jointly with Hans Jensen, for their independent suggestion of the *shell model of the nucleus*). She was born in Kattowitz, Upper Silesia (now Katowice, in Poland), on 18 June 1906. The family moved to Göttingen in 1910, and (as Marya Goeppert) she studied first mathematics (David Hilbert was a friend of her family) and then physics (inspired by Max Born's lectures) at the University of Göttingen, where she received her PhD in 1930. The

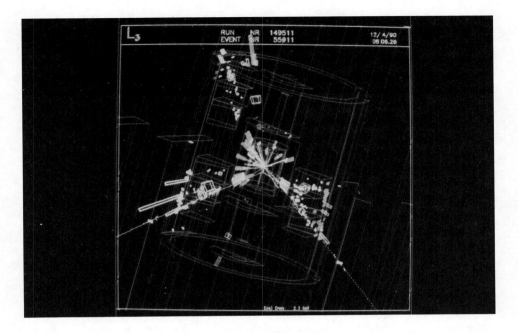

Gluon. A three-jet event seen in the L3 detector at CERN. The particles emerging from the decay (in this case) of a Z particle are tracked by detectors inside L3, and the data converted into this three-dimensional simulation inside a computer.

same year, she married an American chemist, Joseph Mayer, and changed her name to Goeppert-Mayer. She emigrated to the United States in 1931, but found it impossible to get work as a woman scientist in the Depression years. Nevertheless, she worked successively at Johns Hopkins University (tolerated as her husband's assistant, but not part of the male academic establishment), Columbia University in New York (on much the same basis), and then simultaneously at the University of Chicago and the Argonne National Laboratory (welcomed with open arms in the changed post-Second World War climate). It was while at Columbia that she had become respectable in the eyes of the male chauvinists who then ran American science, when the Manhattan Project made her expertise in nuclear physics invaluable. The work for which she received the Nobel Prize was carried out between 1948 and 1950, when she was working in Chicago, where she had been appointed professor of physics in 1946 and worked alongside Enrio Fermi and Edward Teller. In 1960 together with her husband she settled at the University of California, San Diego, where she stayed (as professor of physics, with Joseph Mayer as professor of chemistry) for the rest of her career. She died in San Diego on 20 February 1972.

Goldstone boson   Generic name for any massless particle with spin zero that occurs in any globally symmetric *gauge theory* if the symmetry is broken. Named after Cambridge physicist Jeffrey Goldstone. In order to give the particles mass, an extra field, called the Higgs field, has to be invoked. Without the Higgs mechanism, the W and Z particles would be massless Goldstone bosons.

Goldstone fermion   Supersymmetric counterpart to the Goldstone boson. See *SUSY particles*.

Göttingen   University town in Germany where many important developments in quantum theory were carried out, notably by Max Born and Werner Heisenberg. This became an integral part of what was known (to Born's disgust) as the Copenhagen interpretation of quantum mechanics.

Goudsmit, Samuel Abraham (1902–1978)   Dutch-American physicist who was a member of the first team to introduce the concept of electron spin.

Born on 11 July 1902 in The Hague, Goudsmit took his first degree in Amsterdam and obtained his PhD from the University of Leiden in 1927. While still a student, in 1925 he and George Uhlenbeck explained an experiment involving the splitting of a beam of silver atoms into two streams by a magnetic field (the Stern–Gerlach experiment) by suggesting that the silver atoms possess an intrinsic spin ('up' or 'down'). They suggested that this is because electrons have spin, and since there are an odd number of electrons in a silver atom (47), all but one of the electron spins cancel each other out (23 up and 23 down), leaving one unit of spin (up for half the atoms and down for the other half) to provide a 'handle' by which the magnetic field could get a grip on the silver atoms.

This key insight explains, among other things, how two electrons can occupy the same energy level in an atom, even though they are *fermions* – they are distinguished by their spin. Spin was later put on a secure theoretical foundation by the work of Paul Dirac. Many physicists were baffled that this important discovery was never rewarded with the Nobel Prize.

After completing his PhD, Goudsmit emigrated to the United States, where he worked at the University of Michigan and became professor of physics in 1932; he held the equivalent post at Northwestern University from 1946 to 1948. He worked on radar during the Second World War and was head of a scientific mission sent into Germany with the front-line troops in 1944 to assess the progress of German scientists towards making an atomic bomb. From 1948 to 1970, when he retired, he worked at the Brookhaven National Laboratory. He edited the *Physical Review* from 1951 to 1962, and founded *Physical Review Letters* in 1958. Goudsmit died in Reno, Nevada, on 4 December 1978.

grand unified theories (GUTs)   Any *gauge theory* that attempts to combine the description of the electromagnetic force, the weak nuclear force and the strong nuclear force in one mathematical package is called a grand unified theory. There is no single grand unified theory because all of the candidate theories investigated so far have flaws. So, although the search for a grand unified theory can be said to have begun in the 19th century, when James Clerk Maxwell found one set of equations that describes both electricity and magnetism, uniting them in one theory, it is not yet possible to talk of 'the' grand unified theory in the same way that it is possible to talk of 'the' theory of electromagnetism (or, in its development within the context of quantum mechanics, 'the' theory of quantum electrodynamics, QED). In addition, the grandiose name is slightly misleading because no grand unified theory even attempts to include gravity in the package. There are attempts to bring gravity into a mathematical description of all four of the known *forces of nature*, but these go by different names (see, for example, *supersymmetry*, *string theory*, *theory of everything*).

Electromagnetism (strictly speaking, QED) and the theory of the weak interaction were satisfactorily combined into one theory, the electroweak theory, in the 1960s, as a result of the work of Sheldon Glashow, Abdus Salam and Steven Weinberg. This is a gauge theory which uses the idea of broken symmetry to explain the differences between the electromagnetic force and the weak force, and as a result of the success of this approach in that context all modern attempts to find a grand unified theory follow the same route (gravity and electromagnetism were already known to be gauge theories). There is also a good (but possibly imperfect) description of the way the strong interaction works, in terms of the theory of quantum chromodynamics (QCD), which is consciously modelled on QED. So attempts to construct a grand unified theory are essentially attempts to unify just two successful theories, the electroweak theory and QCD, in one package.

The basic idea underpinning all such theories is that the different forces of nature become indistinguishable from each other (the broken symmetry is made whole) at very high energies. This can be understood in terms of the messenger particles that carry the various interactions – the intermediate bosons.

The electromagnetic force is mediated by photons, which have zero mass. So electromagnetic forces have (in principle) infinite range. The weak force is mediated by three intermediate bosons, the W^+, W^- and Z^0 particles. These all have mass, the two W particles about 80 GeV and the Z about 90 GeV. In order to carry the weak interaction from one particle to the next, the appropriate intermediate boson has to be created out of nothing at all, using energy borrowed from the vacuum. This is allowed by the **uncertainty principle**, but only provided the energy is repaid very quickly (the bigger the mass involved, the more quickly the debt has to be repaid). So the carriers of the weak force are short lived and can travel only a short distance before they disappear. This is why the weak force has short range.

At energies significantly above 100 GeV (say, 1,000 GeV and upwards), however, the W and Z particles can be manufactured without borrowing energy from the vacuum. They become real particles with potentially infinite lifetimes and infinite range, and now behave exactly like photons (except for the fact that the W particles carry charge). Under these conditions, electromagnetism and the weak interaction behave in exactly the same way as each other. All this is explained beautifully by the electroweak theory, which even predicted the masses of the W and Z particles, later found in high-energy experiments at CERN. The other place where there was enough free energy available to make W and Z particles in profusion was in the early history of the Universe, in the Big Bang. So electromagnetism and the weak interaction were on an equal footing when the Universe was very young, and split apart (breaking the symmetry) as it cooled and expanded away from the Big Bang.

It took until the 1960s to find the right way to combine electromagnetism and the weak interaction partly because the weak interaction is more complicated than QED. In QED there is only one kind of intermediate boson to worry about, the photon; but in the weak interaction there are three. Things get even more complicated in quantum chromodynamics, where there are eight distinct gluons, which makes QCD itself that much more complicated to work with, and increases the difficulty of trying to combine QCD with the electroweak theory.

Nevertheless, it is tempting to apply the same trick to the effort to unify the electroweak theory with QCD. Such a grand unified theory would have to treat quarks (which feel the strong force) and leptons (which don't) on an equal footing. There is a strong physical justification for believing that in some sense quarks and leptons belong to the same family (or superfamily) of particles, because there are three *generations* of each kind of particle, with pairs of leptons matched up with pairs of quarks at successively higher masses.

If this is more than a coincidence, in some sense quarks are like leptons, and leptons are like quarks. That means there would have to be a kind of intermediate boson that could transform quarks into leptons, and vice versa, in a way analogous to the way the weak force can transform a neutrino into an electron (see *beta decay*). But the carriers of such an interaction (dubbed X bosons) would have enormous masses, around 10^{15} GeV, and would be as active as photons are now only at energies above about 10^{16} GeV. Worse, it turns out that there would have to be a lot of them – twelve, even in a relatively simple version of a grand unified theory. So although it is possible to assume that QCD, QED and the weak interaction are on an equal footing at such energies, with all the intermediate bosons as real as photons, and it is even possible to construct theories which make predictions about how matter would behave at such energies, there is no way that these theories could be tested in high-energy experiments here on Earth. It would need an accelerator 1,000 billion (10^{12}) times more powerful than the accelerators which detected the W and Z particles, and the whole Solar System is not big enough to encompass the machine needed to do the job. The last time such energies were reached was in the first 10^{-35} sec of the Big Bang – which is why particle physicists are now keenly interested in cosmology, searching the heavens for traces of effects that might result from what went on in that first split-second of creation, in the hope of testing their grand unified theories.

The GUTs do all make one prediction that might have observable consequences here on Earth. Very, very occasionally, even today, a quark inside a proton or a neutron might be able to borrow enough energy from the vacuum to make an X boson and transform itself into a lepton plus a pion. More accurately, since a proton is composed of two up quarks and one down quark, in proton decay an up quark emits a virtual X boson and changes into an antiquark; the X boson is captured by a down quark which becomes a positron; and the leftover quark–antiquark pair forms the pion. The overall effect would be that the proton or neutron had disappeared, transformed into a pion and a positron. The rate at which this happens depends on the mass of the X boson,

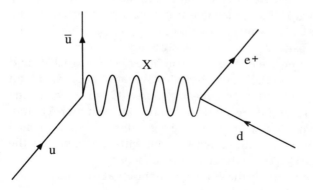

Grand unified theories. The process which must, according to most grand unified theories, produce the decay of the proton.

and on other details of the chosen GUT. It is so difficult to borrow enough energy from the vacuum to do this trick that the average lifetime of an individual proton would be well over 10^{30} years – 100 billion billion times the age of the Universe. But if you had enough protons in one lump, you might expect one of them to decay each year.

One version of a grand unified theory predicts that the lifetime of the proton would be 10^{31} years. In 100 tonnes of matter (anything – water, iron, marmalade) there are about 5×10^{31} protons and neutrons, so if you monitored such a lump for a year, you would, on this version of a grand unified theory, expect to see five decays. Such experiments have been done, and have failed to find that many decays (an average person, by the way, contains about 10^{29} protons, so the chances are that none of the protons in your body will decay during your lifetime). The lifetime of the proton is now known to be at least 5×10^{32} years, and that rules out some GUTs. But it still leaves the door open for others. In particular, it fits in with the idea that the neutrino, originally regarded as a massless particle, has a tiny mass, between 0.001 eV and 10 eV. This would also please astrophysicists and cosmologists – but that is another story (see *Companion to the Cosmos*).

The jury is still out on GUTs and we may never find 'the' grand unified theory. But a lot of physicists are having a lot of fun looking for it.

Further reading: G. D. Coughlan and J. E. Dodd, *The Ideas of Particle Physics*; Yuval Ne'eman and Yoram Kirsch, *The Particle Hunters*.

gravitational constant   A number which indicates the strength of the gravitational interaction. Usually denoted by the symbol G, it has a value of $6.6725985 \times 10^{-11}$ Newton metres² per kilogram². The gravitational force F between two masses M and m a distance r apart is given by the equation $F = GMm/r^2$.

gravitational field   A way of representing the influence of a massive object (that is, any object which has mass, not necessarily one with a lot of mass) on other masses in its vicinity. See *field*.

gravitational force   See *forces of nature*.

gravitational interaction   See *forces of nature*.

gravitational mass   A measure of the amount of matter in a body, determined by its gravitational force. The force of attraction F between two bodies with gravitational masses m and M separated by a distance r is given by the equation $F = GMm/r^2$, where G is the gravitational constant. The gravitational mass of an object is exactly equal to its *inertial mass*; this is a deep truth about the way the Universe works, but nobody can quite explain it. See *Mach's Principle*.

gravitational radiation   Ripples in the fabric of spacetime caused by the motion of objects having mass when they move in certain ways. Gravitational radiation, which is associated with acceleration and orbital motion, is a prediction of Albert Einstein's general theory of relativity, and travels at the speed of light. The theory says that this radiation is completely negligible except where there are strong gravitational fields; although the radiation has yet to be detected directly, its existence was spectacularly confirmed by observations of a system known as the binary pulsar in the 1980s.

The image of matter as solid lumps embedded in a stretched rubber sheet, representing spacetime, makes the origin of gravitational radiation clear. When one of the lumps vibrates, it sends out ripples through the sheet, and these ripples set other lumps of matter vibrating. This is analogous to the way in which a vibrating charged particle sends out electromagnetic radiation in the form of waves which shake other charged particles; but gravitational radiation is very hard to detect because it is only 10^{-40} times as strong as electromagnetic radiation.

One way in which to detect gravitational radiation is to suspend a large bar of material, protected as far as possible from all other sources of vibration, and watch it with sensitive instruments to see if it is disturbed by the passage of gravitational waves. This was tried, using large bars of aluminium, in pioneering experiments in the 1960s and 1970s. Those experiments were so sensitive that they could monitor vibrations in the bar caused by vehicles driving by in the streets outside the laboratory, but they did not succeed in identifying the 'signature' of gravitational radiation. This was no surprise because, if Einstein's theory is correct, any such radiation in the vicinity of the Earth would be too weak to produce detectable shaking of the bars. Nevertheless, it was, of course, worth carrying out the experiments, both to make sure that there was nothing going on that Einstein's theory had not predicted, and to develop techniques that could be used in more sensitive gravitational radiation detectors. Such detectors are now being built, and if Einstein's theory is correct and they work as planned, they may detect gravitational waves around the beginning of the 21st century.

There are two sources of such radiation which ought to produce ripples in spacetime big enough to be detected by the next generation of instruments. The first is the collapse of the core of a large star to form a neutron star or a black hole, when the outer layers of the star explode in a supernova. Such events are rare by human timescales, but a detectable supernova will occasionally occur in our Galaxy – on average, about once every 25 years. When they do occur, they should produce enormous amounts of gravitational radiation in a short time – the energy equivalent (mc^2) of the entire mass of our Sun in a burst of radiation lasting for just 5 microseconds (for comparison, the rate of gravitational radiation produced by the motion of the Earth in its orbit around the Sun is just 200 watts, equivalent in energy terms to the output of an ordinary light bulb).

Even if such an event occurred 10 kiloparsecs away, near the centre of our Galaxy, the amount of gravitational radiation reaching the Earth from it would be equivalent in energy terms to all the electromagnetic energy, across the entire spectrum, reaching us from the Sun in about 100 seconds. Such an outburst would be relatively easy to detect. But because such events are rare, the best bet for the first direct observation of gravitational radiation is to detect the kind of gravitational radiation produced in systems like the binary pulsar, where two very dense stars are in orbit around one another.

Such a system is like an extreme version of a weightlifter's barbell. Viewed in the plane of rotation, this produces gravitational waves which can be visualized in terms of their effects on a circular ring in the same plane. Physicists call this kind of radiation 'quadrupole radiation'.

Quadrupole radiation can be understood most simply in terms of radiation from electric charges. A pair of electrical charges, one positive and one negative, forms a dipole, and when these two charges move (vibrating in and out, or rotating about one another), they produce dipole electromagnetic radiation. A dipole is electrically neutral overall, even though it can radiate in this way.

A pair of dipoles together can make a quadrupole, with two positive charges and two negative charges. When the charges in such an array move in an appropriate way (for example, with one dipole rotating around the other), they produce quadrupole radiation. Unlike electricity, however, mass comes with only one 'sign', so there is no gravitational equivalent of electromagnetic dipole radiation. Two masses which rotate around one another actually behave like a pair of dipoles, producing the gravitational quadrupole radiation which is visualized in terms of its effect on that circular ring.

As the wave passes by, the ring is simultaneously squeezed in one direction and stretched in another direction at right angles, so that it becomes an ellipse. Then the pattern reverses, first returning the circle to its original shape, then distorting it into an ellipse at right angles to the first ellipse. This pattern of alternate squeezing and stretching in two directions at right angles is the characteristic signature of quadrupole radiation. In order to detect such radiation, you need just three test masses, placed in a right-angled 'L' shape, to monitor the distortion of spacetime as the wave passes. And you need some *very* accurate measuring instruments.

Gravitational radiation. The effect of quadrupole gravitational radiation on a circular ring of material. The ring is squeezed into an elliptical shape, first one way and then in a direction at right angles to the first squeeze. This repeats as long as the wave is passing through the ring.

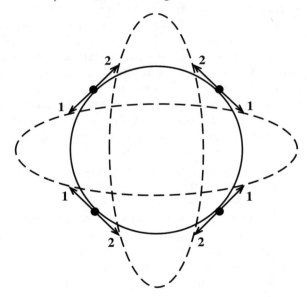

The approach being used in the new generation of gravitational radiation detectors involves placing the three heavy masses in evacuated tubes several kilometres long, built underground. The test masses have mirror surfaces and are monitored using laser beams which bounce to and fro off the mirrors in the evacuated tunnels. The laser beams from each arm of the detector are combined in an interferometer, which measures changes in the positions of the test masses in terms of the wavelength of light. A typical design involves tunnels 3 km long and can measure changes in the separation of the masses at each end of one of these tunnels to an accuracy of 10^{-18} m – less than the size of an atomic ***nucleus***.

Further reading: Kip Thorne, *Black Holes and Time Warps*.

gravitino   The *fermion* that is the counterpart required by the theory of *supersymmetry* to partner the *graviton* (which is itself a *boson*).

graviton   The messenger particle (*intermediate vector boson*) which carries the gravitational force (mediates the gravitational interaction). It plays a part in quantum gravity analogous to the role of the photon in electromagnetic interactions described by QED.

gravity   One of the four fundamental *forces of nature*; a universal force of attraction between all objects that have *mass*. See also *quantum gravity*, *relativity theory*.

Green, Michael (1946–)   British mathematical physicist who played a major role in the development of *string theory* in the context of particle physics.

Born in London on 22 May 1946, Green studied at the University of Cambridge, receiving his first degree in 1967 and his PhD (in elementary particle theory) in 1970. He worked at the Institute for Advanced Study in Princeton from 1970 to 1972, and at the Cavendish Laboratory in Cambridge from 1972 to 1977. After two years in the Department of Theoretical Physics at the University of Oxford, in 1979 he joined the faculty at Queen Mary College (now Queen Mary and Westfield College) in London, becoming a full professor there in 1985. In 1993 Green moved back to Cambridge as John Humphrey Plummer Professor of Theoretical Physics. He has also been a frequent visitor to the Aspen Center for Physics in Colorado, CERN and Caltech. His main research interests are in *superstring* theory and *quantum gravity*.

ground state   The state in which a system has the lowest possible energy. If a system is in any other state, it will tend, if perturbed, to return to the ground state; if it is in the ground state, it will tend to stay there unless it is given a large kick. It is possible for a quantum system to have two (or more) ground states with the same energy but different quantum properties – in the hydrogen atom, for example, the lone electron can occupy either of two states of the lowest energy level, with opposite alignments of the electron's spin. Strictly speaking, the two states corresponding to these two electron orientations do have very slightly different energy, but they are usually regarded as a 'split' ground state of the atom.

The simplest analogy is with a ball on a staircase. When the ball is at the foot of the staircase, it is in the ground state of the system – in this case, literally on the ground. It takes energy to lift the ball up on to a higher step. But it takes only a tiny push to roll the ball over the edge of the higher energy step and send it back down to the ground state.

group   See *group theory*.

group theory   The branch of mathematics that deals with groups and symmetry. In mathematics, a group (or symmetry group) is defined as a collection of elements (a *set*), which are labelled a, b, c and so on, *and* which are related to one another by certain rules.

First, if a and b are both members of the group **G**, then their product, ab is also a member of the group **G**. This process is associative, which means that $a(bc) = (ab)c$, and so on.

Second, there must be an element, called the unit element and usually denoted by e, defined so that $ae = a$, $be = b$, and so on for all elements in the group.

Third, each element has an inverse, written as a^{-1}, b^{-1} and so on, defined so that $aa^{-1} = e$ and so on.

A group for which *ab* = *ba* is an **Abelian group**. The set of ordinary integer numbers (1, 2, 3...) is a simple example of an Abelian group. More generally, groups are made up of elements which are themselves **matrices**. If the smallest object that represents a particular group is a matrix made up of *N* rows and *N* columns (an *N* x *N* matrix), then *N* is the dimension of the group. This is where the number 3 in the group SU(3), which turns out to be important in particle theory, comes from – it is the dimension of that particular group (the 'SU' stands for 'special unitary group').

Group theory was developed in the 19th century by the Norwegian mathematician Sophus Lie (so these groups are sometimes known as 'Lie groups'). Although the theory had been used in mathematical descriptions of the symmetry of crystals, it was a largely obscure branch of mathematics until the second half of the 20th century, when Chen Ning Yang and Robert Mills found a way to describe the strong interaction in terms of Lie groups, and then Murray Gell-Mann and Yuval Ne'eman (working independently of one another) found that SU(3) provided a framework for describing mathematically the relationships between elementary particles. Since then, symmetry groups have been an essential tool used by physicists in their development of gauge theories of the forces of nature. In this context, symmetry groups are sometimes called gauge groups.

A simple example of a group is the set of rotations of a coordinate system (an *x*, *y* graph system) around the point where the *x* and *y* axes meet. If you turn the axes of the graph around, the coordinates of every point measured relative to those axes change, but the relationships between those points do not change – it is simply a relabelling exercise, a gauge transformation. This means, for example, that although the Earth is rotating, the distance between London and Paris (or between any other two points on the globe) stays the same. We are all experiencing a gauge transformation, literally every minute of the day. And if you turn the axes first through an angle *A* and then through an angle *B*, the effect is just the same as turning them through the angle *C*, where *C* = *A* + *B*. Because the angle of rotation can be as small as you like and varies smoothly, as in the example of the rotating Earth, this rotation group is called a continuous group (the SU groups that are so important in particle theory are also continuous groups). The fact that the laws of physics are unchanged by such a rotation implies the law of conservation of angular momentum; in general, whenever a symmetry group describes the behaviour of a physical phenomenon, there must be some conserved quantity associated with that phenomenon (this is sometimes referred to as Noether's theorem, and is a useful feature of group theory which can be used to provide physical insights into the behaviour of particles and forces).

The groups describing the behaviour of particles and fields in the quantum world are, alas, harder to visualize in physical terms, but obey the same mathematical principles. One of the key features of this application of group theory is that, because of the inherent symmetries they predict, there should be certain numbers of particles of a particular type (quarks, say, or gluons), described by a particular symmetry group. SU(3), for example, has 'room' for quarks with just three different varieties of colour charge, and for just eight different varieties of gluon.

GUTs See *grand unified theories*.

h See *Planck's constant*.

hadron Any particle that feels the strong force. All hadrons are composed of quarks. Baryons, which are particles in the everyday meaning of the term, are each composed of three quarks; mesons, which are force carriers, are each composed of a quark–antiquark pair. Baryons and mesons are both members of the hadron family.

Hahn, Otto (1879-1968) German chemist who was awarded the Nobel Prize for Chemistry in 1944, for his role in the discovery of nuclear fission. His colleague Lise Meitner was closely involved in this work, but was shamefully ignored by the Nobel Committee.

Hahn was born in Frankfurt-am-Main on 8 March 1879 (six days before the birth of Albert Einstein). He studied at the University of Marburg and was awarded his PhD in organic chemistry in 1901. After a year's military service and a spell as a junior researcher in Marburg, Hahn intended to take up a career in industry. He was advised that it would be a good career move to become proficient in English, and planned to spend six months in England in 1904. But he got hooked on research and spent two years working in London (1904–5) with William Ramsay, who introduced him to the investigation of radiochemistry, then spent a year working in Canada with Ernest Rutherford before returning to Germany, where he took up a post at the University of Berlin and became professor of chemistry in 1910. In 1907 Liese Meitner joined him there.

In 1912 Hahn joined the Kaiser Wilhelm Institute for Chemistry in Berlin, leaving only for military service in the First World War (where he worked as a chemical warfare expert, not in the trenches) and becoming director of the Institute in 1928. After the Second World War, when the German research institutes were restructured as Max Planck Institutes, Hahn became president of the Institute in Göttingen, which absorbed the old Kaiser Wilhelm Institute for Chemistry.

Hahn's main line of research, often in collaboration with Meitner, was the investigation of radioactive isotopes. In 1934 Enrico Fermi found that when uranium is bombarded with a beam of neutrons, it is transformed into other elements. Hahn and Meitner, together with their colleague Fritz Strassman, carried out further experiments along these lines, bombarding uranium with barium nuclei; but in 1938, in the middle of these investigations, Meitner, who was Jewish, had to move to Copenhagen to avoid persecution by the Nazi regime.

Hahn began to wonder whether the uranium nuclei were actually being broken apart by the bombardment with other particles, the process now known as fission. He lacked the nerve to publish such a seemingly outrageous idea, but Meitner, now working with Otto Frisch, confirmed that this was indeed what was happening, and coined the name 'nuclear fission' to describe the process. The initial impetus for the atomic bomb project by the Allies in the Second World War came from the knowledge that German researchers such as Hahn were well aware of the implications of this discovery. In fact, German progress towards the exploitation of atomic energy during the war was very slow, partly because the leading Nazi politicians did not appreciate its potential, and partly because so many of their best scientists, like Meitner, had fled the country.

Hahn lived to be 89, dying in Göttingen on 28 July 1968. In his postwar writings, he played down the role of Meitner in the discovery of nuclear fission, which fooled nobody and did him no credit. But, very much to his credit, he was a strong opponent of the use of nuclear weapons.

half life The time it takes for half of the radioactive nuclei in a sample to decay. Many processes in the quantum world, notably the radioactive decay of an unstable nucleus, take place in accordance with the statistical rules of chance. This means that, for example, in a collection of thousands or millions of radioactive nuclei in a lump of radioactive material, it is impossible to say which nucleus will decay next, because the process occurs at random. But there is a characteristic lifetime associated with the process (a different lifetime for each radioactive isotope), so that if there is a large enough number of radioactive nuclei in the lump, it is possible to say with very great precision how many of them will decay in a particular interval – the next second, the next hour, the next day or whatever.

The half life is the time it takes for exactly half the nuclei in the sample to decay. It doesn't matter how many you start with: after one half life, half of the original sample will have decayed; after a further half life, half of the rest will also have decayed (leaving one-quarter of the original number); and so on indefinitely. It was this kind of blind obedience of the quantum world to the rules of chance that led Albert Einstein to make his famous comment 'I cannot believe that God plays dice'; but all the evidence is that Einstein was wrong and that these processes do occur entirely at random.

Half lives vary enormously from one radioactive isotope to another. The half life of carbon-14 is just over 5,700 years, which makes it very useful in *carbon dating*; uranium-238 has a half life of just under 4.5 billion years (about the age of the Earth); unnilpentium-258 has a half life of 4 seconds (and exists on Earth only when it is manufactured artificially in particle accelerators). The term is also used to describe the behaviour of unstable particles – the half life of the neutron, for example, is 11.6 minutes (see *beta decay*).

Because a small proportion of the particles or nuclei stay around for much longer than the half life, the mean lifetime is longer than the half life. And because the decay process obeys the precise statistical rules just described, the mean life is always equal to 1.44 times the half life. See also *lifetime*.

Hamilton, William Rowan (1805–1865) Irish mathematician who found a way of restating the equations of motion devised by Joseph Lagrange and whose *Hamiltonian* function became widely used in the variation on the quantum theme known as *wave mechanics*. He also developed the principle of least *action*, which, when applied to motion, leads to Newton's laws of mechanics.

Hamilton was born in Dublin, on 4 August 1805 (legend has it, on the stroke of midnight on the night of 3/4 August). He was the son of a solicitor, but from the age of three was raised by an aunt and uncle in Trim. He was a child prodigy who learned thirteen languages by the age of nine, including Sanskrit and Malay. He mastered Euclid at about the same time and was soon working on Newton. He went to Trinity College in Dublin in 1823, when he was eighteen, but was so outstanding that he was never asked to complete his degree. He had, after all, scored the highest possible grade in intermediate examinations in both Greek and mathematical physics, and during his second year as an undergraduate he had discovered the principle of least action in the context of light paths. He became professor of astronomy, head of the Dunsink Observatory and Astronomer Royal for Ireland in 1827.

Although knighted in 1835 (Ireland was then still a province of Britain), Hamilton developed a drinking problem and spent the last twenty years of his life out of the public gaze, although he did continue his mathematical work, both working and drinking to excess. He died (of gout) at Dunsink on 2 September 1865, shortly after becoming the first foreign member of the US National Academy of Sciences.

Hamilton's early work was on the theory of optics. At the end of the 1820s (beginning in 1827), he found that by using the techniques he had developed to describe the behaviour of light rays, he could modify the equations of motion derived by Lagrange into a set of equations which is in some ways simpler than Lagrange's equations, although you need twice as many of them to describe the same system. Hamilton's version of the equations became widely used in the early days of quantum mechanics, although the Lagrangian approach in fact has many virtues which were first fully appreciated by Richard Feynman. But Hamilton also devised the principle of least action (sometimes known as Hamilton's Principle; see *action*), initially through his work on optics, but soon in a much broader context. This has become a cornerstone of modern ideas about the quantum world.

The rest of Hamilton's work, in pure mathematics, looks at first sight like something far removed from the practicalities of physics, but actually deals with many ingredients, such as complex numbers, non-commutative multiplication and the theory of rotations, which have also proved useful in developing a description of the quantum world. He developed an algebra of four dimensions (quaternion theory) which splits naturally into a one-dimensional component and a three-dimensional component, and coined the term 'vector' in its modern mathematical context.

Hamiltonian   A function (sometimes called the Hamiltonian function) which expresses the energy of a system in terms of momentum and position. In the simplest situations, the total energy is just the sum of the kinetic and potential energies. For more complicated systems, 'the Hamiltonian' becomes a set of differential equations describing the system that is being investigated.

This makes it possible to describe mechanics using Hamiltonian equations, representing interactions in terms of changes in momentum, instead of in terms of forces (the way things are described in Newtonian mechanics, and which was developed further by Joseph Lagrange), and solving those differential equations as necessary. It can be a tedious process, but it has a pedigree dating back to Isaac Newton himself and it is the traditional way that physicists are still taught to tackle such problems. The approach is conceptually similar to using the equations of motion based on Newton's laws to describe the way in which the position of a ball flying through the air changes from one instant to the next as it moves along its trajectory (contrast this with the *Lagrangian* approach; see also *action*). The Hamiltonian of a system may change as time passes, but it is a fundamental property of the system that became an important variable used in the mathematical formulation of wave mechanics, one version of quantum mechanics. This early successful mathematical treatment of the quantum world was built upon a direct analogy with the Hamiltonian approach to classical mechanics; later, however, Richard Feynman showed that the Lagrangian approach is more fundamental and provides a complete model which includes both classical and quantum mechanics.

Hamiltonian equations   See *Hamiltonian*.

Hamiltonian mechanics   Mathematical treatment of mechanics using the *Hamiltonian* approach.

Hamilton's Principle   = principle of least action. See *action*.

hand-waving argument   A vague *qualitative calculation*; it gets its name from the gestures used by physicists while discussing the problem in question. See also *order of magnitude*.

hard scattering   = *deep inelastic scattering*.

Harwell   Home of the *UKAEA*.

Hawking, Stephen William (1942–)   British theoretical physicist who has made important contributions to the study of black holes and the early Universe. His work is relevant to the quantum world largely because of his suggestion that black holes can 'evaporate' through a process in which particle–antiparticle pairs are created near the surface of the black hole (thanks to quantum uncertainty) and one member of the pair falls into the hole while the other carries energy away. The escaping particles are sometimes referred to as Hawking radiation. The existence of this radiation means that any black hole has a characteristic temperature; the work provides a direct link between gravitation theory, quantum physics and thermodynamics.

   Hawking was born in Oxford on 8 January 1942. He completed his first degree at Oxford in 1962 and his PhD in Cambridge in 1965. He has worked in Cambridge ever since, and was appointed Lucasian Professor of Mathematics in 1979.

   Further reading: Michael White and John Gribbin, *Stephen Hawking*.

Hawking radiation   See *Hawking, Stephen William*.

h-cross   See *Planck's constant*.

heat capacity   See *specific heat*.

Heisenberg, Werner Karl (1901–1976)   German physicist who was one of the founding fathers of quantum theory and discovered the *uncertainty principle*. He was, however, awarded the Nobel Prize for Physics (in 1932) not directly for the work he is famous for, but for a subtle explanation of features in the spectrum of molecular hydrogen.

   Born in Duisberg on 5 December 1901, Heisenbeg was the son of a professor of Greek at the University of Munich. Heisenberg studied at the University of Munich, where he was a contemporary of Wolfgang Pauli and was supervised for his PhD (awarded in 1923) by Arnold Sommerfeld. He then worked with Max Born in Göttingen and Niels Bohr in Copenhagen before becoming professor of theoretical physics at the University of Leipzig in 1927. In 1941 he was appointed as director of the Max Planck Institute for Physics and simultaneously professor of physics at the University of Berlin, where he stayed until 1945.

   As the leading physicist to stay in Germany during the Second World War, Heisenberg has been suspected of having Nazi sympathies, and he was the person that the Allies feared might be heading a German effort to develop the atomic bomb. In fact, the limited nuclear research in Germany during the Second World War was largely directed at developing a means for generating power, not weapons. Heisenberg always claimed afterwards that this was largely thanks to his deliberate efforts to divert the Nazis from the potential of atomic weapons; many historians suggest, however, that it was largely because the Nazi leadership was obsessed with new weapons (such as the V2 rocket and jet aircraft) which could be developed more quickly.

None of this obscures Heisenberg's towering achievements in quantum physics. In May 1925, having been struck down with a ferocious attack of hayfever, Heisenberg went to recuperate on the rocky island of Heligoland. It was there, with the hayfever gone and no distractions, that he formulated what was to become known as *matrix mechanics*, the first complete, self-consistent theory of quantum physics. He took the idea back to Göttingen, where Max Born and Pascual Jordan collaborated with Heisenberg in developing it into a complete theory. A copy of this 'three-man paper', passed to Ralph Fowler in Cambridge before publication, was the inspiration for Paul Dirac's formulation of quantum theory. And all of this happened the year before Erwin Schrödinger published the version of quantum physics that became known as wave mechanics. Using the new theory, Heisenberg predicted that hydrogen molecules should exist in two different forms, depending on whether the spins of their nuclei were aligned parallel (in the same direction) or antiparallel (in opposite directions) to each other. It was this work which led to the explanation of the detailed spectrum of hydrogen mentioned in Heisenberg's Nobel citation. Heisenberg discovered the central role of *uncertainty* in the quantum world late in 1926. He was deeply interested in the philosophy of quantum mechanics, and suggested that the wave–particle duality could be explained if entities such as electrons were real little particles that were guided to their destinations by 'pilot waves'. This idea was swamped by the success of the *Copenhagen interpretation*, but bears a close resemblance to the ideas of David Bohm (indeed, it helped to provide the inspiration for Bohm's work), which have become accepted as a valid interpretation of the behaviour of the quantum world.

Heisenberg also explained ferromagnetism and, in the 1930s, made important contributions to particle theory, espousing the idea that the nucleus is made up of protons and neutrons. He suggested that these nucleons could be regarded as different states of the same basic entity, distinguished by a property which he called *isotopic spin*.

After the Second World War, Heisenberg played a major part in setting up the Max Planck Institute for Physics (successor to the Kaiser Wilhelm Institute) in Göttingen, serving as its first director and overseeing its move to Munich in 1955. In 1955 he was also appointed professor of physics at the University of Munich. His later scientific work was largely involved with an unsuccessful attempt to develop a unified field theory. He was also a proponent of the idea of the 'undivided whole', that everything in the world (especially the quantum world) is part of a single system, and therefore it should, for example, be no surprise if closing one slit in the experiment with two holes instantaneously affects the behaviour of an electron passing through the other slit. Although not widely accepted at the time, this idea was developed much more fully by David Bohm. Heisenberg died in Munich on 2 February 1976.

Further reading: David Cassidy, *Uncertainty, Heisenberg*.

Heisenberg's uncertainty principle   See *uncertainty principle*.

Heitler, Walter (1904–1981)   German physicist who, with Fritz London, provided the first quantum-mechanical account of the covalent bond in the hydrogen molecule. This was the birth of *quantum chemistry*.

Born in Karlsruhe on 2 January 1904, Heitler studied first at the University of Karlsruhe, then moved to Munich, where he completed his PhD in 1926. He visited the United States in the following academic year, then took up a junior post at the University of Göttingen, leaving in 1933 when the Nazis came to power and moving to England, where he worked at the University of Bristol until 1941. He then became a professor at the Institute for Advanced Studies in Dublin, and was appointed director there in 1946. In 1949 he left Dublin to become professor of physics at the University of Zurich, where he spent the rest of his career.

helicity   See *spin*.

helium   The second-lightest *element*. Each helium nucleus contains two protons, and each helium atom has two electrons surrounding the nucleus. Along with the two protons (giving it atomic number 2), the helium nucleus must contain at least one neutron. It may contain a single neutron (making it a nucleus of helium-3) or two neutrons (making it a nucleus of helium-4, also known as an alpha particle). Heavier isotopes (containing more neutrons) can be manufactured artificially.

Hertz, Gustav Ludwig (1887–1975)   German physicist who, with James Franck, confirmed that energy is absorbed by atoms only in discrete units – quanta. They shared the Nobel Prize for Physics for this work in 1925; it had been carried out in Berlin in 1914, when they were both junior researchers (the equivalent of modern-day 'postdocs').

Born in Hamburg on 22 July 1887, Gustav Hertz was the nephew of Heinrich Hertz. He studied at Göttingen and Munich before completing a PhD at the University of Berlin in 1911. He was severely wounded in the First World War, and after a spell as a *Privatdozent* at the University of Berlin he worked in industry, for the Philips company in Holland, from 1920 to 1925, when he was appointed professor of physics at Halle University. In 1928 he became a professor at the Technische Hochschule in Berlin. His main work concerned techniques to separate different isotopes of an element. Forced to leave his academic post in 1934 after the Nazis came to power (Hertz was Jewish), he stayed in Germany as director of the research laboratory of the Siemens Corporation. After the Second World War, Hertz worked (not entirely voluntarily) in Russia for ten years, then returned to (East) Germany to become director of the Physics Institute in Leipzig in 1954. He retired in 1961 and settled in what was then East Berlin, where he died on 30 October 1975.

Hertz, Heinrich Rudolph (1857–1894)   German physicist who was the first person to demonstrate the existence of long-wavelength electromagnetic radiation – radio waves.

Born in Hamburg on 22 February 1857, Hertz started an engineering course in Dresden in 1876, but after a year's military service (1876–7) he decided to switch to science and moved to Munich University, initially to study mathematics, then changing to physics. He moved to Berlin in 1878, completing a PhD in 1880, and stayed there for three more years as a *Privatdozent*, then went to the University of Kiel in the same capacity. He became professor of physics at Karlsruhe in 1885, then moved to the University of Bonn in the equivalent capacity in 1889. He died in Bonn on 1 January 1894, when he was in his 37th year.

Hertz's major work was carried out while he was in Karlsruhe. In 1888 he showed that a spark could be induced to jump across a gap in a circuit that was not connected to a source of energy, in response to electrical energy flowing through a similar circuit nearby. He had invented the wireless transmitter and receiver, and proved that electromagnetic waves are generated exactly as the equations discovered by James Clerk Maxwell had predicted.

hertz   Unit of frequency, symbol Hz, equal to 1 cycle per second. Named in honour of Heinrich Hertz.

Hess, Victor Francis (1883–1964)   Austrian-born American physicist who discovered cosmic rays and shared the Nobel Prize for Physics with Carl Anderson in 1936.

Born in Waldstein on 24 June 1883, Hess studied at the University of Graz and was awarded his PhD in 1906. He held various academic posts, and was professor of physics at Innsbruck University in 1938, when Nazi Germany took over Austria. Hess, who had a Jewish wife, moved to the United States, becoming an American citizen in 1944. He was professor of physics at Fordham University, in New York City, until 1956, and died at Mount Vernon, New York, on 17 December 1964.

The work which earned Hess his Nobel Prize was carried out on a series of balloon flights in 1911 and 1912, which showed that penetrating radiation from outer space comes through the atmosphere, some of it to the ground but more being absorbed at high altitudes. Hess actually went up in the balloons, to an altitude of about 5,000 m – a daring adventure in those days.

heterotic string   See *string theory*.

hidden variables   An interpretation (or family of interpretations) of quantum theory based on the assumption that all the usual versions of quantum mechanics are incomplete, and that there is an underlying layer of reality (a kind of sub-quantum world) which contains additional information about the world. This additional information is in the form of the hidden variables. If physicists knew the values of these hidden variables, the argument runs, they could predict the precise outcomes of particular measurements, not just the probabilities of getting particular outcomes.

A helpful analogy can be made with a well-shuffled pack of cards, being turned over one at a time. Probability theory can tell you what the chance is of the cards coming out in any particular order on the deal. But if you had taken a look at the cards before the deal was started, you would be able to 'predict' exactly which card would come next as the deal was being made. The order of the cards is actually well determined, even though you don't know what it is. The *Copenhagen interpretation*, still the way that quantum physics is usually taught, says that at the level of quantum experiments it is as if each card *does not have* a value until it is turned over and examined! There is only a probability of getting any particular card, and the probability changes as you turn over more cards (make more observations).

Of course, we are not saying that real playing cards behave like this. But look at the analogy in more detail, as if playing cards were quantum entities, like electrons. If the first card turned over happens to be a three of diamonds, the probabilities change like this. Before we turned it over, there was a 1 in 52 chance of getting that particular card, a 1 in 2 (26 in 52) chance of getting a red card, a 1 in 4 (13 in 52) chance of getting a diamond, and so on. After we turn over the three of diamonds, there are only 51 possibilities left; there is zero chance that the next card will be the three of diamonds, a 25 in 51 chance of getting a red card, a 12 in 51 chance of getting a diamond, and so on. The Copenhagen interpretation says that these are literally probabilities, and that the value of the next card at the top of the deck, waiting to be turned over, has not been determined in advance. Hidden variables theory says that the card already has a certain value, it just happens that we don't know what it is. Either way, the numbers for the probabilities are the same. Put like this, the hidden variables idea is much more like common sense than the Copenhagen interpretation is. So why isn't it taught more widely?

For many years, hidden variables theories were ignored by most physicists largely because the mathematician John von Neumann had devised a proof that they could not work in the quantum world. In spite of this, David Bohm, in particular, continued to work on hidden variables, convinced that von Neumann was wrong, even though he could not show where he was wrong. Eventually, the mathematician John Bell showed that, although the logic used in von Neumann's argument was correct, the whole argument was based on an unfounded assumption. So there is, in fact, no mathematical or logical reason to discard the idea of hidden variables – although, as we shall see, there is a reason why some people still balk at the idea.

The pedigree of this work goes back to Werner Heisenberg's ideas concerning the hypothetical pilot wave which guides quantum particles to their destination (Louis de Broglie had a similar view of the quantum world, but abandoned it in the face of Niels Bohr's vigorous campaigning for the Copenhagen interpretation); it was kept alive and refined by Bohm and his colleagues. The key feature of hidden variables theories which still disturbs many people is that they are essentially non-local. This means that what is going on in one part of the world affects the entire pilot wave, instantly, everywhere in the world. That is how, on this picture, an electron going through one slit in the experiment with two holes can be 'aware' of the existence of the other slit, and whether or not it is open – because the pilot wave is shaped by the whole experiment. But this non-local aspect of hidden variables theory turns out not to be unique: Bell showed, and the ***Aspect experiment*** confirmed, that non-locality is an integral part of the quantum world. Any version of quantum theory which deals with entities such as electrons as real objects that exist when we are not looking at them is inherently non-local. In the 1990s, these ideas are, therefore, being taken more seriously, by more scientists, than ever before.

Higgs, Peter Ware (1929–) British physicist who developed the first field theory of particle interactions in which the forces of nature are explained in terms of the exchange of gauge bosons. This pointed the way to the development of the electroweak theory. Because the gauge bosons can have mass, this raises the possibility that all particle masses can be explained in terms of a Higgs field which fills the Universe. Everyday particles could acquire their mass, on this picture, by swallowing Higgs bosons. The search for Higgs particles is one of the primary aims of the current generation of particle accelerator experiments.

Higgs was born in Newcastle upon Tyne, and studied at King's College in London, receiving his PhD in 1954. He became a lecturer in mathematical physics at the University of Edinburgh in 1960, and spent the rest of his career there, becoming professor of theoretical physics in 1980.

Higgs boson   See *Higgs particle*.

Higgs field   See *Higgs particle*.

Higgs mechanism   See *Higgs particle*.

Higgs particle   An as yet hypothetical particle invoked to explain why the carriers of the electroweak force (the W and Z bosons) have mass. Quantum electrodynamics requires the photon to have zero mass (which is good because it does, indeed, have zero mass); early attempts to develop an electroweak theory, notably by Sheldon Glashow, seemed to require the equivalent bosons also to be massless (which was bad, since any such massless bosons would be as abundant and obvious in the Universe as photons are, and they are not). Peter Higgs, in Edinburgh, and two Belgian researchers, Robert Brout and François Englert (working together, but independently of Higgs) hit on the same idea for resolving this puzzle, in 1964. If there is an otherwise undetectable field (now called the Higgs field) filling the Universe, it could have associated with it a previously unknown kind of boson, the Higgs particle, which has mass. This would allow any photon-like particle to become massive by swallowing up a Higgs boson. Abdus Salam and Steven Weinberg used this trick – the Higgs mechanism – in their development of the electroweak theory. The whole package was widely accepted by physicists only after Gerard 't Hooft showed, in 1971, that the theory is renormalizable. It is possible, but not proven, that all massive particles get their mass in this way.

The Higgs field differs from the other fundamental fields, such as the electromagnetic field, in being scalar. It has a definite magnitude (a strength), but no direction. This means that the Higgs particle must have zero spin, unlike the particles which carry the forces of nature, such as the photon. Because the Higgs field has no preferred direction and is the same strength everywhere, it has no influence on us, except through the Higgs mechanism (see also *gauge theory*).

There are fierce debates among the experts about the expected properties of the Higgs particle and the workings of the Higgs mechanism. These may be resolved if, as is hoped, the latest generation of particle accelerators succeeds in manufacturing Higgs bosons for investigation.

Hilbert, David (1862–1943)   German mathematician who made many contributions to his subject, including the completion (together with Emmy Noether) of the mathematical theory of symmetry groups, which lies at the heart of the modern picture of quantum field theory.

Born in Königsberg on 23 January 1862, Hilbert studied at the University of Königsberg and in Heidelberg, completing his PhD in 1885. In 1886 he became a *Privatdozent* in Königsberg, and he was appointed professor there in 1892. In 1895 he moved to the University of Göttingen, where he remained as professor of mathematics until he retired in 1930. Hilbert continued to be associated with the university until he died, in Göttingen, on 14 February 1943. As well as his contribution to modern ideas through group theory, Hilbert was an influential figure at Göttingen during the development of quantum mechanics in the 1920s, when his more junior colleagues included (at various times) Max Born, Max von Laue, James Franck and Werner Heisenberg. His work on the theory of *invariants* (entities that are not changed by geometrical changes such as rotation and reflection) provides the mathematical basis for much of modern quantum field theory.

Hofstadter, Robert (1915–1990)   American physicist who won a share of the Nobel Prize for Physics in 1961 for his work on the internal structure of the nucleus.

Born in New York City on 5 February 1915, Hofstadter graduated from City College in New York in 1935, and was awarded his PhD by Princeton University in 1938. He stayed on briefly at Princeton, moving to the University of Pennsylvania in 1939 and back to City College in New York in 1941, but his career was interrupted by war work (on proximity fuses); he returned to Princeton in 1946, and in 1950 moved to Stanford University, where he investigated the way in which electrons are scattered by nuclei. These experiments (picking up the probing of atomic structure where Ernest Rutherford had left off) continued over the next two decades; he found that the proton and the neutron have similar sizes and shapes, and developed an overall picture of the structure of the nucleus.

Hofstadter was also involved in the development of the Crystal Ball detector at SLAC, and promoted the idea of putting gamma ray detectors in orbit, to study high-energy events in the Universe. He died in Stanford on 17 November 1990.

homogeneous   The same everywhere.

Hooft, Gerardus 't (1946–)   Dutch physicist who found a way to renormalize the electroweak theory of Abdus Salam and Steven Weinberg, removing the infinities which plagued it and thereby making it respectable. It was the renormalization of field theory by 't Hooft in 1971 that led to the explosive development of field theory in the 1970s.

Gerardus 't Hooft was born at Den Helder on 5 July 1946. The work which made 't Hooft's name was carried out while he was still a research student at the University of Utrecht. He has spent most of his career there, apart from two years at CERN (1972–4) and short-term visits as a guest professor to various American universities. He became a full professor at Utrecht in 1977.

He first joined the university as an undergraduate in 1964, and began full-time research for his PhD in 1969. His supervisor was Martin Veltman (born in 1931), who had studied at Utrecht and spent five years at CERN before returning to Utrecht as professor of physics. From the middle of the 1960s onwards, Veltman had been trying to develop a theory of the weak interaction, and following a suggestion made by John Bell he was tackling the problem in terms of a gauge theory along the lines of *Yang–Mills theory*. All of the variations on this theme, including the theory developed by Salam and Weinberg, were plagued with infinities. Using the new electronic computers that were just becoming an important tool in physics, Veltman spent years finding ways to cancel out many of these infinities, but was never able to get rid of them all.

Following on from Veltman's work (which, incidentally, used the path integral formalism developed by Richard Feynman), 't Hooft was able to show, in a paper published in 1971, that massless gauge theories are indeed renormalizable. This was impressive stuff from a graduate student, but the real problem was to renormalize theories which allowed for the existence of massive particles like the W and Z bosons. It was early in 1971 that Veltman and 't Hooft had a crucial conversation on the problem (recalled in *Constructing Quarks*, by Andrew Pickering). Veltman told 't Hooft that what was needed was 'at least one renormalizable theory with massive charged vector bosons'. But 't Hooft was unfazed. 'I can do that,' he said. 'What did you say?' replied Veltman, unable to believe his ears. 'I can do that.' Veltman told the confident young man to go away and write it down. To the professor's surprise and delight, he did just that. The resulting paper was also published before the end of 1971, and 't Hooft was awarded his PhD the following year.

The American physicist Benjamin Lee, who had been visiting Utrecht in the summer of 1971, took news of the breakthrough back to America with him, and published a paper translating 't Hooft's ideas into a more conventional mathematical language. The news spread like wildfire. Weinberg's paper on electroweak unification had been published in 1967. The fate of scientific papers is recorded in the *Scientific Citation Index*, which details how many times a particular paper is referred to in other papers. In 1967, 1968, 1969 and 1970 *nobody* (not even Weinberg) referred to the paper at all. In 1971 there were four citations (including those by 't Hooft). In 1972 there were 64 citations, and in 1973 there were 162. The sudden upsurge was entirely a result of the breakthrough made by 't Hooft in renormalizing gauge theory, and indicates precisely when and why gauge theory took centre stage in physics.

Further reading: Gerard 't Hooft, *In Search of the Ultimate Building Blocks*.

Hooke, Robert (1635–1703) British physicist and all-round scientist whose many contributions included ideas about gravity, coining the term 'cell' in its biological context and an early attempt to make a watch powered by a spring. His ideas about light led him into a bitter argument with Isaac Newton, which Newton won by waiting until after Hooke's death to publish his own book, *Opticks*.

Born in Freshwater, on the Isle of Wight, on 18 July 1635, Hooke was educated at Westminster School in London, and then (from 1653) at the University of Oxford, where he became Robert Boyle's assistant. Actually obtaining a degree was of secondary importance, but Hooke completed the formalities for his MA in 1663, three years after he had moved to London and helped to set up what would become the Royal Society. The following year he became lecturer in mechanics at the new Royal Society (chartered in 1662, when Hooke had immediately become the first curator of experiments), and in 1665 he became professor of geometry at Gresham College, in London. He held both these posts until he died, in London, on 3 March 1703; from 1677 to 1683 he was also Secretary of the Royal Society. More than anyone else, he was responsible for transforming the Royal Society from a gentleman's club into a professional scientific organization.

Hoyle, Fred (1915–) British astrophysicist and cosmologist, best known for his promotion of the idea that the Universe is eternal (the Steady State hypothesis).

Born in Bingley, Yorkshire, on 24 June 1915, Hoyle studied in Cambridge, graduating in 1936 and becoming a research student, but he deliberately never completed the formal requirements for a PhD (although he became a Fellow of St John's College in 1939) because he was hard up and there were at the time tax advantages in remaining technically a student. He carried out war work on radar for the Admiralty from 1939 to 1945, then returned to Cambridge to become a lecturer in mathematics. His most inspired piece of work was carried out in the early 1950s, when he realized that in order for carbon (and therefore carbon-based life forms like ourselves) to exist in the Universe, it must be manufactured inside stars, and that this could happen only if a precise piece of 'fine tuning' allowed three helium-4 nuclei to fuse together almost instantaneously to form an excited state of carbon-12 (that is, a carbon-12 nucleus sitting on a particular energy level above its ground state).

At the time, there was no other reason to believe that carbon-12 could exist in such a state. In the face of stiff opposition to the idea, Hoyle persuaded his friend Willy Fowler to carry out experiments in the laboratory to test whether carbon-12 possessed such an energy level. The experiments showed that Hoyle was right, and Fowler received the Nobel Prize for this work in 1983. Hoyle was knighted in 1972.

Further reading: Fred Hoyle, *Home is Where the Wind Blows*; see also *Companion to the Cosmos*.

Huygens, Christiaan (1629–1695) Dutch physicist and astronomer who had the misfortune to be active in science at the same time as Isaac Newton, and whose own brilliant contributions are sometimes overshadowed by those of his contemporary. After Newton, Huygens was the greatest physicist of his time. He explained the motion of a pendulum and designed the first practical pendulum clock (as well as a spring-driven watch), he developed the first scientific wave theory of light, and he was a skilled astronomer who, among other things, provided the first correct interpretation of the rings of Saturn, and discovered Saturn's moon Titan.

Born at The Hague on 14 April 1629, Huygens came from a family with a long tradition of diplomatic service to the House of Orange; his father was a diplomat and a member of the cultural élite in the Netherlands, and René Descartes was a frequent visitor to the family home. Huygens was born with the proverbial silver spoon in his mouth. He spent the years 1645 to 1647 studying law and mathematics at the University of Leiden and the following two years studying law in Breda, with the intention that he should follow the family tradition in his career. Instead, Huygens

Christiaan Huygens (1629–1695).

became fascinated by science, and he spent the next sixteen years investigating exactly what he pleased and studying the way the world worked while living off an allowance provided by his father.

In 1666, four years after the Royal Society had received its charter in London, the French Royal Academy of Sciences was founded in Paris. Huygens, by then an eminent scientist, was invited to work there, and he stayed for fifteen years before returning home. Partly because of ill health, his travel was limited in the later years of his life, but in 1689 he travelled to London and met, among others, Newton. Huygens died at The Hague, after a long illness, on 8 July 1695.

Huygens' most significant achievement in the context of the present book was his wave theory of light – ironically, this was also the part of his work that was most completely overshadowed by Newton for well over a hundred years, because Newton developed a rival corpuscular theory of light and the weight of Newton's reputation was such that for generations few people bothered to take the alternative model seriously.

Huygens' interest in the nature of light was stimulated through his astronomical work and his practical experience of telescope building. Working with his brother Constantijn, Huygens made the best refracting telescopes of their time, and invented an eyepiece made of two thin lenses, instead of a single fat lens, which reduced the problem of chromatic aberration (which causes coloured fringes around the objects being viewed).

The wave theory of light was put forward by Huygens in 1678 and developed fully in his book *Treatise on Light*, published in 1690. But his death five years later left the theory without a champion when Newton's theory of light was promoted in his book *Opticks* in 1704. Huygens' theory treated light as ripples in an all-pervading medium which filled the Universe, but, unlike modern versions of the wave theory, it regarded these as longitudinal (push–pull) waves (like sound waves in air) rather than transverse (side-to-side) waves (like waves on a plucked guitar string). Huygens explained reflection and refraction using this model, and predicted that light would travel more slowly in a more dense medium, like glass, than in air. Newton's rival corpuscular theory (treating light as a stream of tiny cannon balls) also explained reflection and refraction, but predicted that light would travel more quickly in glass. But this seemingly simple way to test the two models could not be carried out accurately enough to distinguish between them until well into the 19th century. By then, the work of Thomas Young and Augustin Fresnel had already revived the wave theory, and Newton's model had become dormant. Today, both models are seen as being correct. Light travels as a wave but arrives as a particle, and the wave–particle duality of light is one of the concepts at the heart of quantum physics.

Huygens' construction = *Huygens' Principle*.

Huygens' Principle The idea that every point on the wavefront of an advancing wave acts as a source of spherical wavelets, which combine with one another to make up the moving wavefront. This image is particularly useful in providing a picture of how waves diffract (bend round corners) and refract (change direction when moving from one medium into another medium of a different density). It was first put forward by Christiaan Huygens, in 1690, in the context of his wave theory of light.

hybrid bond See *quantum chemistry*.

hydrogen The lightest *element*. Each hydrogen nucleus has a single proton, and each hydrogen atom has a single electron surrounding the nucleus. Along with the proton (giving it atomic number 1), the hydrogen nucleus may contain neutrons. If it has no neutron, it is a nucleus of hydrogen-1 (hydrogen); if it contains one neutron, it is a nucleus of hydrogen-2 (deuterium, or heavy hydrogen); if it contains two neutrons, it is a nucleus of hydrogen-3 (the unstable isotope tritium, which has a *half life* of 12.26 years). No other isotopes of hydrogen can exist.

hydrogen bomb See *nuclear weapons*.

hydrogen bond A force of electrical attraction that operates between some molecules. It occurs when hydrogen atoms are bound by ordinary *chemical bonds* to so-called electronegative atoms (notably oxygen, nitrogen and fluorine), which have a tendency to hold on to extra electrons. Because the hydrogen atom has only one electron to shield its positively charged nucleus from the outside world, and this electron participates in the chemical bonding with the electronegative atom, the overall distribution of electric charge on the resulting molecule is dipolar, with the hydrogen's excess positive charge showing at one end, and the excess negative charge of the cloud of electrons surrounding the other atom showing at the other end of the molecule. So the molecules can hold on to each other as a result of the electrostatic attraction between the positively charged end of one molecule and the negatively charged end of the next molecule.

Tempting though it is, this should not be thought of in terms of the hydrogen nucleus giving up its electron entirely and being left bare; *quantum chemistry* explains what is going on much more accurately by treating even a single electron as forming a 'cloud' around its parent nucleus. It is just that the centre of the cloud of charge representing the lone electron shifts towards the electronegative partner, *partially* uncovering the proton that forms the nucleus of the hydrogen atom. It is only by taking quantum effects fully into consideration that the theory can predict the correct form of the hydrogen bond – in particular, its strength.

The hydrogen bond between water molecules is strong enough to keep water in the liquid form at room temperature, while comparable compounds such as hydrogen sulphide are gaseous under the same conditions (H_2S molecules are actually much heavier than H_2O molecules, so without hydrogen bonding you would expect water to vaporize at a much lower temperature than hydrogen sulphide). The hydrogen bond is an important factor in the chemistry of life – among other things, it is hydrogen bonding that holds the paired strands of nucleic acid in a molecule of deoxyribonucleic acid together in the famous double-helix structure of DNA, allowing life forms like us to exist. Hydrogen bonds are about one-tenth as strong as ordinary covalent bonds.

hypercharge A number used to identify some of the properties of an elementary particle. The hypercharge is equal to the sum of the *baryon number* and the *strangeness* of the particle. The amount of hypercharge is conserved in interactions involving electromagnetism and the strong force, but not in those involving the weak force. See *hyperon*.

hyperfine splitting The production of closely spaced lines in the spectrum of an atom, caused by the influence of the nucleus in disturbing the allowed quantum energy states of the atom. The magnetic moment of the nucleus interacts with the (tiny) magnetic moment of the electron to split one energy level into several states with almost equal energy.

hyperfine structure   See *fine structure*.

hyperon   Any *baryon* which has non-zero *strangeness* is a hyperon. Hyperons are, therefore, 'strange' particles (in the quantum meaning of the term, indicating that they contain at least one strange quark), each of which eventually (within about 10^{-10} to 10^{-8} sec) decays into a proton or a neutron (which itself then decays into a proton). They have half-integer spin and are all baryons, so they each have a baryon number of 1. See also *kaon*. It was the discovery of most of these particles in the 1950s that led (eventually) to the *eightfold way* system of classification, and then to the quark model. All hyperons are hadrons, and they are each heavier than the neutron.

IAT   See *International Atomic Time*.

ideal gas (perfect gas)   A (hypothetical) gas made up of molecules which each take up no space, and which do not interact with one another or with the walls of their container in any way, except by perfectly elastic collisions. By comparing the behaviour of real gases with the predicted behaviour of an ideal gas, it is possible to work out some of the properties of real molecules (for example, their average sizes).

imaginary number   A number which involves the square root of -1 (i) in a simple way. For example, $4i$ is an imaginary number. See *complex number*.

induced radioactivity   Radioactivity induced in a previously non-radioactive sample of material by bombarding it with particles such as *neutrons* and *protons*. The particles interact with the *nuclei* of the *atoms* of the non-radioactive substance, converting them into radioactive *isotopes*.

inelastic collision   A collision in which there is a change in the kinetic energy of the particles involved – the total kinetic energy of the particles going towards the collision is not the same as that of the particles heading away from the collision. In the everyday world (for example, when two billiard balls collide), some of the kinetic energy is converted into heat, so the collision is inelastic. In some quantum scattering events (for example, when a proton approaches an atomic nucleus and is repelled because both the proton and the nucleus have positive charge), the collision can be essentially elastic. But in others, kinetic energy may be converted into 'new' particles, making the scattering highly inelastic.

inelastic scattering   See *inelastic collision*.

inertia   The resistance of a body to acceleration. See *Mach's Principle*.

inertial frame   See *relativity theory*.

inertial mass   A measure of the amount of matter in a body, determined by its resistance to acceleration. The acceleration a produced in a body with inertial mass m by a force F is given by the equation $a = F/m$. The inertial mass of an object is exactly equal to its *gravitational mass*; this is a deep truth about the way the Universe works, but nobody can quite explain it. See *Mach's Principle*.

infinity   There is more (sometimes less) to infinity for a mathematician than to the person in the street. Science-fiction fans and amateur philosophers may be familiar with the idea that, if the Universe is infinite, then not only must anything that is possible happen somewhere in the Universe, but anything that is possible will happen an infinite number of times, in an infinite number of places. In that case, all the weirdness of the quantum world could be explained as just one huge statistical fluke affecting our corner of an infinite Universe. But the catch (apart from the mind-boggling nature of such a statistical fluke) is that this requires a special kind of infinity, called an exhaustively random infinity. It is quite possible to have an infinity that does not include everything – a trivial example is the set of all the even numbers. It is certainly infinite, but it is not exhaustive (or random) because it does not contain any of the odd numbers. Nobody knows whether or not the Universe is infinite, let alone whether or not it is an exhaustively random infinity.

inflation   See *quantum cosmology*.

infrared radiation *Electromagnetic radiation* with wavelengths longer than those of visible light, beyond the red end of the visible *spectrum*. Usually divided into three bands: the near infrared, from 0.8 to 8 micrometres wavelength; the mid infrared, from 8 to 30 micrometres; and the far infrared, from 30 to 300 micrometres. Infrared radiation is the warmth you feel when you hold your hand near, but not touching, a hot object.

infrared slavery   See *asymptotic freedom*.

inos   The fermionic counterparts to ordinary bosons required by the theory of *supersymmetry* are given names by altering the names of their everyday bosonic counterparts so that the names end in 'ino': *photino*, gluino, Wino, Zino and so on.

intensity   The amount of energy per second (power) carried across a unit area by any form of radiation. The intensity of light, or other electromagnetic radiation, diminishes as the distance from the source increases, following an inverse square law. So the light is one-quarter as bright at twice the distance, one-ninth as bright at three times the distance, and so on. For a sinusoidal wave (the archetypal 'pure' wave), the intensity is proportional to the square of the *amplitude* of the wave. In the *wave mechanics* version of quantum physics, this can also be regarded as an indication of the probability of finding the particle associated with a wave at any particular place.

interaction   See *forces of nature*.

interference   The way in which two (or more) sets of waves interact with one another to produce an overall pattern in which there are characteristic regions of high intensity and low intensity. The archetypal example is the way in which light waves passing through the two holes in the *double-slit experiment* interfere with one another to produce a pattern of light and dark stripes on a screen (an interference pattern). But interference can occur with other kinds of wave as well, including sound waves in air, ripples on a pond and the *probability waves* invoked by some interpretations of quantum theory to describe what is going on in the quantum world. The study of interference in light was pioneered by Thomas Young and by Augustin Fresnel, early in the 19th century. Where two waves combine to form an extra-strong peak of intensity, it is called constructive interference; where they cancel each other out to produce a trough in intensity, it is called destructive interference.

Thin films, such as the skin of a soap bubble or a layer of oil on water, produce interference patterns because light waves reflected from the back of the film and light waves reflected from the front of the film are out of step with each other. White light contains a mixture of different wavelengths, corresponding to different colours. Depending on the angle at which you view the film, and its thickness, different wavelengths of light (different colours) are either exaggerated by constructive interference or removed by destructive interference, which is why such films show a shifting pattern of swirling colours.

interference pattern   See *interference*.

interferometer   An instrument which makes use of the *interference* between waves to measure properties of the waves (such as their wavelength) or of the medium through which the waves have travelled (such as the extent to which a supposedly flat surface really is flat). The basic principle on which an interferometer works is to split a beam of light into two parts, which travel by two different routes through the instrument before being recombined to make an interference pattern. The interference occurs if one of the two light beams has travelled a slightly different distance from the other before they are recombined, so that the two beams have got out of step.

Using laser beams, this technique can be used to measure distances with very high accuracy. It also lies at the heart of the **Michelson–Morley experiment**, which showed that the measured speed of light is not affected by the Earth's motion through space.

On a much larger scale, in radio astronomy, data from two (or more) radio telescopes that observe the same source simultaneously can be combined by a computer which is programmed to analyse the interference between the 'signals' arriving at the different antennae. For a uniformly bright disc source, for example, at some critical spacing of the antennae the interference pattern disappears, because bright fringes made by waves from one side of the source coincide with dark fringes produced by waves from the other side of the source. So the nature of the interference pattern (actually much more complicated than in this idealized example) can be used to provide information about the structure of the source of the radio waves. The astronomical interferometer technique was previously used (notably by Albert Michelson) with visible light, to measure the diameters of stars. Although only six stars are close enough and bright enough for the technique to work, nevertheless it means that we know that the star Betelgeuse, for example, is 800 times the size of the Sun.

intermediate vector bosons   The carriers of the weak force, the W and Z particles. See **forces of nature**.

International Atomic Time (IAT or, from the French, TAI)   The standard of international timekeeping, established in 1972 and maintained by the Bureau Internationale de l'Heure in Paris. The time is measured by 80 atomic clocks in 24 countries, which each send their information to Paris where it is coordinated into one definitive time signal, correct to 1 millisecond, which is the basis for all civil timekeeping. Because there is not a whole number of seconds in a day, and the Earth's rotation changes slightly, an occasional 'leap second' is introduced into broadcast time signals to ensure that the Sun is always at its highest in the sky at local noon by our watches and clocks (except, of course, for the local changes we make to keep the same time across each time zone, and for summer time!).

International Practical Temperature Scale (IPTS)   For all everyday purposes, this is the same as the **Kelvin scale**. The IPTS also, however, defines the precise temperatures corresponding to eleven fixed points, including the boiling point of neon and the freezing point of gold, with precise rules for measuring temperatures in between the fixed points. This means that, when a scientist reports that something happened in a certain experiment at a certain temperature, there is absolutely no doubt about the temperature that is referred to, and other scientists can repeat the experiment at precisely the same temperature.

interpretations   See *models*.

invariance   The property of remaining unchanged under the action of a group transformation (see *group theory*). Every symmetry operation leaves something unchanged and so has a conservation law associated with it, a property of symmetry groups known as Noether's theorem. Conservation laws are among the most important guides to what goes on in the Universe, because they specify which kinds of interaction *cannot* occur. It is a reliable rule of thumb in physics that anything which is not expressly forbidden will happen. In the realm of classical mechanics, the invariants which define the way the world works are the conservation of energy and mass (or just the conservation of energy, since the two are related by Einstein's equation $E = mc^2$), the conservation of linear momentum, the conservation of angular momentum and the conservation of electric charge. In the quantum world, there are also other invariants, such as lepton number and baryon number (almost always!: see *grand unified theories*), and other properties which are conserved in some kinds of interaction but not in others – for example, *isotopic spin* is conserved during strong interactions, but need not be conserved during interactions involving the electroweak force.

invariant   See *invariance*.

inverse beta decay   See *beta decay*.

inverse Compton effect   A process whereby low-energy photons gain energy when they scatter from electrons. See *Compton effect*.

ion   An atom which has either lost one or more of its electrons (becoming a positive ion because it has lost negative charge) or gained one or more electrons (becoming a negative ion because it has gained negative charge). In chemistry, a positive ion is also known as a cation because in a solution it will be attracted towards the cathode during electrolysis. Similarly, a negatively charged ion is also known as an anion because it is attracted to the anode. The term ion is sometimes used to refer to an electrically charged molecule.

ionic bond   See *chemical bond*.

ionization   Any process which produces *ions*. This may be as simple as dissolving a so-called electrolyte in a suitable solvent. For example, when common salt (NaCl) is dissolved in water, it forms a sea of negatively charged chlorine ions (Cl^-) and positively charged sodium ions (Na^+). Gases can be ionized when electrons are knocked out of their atoms by alpha, beta, gamma or X-rays, which is why these forms of radiation became known historically as ionizing radiation.

isospin   See *isotopic spin*.

isothermal process   A process that occurs at a constant temperature. This may mean that heat has to be put in from outside, or carried off, at just the right rate. For example, if gas is expanding in a cylinder and pushing a piston out of the way, heat has to be put in to maintain a constant temperature even though the gas is doing work by pushing the piston. Contrast with *adiabatic process*.

isotopes Atoms of the same element (which therefore each have the same number of protons in the nucleus) which have different numbers of neutrons in the nuclei (but the same number for each atom of a particular isotope) and therefore have different atomic weights, but otherwise identical chemical properties. Each atom of every isotope of the same element has the same number of electrons.

isotopic spin A label introduced by Werner Heisenberg in 1932 as a way of distinguishing between protons and neutrons, while emphasizing their similarities. It has nothing to do with spin or angular momentum, but there is a slightly contrived mathematical analogy between the way isotopic spins add up and the way quantum spins add up, which is how it got its name. The proton and the neutron are regarded, on this picture, as different quantum states of the same underlying entity (the nucleon), distinguished because the proton has isotopic spin ½ ('pointing upwards') and the neutron has isotopic spin –½ ('pointing downwards'). This provides a way to distinguish between protons and neutrons when they are taking part in interactions mediated by the strong force alone, when their electric charge (or lack of it) can be ignored.

The concept has since been extended to other hadrons. The pions, for example, come in three varieties, with positive, zero and negative electric charge; these are allocated isotopic spins of 1, 0 and –1, respectively.

In an analogous way, the weak force cannot tell the difference between an electron and a neutrino (or between their heavier counterparts). It 'sees' only a lepton. So it is possible to define a label 'weak isospin', which distinguishes between leptons in the same sort of way that isotopic spin distinguishes between nucleons.

The importance of all this is that treating nucleons in this way was an early application of **group theory** to particle physics, although it was not thought of in those terms in the 1930s. The neutron and the proton together form an isospin group, and the electron and the neutrino form a weak isospin group. If we regard the proton as having isospin pointing upwards, and the neutron as having isospin pointing downwards, we can imagine turning a proton into a neutron by rotating the isospin. This is a symmetry operation (see **gauge symmetry**) called an isospin rotation. We can imagine changing the isotopic spin of every nucleon in the Universe simultaneously, by rotating the orientation of the isotopic spin of every

Isotopic spin. Protons and neutrons can be regarded under many circumstances as different quantum states of the same kind of particle, one with 'isotopic spin' pointing upwards and the other with 'isotopic spin' pointing downwards. On this picture, you convert one into the other by rotating the isotopic spin. This is an example of a symmetry operation.

nucleon through 180 degrees. The strong force would be completely unaffected by such a transformation. This is an expression of the underlying symmetry of the isotopic spin 'field'.

We can also imagine the nucleon existing in an intermediate state, a superposition of proton and neutron (say, 30 per cent proton and 70 per cent neutron) corresponding to the isospin 'pointing' in some intermediate direction. The angle corresponding to a particular mixture of such quantum states is called the **mixing angle**. The symmetry implicit in isotopic spin is associated with a field, called the Yang–Mills field. The Yang–Mills field possesses three quantum field particles, which can be identified with the pions, the carriers of the strong force. It was this example of the power of symmetry in the context of field theory that led (eventually!) to the modern field theory interpretation of the quantum particle world.

For example, the colour charge of quarks, which comes in three varieties, can be thought of in terms of a pointer which can point in one of three directions at 120 degrees to each other, not just up or down. Different coloured quarks can be distinguished in this way by a kind of colour isospin, as members of a symmetry group with each colour representing a different quantum state of the same basic entity; and quarks in mixed states can be described in terms of the appropriate mixing angle for this pointer. Carrying through the equivalent mathematical analysis as for the Yang–Mills description of the strong force, this leads to the requirement that there are exactly eight gluons associated with the glue force.

isotropic   The same in all directions.

Jacobi, Karl Gustav Jacob (1804–1851)   German mathematician who was one of the first people to appreciate the importance of symmetry and invariance in physics. A contemporary of William Hamilton, who also made major contributions to this field.

Born in Potsdam on 10 December 1804, Jacobi was a child prodigy who was thought by his schoolteachers to be ready for university at the age of twelve, but was not allowed to become a student at the University of Berlin until 1821, in his seventeenth year. He graduated in the same year! He qualified as a teacher at the age of nineteen, and was awarded his PhD a year later. He taught at Berlin University as a **privatdozent** for a year, then in 1826 he moved to Königsberg University. He was appointed professor there in 1827. After an illness in 1843 and a long convalescence in Italy, he returned to Berlin as a full professor in 1844. He died of smallpox, in Berlin, on 18 February 1851.

Jacobi's work embraced many areas of mathematics, the most relevant to the present book being his investigation of symmetry. His Jacobian determinants, which developed further some of the ideas of Hamilton, became an important feature of quantum mechanics.

Jacobian   The value of a square **matrix** (that is, one which has the same number of rows as it has columns), calculated in accordance with a set of rules laid down by Karl Jacobi. The values of the various elements in the matrix are multiplied together in pairs and the resulting products added to and subtracted from one another in a consistent way to produce a single number which is characteristic of the matrix. Also known as the determinant.

Janssen, (Pierre) Jules Cesar (1824–1907)   French astronomer who, with Norman

Lockyer, identified the existence of helium in the Sun, by spectroscopy, in 1868. It was not identified on Earth until 1895.

Born in Paris on 22 February 1824, Janssen worked in a bank while studying part time at the Faculty of Sciences of the University of Paris, graduating in 1852. He then worked as a teacher before going on a scientific expedition to Peru in 1857, to measure the location of the magnetic equator. On his return, he was awarded his PhD (from the Faculty of Sciences) in 1860, and held a succession of scientific posts, culminating in his appointment as director of the new astronomical observatory at Meudon (near Paris) in 1875. He stayed in this post until he died, in Paris, on 23 December 1907.

Janssen's interest in the possibilities of astronomical spectroscopy was fired by the pioneering work of Gustav Kirchoff at the end of the 1850s. He developed a technique for observing the spectrum of prominences on the Sun even when it is not in eclipse, and tested this following the eclipse of 1868, which he observed in India. (Incidentally, Janssen was such a keen eclipse follower that in 1870 he left the besieged city of Paris by hot-air balloon in order to see the eclipse visible that year from Algiers.) Lockyer observed the solar atmospheric spectrum using an identical technique at about the same time. Janssen noticed the presence of a line in the solar spectrum that he had never seen before, and passed the news on to Lockyer, who confirmed its presence from his own observations and suggested that it must be caused by the presence of a previously unknown element, which he called helium, in the atmosphere of the Sun.

Janssen and Lockyer became good friends as a result of this joint discovery, and the French Academy of Sciences struck a commemorative medallion with their profiles on either side.

Jeans, James Hopwood (1877–1946) British mathematician and astrophysicist who became well known for his ideas about the Universe, and as a popularizer of science, but who earlier in his career contributed to the development of the kinetic theory of gases, and made early applications of the idea of quanta to the investigation of specific heats and the nature of ***black body radiation.***

Jeans was born at Ormskirk, in Lancashire, on 11 September 1877. His family moved to Tulse Hill, in London, when he was three, and he was educated at the Merchant Taylors' School before going on to Cambridge in 1896. He graduated in 1900 and was made a Fellow of Trinity College the following year, completing the requirements for his MA in 1903. Between 1905 and 1909 he was professor of applied mathematics at Princeton University, then he returned to Cambridge as a lecturer in applied mathematics from 1910 to 1912. After 1912 he had no paid academic post, but carried out research privately and wrote a series of books and articles on science, turning to broadcasting as well in the late 1920s. He did, though, enjoy the status of being a research associate of the Mount Wilson Observatory, in California, from 1923 to 1944. In 1928 Jeans suggested that matter is continuously being created in the Universe (a forerunner of what became known as the Steady State hypothesis), but never developed the idea. He was also knighted in 1928. He died at Dorking, in Surrey, on 1 September 1946.

Jeans' early work, up to 1912, largely dealt with problems in quantum physics. The most important of these concerned the nature of black body radiation, where he

James Jeans (1877–1946).

was able to make a correction to earlier work by Lord Rayleigh, and came up with an improved mathematical formula for the variation of black body radiation in terms of wavelength and temperature. This applies to the long-wavelength end of the black body spectrum; it became known as the ***Rayleigh-Jeans Law***.

Jensen, (Johannes) Hans (Daniel) (1907–1973) German physicist who was awarded the Nobel Prize for Physics in 1963, jointly with Marya Goeppert-Mayer, for their suggestion (made independently of each other at the end of the 1940s) that the structure of the nucleus could be described in terms of a series of shells (see ***shell model of the nucleus***). Born in Hamburg on 25 June 1907 (the son of a gardener), Jensen's ability was noticed by a schoolteacher, who helped him obtain a scholarship for higher education. He studied at the University of Hamburg and in Freiburg, where he received his PhD in 1932. He worked in Hamburg from 1932 to 1941, and then at the Institute of Technology in Hanover, before being appointed professor of physics at Heidelberg in 1949. He stayed there for the rest of his career, although spending various spells as a visiting professor at American universities, and died (in Heidelberg) on 11 February 1973.

When they discovered that they had been thinking along the same lines, Goeppert-Mayer and Jensen collaborated in the early 1950s and wrote a book together, *Elementary Theory of Nuclear Shell Structure*, published in 1955. At one time, when Goeppert-Mayer felt that their ideas had fallen on stoney ground, Jensen wrote to her, saying, 'You have convinced Fermi, and I have convinced Heisenberg. What more do you want?'

jet In particle physics, the term 'jet' is used specifically to refer to a shower of particles moving in the same general direction, generated from a quark, antiquark or gluon

Jet. A two-jet event, recorded by DELPHI, a detector at CERN.

that has either been knocked out of its 'parent' particle (for example, a proton) or created out of energy (in line with Einstein's equation $E = mc^2$) in a particle accelerator experiment. Because of **confinement**, these particles cannot exist in isolation, but must use some of their kinetic energy (ultimately provided by the accelerator) to create more quarks, antiquarks and gluons, which group together to make particles such as protons and pions, which are detected moving together in the jet (actually a cone-shaped beam, rather like the bristles on a witch's broom). The effect was first observed at SPEAR, in electron–positron collisions, in the 1970s. The jet is the closest we can come to observing an individual quark, and the properties of the jet (including its overall energy and the direction in which it is moving) help to pin down the properties of the quarks themselves (for example, their masses).

The characteristic signature of quarks produced in this way is a pair of jets moving in opposite directions, corresponding to the creation of a quark–antiquark pair out of energy. At energies greater than about 20 GeV, one member of the pair may have so much energy that it can create a gluon (a real gluon, that is, not a **virtual particle**), which moves away at an angle to form a third jet. If enough energy is available, the quark and the antiquark may each emit a gluon, making a four-jet event. Two-jet events are quite common; three-jet events occur about one-tenth as often as two-jet events; and four-jet events are one-tenth as common as three-jet events.

The term 'jet' is also sometimes used to refer to the shower of particles produced in an energetic cosmic ray event.

JET Acronym for Joint European Torus, an experimental reactor used for research into *nuclear fusion*. JET is a tokamak machine located at Culham, in England (the name 'tokamak' comes from a Russian acronym). On 9 November 1991, it generated 1.7 megawatts of power in a pulse lasting two seconds; this was the first sustained burst of power on a large scale from a controlled fusion reaction on Earth.

Joliot-Curie, (Jean) Frédéric (1900–1958) and Irène (1897–1956) French husband-and-wife team who shared the Nobel Prize for Chemistry in 1935 for their work on artificial radioactivity. Irène was the daughter of Pierre and Marie Curie; the scientific tradition continued into a third generation with Paul Joliot-Curie, who became a physicist, and Hélène Joliot-Curie, their daughter, who became a nuclear physicist and married a physicist who was himself a grandson of Paul Langevin.

Frédéric Joliot was born in Paris on 19 March 1900. He studied at the School of Physics and Chemistry in Paris and, after a spell of military service, worked briefly in industry before joining Marie Curie's Radium Institute as an assistant in 1925. The following year, he married Irène Curie and they both changed their name to Joliot-Curie. In 1927 he received his BA from the University of Paris, and in 1930 he was awarded a PhD for his study of the chemistry of radioactive elements. He worked at the Sorbonne from 1935 to 1937, and then became professor of nuclear chemistry at the Collège de France. In 1939, after hearing of the work by Otto Hahn and Fritz Strassman which suggested the possibility of *nuclear fission*, he quickly carried out tests which proved the reality of fission, independently of (but just after) the similar work by Otto Hahn and Lise Meitner. He also calculated the possibility of a runaway chain reaction (of the kind later used in the atomic bomb); but with the outbreak of war he decided not to publish these results, and instead the Joliot-Curies recorded them in a sealed letter deposited with the French Academy of Sciences. It was opened in 1949.

During the war, both the Joliot-Curies initially stayed in France, with Frédéric working on the use of isotopes in biology (he pioneered, with Antoine Lacassagne, the use of radioactive iodine as a tracer in investigation of the thyroid gland) and Irène concentrating on the upbringing of their children. Frédéric was active in the Resistance, and managed to prevent the equipment in his laboratory being removed to Germany. He had also been instrumental in obtaining supplies of heavy water (useful as a *moderator*, and a valuable resource in any atomic energy programme) from Norway in 1939, and sending the entire stock (just under 200 kg) on to England in 1940. In May 1944 it became too dangerous for Irène and the children to stay in Paris, and they escaped to Switzerland. Frédéric stayed on, under the name Jean-Pierre Gaumont, and supervised the production of a large amount of explosives at the Collège de France, used during the liberation of Paris. He received the Croix de Guerre for his work in the Resistance.

After the Second World War, Frédéric became the head of the French atomic energy programme and director of the atomic synthesis laboratory at the Centre National de la Recherche Scientifique (CNRS). But in 1950 he was removed from the former post because of his left-wing politics (he had joined the Communist Party during the war, and had an intense dislike of establishment institutions), and Irène was dismissed from her post as an atomic energy commissioner soon after. Frédéric returned to the Collège de France, and in spite of ill health he and Irène were active in the peace movement over the next few years, alongside their teaching and efforts to establish a new nuclear physics laboratory at the University of Orsay, south of Paris. When Irène died, Frédéric succeeded her as professor at the Sorbonne and head of the Radium Institute. He died of cancer, in Paris on 14 August 1958 – almost certainly a result of his prolonged exposure to radioactive materials.

Irène Curie was born in Paris on 12 September 1897. She learned physics privately from her mother and her mother's colleagues at the Sorbonne (including Langevin and Jean Baptiste Perrin), and worked with Marie Curie as a radiographer during the First World War. She graduated from the University of Paris in 1920, and studied for a PhD at the Sorbonne. This was awarded in 1925 for her investigation of the behaviour of alpha radiation from polonium, conducted while working as an assistant at her mother's Radium Institute, where she met Frédéric Joliot.

The work for which the Joliot-Curies gained the Nobel Prize was carried out in the early 1930s. In the crucial experiment, aluminium was bombarded with alpha radiation. After the source of alpha rays was removed, the aluminium emitted positrons for several minutes. Some of the aluminium nuclei had each absorbed an alpha particle and been transformed into nuclei of a radioactive form of phosphorus, which decayed (emitting the positrons) into silicon with a *half life* of about 3.5 minutes. This was the first recognized production of artificial radioactivity, and clear evidence that the transmutation of elements was, at least in some cases, a practical possibility. So-called radioisotopes created in this way became a valuable tool for research, and in some cases found applications in medicine. A few years later, Irène and her colleagues investigated the response of uranium to bombardment with slow neutrons, and they found that some of the uranium seemed to be transformed into lanthanum. This led to the work by Hahn, Strassman, Meitner and Frisch (and Frédéric Joliot-Curie) which demonstrated nuclear fission.

In 1936 Irène was appointed Under-Secretary of State for Scientific Research (one of only three women in the Popular Front government of the time), and helped to create what became the Centre National de la Recherche Scientifique (CNRS). Both the Joliot-Curies supported and campaigned for the Republicans in the Spanish Civil War. She became a professor at the Sorbonne in 1937, but, as we have mentioned, had to leave France in 1944. On her return, she took up her old post at the Sorbonne in 1946, and became director of the Radium Institute; she also served for five years (1946–51) as a commissioner on the French atomic energy programme, in spite of her left-wing sympathies (she was, however, refused membership of the American Chemical Society on political grounds). She died from leukaemia in Paris on 17 March 1956 – almost certainly a result of her prolonged exposure to radioactive materials.

Jordan, (Ernst) Pascual (1902–1980) German physicist who worked with Max Born and Werner Heisenberg in Göttingen on the formulation of the *matrix mechanics* version of quantum theory, and who was, together with them, one of the authors of the 'three-man paper'.

Jordan was born in Hanover on 18 October 1902 and studied at the Hanover Institute of Technology and at the University of Göttingen, where he received his PhD in 1924 and stayed on until 1928 as Born's assistant. He then moved to the University of Rostock, becoming professor of physics there in 1935, to the University of Berlin in 1944, and to the University of Hamburg in 1951 (each time as professor of physics), retiring in 1970. Together with Paul Dirac and Wolfgang Pauli (but often working independently), he was also one of the pioneers of the theoretical work in the 1920s and 1930s which paved the way for the theory of *quantum electrodynamics*. He also carried out important work on the theory of gravity. He died in Hamburg on 31 July 1980.

Josephson, Brian David (1940–)   Welsh physicist who was awarded the Nobel Prize for Physics in 1973, for work that he carried out while still a student.

Born in Cardiff on 4 January 1940, Josephson graduated from the University of Cambridge in 1960 (having published an important paper on the *Mössbauer effect* while still an undergraduate), and he has worked there (apart from a year at the University of Illinois, 1965–6) ever since. He has been a Fellow of Trinity College since 1962. It was also in 1962, while he was working for his PhD (which was awarded in 1964), that he discovered (or invented) the *Josephson junction*, the work for which he received the Nobel Prize. He became a professor of physics in Cambridge in 1974. Having achieved such a fantastic breakthrough so early in his career as a physicist, he later shifted his attention to the study of intelligence and the mind, and espoused ideas relating to phenomena such as extrasensory perception that are regarded as somewhat beyond the pale by many of his physicist colleagues.

Josephson effect   See *Josephson junction*.

Josephson junction   The join between two pieces of superconducting material separated by a thin layer of insulating material (typically, a layer of oxide less than 1 millionth of a centimetre thick). If the material either side of the barrier was a normal conductor, a small electric current could still flow across such a thin barrier because of the *tunnel effect*. But in 1962 Brian Josephson realized that with superconductors, quantum effects allow a supercurrent to flow through the barrier, encountering no resistance at all.

But this is only the beginning of the story. If the current is gradually increased, at some critical value the superconducting ability of the barrier is lost, and the flow through the barrier drops back to the trickle associated with the 'normal' tunnel effect. And if a magnetic field is applied to the junction, starting small and gradually being increased, the current increases from zero to a maximum value; then, as the magnetic field continues to increase, the current drops back to zero again, increases

Josephson junction 1. Quantum effects will allow a current to flow from one superconductor to the other, in spite of the presence of the insulator, encountering no resistance at all.

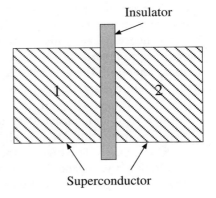

Insulator

Superconductor

to a (lower) maximum, falls to zero again, and so on. At a critical value of the magnetic field, the superconductivity of the barrier disappears.

There's more. If a *steady* potential difference is applied across the junction, a high-speed *alternating* current flows through it (*not* a direct current!). The frequency of this current is determined by the size of the voltage difference across the junction, and also depends on the ratio of the charge on an electron to Planck's constant (*e/h*). This makes it possible to measure this ratio with very great accuracy.

This whole package of phenomena is known as the Josephson effect, or effects. The behaviour of Josephson junctions is very sensitive to the magnetic fields or voltages applied to them, so they can be used to measure the strength of magnetic fields with great accuracy, or in defining the size of the volt (precise to better than one part in 100 million). Because their behaviour switches sharply from allowing current to flow to not allowing current to flow at critical values of the applied magnetic field (in a matter of picoseconds), they can also be used as sensitive switching devices, with applications in the development of computers. A computer is simply a lot of on/off switches wired together in the right way, and 'Josephson switches', as they might be called, would use very low power (so many switches could be packed closely together without generating excessive amounts of heat) and operate very quickly (with switching times tens of times faster than conventional computers). The snag is that in order to remain superconducting, the switches have to be kept very cold (see *superconductivity*). Josephson junctions can also be used to demonstrate quantum phenomena on a large scale, with further potential practical applications (see *SQUID*).

J/psi particle   A massive meson, with a mass of 3,097 MeV (more than three times as heavy as a proton), composed of a charmed quark and an anticharmed quark, and therefore with zero charm overall. It has a lifetime of 10^{-20} sec. Evidence for the existence of the particle was found by two groups in the United States, working independently of each other, in 1974. Samuel Ting and his colleagues at the Brookhaven Laboratory called it the J particle because the letter J resembles the Chinese character that means 'Ting'. Burton Richter and his colleagues at Stanford called it the psi particle because the tracks in their detector at SPEAR which provided the evidence of the particle resembled the Greek letter psi (Ψ). So it is usually referred to now as the J/psi (the neat combination of the two names to make the word 'gypsy' has not, alas, caught on as a name for the particle). Its existence was predicted by the developing quark model in the early 1970s, but at that time quarks were still not fully accepted. It was the discovery of the J/psi that made most physicists sit up and take notice of quark theory, ushering in a new era in physics.

jump   See *quantum leap*.

Kaluza, Theodor (1885–1954)   See *Kaluza–Klein theory*.

Kaluza–Klein theory   Strictly speaking, the Kaluza–Klein model is the unification of gravity and electromagnetism in one five-dimensional model, first proposed by the German Theodor Kaluza in 1919, and refined to take account of the requirements of *quantum theory* by the Swedish physicist Oskar Klein in 1926 (Kaluza and Klein never worked together). This model – the equivalent in five dimensions of Albert Einstein's equations of the general theory of relativity, which describe gravity in terms of four-dimensional spacetime – yields not only Einstein's equations for gravity, but also

James Clerk Maxwell's equations that describe electromagnetic radiation. Electromagnetism is seen as a 'ripple in the fifth dimension', analogous to gravity being a 'ripple in the fourth dimension'. But where is the fifth dimension?

The standard explanation is that the fifth dimension is hidden from us by compactification. You can make an analogy with a hosepipe. The pipe is made of a sheet of two-dimensional material rolled round in the third dimension; but from a distance it looks like a one-dimensional line. It is possible to describe every point in spacetime as being made up of a whorl of five-dimensional spacetime wrapped up so that it looks four-dimensional – provided that the 'wrapping up' happens on a scale much smaller than that of an atomic *nucleus*.

With the discovery of the two forces that operate within the nucleus (see *forces of nature*) and the development of *grand unified theories* involving many dimensions, the term 'Kaluza-Klein model' has come to be applied to any version of grand unification which operates in many dimensions and requires compactification. The theory of *superstrings*, for example, seems to require a total of ten dimensions, six more than the three of space plus one of time that we experience directly. The 'extra' dimensions are disposed of by compactification in just the same way as the one extra dimension is in the original Kaluza–Klein theory.

Kamerlingh Onnes, Heike   See *Onnes, Heike Kamerlingh.*

kaons   Two electrically charged mesons, the K$^+$ and the K$^-$, and the neutral K^0 particle (also known collectively as K mesons). They all have masses of about 500 MeV, roughly half the mass of the proton. The K$^-$ has a *strangeness* of –1 and *isotopic spin* of –½, and is the antiparticle of the K$^+$, which has a strangeness of +1 and an isotopic spin of +½. The K^0 has a strangeness of +1 and an isotopic spin of –½, and its antiparticle has a strangeness of –1 and an isotopic spin of +½, making a neat symmetry between the four particles. Like all mesons, each kaon can be regarded as made up of a quark–antiquark pair, held together by the exchange of gluons. The K$^+$ contains an up quark and an antistrange quark; the K$^-$ contains a strange quark and an antiup quark; and the K^0 contains a down quark and an antistrange quark, while its antiparticle counterpart contains a strange quark and an antidown quark. Subtle differences between the way the K^0 and its antiparticle counterpart decay showed in the mid-1960s that the so-called CP symmetry is violated by the weak interaction to a tiny degree (see *CP conservation*).

Kapitza, Piotr Leonidovich (1894–1984)   Russian physicist who made major contributions to the understanding of low-temperature physics, and was awarded the Nobel Prize for Physics as a result, in 1978.

Born in Kronstadt on 8 July 1894 (26 June, Old Style), Kapitza studied at the Polytechnical Institute in St Petersburg, graduating in electrical engineering in 1918, by which time the city had been renamed Petrograd. He stayed in Petrograd, carrying out research and teaching under very difficult circumstances – his wife and two children died in the famine of 1919. In 1921 he was a member of a Soviet commission sent to renew scientific links with other countries after the turmoil of the Russian Revolution and Civil War. He went to the Cavendish Laboratory on what was intended to be a short visit, but which lasted for thirteen years. Kapitza, happy to leave the horrors of his life in Petrograd behind, intended to spend his career in Cambridge. He completed his PhD in 1923 and worked on techniques for generating strong magnetic fields. He was interested in the way metals respond to intense magnetic fields at low temperatures, and this led him to design an improved system for liquefying helium. His work was so highly regarded that the Royal Society Mond Laboratory was built, in Cambridge, to create a new centre for the study of low temperature and solid-state physics largely because of his presence, and he became its director in 1930, although the building was not completed until a couple of years later (in 1929 Kapitza was the first foreigner for 200 years to be elected a Fellow of the Royal Society). He had, however, made regular visits to Russia for scientific meetings and to see his mother, and when he did so again in 1934, he was not allowed to leave.

In 1936 a new Institute for Physical Problems was set up for him in Moscow, and it was here that his investigations of the behaviour of liquid helium led to the discovery of *superfluidity*, which he named in 1938. During the Second World War, he worked on techniques for producing liquid air and oxygen, of immense value to the war effort, but refused to work on nuclear weapons research. In 1946 Kapitza wrote to Stalin criticizing the activities of the notorious NKVD (the secret police). He was dismissed from his posts and restricted to house arrest at his dacha, where he set up a small laboratory and invented a technique for generating high-power microwaves. When Stalin died in 1954, Kapitza was restored to his post as director of the Institute (he held it until he died), and used this high-energy microwave source to heat plasmas to very high temperatures, a step on the road to nuclear fusion.

His rehabilitation allowed Kapitza to play a part in the development of the Soviet space programme, and in the establishment of the science city, Akademgorodok, near Novosibirsk, in western Siberia. As an elder statesman of Soviet science, late in life Kapitza was allowed to travel abroad again (he visited Cambridge in 1966), even though he spoke out on behalf of colleagues, including Andrei Sakharov, who were victimized by the Soviet regime. He died in Moscow on 8 April 1984.

KEK Japanese particle physics research laboratory in the science city of Tsukuba.

Kelvin, Lord (William Thomson, Baron Kelvin of Largs) (1824–1907) British physicist who devised the absolute scale of temperature (now known as the *Kelvin scale*), made pioneering contributions to the study of thermodynamics and to the understanding of electricity and magnetism, and was the man who made sure that the first successful transatlantic telephone cable was indeed a success.

Born in Belfast on 26 June 1824, William Thomson was the son of James Thomson, professor of mathematics at the Academical Institution in Belfast (a forerunner of Belfast University). William and his brother James (1822–92), who also became a physicist, were educated at home by their father. The family moved to Glasgow in 1832, when the boys' father became professor of mathematics at the University of Glasgow, and in 1834 William enrolled at the university (at the age of ten!) to study what was then known as natural philosophy. This early entry into university was not quite as unusual then as it would be today, especially given the family connection, but it was not exactly commonplace either. He did not graduate from Glasgow, but went on to Cambridge in 1841, at the ripe old age of sixteen, and graduated in 1845.

After a brief visit to Paris, where he worked for a year, Thomson became professor of natural philosophy at Glasgow in 1846, where he established the first dedicated physics laboratory in any British university.

Lord Kelvin (1824–1907).

While he was still an undergraduate in Cambridge, Thomson began a study of electricity and magnetism which led him, in 1847, to put forward the idea that electric and magnetic fields are distributed throughout space, a step towards the comprehensive field theory of electromagnetism developed by James Clerk Maxwell a little later (Thomson was the scientific link, in this connection, between Michael Faraday and Maxwell). While he was in Paris, Thomson was introduced to the ideas of Sadi Carnot (1796–1832) on the relationship between work and heat, which led him into his investigation of thermodynamics, and his definition of the absolute scale of temperature in 1848. He also formulated the *second law of thermodynamics*, in 1851, and introduced the concept of energy (in its modern sense) into physics. His interests extended into astronomy, where he estimated the ages of the Earth and the Sun by arguing that the Sun released energy steadily as it gradually shrank through gravitational collapse; this gave an age of only about 25 million years, which led to a fierce debate with the geologists and evolutionists (including Charles Darwin), who needed a much longer timescale. Neither side in this 19th-century debate knew, of course, of the possibility of generating heat by nuclear fusion inside the Sun, which increases its lifetime to billions of years, ample for evolution to do its work.

In spite of his impressive achievements in physics, Thomson made his biggest impact in Victorian Britain in a much more practical way. After two unsuccessful attempts (by other people) to establish a transatlantic telegraph cable link, a third attempt was made in 1866 using principles proposed by Thomson, and this time it worked. He was knighted for this work in 1866 and made a peer in 1892, when he chose the title Baron Kelvin of Largs (he took the name Kelvin from that of a little stream that ran through the site of the University of Glasgow). He also made a fortune from the patents on the cable system and other electronic devices, and owned a private yacht, the *Lalla Rookh*, equipped as a floating laboratory. He retired from his chair in Glasgow in 1899, at the age of 75, and promptly enrolled as a research student at the university to keep his hand in. So Thomson was both one of the youngest and one of the oldest students ever to attend the University of Glasgow. He died at Largs, in Ayrshire, on 17 December 1907, and was buried next to Isaac Newton in Westminster Abbey. He was not related to J. J. Thomson and George Thomson.

kelvin   Unit of temperature on the *Kelvin scale*. 1 K represents the same increment in temperature as 1 °C; but note that it is written without a 'degrees' symbol.

Kelvin scale   Temperature scale devised by William Thomson (later Lord Kelvin) and known originally as the absolute scale of temperature; sometimes referred to as the thermodynamic temperature scale. From thermodynamic principles, in 1848 Thomson calculated that there must be an absolute lower limit to temperature, –273.16 °C (now written as 0\K). In quantum terminology, this is the temperature at which atoms and molecules would have the minimum possible energy, their *zero-point energy*.

Kendall, Henry Way (1926–)   American physicist who received a share of the Nobel Prize in 1990 for his contribution to the experimental work which confirmed the existence of particles within neutrons and protons – the quarks.

Kendall was born in Boston on 9 December 1926, and studied at Amherst College, where he graduated in 1950, and MIT, where he obtained his PhD in 1955. He worked at Stanford University from 1956 to 1961 and then moved to MIT. At the end of the 1960s, although still based at MIT, he worked at SLAC (with Jerome Friedman and Richard Taylor) on experiments which probed the structure of nucleons (protons and neutrons) using beams of high-energy electrons. Together with the theoretical interpretation of these experiments provided by James Bjorken and Richard Feynman, they showed that the structure within the nucleons is caused by the presence of point-like particles. In the 1970s, further experiments confirmed that the particles could be identified with quarks. See *partons*.

keV   1,000 *electron volts*.

kinetic energy   The energy an object possesses by virtue of its motion. See *energy*.

Kirchoff, Gustav Robert (1824–1887)   German physicist who was a pioneer of spectroscopy.

Born in Königsberg on 12 March 1824, he studied at the University of Königsberg and became professor of physics at the University of Breslau, where he first worked with Robert Bunsen, in 1850. In 1854 Kirchoff followed Bunsen to Heidelberg, and in 1875 he became professor of mathematical physics at the University of Berlin, a post he held until his death, on 17 October 1887.

It was Kirchoff who, in 1859, explained the lines observed in the spectrum of the Sun by Josef von Fraunhofer, and who, with Bunsen, pioneered the technique of spectroscopic analysis over the next few years. He also investigated dynamics, the theory of sound, electrical networks and the diffraction of light. His law of emission

(*Kirchoff's Law*) was to lead to the concept of a *black body*, and played an important part in the early development of quantum theory.

Kirchoff's Law At a given temperature, the rate of emission of electromagnetic energy by an object is equal to the rate at which the object absorbs electromagnetic energy of the same wavelength (frequency). This law was first stated by Gustav Kirchoff in 1859, and proved by him in 1861. It led him to develop, in 1862, the idea of a *black body*, and *black body radiation*, which in turn led Max Planck to introduce the idea of quanta into physics.

Klein, Oskar (1894–1977) See *Kaluza–Klein theory*.

Klitzing, Klaus von (1943–) German physicist who was awarded the Nobel Prize for Physics in 1985 for his work on the quantum Hall effect. This is not something you need to worry about, but deals with the behaviour of electrons moving in a very thin layer of material: for example, at the semiconductor surface of a so-called field effect transistor. This has important applications in electronics. Von Klitzing was born in Schroda on 23 June 1943, studied at the Technical University in Brunswick and at the University of Würzburg, and was the first German to win the Nobel Prize for Physics since 1963. He works at the Max Planck Institute for Solid State Research, in Stuttgart.

Kramers, Hendrik Anthony (1894–1952) Dutch physicist who made many solid but unspectacular contributions to the development of quantum physics, and predicted (with Werner Heisenberg) *Raman scattering*. He also introduced the idea of *renormalization*, but did not follow through with it; Hans Bethe was the first person to make significant use of the idea, in his early work on the *Lamb shift*, in 1947. Born in Rotterdam on 17 December 1894, Kramers became professor of physics at the University of Leiden, and died at Oegstgeest on 24 April 1952.

Kronig, Ralph de Laer (1904–) Physicist who came up with the idea of electron *spin* in 1925, but was dissuaded from publishing it after discussions with Wolfgang Pauli, Werner Heisenberg and Hendrik Kramers, all of whom were sceptical about the idea. This left the field clear for Samuel Goudsmit and George Uhlenbeck, and shows that you shouldn't always take the advice of experts. It has been suggested that it was because Kronig thought of the idea first (even though he failed to publish it) that the Nobel Committee could never quite bring themselves to recognize the work of Goudsmit and Uhlenbeck in the way that it deserved.

Kusch, Polykarp (1911–1993) German-born American physicist who shared the Nobel Prize for Physics in 1955 with Willis Lamb, for his accurate measurement of the *magnetic moment* of the electron. This was a key piece of observational evidence on the road to the development of the theory of quantum electrodynamics.

Born at Blankenburg on 26 January 1911, Kusch was taken to America by his parents when he was only one year old, and became a US citizen in 1922. After a few years moving around, the family had settled in Cleveland, Ohio. He took his first degree at the Case Institute of Technology, graduating in 1931, and received his PhD from Illinois University in 1936. After a year at the University of Minnesota, in the late 1930s he worked at Columbia University with Isidor Rabi, investigating the way beams of atoms are affected by magnetic fields, then worked on radar during the Second World War, before returning to Columbia as professor of physics in 1946. He remained in this post until 1972, when he moved to the University of Texas, Dallas, finally retiring in 1982. He died in Dallas on 20 March 1993.

The work for which Kusch received his Nobel Prize was carried out in 1947. His extremely precise measurements showed that the magnetic moment of the electron did not match the predictions of existing versions of quantum theory (the error was only about 0.1 per cent); one of the triumphs of QED is the precise agreement between measurements like those carried out by Kusch (and since refined still further) and the predictions of the new theory.

Lagrange, Joseph Louis (1736–1813)   Italian-born mathematician (christened Giuseppe Luigi Lagrangia) who developed techniques that proved invaluable in the formulation of *group theory*, and whose *Lagrangian* function provides a simple way to describe the behaviour of entities as diverse as light rays, planets in their orbits, and subatomic particles scattering from one another in particle accelerators.

Born in Turin on 25 January 1736, Lagrange came from a family of French and Italian descent. He became interested in mathematics when he read, at the age of seventeen, Edmund Halley's discussion of the use of algebra in optics. In 1755, at the age of nineteen, Lagrange became professor of mathematics at the Artillery School in Turin (he had already been teaching there for about a year; the exact dates are not recorded). In 1759 he sent some of his work to Leonard Euler, who encouraged him to publish. By the early 1760s he was regarded as the ablest mathematician of his time. He suffered from ill health and bouts of depression brought on by overwork, but moved to Berlin in 1766 (where he succeeded Euler as director of mathematics at the Berlin Academy of Sciences) and to Paris in 1787, where he worked at the Academy of Sciences. Although he had originally been invited by Louis XVI, he was held in high regard by the revolutionary regime which came to power a couple of years later (other scientists were not so lucky; the chemist Antoine Lavoisier, for example, was guillotined). He did little original work in mathematics after he was 40, but he developed the metric system for the new Republic, and in 1797 he became professor of mathematics at the new Ecole Normale. He also contributed to astronomy. His great book *Analytical Mechanics* was published in 1788, soon after he moved to Paris, although it had largely been written several years earlier. It did not contain a single diagram, but used entirely algebraic techniques, reflecting the inspiration he had drawn from Halley's work on optics as a young man (ironically, one of the people most profoundly influenced by the Lagrangian method was Richard Feynman, the man who put diagrams back at the heart of physics). Lagrange was made a Count by Napoleon.

He died in Paris on 10 April 1813, and was buried in the Panthéon.

Lagrangian   Sometimes referred to as the Lagrangian function. A parameter which characterizes the path (trajectory) of an object. The Lagrangian is equal to the *kinetic energy* minus the *potential energy* of the object at any point along the path. The moving object always moves in such a way that the sum of the values of the Lagrangian along the entire path (the integral of the Lagrangian) is a minimum. For a light ray, this is equivalent to *Fermat's Principle*, that the path followed by the light ray is the one that takes least time. More generally, it is an expression of the principle of least *action*, and by using the Lagrangian in the context of quantum physics Richard Feynman was able, in his PhD thesis, to reformulate quantum mechanics into a single, complete system of mechanics that includes all of classical mechanics as well. Feynman's mechanics, based on the Lagrangian, is all you need to explain all of mechanics, from the motions of the stars to the motions of electrons. For obvious reasons, this is often known as the path integral formalism of quantum theory.

It is actually much easier, in terms of the mathematics, to work with the Lagrangian than with the alternative *Hamiltonian* approach (which, through a historical accident, is the way most people are introduced to mechanics); John Wheeler, who was Feynman's PhD supervisor, says that his thesis, presented in 1942, marked the moment 'when quantum theory became simpler than classical theory'. If only teachers of physics in schools had the sense to teach mechanics from the beginning using the Lagrangian formalism, students could learn both classical and quantum mechanics at once, using equations that are easier to manipulate. One of the main reasons why quantum mechanics often seems difficult when students do encounter it is that they have to unlearn all the old stuff first.

Further reading: John and Mary Gribbin, *Richard Feynman: A life in science.*

Lamb, Willis Eugene, Jr (1913–)   American physicist who received the Nobel Prize for Physics in 1955 for his precision measurements of details of the spectrum of hydrogen. These included studies of *fine structure* and measurements of the *Lamb shift*, a key observational step on the road to the development of the theory of *quantum electrodynamics.*

Born in Los Angeles on 12 July 1913, Lamb studied at the University of California, Berkeley, where he obtained his first degree (in chemistry) in 1934 and his PhD (in physics, under the supervision of Robert Oppenheimer) in 1938. He then took up a post at Columbia University in New York, where he became a professor in 1948. Between 1943 and 1951 he also worked at the Columbia Radiation Laboratory. The key work which led to the award of his Nobel Prize was carried out at Columbia immediately following the end of the Second World War.

In 1951 Lamb moved to Stanford University, as professor of physics. There he studied the hyperfine structure of the spectrum of helium. In 1956 he moved to the University of Oxford, returning to the United States in 1962 to become professor of physics at Yale University. Since 1974 he has been professor of physics and optical sciences at the University of Arizona.

lambda   A *hyperon* carrying zero charge, a *strangeness* of –1 and with a mass of 1,115.6 MeV, about 20 per cent more than that of the proton, making it the lightest strange *baryon*. It has a lifetime of 2.63×10^{-10} sec. Although strangeness is conserved during strong interactions, the lambda can decay through the weak interaction, transforming into a proton, an electron and an antineutrino, in a process similar to *beta decay*; the strangeness disappears in this process. The lambda is composed of one up quark, one down quark and one strange quark; in effect, it is a neutron in which one of the down quarks has been replaced by a strange quark. When the lambda decays, the strange quark is transformed into an up quark.

Lamb shift   A small difference in the energy levels of two possible quantum states of the hydrogen atom, revealed by the splitting of a line in the spectrum of hydrogen into two components. It is caused by the *self-interaction* of the electron, and was discovered by Willis Lamb and measured in 1947. The original theory of the electron in quantum mechanics, developed by Paul Dirac, said that these two states of hydrogen should have exactly the same energy. But the idea of self-interaction suggested that there should be an infinite difference between the two energy levels!

The Lamb shift was explained, before the end of the 1940s, with the aid of **renormalization**, in the context of **quantum electrodynamics**, as a result of the way charged particles such as electrons interact with electromagnetic fields.

Landau, Lev Davidovich (1908–1968)   Soviet physicist who predicted the existence of neutron stars, developed a theory of superfluidity and received the Nobel Prize for Physics in 1962, chiefly for his theoretical studies of the behaviour of liquid helium.

Landau was born on 22 January 1908, in Baku, Azerbaijan, where his father was working as a petroleum engineer. He studied at the University of Baku from 1922 (when he was fourteen) to 1924, then transferred to the University of Leningrad (as it then was), graduating in 1927. The following year he joined the faculty of the Leningrad Physical-Technical Institute, a position he held until 1931. At the end of the 1920s he visited several centres of scientific research in Europe, including Niels Bohr's institute in Copenhagen, returning to the USSR in 1931. He visited Copenhagen again in 1933 and 1934, the year he was awarded his PhD. In 1932 he moved from Leningrad to Kharkov, initially as director of the theoretical physics department of the Physical-Technical Institute there, and in 1935 he became in addition professor of physics at Kharkov University. In 1937 he moved to Moscow (at the request of Pyotr Kapitza), became director of theoretical physics at the Institute of Physical Problems, and turned his attention to superfluidity, an area of research pioneered by Kapitza. At the height of a Stalinist purge of scientists and other intellectuals, Landau was arrested and imprisoned on suspicion of being a German spy (a ludicrous trumped-up charge) in 1938. Only strenuous intervention by Kapitza saved him from being sent to Siberia, or worse; many of his contemporaries disappeared without trace in those troubled times. In 1943 Landau was appointed professor of physics at Moscow State University.

Landau made important contributions to almost every branch of theoretical physics, from quantum electrodynamics to astrophysics. He also wrote a series of influential books, mostly in collaboration with Evgeny Lifshitz (1917–69), and established a thriving research centre for theoretical physics in Moscow. In 1962 he was seriously injured in a car accident and never fully recovered (the timing of the award of his Nobel Prize was undoubtedly linked to his fragile state of health, and the fear that he might not live to receive it if it were delayed much longer). He died in Moscow on 1 April 1968.

Large Hadron Collider   See **LHC**.

laser   A system for producing a powerful beam of monochromatic light, with waves from many different atoms marching precisely in step with one another. The name is an acronym for Light Amplification by Stimulated Emission of Radiation. It works by preparing the source (atoms, molecules or ions) in an excited state, in which very many atoms in the source each occupy a particular energy level (each with an electron on a specified rung of the energy ladder; see **Bohr model**). A weak light wave of the right frequency moving through the system then triggers each of the excited atoms to fall back to its **ground state**, adding more photons with precisely the same wavelength, and all moving in phase with one another, to the growing pulse of radiation.

latent heat   See **specific heat**.

Laue, Max Theodor Felix von (1879–1960)   German theoretical physicist who predicted the **diffraction** of **X-rays** by crystals.

He was born at Pfaffendorf (near Koblenz) on 9 October 1879, when the family name was still plain 'Laue'. He began studying physics at the University of Strasbourg in 1898, moved to the University of Göttingen in 1899, and completed his PhD (under the supervision of Max Planck) at the University of Berlin in 1903. He then spent two years studying art in Göttingen, and obtained a teaching certificate, but was brought back into physics by Planck, who found a post for him as an assistant in Berlin from 1905 to 1909. During that period, Laue's work mainly involved the applications of the then new special theory of relativity. He then moved to the University of Munich, (initially as a *Privatdozent*), where he stayed until 1912. As well as continuing his interest in relativity theory, he began to study the theory of optical waves (light), and through this he developed the realization that if X-rays (then often referred to as Röntgen rays) were short-wavelength electromagnetic waves, as was beginning to be suspected, they should be diffracted by the regular arrangement of atoms in a crystal lattice. In 1912 an experiment to test the prediction was carried out by Laue's assistant, Walter Friedrich, and a research student, Paul Knipping, confirming the wave nature of X-rays. As a result of his insight, Laue received the Nobel Prize for Physics in 1914, the same year that he became professor of physics at the University of Frankfurt, after two years as an associate professor in Zurich. It was also in 1914 that Laue's father, Julius (a military legal official), was elevated to the peerage, and the family added the 'von' to their name. The discovery of X-ray diffraction was picked up and developed by William and Lawrence Bragg as a tool for the study of crystal structure.

During the First World War, von Laue carried out war work on military communications systems, at the University of Würzburg, before in 1919 becoming professor of theoretical physics at the University of Berlin, where he stayed for the rest of his career. He was a vociferous opponent of the Nazi regime, courageously speaking out against the dismissal of Albert Einstein and doing what little he could to oppose the Nazification of German science; eventually, he resigned his post in protest at the racism of the regime in 1943. But von Laue was active in the rebuilding of German science after the war, serving as deputy director of the Kaiser Wilhelm Institute for Physics (before it became the Max Planck Institute), and from 1951 as director of the Max Planck Institute for Physical Chemistry. He died, following a motoring accident, in Berlin on 24 April 1960.

Lavoisier, Antoine Laurent (1743–1794)   French chemist who, more than anyone else, can be regarded as the father of modern chemistry. His single most important contribution was to disprove the phlogiston theory, realizing that burning involves a substance combining with oxygen (which he named) from the air, not losing 'phlogiston'.

Born in Paris on 26 August 1743, Lavoisier came from an affluent family. He studied at the College Mazarin, and in 1768 he was elected to the French Academy of Sciences. In the same year, he became an assistant tax collector, and later a full member of the *ferme générale* of tax 'farmers', a body held in roughly the same esteem in those days as Lloyd's 'names' are today. In 1772 his father purchased a title for Lavoisier. All of this activity went on alongside his scientific work, and financed it; but it brought that work to an untimely end when, during the Terror following the French Revolution, Lavoisier was guillotined, in Paris, on 8 May 1794. This was doubly unfortunate because Lavoisier was in fact a reformer who had supported the initial stages of the revolution. His wife Marie, whom he had married in 1771 when she was fourteen and who had acted as his assistant in his chemical investigations, retained her interest in matters scientific, and later (in 1805) married the American-born English chemist Count Rumford.

Lawrence, Ernest Orlando (1901–1958)   American physicist who invented the cyclotron.

Born at Canton, South Dakota, on 8 August 1901, Lawrence studied at the University of South Dakota (graduating in 1922), the University of Minnesota (MA 1923), started his PhD studies in Chicago (1923–4) and completed them at Yale (1925). After two years as an assistant professor at Yale, he joined the University of California, Berkeley, in 1927, and stayed there until he died – from 1936 as director of the Radiation Laboratory there. Lawrence was interested in the way nuclear reactions might generate energy inside stars, and devised the cyclotron in order to accelerate particles to sufficiently high energies to study some of the relevant nuclear processes. Once such machines existed, they were used for many purposes, including the production of radioactive isotopes for use in medicine, and the synthesis of plutonium and neptunium (both manufactured for the first time in Lawrence's laboratory in 1940). Lawrence received the Nobel Prize for Physics in 1939 for his work (especially the invention of the cyclotron), and was an important member of the Manhattan Project during the Second World War. After the war, he worked with his brother, John (a medical man) on the application of physics to medicine; they invented the use of neutron beams to treat cancer. Lawrence died at Palo Alto, in California, on 27 August 1958; in 1961, when element number 103 was synthesized at the Radiation Lab (now known as the Lawrence Berkeley Radiation Laboratory), it was named lawrencium in his honour.

least action   See *action*.

Lebedev, Pyotr Nikolayevich (1866–1912)   Russian physicist who was the first person to measure the pressure produced by light, an effect predicted by James Clerk Maxwell.

Lebedev was born in Moscow on 24 February 1866 (8 March, New Style). He studied business and engineering, but decided he was more interested in physics and in 1887 went to the University of Strasbourg to work with August Kundt (1839–94; German physicist who devised an accurate technique for measuring the speed of sound). Sources disagree on details of how Lebedev's career developed next. When Kundt moved to Berlin in 1888, Lebedev seems to have followed him, but may not have had the qualifications needed to enrol at the University of Berlin. So he did not formally complete his PhD studies, although he had done all the work, and returned to Russia in 1891, where he joined the physics department at the University of Moscow. In 1892 he became a junior member of the physics department there, but only completed a 'proper' PhD (for work on electromagnetic, sound and water waves) in 1900. He then became a full professor of physics. In 1911 he was involved in a mass protest at government interference in the way universities were run, and resigned; he died in Moscow on 1 March 1912 (14 March, New Style).

The studies of light pressure, for which Lebedev is best remembered, were carried out intermittently from 1898 until 1910. Using very light apparatus in an evacuated chamber, he actually measured the pressure of light, and suggested that this pressure explains why comet tails always point away from the Sun (he was partly right; there is an additional pressure, caused by the solar wind of particles (similar to particles in cosmic rays), which also pushes comet tails outward). The P. N. Lebedev Institute in Moscow is named after him.

LED   Acronym for light emitting diode. A simple semiconductor device that converts electrical energy into visible light. LEDs use very little electricity, and were widely used in numeric displays for electronic calculators, watches and so on, where this meant that they were not going to drain the batteries quickly. They have largely been replaced by liquid crystals in these applications, but are still used as indicators or warning lights where power is at a premium.

Lederman, Leon Max (1922–)   American physicist who investigated muon decay and confirmed, in 1962, the existence of the muon neutrino. Lederman was also involved in the discovery of evidence for the bottom quark (see *J/psi particle*), and shared the 1988 Nobel Prize for Physics with his colleagues Melvin Schwartz and Jack Steinberger. (Curiously, this award came seven years before Frederick Reines was similarly honoured for the experiment which first detected the electron neutrino!) This was a fundamental step towards the idea that particles come in *generations*.

Born in New York City on 15 July 1922, Lederman took his first degree at the City College of New York (graduating in 1943) and received his MA (1948) and his PhD (1951) from Columbia University (also in New York). He stayed on at Columbia, where he became a professor of physics in 1958. In 1979 he became director of Fermilab, retiring from that post in 1989. The experimental collaboration which resulted in the identification of the muon neutrino was carried out from 1960 to 1962 at the Brookhaven National Laboratory, on Long Island, where Lederman was a 'guest scientist'. The team of Lederman, Schwartz and Steinberger were so quiet and unassuming that they were jokingly referred to by their colleagues as 'Murder Incorporated'.

Lee, David M. (1931–)   American physicist who shared the Nobel Prize for Physics in 1996 for the discovery of *superfluidity* in liquid helium-3.

Born in Rye, New York, on 20 January 1931, Lee studied at Harvard University (where he received his first degree in 1952), the University of Connecticut (MSc 1955) and at Yale (PhD 1959). He then joined the staff at Cornell University, where he has stayed apart from spells as a visiting researcher at the Brookhaven National Laboratory (1966–7), the University of Florida (1974–5) and the University of California, San Diego (1988). He is now professor of physics at Cornell. The work for which Lee received the Nobel Prize was carried out in 1971 and 1972 at Cornell, after Douglas Osheroff (then a research student) noticed the peculiar behaviour of helium-3 at very low temperatures. Together with Robert Richardson, Lee and Osheroff established that this was a result of superfluidity.

Lee, Tsung Dao (1926–)   Chinese-born American physicist, usually referred to by his initials as T. D. Lee. He shared the Nobel Prize for Physics in 1957 with Chen Ning Yang for their joint theoretical work which predicted (following on from a suggestion by Martin Block and Richard Feynman) the non-conservation of *parity* (see *CP violation*) under certain conditions. Unusually, the prize (the first Nobel Prize for Physics awarded to Chinese researchers) was awarded in the year following the work for which it was awarded; this was because the prediction had been confirmed almost immediately by an experiment carried out by Chien-Shiung Wu and her colleagues at Columbia University.

Lee, who was born in Shanghai on 24 November 1926, was also the second-youngest recipient of the Physics Prize, after Lawrence Bragg. He studied at several Chinese universities, including Checkiang and the National Southwest Associated University, where he met Yang. He had had to move repeatedly to keep away from the invading Japanese army. Because of these difficulties, he never formally completed his degree studies. But in 1946 Lee and Yang were both awarded government scholarships to the University of Chicago, where Lee was allowed to enter graduate school even though he was not technically a graduate, and received his PhD in 1950, for a study of the theory of white dwarf stats.

He moved on briefly to the University of California and the Yerkes Observatory, still working in astronomy, then to the Institute for Advanced Study in Princeton in 1951, where his interests shifted to particle physics, and to Columbia University in 1953. Apart from another spell at the Institute for Advanced Study (1960–3), he spent the rest of his career at Columbia. His collaboration with Yang continued even when they were at different research centres, and together they were among the theorists who argued in 1960 that there was more than one kind of neutrino (see *Lederman, Leon Max*); they also predicted the existence of the charged W boson and the existence of weak neutral currents, carried by what is now known as the Z particle.

Lenard, Philipp Eduard Anton (1862–1947) German physicist who studied the photoelectric effect, providing the experimental evidence which Albert Einstein explained in terms of *photons*.

Lenard was born at Pozsony (then part of the Austro-Hungarian Empire; now Bratislava, in the Czech Republic) on 7 June 1862. He was the son of a wine merchant, and this affluent background enabled him to study at many of the best universities, moving from Budapest to Vienna, Berlin and finally Heidelberg, where he was awarded his PhD in 1893. The grand tour continued with brief teaching spells in Bonn, Breslau and Aachen, before he returned to Heidelberg in 1896, where he stayed for two whole years before becoming professor of experimental physics in Kiel. In 1907 he returned once again to Heidelberg, and stayed there until he retired in 1937. He died in Messelhausen on 20 May 1947.

Lenard's investigations of the photoelectric effect began in 1899, while he was at the University of Kiel. He also studied the nature of cathode rays (electrons), and was awarded the Nobel Prize for Physics for this work in 1905. Among other things, he found that cathode rays could pass right through thin sheets of aluminium foil, and inferred that atoms must be mostly empty space, ten years before Ernest Rutherford came up with his model of the atom. Lenard came close to discovering X-rays (and might have done so if he hadn't moved around so much in the 1890s), and later felt that he had not received due credit for his work in this area and on electrons. This almost paranoid reaction to the credit given to Wilhelm Röntgen and J. J. Thomson developed into full-blown crankiness, and after 1919 Lenard was a leading proponent of the desire to establish a proper 'German' physics, meaning one which ignored the contributions of Jewish researchers (including Einstein). This was seized upon by the Nazis as a basis for their own propaganda, and Lenard, as the only

LEP. Even before the particles (such as electrons) go into a large particle accelerator such as LEP, they have to be boosted to energies that previous generations of physicists could only dream about. This long tube is only the pre-injector accelerator for LEP, and boosts electrons and positrons to energies of a mere 600 million electron volts. Compare this with the original **Cockcroft–Walton machine,** developed less than 60 years before LEP.

leading scientist who openly and enthusiastically supported the Nazis, was directly responsible for the exodus of many German scientists from the country in the 1930s. Ironically, this ripped the heart out of German physics, and it took decades to recover. Happily, it was a major reason why the Nazis never obtained nuclear weapons.

LEP Acronym for Large Electron–Positron collider, a circular particle accelerator 27 km in diameter, at CERN, which can reach energies of more than 100 GeV. Started working in 1989; the best 'factory' for the production of Z^0 particles in the world.

lepton Any of the six *elementary particles* (and their antiparticle counterparts) that are not quarks – the electron, the muon, the tau particle and their associated neutrinos. Leptons do not feel the strong force and have no substructure, behaving as point-like particles. They are all *fermions*. See also *generation*.

lepton number   A label used to denote which particles are leptons and which ones are not. Each lepton has a lepton number of 1, each antilepton has a lepton number of –1. Every other particle has a lepton number of 0. The total lepton number is conserved in all particle reactions that have been carried out on Earth (so if you make a lepton, you have to make an antilepton to balance it), but a slight asymmetry in the laws of physics allowed leptons to be created in the Big Bang. The same asymmetry implies that baryons will eventually decay into leptons and photons. See *grand unified theories*.

lepton–quark symmetry   The observation that there is the same number of pairs of leptons (three) as there is pairs of quarks (also three). See *generation*.

Leucippus (about 500–450 BC)   Greek philosopher who was a disciple of Zeno and the teacher of Democritus, and who is credited with inventing the atomic theory, although very little is known about his life and work.

Lewis, Gilbert Newton (1875–1946)   American physical chemist who invented the idea of the covalent bond (see *chemical bond*).

   Born in Weymouth, Massachusetts, on 25 October 1875, Lewis was the son of a lawyer. He studied at the University of Nebraska, preparing for entry to Harvard, where he graduated in 1896, was awarded an MA in 1898 and received his PhD in 1899. He stayed on at Harvard for a year, then worked at Göttingen and Leipzig (on a travelling scholarship). He returned to the United States to take up a post at Harvard in 1901, and technically held this post until 1906, but went to the Philippines to work as a chemist for the Bureau of Science in 1904–5. He followed this with a spell at MIT (1907–12). In 1912 he became professor of physical chemistry at the University of California, Berkeley, where he stayed for the rest of his career (apart from war service in 1917–18). It was there (in 1916) that he published the idea of the covalent bond, in which atoms in a molecule are envisaged as sharing pairs of electrons, but the idea had been gradually developed over about fifteen years. It was refined further by Irving Langmuir (1881–1957) in the United States, and by Nevil Sidgwick (1873–1952) in England.

   Lewis also carried out important work in thermodynamics, and in the theory of acids and bases. He died in Berkeley on 23 March 1946.

LHC   Abbreviation for the Large Hadron Collider, a particle accelerator being built in the tunnel of the LEP machine at CERN. Construction began in 1995. The machine should begin operating, at energies as high as 14 TeV, in 2003; it may be able to produce direct evidence for the existence of the *Higgs particle*, which is predicted to have a mass of a few TeV.

lifetime   The time it takes for a certain fraction of the number of radioactive particles (or nuclei) in a large number of identical particles (nuclei) to decay, leaving a fraction $1/e$ of the original population, where e is the number that is the base for so-called natural logarithms, 'exponential e', 2.71828. Also known as the mean life. See *half life*.

light   Usually regarded simply as the range of *electromagnetic radiation* which human eyes are sensitive to, with *wavelengths* in the range from about 380 nanometres to 750 nanometres. This part of the electromagnetic *spectrum* is bounded by *ultraviolet radiation* (at shorter wavelengths) and by *infrared radiation* (at longer wavelengths). It corresponds, in terms of colours, to the rainbow range from red (at the long-wavelength end) through orange, yellow, green, blue and indigo to violet (at the short-wavelength end).

Our eyes have evolved and adapted to be sensitive to light because it is there – radiation in this part of the spectrum is produced copiously by the Sun and is not absorbed by the atmosphere of the Earth, so it penetrates to the ground. There is no essential difference between light and other forms of electromagnetic radiation such as X-rays and radio waves; they simply have different wavelengths.

Since the development of quantum theory in the first quarter of the 20th century, light has also been regarded as having particle-like properties. In appropriate circumstances, the energy in light can be thought of as being carried by a stream of particles, called *photons*. Many modern detectors, such as the charge coupled devices widely used in astronomy, are essentially photon counters that record the arrival of individual photons from faint sources and build them up into images over long periods of time.

light emitting diode   = *LED*.

lightest supersymmetric partner (LSP)   The lightest of the 'new' particles predicted by the theory of *supersymmetry*. Although other *supersymmetric partners* may be unstable, the lightest member of the family must be stable (because there is no lighter member of the family for it to *decay* into), and therefore in principle detectable. The best candidate for the LSP is the *photino*.

light quantum   = *photon*.

linear accelerator   Any particle *accelerator* which does its job by accelerating particles in straight lines, not round in circles or spirals. This includes the pioneering prototype built by John Cockcroft and Ernest Walton, which fitted comfortably inside their lab, and the Stanford Linear Accelerator (SLAC), an electron 'gun' 3 km long.

lines of force   Imaginary lines which indicate the direction and strength of a *field*.

liquid crystal   A substance which behaves like a liquid, but in which there is some order in the arrangement of the molecules – for example, a liquid crystal may be made up of long molecules which tend to align with one another. The molecules can be rotated using magnetic fields, and a liquid crystal can be set up so that in one orientation the molecules allow polarized light to pass through, while in another orientation they do not. This is the basis of the liquid crystal displays widely used in watches and pocket calculators.

liquid drop model   A model of the nucleus which treats it as a drop of liquid, instead of regarding the individual neutrons and protons in the nucleus as completely discrete particles, like marbles in a bag. The individual nucleons are, rather, regarded as behaving like molecules in a liquid, with the whole thing being held together by a kind of nuclear surface tension, maintaining the roughly spherical shape of the nucleus. The model was dreamed up by George Gamow, while he was a visitor at Niels Bohr's Institute in Copenhagen, towards the end of 1928. But Gamow's original liquid drop model differed from the modern version because in 1928 the neutron had not been discovered; he thought of the nucleus as being made up mainly of alpha particles (not at all a ridiculous idea; see *nucleosynthesis*), with a few extra protons and electrons. The modern version of the liquid drop model works quite well in describing some features of what is going on in heavier nuclei – in particular, nuclear fission can be envisaged in terms of the liquid drop breaking into smaller droplets, and this can be modelled mathematically to predict how different nuclei will behave.

Gamow's idea was taken up and developed by other researchers, including Niels Bohr, after the discovery of the neutron in 1932. Because of this, and the fact that it came out of the Copenhagen Institute in the first place, it is often mistakenly credited to Bohr.

Further reading: A. P. French and P. J. Kennedy (eds), *Niels Bohr: A centenary volume.*

locality   See *local reality*.

local reality   A view of the way the world works which combines common sense with the basic postulate of the special theory of relativity. That says that the speed of light is the ultimate limit at which interactions between different locations can take place (this is called 'locality'). The commonsense part is the assumption that there is a real, physical world (made up of real particles, like electrons) which exists even when nobody is looking at it (this is called 'reality'). Alas for common sense, the **Aspect experiment** and **Bell's inequality** combine to show that the world does not obey local reality. But the experiment and the inequality do not tell us which part of the package we have to abandon.

You have a choice – either you can believe that there is a real world out there even when nobody is looking at it (in which case you must accept that what Albert Einstein called 'spooky action at a distance' operates instantaneously between pairs of correlated particles, even on opposite sides of the Universe); or you can believe that no interaction of any kind can operate faster than light (in which case you must accept that things exist in a fuzzy indeterminate state, like **Schrödinger's cat**, until they are observed, triggering the **collapse of the wave function**).

We emphasize that there is nothing in the physics to tell you which choice to make. For more than half a century, under the influence of the **Copenhagen interpretation**, the 'standard' choice would have been to keep locality and abandon reality. Our own preference is to keep reality and abandon locality. See **hidden variables**, **transactional interpretation**.

local symmetry   See *gauge symmetry*.

Lockyer, (Joseph) Norman (1836–1920)   British astronomer and physicist who discovered the existence of helium in the Sun, by spectroscopy, almost 30 years before it was identified on Earth.

Born in Rugby on 17 May 1836, Lockyer worked as a civil servant at the War Office, in London, but established such a reputation as an amateur astronomer that in 1869 he was elected a Fellow of the Royal Society; the same year, he helped to found the science journal *Nature*, which he then edited for 50 years. In 1890 he became director of the Solar Physics Observatory in South Kensington, staying in the post until 1911, when the observatory moved to Cambridge. Lockyer, now 75, 'retired' to Salcombe Regis, in Devon, built his own observatory nearby, and remained an active observer until his death, on 16 August 1920.

Spectroscopy was still a new science in the 1860s, having been developed by Gustav Kirchoff and Robert Bunsen at the end of the 1850s. It was relatively easy to combine a spectroscope with a telescope to observe the spectrum of sunlight during a solar eclipse (actually the spectrum of light from prominences, jets of material stretching out from the surface of the Sun, and therefore not obscured by the Moon's disc), but much harder to observe the atmosphere of the Sun when it was not in eclipse, because the light from the main body of the Sun tends to swamp the detector. In 1868 Lockyer had developed a technique for making these observations in ordinary daylight. The French astronomer Jules Janssen had the same idea as Lockyer, independently, and both Lockyer and Janssen tested the technique within a few weeks of each other in 1868. Janssen noticed a feature in the spectrum that he had never seen before, and sent his observational data to Lockyer, who confirmed the feature from his own observations and decided that the line was caused by the presence in the atmosphere of the Sun of an element that had never been found on Earth. He called it helium, from the Greek word for the Sun, *helios*. Helium was identified on Earth (from its unique spectrum) only in 1895, by William Ramsay and William Crookes; Lockyer was knighted two years later.

London, Fritz Wolfgang (1900–1954) German-born American physicist best known for his collaboration with Walter Heitler, which provided the first quantum-mechanical description of the covalent bond in the hydrogen molecule. This was the birth of *quantum chemistry*.

Born in Breslau (now Wroclaw, in Poland) on 7 March 1900, London studied in Frankfurt, Munich and Bonn, where he received his PhD in 1921. But this education had been in classics and philosophy, and it was only after working as a teacher for four years that he turned to physics in 1925, working at several German universities before 1933, then leaving the country after the Nazis came to power. London worked in Oxford and Paris in the 1930s, before becoming professor of theoretical chemistry at Duke University, North Carolina, in 1939. He stayed in that post until he died, in Durham, North Carolina, on 30 March 1954.

The collaboration with Heitler was carried out in 1927, while London was working with Erwin Schrödinger, in Zurich. He also calculated (in 1930) the strength of what are now known as *van der Waals forces*, and investigated superconductivity (initially in collaboration with his brother, Heinz, and later on his own).

London, Heinz (1907–1970) German-born British physicist who studied superconductivity, superfluidity and techniques for separating isotopes.

Born in Bonn on 7 November 1907, London studied in Bonn, Berlin, Munich and Breslau, where he completed his PhD (dealing with aspects of superconductivity) in 1933. Forced to leave Germany almost immediately because of the rise of Nazism, London worked for a time with his brother Fritz, at the Clarendon Laboratory in Oxford. In 1936 he moved to the University of Bristol, and (after war work on the atomic bomb project) from 1946 onwards he worked at the Atomic Energy Research Establishment in Harwell, near Oxford. He died in Oxford on 3 August 1970.

Lorentz, Hendrik Antoon (1853–1928) Dutch physicist who received the Nobel Prize for Physics in 1902, for his work on the theory of electromagnetism.

Born in Arnhem on 18 July 1853, Lorentz studied at the University of Leiden, graduating in 1873 and leaving to become a teacher while working on his PhD, which was awarded in 1875. The authorities were so impressed with this piece of work (titled *The Theory of Reflection and Refraction of Light*, it showed how to solve

Maxwell's equations at a boundary between two materials) that in 1877 he was appointed professor of theoretical physics at the University of Leiden. He stayed in the post for 39 years before moving on his retirement to become director of the Teyler Institute, in Haarlem. He stayed in that post until he died, in Haarlem, on 4 February 1928.

Lorentz developed James Clerk Maxwell's theory of electromagnetism, providing the bridge between Maxwell's work and Albert Einstein's special theory of relativity. He explained light as being produced by the oscillations of charged particles (which would now be identified with electrons) within atoms. Lorentz coined the name 'electron' in 1899, and he developed (independently of George Fitzgerald (1851–1901)) the so-called Lorentz transformation equations, which describe the way space and time are distorted for objects travelling at a sizeable fraction of the speed of light. See *relativity theory*.

Loschmidt, Johann Joseph (1821–1895)   German physical chemist who made the first reasonably accurate calculation of the size of the molecules present in air.

Loschmidt was born in Putschirn, Bohemia (now part of the Czech Republic), on 15 March 1821. He studied in Prague and Vienna, failed as a businessman (he went bankrupt in 1854) and became a science teacher in Vienna. In 1865 he calculated the number of molecules of gas in a cubic centimetre of air (see *Loschmidt's number*), obtaining a figure about 30 times smaller than modern estimates. In 1866 he became a *Privatdozent* at the University of Vienna, and two years later an assistant professor, securing his future. He died in Vienna on 8 July 1895.

Loschmidt constant   The number of molecules in 1 cubic metre of an ideal gas under standard conditions of temperature and pressure (STP). This is 1 million times *Loschmidt's number*, which refers to the number of molecules in 1 cubic centimetre, not in 1 cubic metre.

Loschmidt's number   The result of a calculation, originally carried out in 1865 by Joseph Loschmidt, of the number of molecules in 1 cubic centimetre of air; see also *atom, Avogadro's number*.

The nub of Loschmidt's approach is that he used two sets of equations to determine simultaneously two properties of molecules – their sizes and Avogadro's number. This is a standard technique, the memory of which may be familiar even to non-mathematicians from school days. If you have one unknown quantity, and one equation in which that quantity appears, you can solve the equation to find the unknown quantity. If you have two unknown quantities, you need two equations each involving both quantities before you can solve the equations to find out both the unknown numbers. With three unknowns, you need three equations, and so on.

Loschmidt's calculations involved the average distance that a molecule travels between collisions in a gas – the 'mean free path' – and the fraction of the volume of the gas actually occupied by the volume of all the molecules added together. He assumed that in a liquid all the molecules are touching each other, which gave him a handle on the volume occupied by all the particles (molecules) in the liquid when they are closely packed together. Then, when the same liquid was heated to become a gas, he knew that the volume of gas actually occupied by the molecules must be the same as the volume of the liquid that had been evaporated, and that the rest of the volume of the gas is simply the empty space that the molecules whiz through. 'Whiz' really is the operative word, here. At 0 °C, the molecules in air are moving at several hundred metres per second, something that the pioneers of kinetic theory such as Lord Kelvin fully appreciated, in the second half of the 19th century, from their studies of the way the pressure exerted by a gas changes when it is squeezed into a smaller volume.

Since he actually carried out his calculations for air, which is mainly a mixture of nitrogen and oxygen, Loschmidt had to use estimates of the densities of liquid nitrogen and liquid oxygen which were not as accurate as modern measurements, but he still came up with answers to his calculations that stand up very well even today. Loschmidt said that the diameter of a typical molecule of air must be measured in millionths of a millimetre, and in 1866 he gave a value for Avogadro's number of 0.5×10^{23}. The modern value of Loschmidt's number is 2.686763×10^{19}.

Using modern data, incidentally, the mean free path of molecules of air turns out to be just 13 millionths of a metre at 0 °C, and an oxygen molecule in air at that temperature will be travelling at just over 461 metres per second. So it undergoes more than 3.5 billion (thousand million) collisions every second.

LSP = *lightest supersymmetric partner.*

Lucretius (Titus Lucretius Carus) (about 95–55 BC) Roman philosopher/poet who wrote a poem, *De rerum natura*, which propounded the ideas of Epicurius – including the notion of atoms moving in an infinite void.

luminescence Emission of light from a substance after it has been stimulated by some outside influence other than a rise in temperature. The influence could, for example, be the effect of *X-rays*, or bombardment with electrons. See *fluorescence*, *phosphorescence.*

Lyman, Theodore (1874–1954) American physicist who discovered the *Lyman series* of lines in the spectrum of hydrogen, in 1914.

Born in Boston on 23 November 1874, Lyman studied at Harvard, where he was awarded his PhD in 1900. He then studied at the universities of Cambridge and Göttingen, before returning to Harvard, where he became an instructor in 1902 and an assistant professor in 1907. Alongside a succession of professorial posts, he became the director of the Jefferson Physical Laboratory in 1910, and held the post until he retired in 1947 (he also served in the Signal Corps in the First World War). From 1926 onwards, he had the status of professor emeritus. All of Lyman's scientific work concentrated on investigations of the ultraviolet part of the spectrum, where he made his most famous discovery. He died in Brookline, Massachusetts, on 11 October 1954.

Lyman alpha See *Lyman series.*

Lyman series A series of lines in the ultraviolet part of the spectrum of hydrogen, predicted empirically by Johann Balmer and identified (in 1914) by Theodore Lyman. This was an important confirmation of the accuracy of the *Bohr model* of the atom. The first, and brightest, line in the series is called the Lyman alpha line and occurs at a wavelength of 121.57 nanometres; the other lines in the series all have shorter wavelengths. See also *Balmer series.*

Mach, Ernst (1838–1916) Austrian physicist whose name is most familiar today from the Mach number, which gives the speed of an object relative to the speed of sound in the medium through which the object is travelling (Mach 1 is the speed of sound, Mach 0.5 half the speed of sound, Mach 2 twice the speed of sound, and so on). But Mach also made profound contributions to the way scientists think about the Universe and the **models** they use to describe it, both on the large scale and the small. These ideas influenced both Albert Einstein, when he was working on the general theory of relativity, and some of the quantum pioneers when they were developing the first versions of quantum mechanics. Mach's ideas have recently gained renewed attention from both cosmologists and quantum physicists (see **Mach's Principle**).

Born in Turas, in Moravia (then part of the Austro-Hungarian Empire, but now in the Czech Republic), on 18 February 1838, Mach moved with his family to Unter Siebenbrunn, near Vienna, in 1840, where he was educated at home (by his parents) until he was fifteen. After two years attending the local Gymnasium (high school), where he first became interested in science, he entered the University of Vienna in 1855, and received his PhD, for studies of electricity, in 1860.

Mach then taught at the University of Vienna as a **Privatdozent**, gave public lectures which brought in a little extra money, and wrote two books, before moving to the University of Graz in 1864, initially as professor of mathematics and then, from 1866 onwards, as professor of physics. But this appointment was short lived and in 1867 Mach moved once again, this time to Prague, where he became professor of experimental physics, and stayed for 28 years, before returning to Vienna in 1895 as professor of history and theory of inductive science.

In his early days of research in Vienna, Mach had worked on the study of sound, following up the work of Johann Doppler (1803–53), who had predicted the relationship between the pitch of a note and the motion of the source of the note (the Doppler effect) while working in Prague in the 1840s. Doppler later moved to Vienna, so Mach's work was directly descended, so to speak, from Doppler's. But alongside his practical work in experimental physics (only a tiny fraction of which we have mentioned here), Mach was interested in more abstract problems, such as the nature of perception, from the beginning of his career. This developed into what can best be described as a philosophy of science, although Mach never regarded himself as a philosopher. He argued that the nature of scientific discovery itself alters the way that we envisage the world, and that the interpretation of later discoveries is coloured by our knowledge of the discoveries that have preceded them, so that we might have a different scientific world-view if the discoveries had been made in a different order (this is an important issue in the context of the meaning of scientific models today; see Martin Krieger, *Doing Physics*, and Andrew Pickering, *Constructing Quarks*).

In particular, Mach said that anything which could not be directly perceived by our senses should not be regarded as a real entity. This led him to reject the notion of atoms, at a time when the atomic theory was developing rapidly and being taken increasingly seriously by his colleagues.

In his lifetime, such views were regarded by most of Mach's colleagues as rather cranky; but when theorists such as Niels Bohr developed quantum theory in the 1920s, they found Mach's ideas to be useful intellectual justification for the idea of entities which exist only as waves of probability until they are observed, at which point there is a collapse of the wave function which (temporarily) makes the electron (or whatever) real. In the strict version of the **Copenhagen interpretation**, the only real events are indeed the ones we perceive directly with our senses – the movement of a pointer across a dial, the click of a geiger counter, the pattern of light and dark made by the experiment with two holes. Everything else, including the idea that the click was made by an electron passing through the tube of the counter, or that the pattern of light and shade is made by interfering waves, is mere inference.

Mach's other enduring contribution came in one of his earliest works, the book *Die Mechanik*, published in 1863 when he was just 25, still a humble Privatdozent in Vienna. It was there that he spelled out the idea now known as Mach's Principle, which was widely debated at the time and has been, off and on, ever since. The principle says that an object possesses inertial mass only by virtue of the presence of all the other masses in the Universe. Einstein was strongly influenced by the idea and intended to incorporate the principle into what became the general theory of relativity. He only partly succeeded. He did, though, succeed in annoying Mach, who did not like Einstein's theory and was in the early stages of preparing a book criticizing Einstein's work when he died.

By then Mach was 78. Even after suffering a stroke in 1897, he had stayed on as a professor in Vienna until 1901, and when he finally did step down he was appointed as a member of the Austrian parliament, where he served for another twelve years. In 1913 he left public service and moved to live with his son in Vaterstetten, near Munich, where he died on 19 February 1916.

Mach's Principle   The idea that the inertia of an object depends on its relationship with all of the other mass in the Universe. Developed by Ernst Mach in the 1860s, the idea goes back to discussions of inertia made by Galileo and by Isaac Newton in the 17th century, but was given this name by Albert Einstein, who was strongly influenced by Mach's work, in 1918.

Galileo seems to have been the first person to realize that it is not the velocity with which an object moves but its acceleration that reveals whether or not forces are acting upon it. On Earth, there are always external forces (such as friction) at work, so we have to keep pushing an object just to keep it going at a constant velocity. But the natural tendency is to keep moving in the same direction at constant speed unless an external force acts (see **Newton's laws of motion**). But what do you measure velocities and accelerations against?

Newton thought that there was a preferred frame of reference in the Universe, defined by absolute space. Space is a tricky thing to pin down – you can't hammer a nail into it and measure your velocity relative to the nail. But Newton thought that the existence of the preferred frame of reference could be demonstrated by experiments on rotating objects – specifically, a bucket of water. He described the experiment in his great book the *Principia*, published in 1686:

> The effects which distinguish absolute motion from relative motion are, the forces of receding from the axis of circular motion … if a vessel, hung by a long cord, is so often turned about that the cord is strongly twisted, then filled with water, and held at rest together with the water; thereupon, by the sudden action of another force, it is whirled about the contrary way, and while the cord is untwisting itself … the surface of the water will at first be plain, as before the vessel began to move; but

after that, the vessel, by gradually communicating its motion to the water, will make it begin sensibly to revolve, and recede by little and little from the middle, and ascend to the sides of the vessel, forming itself into a concave figure (as I have experienced), and the swifter the motion becomes, the higher the water will rise.

Newton is talking about what we now call centrifugal force, and his comment 'as I have experienced' is pertinent because, unlike many of his predecessors and contemporaries, Newton actually did experiments – he didn't just imagine how things 'ought' to work in an ideal world. You can experience the same thing, on a smaller scale, by stirring your cup of coffee. The liquid is pushed to (and up) the sides by centrifugal force, leaving a dent in the middle. But what Newton pointed out is that it is not motion relative to the container that matters, but, in some sense, the absolute motion of the liquid.

At the start of the experiment, the bucket begins to move, but the surface of the liquid stays flat, even though there is relative motion between the liquid and the bucket. Then, as friction makes the liquid rotate, the concave depression builds up (or down), even though there is now no motion of the liquid relative to the bucket. Finally, you can grab the bucket to stop it rotating; now, the liquid inside keeps on rotating, with a concave depression, just like the coffee stirred around in your cup. Somehow, the liquid 'knows' it is rotating, and behaves accordingly. But rotating relative to what?

Newton said it was rotating relative to fixed (or absolute) space. But 30 years later the Irish philosopher and mathematician (and Bishop) George Berkeley (1685–1753) argued that all motion is relative and must be measured against something. Since 'absolute space' cannot be perceived, that would not do as a reference point, he said. If there were nothing in the Universe but a single globe, he continued, it would be meaningless to talk about any movement of that globe. Even if there were two perfectly smooth globes in orbit around one another, there would be no way to measure that motion. But 'suppose that the heaven of fixed stars was suddenly

Mach's principle. When we are whirled around in a circle by a fairground ride, 'centrifugal force' makes us fly outwards. According to Mach's principle, this is because we are rotating relative to the distant stars and galaxies.

created and we shall be in a position to imagine the motions of the globes by their relative position to the different parts of the Universe'. In effect, Berkeley argued that it is because the coffee in your cup knows that it is rotating relative to the distant stars that it rises up the sides of the cup in protest.

The same argument applies to accelerations in straight lines; Berkeley's reasoning would say that the push in the back you feel when you are in a car that accelerates away from a standing start is because your body knows that it is being accelerated relative to the distant stars and galaxies. But Berkeley was 150 years ahead of his time. Although there was some discussion of his ideas in the 18th century, they were largely ignored. Interest in the idea only really developed in the 1860s, when it was taken up by Ernst Mach.

Mach added very little to the ideas put forward by Berkeley, although he did make the intriguing suggestion that, if we want to explain the equatorial bulge of the Earth as due to centrifugal forces, 'it does not matter if we think of the Earth as turning round on its axis, or at rest while the fixed stars revolve around it'. It is the *relative* motion that is responsible for the bulge.

When Einstein set out to develop his general theory of relativity, he intended to come up with a theory that would incorporate Mach's Principle as a natural conse-quence. He was only partially successful – the equations of the general theory incorporate this feedback between distant objects and accelerated motion only if the Universe is closed, meaning that it contains enough matter to ensure that its present expansion will one day be reversed, and it will collapse back into a 'Big Crunch', mirroring the Big Bang in which it was born. But since the currently favoured theory of how the Universe was born, called inflation, suggests that the Universe is indeed closed, this is not much of a drawback. The link between Mach's Principle and the fate of the Universe is the subject of continuing research in the 1990s; Donald Lynden-Bell, of the University of Cambridge, has been a leading light in establishing this link in recent years.

Evidence that the Universe is closed also favours a non-mainstream view of how electromagnetic interactions work, called the **Wheeler–Feynman absorber theory**. This sees such interactions (for example, between one electron and another) in terms of waves which move out from one particle, disturb the other and trigger a reaction which travels backwards in time back to the first particle, so that the effect instanta-neously produces a kind of electrical inertia, known as radiation resistance. One interpretation of Mach's Principle involves the same kind of forward- and backward-in-time interaction between an object here on Earth (such as my computer) and all of the other mass in the Universe, so that the computer 'knows', instantaneously, how much resistance it should offer when I try to move it.

Bizarre though it may sound, all of these ideas have been incorporated into a model of gravity and inertia, operating at the quantum level, by Shu-Yuan Chu of the University of California. In a development of **string theory**, Chu found that gravity can be explained in terms of the time-symmetric exchange of interactions between entities suggested by the Wheeler–Feynman theory. Gravity emerges as an average of all the interactions between strings, which weave a kind of tapestry (spacetime) that looks smooth only over distances much larger than the strings themselves – which doesn't pose any conflict with the smoothness of spacetime in the everyday world, since strings are only about 10^{-35} m long. In Chu's words, 'classical mechanics describes the equilibrium condition (hence the absence of any probabilistic statements in classical mechanics); quantum mechanics describes the fluctuations; and the *path integral* formalism follows from summing over the huge number of strings in the system'. Chu died in 1998, while still working on these ideas.

We stress that this is not, at present, the received wisdom about how the Universe works, but the model does hang together, linking the Universe at large with the smallest entities discussed by physicists today, and unifying quantum theory and gravity in one package. See also *transactional interpretation*.

Further reading: John Gribbin, *Schrödinger's Kittens*.

macroscopic system   See *correspondence principle*.

magic numbers   Numbers of neutrons and/or protons in atomic nuclei that correspond to nuclei with particularly stable structure. This is reminiscent of the way atoms in which the electrons form closed 'shells' are very unreactive (see *Bohr model*, *quantum chemistry*). For both neutrons and protons, the set of magic numbers is 2, 8, 20, 28, 50, 82 and 126. So, for example, the nucleus of helium-4 (2 protons + 2 neutrons) and the nucleus of oxygen–16 (8 protons + 8 neutrons) are both very stable. Just having one kind of nucleon match a magic number is also desirable, in energy terms, so that, for example, there are six stable isotopes of calcium, each with 20 protons in the nucleus. This pattern has led to the development of a *shell model* for nuclei analogous to the shell model of electrons in atoms.

magnetic constant   Also known as the permeability of free space. In any sensible system of units, this should be 1, since free space has less effect on a magnetic field than anything else has. In some systems of units, including the SI system, the magnetic constant has a value of $4\pi \times 10^{-7}$. This nonsense is then removed by dividing all permeabilities by this number to give relative permeabilities, in which the value of the magnetic constant is restored to 1. See *permeability*.

magnetic moment   Also known as the magnetic dipole moment. A measure of the turning force that a magnetic dipole experiences in a magnetic field, because the north pole at one end is pushed one way while the south pole at the other end is pulled the other way. All charged particles, such as the electron and the proton, moving in a magnetic field also have a magnetic moment because of the way their (moving) electric charge interacts with the magnetic field. If there is a distribution of electric charge within a particle, even if the overall charge of the particle is zero, it will still experience this turning effect in a magnetic field. Even the neutron has a magnetic moment, indicating that it has a complex internal structure involving a distribution of charge (see *quarks*). See also *quantum electrodynamics*.

magnetic monopole   = *monopole*.

magnetic permeability   See *permeability*.

magnetism   One component of the *electromagnetic force*; see also *ferromagnetism*.

Maiman, Theodore Harald (1927–)   American physicist who developed the *laser*, while working at the Hughes Research Laboratories, in Miami, in 1960.

Born in Los Angeles on 11 July 1927, Maiman was the son of an electrical engineer, and (after military service in the US Navy) studied at the University of Columbia (graduating in 1949) and at Stanford University (MSc 1951; PhD 1955). He then joined Hughes. After developing the laser (building on the principles of the *maser*, which had already been invented), Maiman left Hughes and founded his own company, Korad Corporation, in 1962, to manufacture high-power lasers. His business interests developed through Maiman Associates in 1968 and the Laser Video Corporation in 1972, and in 1977 he joined TRW Electronics, Los Angeles, as a vice president and director of advanced technology.

Malus, Etienne Louis (1775–1812)   French physicist who discovered the polarization of light.

Malus was born in Paris on 23 June 1775. After a year of military service at the age of eighteen, he was one of the first students to attend the new Ecole Polytechnique, in Paris, where he stayed for only two years before becoming a junior officer in the engineering corps in 1796. He went on Napoleon's Egyptian campaign, but became ill and was sent back to France in 1801. Over the next few years, while still serving as an army engineer in various parts of France, Malus was able to carry out scientific experiments, and he acted as an examiner for the Ecole Polytechnique, which meant that he often had to visit Paris, enabling him to keep in touch with other scientists. He was posted to Paris in 1808 and promoted to major, but ill health prevented him from making the most of this opportunity, and he died of tuberculosis in Paris on 23 February 1812.

Malus became interested in the properties of light during his time in Egypt, but the work for which he is remembered was carried out in 1807 and 1808. It was known that a beam of light is split into two when it passes through a crystal of Iceland spar (or certain other crystals), and although Christiaan Huygens had found rules to describe the behaviour of this 'double refraction' his ideas had largely been ignored because they were based on the wave model of light. Malus found that, when light reflecting from glass passed through Iceland spar, it emerged only as one beam, not two. He also discovered that, if the two beams from a double refraction were directed at the surface of water, at certain angles one would be reflected and the other refracted into the water. He explained all this in terms of a phenomenon he called polarization.

According to Malus, light consisted of a stream of particles, but each particle had a specific orientation (think of a soldier carrying a spear, either held vertically or horizontally across his chest). In double refraction, there are two routes that light can follow through the crystal, but one route only permits light with one orientation to pass, while the other route only permits light with the other orientation to pass. So the two beams that emerge are made up of particles which have the same orientation within each beam, but with the particles in one beam oriented at right angles to the particles in the other beam ('up' and 'across'). Reflected light is polarized and contains particles already lined up in one of these orientations, so they all follow the same route through the crystal and emerge as a single beam.

Shortly after Malus died, the particle model of light was replaced by the wave model, thanks to the work of Thomas Young and Augustin Fresnel. In 1821 Fresnel explained polarization in terms of transverse waves, but the image of vertically and horizontally oriented beams is still a good one.

Manhattan Project   The code name for the project to build the first atomic bomb, in the United States during the Second World War. Many of the leading quantum physicists of the day, including Hans Bethe, Enrico Fermi, Robert Oppenheimer and Richard Feynman, were involved with the project.

many histories interpretation   See *many worlds interpretation.*
many minds interpretation   See *many worlds interpretation.*
many worlds interpretation   The idea that, whenever the world is faced with a choice at the quantum level (for example, if an electron has a choice of which hole to go through in the experiment with two holes), the universe divides into two (or as many parts as there are choices), so that all possible options are followed (in this case, in one world the electron goes through hole A, in the other world the electron goes through hole B).

It is not always appreciated that the roots of this idea can be traced back to the development of the Copenhagen interpretation by Niels Bohr in 1927. Bohr actually suggested that we might think of the experiment with two holes in terms of two different realities, in one of which the electron goes through hole A while in the other it goes through hole B. But he saw our world, the world of our experience, as a hybrid combination of the two possibilities, producing interference between the two worlds. When we look to see which hole the electron goes through, we make one world real while the other disappears, so there is no interference. Many worlds are a feature of the Copenhagen interpretation – but they are *ghost* worlds, which are not supposed to have any physical reality. On Bohr's picture, if we look at the experiment and find the electron going through hole A, that is the end of the story. It does not mean that in some alternative reality an equivalent experiment sees the electron going through hole B.

But why not? In the early 1950s, a graduate student at Princeton University, Hugh Everett, puzzled over the Copenhagen interpretation and the magical collapse of the wave function, and decided that it made more sense to treat each outcome of every possible quantum event as existing in a real world. In the classic example of Schrödinger's cat, this means that if the experiment really were carried out, the Universe would divide into two worlds, in one of which the experimenter opened the box to find a dead cat, and in the other of which the experimenter opened the box to find a live cat. Encouraged by his thesis supervisor, John Wheeler, Everett developed his idea into a fully worked-out interpretation of quantum theory, and showed that the assumption that all of the quantum possibilities are real leads to exactly the same predictions for the outcome of experiments as the Copenhagen interpretation.

This is both good and bad. It is good because every experiment that had been carried out agreed with the predictions of the Copenhagen interpretation, so the many worlds interpretation was, in that sense, not wrong; it is bad because there is no way to test, by experiment, which of the two rival interpretations is 'right'. It is simply a matter of your own personal choice which one you prefer.

Everett's work was published in the journal *Reviews of Modern Physics* in 1957 (vol. 29, p. 454) alongside a paper by Wheeler drawing attention to it (vol. 29, p. 463). In spite of this, it was largely ignored until the idea was taken up by Bryce DeWitt, of the University of North Carolina, at the end of the 1960s. This was largely because of the mind-boggling implications of accepting that there is an infinity of alternative realities 'out there', existing in some sense alongside our own, in which every possible outcome of every possible quantum choice is realized in one world or another. DeWitt himself, writing in *Physics Today* (September 1970, p. 30) described the shock he felt on first encountering the 'idea of 10^{100} slightly imperfect copies of oneself all constantly splitting into further copies'. But recently one group of scientists has taken the idea very seriously indeed.

These are the cosmologists, who find that by using the many worlds interpretation they can get round the puzzle, which is insurmountable in the Copenhagen interpretation, of explaining what observation can collapse the wave function of the entire Universe and bring it into reality. On Bohr's picture, since there is no outside observer to observe the Universe, it should stay for ever in a ghostly superposition of states.

Apart from the cosmologists, the leading champion of the many worlds interpretation today is David Deutsch of the University of Oxford, who brings back on board some of Bohr's ideas. On Deutsch's picture, although two worlds exist while an electron is going through the experiment with two holes, the creation of the interference pattern involves electrons from both worlds somehow getting back together, with the two worlds fusing to make one reality, complete with an interference pattern. So the proliferation of universes is much less extreme than DeWitt imagined in 1970. Deutsch even suggests that it might be possible to build a quantum computer which could experience directly the splitting of worlds during an experiment like the one with two holes, and report back to us that it had indeed divided into two and then been fused back into one.

The many worlds idea has spawned a subset of interpretations of quantum reality, all variations on the same basic theme. One idea is that the world behaves as though it is classical (see **classical mechanics**) because there are many other universes that we are ignorant about. This is sometimes known as the many histories approach (see also **decoherence**). Another idea is that the conscious mind is something 'outside' physics, which does the selecting of realities for us. This has led to the development of a many minds interpretation of quantum reality, so that every quantum possibility is observed by one or other of our minds, even though they do not occupy separate physical realities. On this picture, you would literally be in two minds about the outcome of the Schrödinger's cat experiment!

See **quantum interpretations**.

Further reading: John Gribbin, *In Search of Schrödinger's Cat.*

maser A device which produces an intense beam of microwaves from the stimulated emission of radiation by excited atoms (the name comes from Microwave Amplification by Stimulated Emission of Radiation). The process is exactly equivalent to the way an intense light beam is produced in a *laser*, but the maser was invented first, by Charles Townes in 1951 (and independently by Nikolai Basov and Alexander Prokhorov).

mass A measure of the amount of stuff there is in an object. There are two ways to define mass, either in terms of the strength with which a lump of stuff interacts gravitationally (see **gravitational mass**) or in terms of its resistance to being pushed around (see **inertial mass**). The two masses are identical, which presumably represents a deep truth about the way the Universe works, but this has not yet been fully explained. See **Mach's Principle**.

Mass and *energy* are interchangeable, in line with Albert Einstein's famous equation $E = mc^2$, but this effect is not important in everyday life – the amount of energy required to raise the temperature of 1 kg of water from 273 K to 373 K is equivalent to a mass of just 4×10^{-12} kg. But the measured mass of an object does depend on the speed with which it is moving relative to the observer. An object (such as an electron) moving at 99 per cent of the speed of light has seven times as much mass as when it is at rest. See *relativity theory*.

mass decrement   The difference between the rest mass of a radioactive nucleus and the rest mass of all of its decay products added together. It is because there is less mass in the decay products that the nucleus does decay, with the excess energy (equivalent to the mass decrement, from $E = mc^2$) being liberated in the process.

mass defect   The difference between the rest mass of a nucleus and the rest mass of all its components (the protons and neutrons) taken separately. It is because the mass of the nucleus is less than the sum of the masses of its parts that nuclei are stable. The mass deficit is equivalent to the binding energy, from $E = mc^2$.

mass–energy equation $E = mc^2$. See *relativity theory*.

mass number   The total number of protons and neutrons in the nucleus of a particular isotope. Also known as the nucleon number.

mass spectrometer   See *mass spectroscopy*.

mass spectroscopy   Technique for separating ions with different masses, using a combination of electric and magnetic fields. The extent to which a charged particle moving through such fields is deflected depends both on its mass and on its electric charge, so that a mass spectrometer directly measures the ratio of charge to mass for the ions. In most cases, however, mass spectroscopy is carried out using singly ionized atoms, all with one unit of electric charge, so that in effect it directly measures the mass spectrum.

In one standard form of mass spectrometer, the ions are first accelerated by an electric field and then pass between the poles of a magnet, so that they are bent around part of a circular arc whose radius depends on their mass. Heavier ions are deflected less, lighter ions are deflected more. On the other side of the magnet, the ions continue in straight lines at the appropriate tangent, with the tracks for ions of different mass diverging from one another and fanning out. They are then picked up by a detector spread out across the line of flight (perhaps something as simple as a photographic film), and the position in which they are recorded indicates the mass. The technique provides one way of identifying and separating different *isotopes* of the same *element*.

mass spectrum   See *mass spectroscopy*.

matrices   Two-dimensional arrays of numbers or other mathematical expressions (in a rectangular grid) which represent the elements of a particular mathematical set, or group. Square matrices, which have the same number of rows as there are columns, are a special case of the more general situation in which the number of rows is different from the number of columns. There are well-defined rules which determine the way that matrices can be added together or multiplied; one of the key features of their behaviour is that multiplication of matrices is not *commutative*. It turns out that matrices are a convenient way of expressing the properties of quantum entities and their interactions (see *matrix mechanics*).

matrix mechanics   The first complete, self-consistent description of quantum mechanics, developed largely by Werner Heisenberg in 1925.

At the beginning of 1925, the understanding of the quantum world was confused and muddled. Every problem involving quantum entities had to be solved first by working out the equivalent pattern of behaviour in classical mechanics, then adjusting the 'answer' by adding in bits and pieces of quantum behaviour by hand, fiddling about with the equations until they matched the results of experiments. In his book *The Conceptual Development of Quantum Mechanics*, Max Jammer describes the situation as 'a lamentable hodgepodge of hypotheses, principles, theorems and computational recipes'. Nobody had the faintest idea how to construct a coherent theory to clear up the mess.

Heisenberg had completed his PhD, at the University of Munich, in 1923, when he was just 22. He was one of the first physicists to be brought up on quantum theory, and after a few months working with Niels Bohr in Copenhagen, in 1924 Heisenberg became Max Born's assistant in Göttingen. The key to the breakthrough he achieved was based on an idea that he picked up almost immediately on his arrival in Göttingen – nobody could recall, later, exactly who expressed it first. The idea was that a physical theory should concern itself only with things that could actually be observed by experiments. It sounds trite, but this is actually a very important insight. We cannot see electrons as little balls moving in elliptical orbits around atomic nuclei, and the only things we know about are the wavelengths of the spectral lines produced when electrons move from one energy state to another. The important, and relevant, point is that all the observable features of atoms and electrons deal with *two* states, and the transition of the atom (or electron, or whatever) from one state to another. We have no picture of what is going on during the transition itself, and all the business about orbits is really just something tacked on from our classical image of the behaviour of objects like planets. Heisenberg deliberately abandoned the classical picture of particles and orbits, and took a long, hard look at the mathematics that describes the associations between pairs of quantum states, without asking himself how the quantum entity gets from state A to state B.

The place where he did this was not in Göttingen, but on the rocky island of Heligoland, where he had gone in May 1925 to recover from a severe bout of hayfever. In his autobiographical memoir *Physics and Beyond* (Harper and Row, New York, 1971), he described his feelings as everything began to fall into place. At 3 a.m. one night he:

> could no longer doubt the mathematical consistency and coherence of the kind of quantum mechanics to which my calculations pointed. At first, I was deeply alarmed. I had the feeling that, through the surface of atomic phenomena, I was looking at a strangely beautiful interior, and felt almost giddy at the thought that I now had to probe this wealth of mathematical structures nature had so generously spread out before me.

There were some very peculiar features about the mathematical relationships that Heisenberg had discovered. Because he was describing relationships between two states, Heisenberg had not been able to work with ordinary numbers. He had to use arrays of numbers, which he laid out as tables that contained information about both states associated with a transition. Among other things, Heisenberg found that these tables did not commute. When two of the arrays were multiplied together, the answer you got depended on the order in which the multiplication was carried out – A x B was *not* the same as B x A.

Back in Göttingen, Born realized immediately what Heisenberg had discovered. Unlike Heisenberg, Born already knew about an obscure branch of pure mathematics dealing with entities known as **matrices**. He had studied them more than twenty years before; but the one thing that sticks in the mind of anyone who has ever studied matrices is that they do not commute!

In the summer of 1925, working with Pascual Jordan, Born translated Heisenberg's mathematical insight into the formal language of matrices, and Born, Heisenberg and Jordan together published a full account of the work, in what became known as the 'three-man paper'. The equations of Newtonian (classical) mechanics were replaced by similar equations involving matrices, and many of the fundamental concepts of classical mechanics – such as the conservation of energy – emerged naturally from the new equations. Matrix mechanics seemed to include Newtonian mechanics within itself, in much the same way that the equations of the general theory of relativity include the Newtonian description of gravity as a special case.

Unfortunately, few people appreciated the significance of this work. The mathematics was not so much difficult as unfamiliar, and it was not seized upon with the cries of delight that, with hindsight, you might expect. The one exception was in Cambridge, where Paul Dirac picked up the idea and developed it further almost before the ink was dry on the three-man paper. Dirac also found, independently of the Göttingen group, that the equations of matrix mechanics have the same structure as the equations of classical mechanics, with Newtonian mechanics included within them as a special case. Indeed, Dirac's formulation (**quantum algebra**) went even further than matrix mechanics, and included matrix mechanics within itself as a special case.

Some mathematicians appreciated the importance of this work, but most physicists were unhappy about its abstract, theoretical nature. They liked the idea of particles in orbits, and were baffled by a theory which deliberately did away with any physical picture of what was going on inside atoms. So when, just a year later, Erwin Schrödinger came up with a version of quantum mechanics based on the familiarity of waves (see **wave mechanics**), they did seize upon it with delight, and that, not matrix mechanics, became the standard way for physicists to think about the quantum world. This is, perhaps, unfortunate because the one thing that is now absolutely clear about the quantum world is that it is *not* like the everyday world, and although images like waves and orbits may be appealing and comforting, they do not actually describe quantum reality (see **models**).

Nevertheless, we can use a simple pictorial analogy to help get an idea of what matrix mechanics is all about. Think of a chess board. There are 64 squares on the board and you could identify them by a series of numbers, from 1 to 64. But chess players prefer to use a notation in which the 'columns' of squares across the board are labelled a, b, c, d, e, f, g and h, while the 'rows' up the board are numbered 1, 2, 3, 4, 5, 6, 7 and 8. Now, each square is labelled by two coordinates – the bottom left-hand square is a1, the home square of one particular knight's pawn is g2, and so on. At any point in a game of chess, the 'state' of the game can be represented by

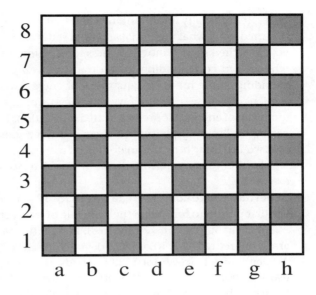

Matrix mechanics 1. Squares on a chessboard can be represented by unique coordinates – a1, b7, and so on. This is a matrix notation.

noting which piece occupies which square. The figure below shows the state at the start of a game; 1 represents a pawn, 2 a rook, and so on. Positive numbers indicate white pieces, negative numbers black. Instead of describing a move in the game in words (such as 'pawn to queen four'), we can use algebraic notation (in this case, e2–e4). We could actually write down the state of the entire board before the move, as a matrix, and the state of the entire board after the move, as another matrix, although in this case there is no need because only one piece has been moved.

Quantum transitions are described in a similar notation linking initial and final

Matrix mechanics 2. The 'state' of the chess game can be represented in matrix notation by using numbers to indicate which piece occupies each square. This particular state corresponds to the start of a game. It could be written out as a1(2), b1(3), c1(4) . . . and so on.

-2	-3	-4	-5	-6	-4	-3	-2
-1	-1	-1	-1	-1	-1	-1	-1
0	0	0	0	0	0	0	0
0	0	0	0	0	0	0	0
0	0	0	0	0	0	0	0
0	0	0	0	0	0	0	0
1	1	1	1	1	1	1	1
2	3	4	5	6	4	3	2

states. It is crucially important, though, that we do not know how the transition has occurred, only what the initial and final states are – a point brought out neatly if you consider the knight's move in chess, or castling. In this analogy, the advance of a white pawn correponding to the transition e2–e4 might be thought of as corresponding to the input of a quantum of energy, while the reverse transition e4–e2 (which, of course, is not allowed in chess) would correspond to the emission of a quantum of energy. More complicated interactions between quantum entities would correspond to multiplying two chess board representations together – but perhaps that would be pushing the analogy too far.

Further reading: John Gribbin, *In Search of Schrödinger's Cat;* Heinz Pagels, *The Cosmic Code.*

Maupertuis, Pierre Louis Moreau de (1698–1759)   French mathematician who was the first person to formulate the principle of least action.

Born at St Malo on 17 July 1698, Maupertuis served as an officer in the French army, but retired early in order to devote himself to science. He taught at the French Academy of Sciences from 1723 onwards, travelling to England in 1728, where he learned about Isaac Newton's work on gravity, and bringing Newton's ideas back to France with him. In 1736–7 he led an expedition to Lapland to measure the length of a degree of longitude, and this work caught the attention of Frederick the Great. After service in the Prussian army, where he was captured by the Austrians at the Battle of Molwitz in 1741, Maupertuis worked at the Berlin Academy, where he stayed from 1741 to 1753. The principle of least action was expressed while he was working in Berlin, initially in 1744, and was published in his *Essai de Cosmologie*, published in 1750. It developed from the work of Pierre Fermat, who formulated the principle of least time, which applies to the way light moves. Maupertuis thought that the principle of least action was the fundamental law of mechanics, and that everything else in mechanics ought to be derived from it; he was not far wrong, as the work of Richard Feynman in the 20th century shows. But he perhaps went too far in his attempts to prove the existence of God starting from the principle of least action.

Following a heated argument about priority, in which the German mathematician Samuel König accused Maupertuis of stealing the idea from Gottfried Leibnitz, and Maupertuis was mercilessly satirized by Voltaire, he left Berlin in 1753 and went back to France. He died in Basle on 27 July 1759.

Maxwell, James Clerk (1831–1879)   Scottish physicist who discovered the equations that describe the behaviour of electromagnetic waves (including light), was one of the founders of the kinetic theory of gases, explained colour vision and was instrumental in establishing the Cavendish Laboratory, in Cambridge, as a centre of excellence in physics.

Born in Edinburgh on 13 June 1831. Maxwell's father was a lawyer with a keen interest in science, who was a Fellow of the Royal Society of Edinburgh. He owned an estate near Dalbeattie, in Galloway, but James was born in Edinburgh because his mother had gone there to ensure the best medical attention for the birth. The young Maxwell spent the first ten years of his life in Galloway, where he was educated at home. His mother died (of cancer) when he was eight, and she only 48, and for the next two years he suffered under an old-fashioned and unsympathetic tutor. When he was sent to the Edinburgh Academy at the age of ten (living with one of his aunts in Edinburgh during term time), he stuck out as something of a country bumpkin, with a different accent and different clothes from the other boys. He quickly acquired the nickname 'Dafty', which stuck until he left the Academy in 1847 and entered the University of Edinburgh.

After three years of the four-year course in Edinburgh, Maxwell moved on to Cambridge, where he graduated (in mathematics) in 1854. After a short spell as a Fellow of Trinity College, Cambridge, in 1856 he became professor of natural philosophy at Marischal College, Aberdeen. Maxwell married the daughter of the principal of the college, but when it merged with King's College, Aberdeen, to become the University of Aberdeen (in 1860), he lost his job and became professor of natural philosophy and astronomy at King's College, London. When his father died, in 1865, Maxwell resigned his post and returned home to look after the family estate, working as an amateur scientist and writing up his great work on electricity and magnetism in book form (*Treatise on Electricity and Magnetism*, eventually published in 1873). In 1871 he was persuaded to return to Cambridge, where he became Cavendish Professor of Experimental Physics and the first head of the Cavendish Laboratory, which opened in 1874 and which he put on a sound footing before he died, in Cambridge, on 5 November 1879 – at the same age as his mother had been when she died, and from the same illness, cancer.

Maxwell was interested in just about all aspects of 19th-century physics. In 1857 he published a paper which proved that the rings of Saturn could not be solid objects, but must be made up of a myriad of small moons in orbit around the planet; at the

James Clerk Maxwell (1831–1879).

beginning of the 1860s, he developed a statistical treatment of the behaviour of gases, based on the idea that a gas is made up of large numbers of atoms and/or molecules in rapid, random motion (this finally established that heat is related to the motion of atoms and molecules); he invented colour photography (using a technique involving three black and white images taken through different colour filters, which was used, more than a century later, by the spaceprobes which sent back colour images of the rings of Saturn); and he came up with the idea of **Maxwell's demon**. But his greatest work was undoubtedly the discovery of the equations of electromagnetism. They built from the idea of lines of force introduced by Michael Faraday, and in turn influenced Albert Einstein to develop the special theory of relativity.

Maxwell's major investigation of electricity and magnetism began just after he graduated from Cambridge in 1854, and culminated in his work *A Dynamical Theory of the Electromagnetic Field*, published in 1864. After a great deal of work, it was in 1862 that Maxwell found that the electromagnetic waves which he had invoked to explain the behaviour of electric and magnetic fields must travel at the speed of light. In a paper published that year, under the title *On Physical Lines of Force*, his excitement shone through in his own words, where the italics are in the original: 'We can scarcely avoid the inference that *light consists in the transverse undulations of the same medium which is the cause of electric and magnetic phenomena.*' The point is that Maxwell did not introduce the constant which turns out to be the speed of light into the equations; when he set the equations up to describe moving electromagnetic waves, using parameters derived from experiments involving stationary electric charges and magnets, the constant emerged naturally from the equations as a fundamental feature of the Universe.

The 1864 paper summed up everything there was to say about electric and magnetic phenomena in a set of just four equations, now known as Maxwell's equations. *Every* problem involving electricity and magnetism (at the level of classical physics) can be solved by using Maxwell's equations, just as every problem involving classical mechanics can be solved by using Newton's equations (that is, **Newton's laws of motion**). Maxwell's work was, indeed, the greatest step forward in physics since Newton's work.

As well as explaining light as a form of **electromagnetic radiation**, Maxwell also realized that there must be longer-wavelength radiation of the same kind, and he predicted the existence of radio waves, which were discovered by Heinrich Hertz not long after Maxwell died. Maxwell himself still thought in terms of these waves being transmitted by the 'ether'; but it was his discovery that the speed of the waves is an intrinsic constant, an integral part of Maxwell's equations, that was, within a few decades, to sound the death knell of the ether.

Further reading: C. W. F. Everitt, *James Clerk Maxwell*.

Maxwell's demon   A thought experiment dreamed up by James Clerk Maxwell (but given the 'demon' name by Lord Kelvin) to highlight an imperfection in the 19th-century understanding of thermodynamics. Maxwell suggested that it would be possible to violate the **second law of thermodynamics** (in a sense, making time run backwards) if a tiny intelligent demon operated a trap door in a wall between two halves of a box, each initially filled with gas at a uniform temperature (the same temperature on both sides of the wall). Every time the demon sees a faster than average molecule approaching from one side of the box, he lets it through; every time he sees a slower than average molecule arriving from the other side of the box, he lets that one through. But all the other molecules have to stay in the half of the box where they started. After a while, one side of the box contains most of the fast-moving molecules (so it has a higher temperature), while the other half of the box contains most of the slower-moving molecules (so it has a lower temperature). It seems as if heat has flowed from a cooler object to a hotter object, without any work being done, increasing the amount of order in the Universe (and therefore decreasing *entropy*), in violation of the second law.

The puzzle was only resolved in 1929, by Leo Szilard. The point is that the demon, no matter how efficient he is at his job, must identify which molecules are moving faster than average and which ones are moving slower than average before he decides whether or not to open the trap door. He might, for example, shine a light on them and measure the velocities from the Doppler shift in the reflected light. When this need for a real, physical interaction between the demon and the molecules is allowed for, it turns out that the cost (in terms of an increase in entropy associated with his activity) is always greater than the benefit (in terms of the decrease in entropy associated with the sorting of the molecules).

This now looks a fairly obvious solution to the puzzle; the interesting, and relevant, point is that Szilard realized this at the end of the 1920s, when quantum mechanics was just becoming established and ideas concerning the interactions between observers and what they are observing were of paramount importance to physicists (see *Copenhagen interpretation*). Only a classical physicist would ever have been baffled by the puzzle of Maxwell's demon. The fact that even physicists as good as Maxwell and Kelvin *were* baffled by it highlights the gulf between the classical world-view and the quantum world-view.

Maxwell's equations A set of four differential equations which together say everything there is to say about the classical (that is, non-quantum) behaviour of electricity, magnetism, electromagnetic fields and electromagnetic radiation (including light). The equations include a fundamental constant, the speed with which electromagnetic waves travel, which is the speed of light and is the same for all electromagnetic waves, wherever they are observed from. See *Maxwell, James Clerk*.

Mayer, Marya (1906–1972) See *Goeppert-Mayer, Marya*.

mean free path Originally, the average distance that a molecule or atom in a gas travels before colliding with another molecule or atom. Also used for the distance travelled between collisions for electrons moving through a metallic crystal, neutrons moving through the *moderator* of a nuclear reactor, particles involved in inelastic or elastic *scattering*, and so on. The mean free path of a molecule in the air, at sea level and at a temperature of 0 °C, is 13 millionths of a metre. See *atom*.

mean free time The average time between collisions of, for example, molecules in a gas. The mean free time for a molecule of air at sea level and at 0 °C, is roughly 3 hundred-millionths (3×10^{-8}) of a second. See *mean free path*.

mean life See *half life, lifetime*.

mechanics The branch of science concerned with interactions between matter and forces. It deals with both the motion of objects and the way they interact while moving (for example, in collisions), and with their equilibrium (static) configurations. See *classical mechanics, relativistic mechanics, quantum mechanics, statistical mechanics.*

Meitner, Lise (1878–1968) Austrian-born physicist who was one of the first people to investigate nuclear fission, and was disgracefully overlooked by the Nobel Committee (not the most perspicacious people in the world) when the prize for this work was given to her colleague Otto Hahn.

Born in Vienna on 7 November 1878, Meitner studied at the University of Vienna (where Ludwig Boltzmann was one of her teachers), and received her PhD in 1906 – only the second woman to be awarded a doctorate in a scientific subject by that university. She then went to Berlin, where the attitude of the university authorities towards women was no more enlightened than the attitude of the authorities in Paris had been to Marie Curie a few years earlier. She attended lectures by Max Planck (which posed no problem for the authorities) and started doing research (which did pose problems). For the first two years, Meitner was not allowed to enter laboratories where men were working, and had to carry out her research in an old carpentry workshop in the basement. It was under those difficult conditions that her long collaboration with Hahn began. In 1912 both Hahn and Meitner moved to the new Kaiser Wilhelm Institute for Chemistry, in the Dahlem quarter of Berlin, where conditions were much easier. In the same year, Meitner became one of Planck's

Lise Meitner (1878–1968).

assistants at the Berlin Institute of Theoretical Physics. But their work was soon interrupted by war.

During the First World War, Meitner served as a radiologist in the Austrian army, continuing her research when on leave. Together with Hahn, she discovered the radioactive element protactinium at the end of the war.

In the changed social climate after the war (and now with a solid track record in research), Meitner became the head of the department of radiation physics at the Kaiser Wilhelm Institute in Berlin, then joint director (with Hahn) of the Institute, and was appointed a full professor at the University of Berlin in 1926. She studied atomic radiation and in 1934 (with Hahn) began to investigate the way uranium nuclei behaved when they were bombarded with neutrons, following up puzzling results obtained by Enrico Fermi. Just as Meitner and Hahn were beginning to think that the uranium nuclei actually split apart under these conditions, Meitner (who was Jewish) had to leave Germany in 1938, to escape persecution by the Nazis in the wake of the German annexation of Austria. She moved to Stockholm, where she worked at the Nobel Institute (ironically) and collaborated with her nephew Otto Frisch, who was based in Copenhagen. Together, they published the first explanation of the uranium experiments as indicating the occurrence of nuclear fission, and gave the phenomenon that name. She did not receive proper recognition for this at the time, partly because of the confusion caused by the war, partly through the incompetence of the Nobel Committee, and partly because of a disgraceful attempt by Hahn to write her out of the story.

Meitner became a Swedish citizen in 1949, at the age of 71, although retaining her Austrian citizenship as well; she remained active in research until 1960, when she retired and moved to England. She died in Cambridge on 27 October 1968.

Further reading: Ruth Sime, *Lise Meitner: A life in physics*.

membranes The two-dimesional equivalent of strings in particle physics (see *string theory*). The idea of replacing the image of fundamental entities as point-like objects with linear objects (strings) surfaced in the 1970s, but was not taken seriously by many researchers for a long time. In the late 1980s, a few theorists, including Michael Duff of Texas A&M University, raised the possibility that we ought not to be dealing with strings at all, but ought to add in another dimension, making them resemble two-dimensional sheets (membranes) rather than one-dimensional lines. The extra dimension brings the total number up to eleven, but one of these dimensions is immediately rolled up so that the membrane behaves like the ten-dimensional string of string theory. The idea was more a speculation than a fully worked out theory, and it was laughed out of court at the end of the 1980s. But it was revived, as a much more complete theory, in the mid-1990s, and today the membrane idea is just about the hottest game in town. John Schwarz coined the term 'M-theory' to describe the idea, saying that the 'M' can stand for 'magic, mystery or membrane, according to taste'.

The reason why M-theory is causing excitement at the end of the 1990s is that it offers, at last, a *unique* mathematical package to describe all of the forces and particles of nature. String theory itself comes in several different varieties, which each have their good points and their bad points. In fact, there are exactly five variations on the theme. These are the Type I theory of John Schwarz and Michael Green, two versions of their Type II theory, and the two versions of heterotic strings. In addition, there is a wild card, the eleven-dimensional supergravity. It can be shown mathematically that these are the only viable variations on the theme; all of the other possibilities involving supersymmetry are plagued by infinities.

At first sight, six rival contenders for the title 'theory of everything' looks a lot. But in fact this is a remarkably short short-list. The old-fashioned particle physics approach to grand unified theories gives you a plethora of possibilities, any of which are just as good as any of the others. To have only half a dozen theories to choose from seemed miraculous in the 1980s. The dramatic new discovery made in the mid-1990s, however, was that all six theories are related to one another. Specifically, they are all different manifestations of a single M-theory. In a manner reminiscent of the way that the electroweak theory is a single theory that describes what seem to be two separate interactions at lower energies (*electromagnetism* and the **weak interaction**), M-theory is a single theory at even higher energies, and describes what seem to be six different models at lower energies. Specifically, the differences between the six models appear at the level of the weak interaction, and the unity is clear at the level of the strong interaction.

We may not have to wait too long to find out just how good a theory M-theory really is, and whether it is indeed the long-sought theory of everything. The kind of energies needed to probe the predictions of M-theory should be achieved at the latest high-energy particle accelerator, the Large Hadron Collider (LHC), which is expected to begin operating at CERN in the middle of the first decade of the 21st century.

mendelevium-101   See *Mendeleyev, Dmitri Ivanovich*.

Mendeleyev, Dmitri Ivanovich (1834–1907)   Russian chemist who came up with the idea of the periodic table of the *elements* and correctly predicted the existence of 'new' elements on the basis of gaps in that table.

Born in Tobol'sk, in Siberia, on 7 February 1834 (27 January, Old Style), Mendeleyev was the youngest of a family of fourteen children; their father was the head of the local school, but went blind when Dmitri was still a child, after which the family was largely supported by their mother, who set up a glass works to obtain income. The spelling of the family name in the western alphabet has been to some extent a matter of choice down the years; we have chosen a common variation on the theme which conveys the way the name should be pronounced.

Mendeleyev was initially educated in Tobol'sk, but his father died in 1847, and a year later, when he was fourteen, his mother's factory was destroyed by fire. A determined woman, she wanted her youngest child to have a good education in spite of these difficulties, and took him to St Petersburg. Because of prejudice by the central authorities against students from the provinces, he was unable to gain admission to any university, but was enrolled in 1850 as a student teacher at the Pedagogical Institute in St Petersburg, where his father had qualified. His mother died just ten weeks later. Having qualified as a teacher and thus proved his ability, after a year teaching in Odessa he was allowed to take a master's degree in chemistry at the University of St Petersburg.

In 1859 Mendeleyev went on a government-sponsored study programme to Paris and Heidelberg, where he worked under Robert Bunsen and Gustav Kirchoff, met Stanislao Cannizzaro and learned of the important distinction between molecular and atomic weights. He returned to St Petersburg in 1861, to become professor of general chemistry in the city's Technical Institute. While working there, he completed his PhD (awarded in 1865), and in 1866 he became professor of chemistry at the University of St Petersburg. Because there was no up-to-date textbook suitable for the course he wanted to teach, he wrote his own, *Principles of Chemistry*, which was widely translated and made his name well known among the international community of chemists. It was during this period in St Petersburg, from the second half of the 1860s onwards (and partly through his work on the textbook), that Mendeleyev worked on his periodic law, alongside other research now seen as of lesser importance. In 1890 he took up the case of students who were protesting about conditions in the Russian academic system, and as a result of this he was removed from his post. After three years he was deemed to have purged his guilt, and in 1893 he became controller of the Bureau of Weights and Measures, a post he held until he died. In 1906 he was nominated for the Nobel Prize for Chemistry, but lost out by one vote to Henri Moissan, who was honoured for his success in being the first person to isolate fluorine. Mendeleyev did not get another chance at the Prize because he died in St Petersburg on 2 February 1907 (20 January, Old Style). Element number 101 was named mendelevium, in his honour, in 1955.

mesons Members of a family of particles, each of which is composed of a quark and an antiquark, bound together by the exchange of gluons (see *colour force*). These are all strongly interacting particles (that is, hadrons), with an integral amount of spin, so they are members of the larger *boson* family. They all have zero *baryon number*. Some have one unit of negative charge, some have one unit of positive charge and some are uncharged. Pions and kaons are examples of mesons. Mesons are involved in the interactions which hold nucleons together in an atomic nucleus (the *strong interaction*). It was the discovery of mesons, and the interpretation of their role in particle interactions, that marked the beginning of modern particle physics in the second half of the 1930s. See also *Yukawa, Hideki*.

messenger particle Another name for the force carriers that mediate interactions between particles. See *intermediate vector boson*.

metre Standard unit of length, originally defined (by the National Assembly in revolutionary France in the 1790s) as 1 ten-millionth of the distance from the North Pole to the equator. Since 1983, the definition of the metre has been as the distance travelled by light in 1/299,792,458 seconds. Several people have suggested redefining the metre so that the speed of light is exactly 300,000,000 metres per second (an adjustment of less than 0.07 per cent), but as yet, unfortunately, this has not been done.

MeV 1 million *electron volts*.

Meyer, (Julius) Lothar (1830–1895) German chemist who discovered the periodic pattern of the chemical *elements*, independently of Dmitri Mendeleyev.

Born at Varel on 19 August 1830, Meyer studied in Zurich and Würzburg, qualifying as a doctor in 1854, but switched to chemistry and obtained his PhD from the University of Breslau in 1858. He then worked as a *Privatdozent* in Breslau before taking up a succession of academic appointments (interrupted by work as a surgeon in a military hospital during the Franco-Prussian War of 1870–1), culminating in becoming the first professor of chemistry at the University of Tübingen, in 1876. He remained there until he died, on 11 April 1895.

Like Mendeleyev, Meyer came to the attention of other chemists through writing a textbook, *The Modern Theory of Chemistry*, published in 1864. The first outlines of his ideas about a periodic law appeared in that book, and in 1868 Meyer prepared a more detailed periodic table, probably intended originally for a second edition of the book. The second edition was not produced immediately, and Mendeleyev's version of the periodic table was published in 1869, a year before Meyer's version appeared in print. By then Meyer had seen Mendeleyev's work and did not claim priority, even though he had reached many of the same conclusions independently of Mendeleyev. Crucially, though, unlike Mendeleyev, Meyer did not point to the need for 'new' elements to fit into the gaps in the periodic table. Nevertheless, he shared the award of the Davy Medal of the Royal Society with Mendeleyev in 1882. He also made important contributions to the study of organic chemistry and to the understanding of the physiology of respiration.

Michelson, Albert Abraham (1852–1931) German-born American physicist who made many determinations of the speed of light, and carried out an experiment with Edward Morley which famously failed to find any evidence for an 'ether' through which light was propagated. He was awarded the Nobel Prize for Physics in 1907 for the development of precision optical instruments and the scientific measurements (including those of the speed of light) that he carried out with them. He was the first American to be awarded a Nobel Prize.

Born in Strelno (now Strzelno, in Poland) on 19 December 1852, Michelson emigrated with his family to the United States when he was four, and went to school in San Francisco. He entered the Naval Academy at Annapolis in 1869, and after graduating in 1873 spent two years at sea before becoming an instructor at the Academy. From 1880 to 1882 he was given leave to study in Paris and Berlin, and on his return to America he left the navy and became professor of physics at the Case School of Applied Science, in Cleveland, Ohio. In 1889 he moved to Clark University, at Worcester, in Massachusetts; in 1892 he became professor of physics at the University of Chicago, staying there until he retired in 1929. He died in Pasadena on 9 May 1931. The famous collaboration with Morley took place while Michelson was in Cleveland, in 1887.

Michelson–Morley experiment An experiment carried out in the 1880s, initially by Albert Michelson and later in collaboration with Edward Morley, in an attempt to detect the motion of the Earth through the 'ether', by measuring differences in the speed of light determined along the line of the Earth's motion and at right angles to that line. Equivalent experiments have been carried out many times since by other experimenters, and all come up with the same result – no effects attributable to the motion of the Earth can be seen in any measurements of the speed of light. The experiments show that there is no ether, and that the measured speed of light does not depend on how the measuring apparatus is moving.

The Michelson–Morley experiment is often regarded as having provided the impetus for Albert Einstein to develop the special theory of relativity, which says that the speed of light is an absolute constant, *c*, and will be measured to have the same value by all observers, in any inertial frame, regardless of how they are moving through space, or at what speed the source of the light is moving through space. But Einstein always told enquirers that he had been unaware of the Michelson–Morley experiment at the time he developed the special theory, in 1905. The impetus for his work at that time came in part from James Clerk Maxwell's equations describing the motion of electromagnetic waves (including light waves); those equations include the speed of light as an absolute constant.

microscopic system   Actually very much smaller than anything that can be seen by an ordinary microscope; physicists use the term to refer to systems small enough to be dominated by quantum processes, not by the laws of classical mechanics. See *correspondence principle*.

microwaves   Electromagnetic radiation with wavelengths in the range from 1 to 30 cm (short-wavelength radio waves).

Millikan, Robert Andrews (1868–1953)   American physicist who received the Nobel Prize for Physics in 1923 for measuring the charge on the electron. The citation also mentioned his work on the photoelectric effect; Millikan had carried out the experiments which proved (against Millikan's initial expectations!) that Albert Einstein's theory of the photoelectric effect was correct. He later studied cosmic rays.

Born in Morrison, Illinois, on 22 March 1868, Millikan was the son of a Congregational minister and initially studied classics, before obtaining his PhD in physics from Columbia University (where he was the only graduate student in physics at the time) in 1895. After a year of postgraduate studies in Germany (including a spell with Max Planck in Berlin), he joined the University of Chicago in 1896 (working with Albert Michelson), and became a full professor there in 1910. In his first dozen years at Chicago, Millikan concentrated on teaching and on writing a series of influential textbooks, but did little research, which is one reason why his greatest scientific achievements came relatively late in life. During the First World War, Millikan was director of research for the National Research Council (the body organizing defence research), and worked directly on anti-submarine devices and in meteorological instrumentation; it was through his involvement with a scheme to use unmanned balloons to carry propaganda behind enemy lines that he became interested in using balloon-borne detectors to study cosmic rays (which he gave that name to, in 1925).

In 1921 Millikan moved to Caltech, where he became director of the Norman Bridge Laboratory and chairman of the executive council of Caltech itself; he was instrumental in establishing the reputation of the Institute over the next few decades. He retired in 1945 and died in San Marino, California, on 19 December 1953.

The work for which Millikan received the Nobel Prize was carried out between 1909 and 1912, while he was in Chicago. Important though this was, it was his next project which was of key significance in the development of quantum physics. Millikan did not accept Einstein's idea of light quanta when it was published in 1905, and in 1912 he set out to prove that Einstein's interpretation of the photoelectric effect in these terms was wrong. Over the next four years, in a series of painstaking experiments, he succeeded in proving that Einstein was right – and, being a good scientist, then accepted the evidence, instead of trying to fight a futile rearguard action against the notion of light quanta. In the course of these experiments, Millikan also obtained a very accurate measurement of Planck's constant, as 6.57×10^{-27}. It was this experimental proof of the reality of light quanta, obtained by a sceptic who had persuaded himself that his initial prejudice against the idea was wrong, that led to Einstein being awarded the Nobel Prize for his theory of the photoelectric effect (the citation for Einstein's prize specifically mentioned Millikan's work), and to the notion of light quanta becoming firmly established as respectable physics. Looking back more than a quarter of a century later, Millikan commented ruefully: 'I spent ten years of my life testing that 1905 equation of Einstein's and contrary to all my expectations, I was compelled in 1915 to assert its unambiguous verification in spite of its unreasonableness' (see *Reviews of Modern Physics*, vol. XXI, p. 343, 1949).

mirror matter   = *shadow matter*.

mirror symmetry   See *parity*.

missing mass   See *dark matter*.

mixed state   A state which is a superposition of two or more quantum states. See *eigenstate, Schrödinger's cat, Copenhagen interpretation*.

mixing angle   A measure of the extent to which two different eigenstates contribute to a mixed state. The idea of rotation is used as a metaphor to describe the mixing process (see, for example, *isotopic spin*), as if the position of a pointer on a dial indicated the amount of each eigenstate in the mixture. For an angle Θ, the sum of the functions $\sin\Theta$ and $\cos\Theta$ is always 1, so the proportion of the mixed state contributed by each of two components can be represented as the sum of a sine and a cosine. If the mixed state is denoted by C and the amplitudes of the two states in the mixture are denoted by l and m, then

$$C = l\sin\Theta + m\cos\Theta$$

and any mixture of l and m can be represented by a suitable choice of the single parameter Θ, the mixing angle.

models   Physicists studying the quantum world cannot make cardboard or wooden scale replicas of the things they are interested in (such as a photon), so their models are a combination of mathematical equations and physical insights which are used to provide some kind of image of what is going on in the quantum world. Some of these models are very precise representations of quantum phenomena, described in terms of equations that can be run on a computer to simulate the way in which that quantum system will respond to different stimuli. Others are much more vague and 'hand waving', designed only to help the limited human imagination picture what is going on.

　　One of the most important things to appreciate about models is that they are not (any of them!) 'the truth'. So even though one particular model may be very successful as a description of what is going on in one context, a completely different model may be equally successful in describing the behaviour of the same quantum entity under different circumstances. Both models can be equally valid. The classic example of this is, of course, the wave–particle duality of quantum entities. It is sometimes appropriate to describe light in terms of particles (photons), and sometimes appropriate to describe it in terms of waves (described by Maxwell's equations). It is *not* the case that light always behaves as a wave, nor is it the case that light always behaves as a particle. Indeed, you should not imagine that light 'really is' *either* a wave *or* a particle. It is something for which there is no analogy in the everyday world of our senses, a something which under certain circumstances seems *like* a wave, and under other circumstances seems *like* a particle.

Another example brings this use of models out with more force. Historically, the idea of atoms as hard, indivisible spheres came before the realization that atoms are composed of other particles. Using this 'billiard ball' model of the atom, physicists were able to provide a very accurate mathematical description of the behaviour of gases – for example, the relationship between pressure and temperature for a box full of gas. Later, the *Bohr model* of the atom, with electrons regarded as tiny billiard balls in orbit around a billiard-ball nucleus, proved extremely successful in explaining the origin of spectral lines. And later still the nature of the chemical bond was explained using the model of electrons as clouds (probability distributions) around the nucleus. Although there is a clear path of historical development in these ideas, this does not mean that the later models are 'right' and that earlier models are 'wrong'. Physicists today still use the billiard ball model of the atom if they are calculating gas pressures, and chemists today still use what is essentially the Bohr model if they are studying spectra. Each model is correct in its own area of application, even though the different models may seem to be quite incompatible with one another.

The best way to think of the models used by physicists is as a set of tools that can be used for different jobs. A carpenter may sometimes need a screwdriver, and other times a hammer. As long as the carpenter uses the right tool for the job, there is no problem with the fact that they are different tools. It is when you apply a screwdriver to a nail (or try to explain spectroscopy using the billiard ball model of the atom) that you run into trouble.

This is particularly important because all of the so-called interpretations of quantum mechanics (for example, the *Copenhagen interpretation*) are really models of this kind. *None* of them represents the ultimate truth about the quantum world, and very probably there is no way that a human brain could ever comprehend the ultimate truth about the quantum world. All of the interpretations are simply aids to help you get a feel for what is going on. A good physicist should carry every quantum interpretation in his or her toolkit, and should apply the right one for the job in hand when confronted with a particular quantum puzzle. For example, the Copenhagen interpretation is clearly not the right model to apply to the puzzle of *Schrödinger's cat*; but the *many worlds interpretation* works very well when applied to this particular problem. But nobody knows what the quantum world 'is'; all we can know is what it is 'like'. Sometimes it is like one model, and sometimes it is like another model. And that's reality.

Further reading: John Gribbin, *Schrödinger's Kittens*.

moderator   Material used in nuclear reactors to slow down fast neutrons (produced by nuclear fission) by scattering, so that they have the appropriate energies to sustain a controlled nuclear reaction by triggering further fission at the rate required. Commonly used moderators include deuterium (in the form of heavy water), graphite and beryllium. See *reactor*.

molecular beam   A beam of molecules, atoms or ions at low pressure travelling in the same direction, so that there are very few collisions between the particles in the beam. The molecular beam technique was used in the *Stern–Gerlach experiment* which provided evidence for electron spin.

molecular biology   Branch of (quantum) physics dealing with the nature of life. Life molecules are complex compounds, usually including carbon and known as organic compounds, which are described by *quantum chemistry*. These include substances such as proteins, and the deoxyribonucleic acid (DNA) that forms the fundamental molecule of life, the famous double helix, and carries the genetic code.

The central role of quantum physics in life processes is best illustrated by the example of DNA itself. Each strand in the double helix of a DNA molecule is a long string of subunits, each one a sugar compound (a carbohydrate), linked to its neighbours on either side by groups known as phosphates. In DNA, the particular kind of sugar in these units is called deoxyribose, which is where DNA gets its name. Each of the deoxyribose sugar units in the chain has attached to it another compound, sticking out from the chain, called a base. Each unit in the molecule consists of a base, plus sugar, plus phosphate group, and is known as a nucleotide. There are four different kinds of base in these subunits, called adenine, cytosine, guanine and thymine, but usually referred to simply by their initials, as A, C, G and T. It is the order of these bases along the chain of a DNA molecule that spells out the biological message carried by the DNA, in a four-letter code (it was, incidentally, George Gamow who first hit on this idea of a genetic code). This is precisely equivalent to the way the words you are reading convey information spelled out in a 26-letter alphabetic 'code', or the way in which information is stored in a computer using a two-digit binary code.

The important feature of the genetic code (like the code in a string of words or in binary form in a computer) is that it can be copied and passed on to later generations. In DNA this is possible because each molecule is double stranded, forming the famous double helix. The bases that stick out from the spines of each strand of DNA have a distinctive physical structure. The structures of adenine and thymine match up, so that if the two molecules are put close together, they will naturally form two hydrogen bonds across the gap; similarly, the structures of guanine and cytosine match up, so that if they are put close together, they will naturally form three hydrogen bonds across the gap. Right down the length of a DNA double helix, everywhere there is a C on one strand of the helix there is a G on the other strand, and everywhere one strand has an A the other has a T. The As and Ts fit together like two-pin plugs in their appropriate sockets, while the Gs and Cs fit together like three-pin plugs in their appropriate sockets. But you *never* find A paired with G or C, or T paired with G or C, because you can't get a three-pin plug in a two-hole socket. So each strand of the double helix in a particular molecule is effectively a mirror image of the other.

When the molecule reproduces (as it does when cells divide, both during ordinary growth and in the process that leads to the production of eggs and sperm that carry the genetic message into the next generation), the double helix unzips, and each separate strand attracts the right chemical counterparts from the chemical soup inside the cell to form another mirror image strand, with A latching on to T and C latching on to G, producing two new double helices, each identical to the original.

But notice that all of this depends crucially on the properties of the hydrogen bond, a phenomenon which can be described only in terms of the quantum behaviour of electrons. Quite apart from the broader need for quantum chemistry to

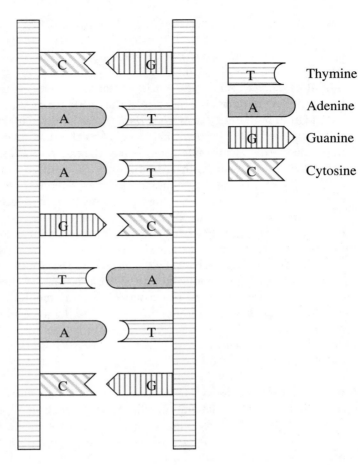

Molecular biology. The components which form the chemical bonds which hold the two strands of a DNA molecule together have distinctive shapes, which only allow matching pairs of molecules to join, like the fit of a key in a lock. But in each case, the actual link is completed by hydrogen bonds.

T Thymine

A Adenine

G Guanine

C Cytosine

explain what is going on in the molecules of life in general, this particular quantum phenomenon lies at the very centre of life itself, holding DNA molecules together and ensuring that each strand of the double helix can act as a template for the other.

It might be possible to imagine other forms of life, and other kinds of genetic code. But the fact is that life as we know it, including the life processes at work in the cells of your own body, depend crucially on quantum interactions. Without the hydrogen bond, you would not exist.

Further reading: John Gribbin, *In Search of the Double Helix.*

molecular spectroscopy   See *spectroscopy.*

molecular weight (molecular mass)   The weight of a *molecule* expressed in the same units as *atomic weight.*

molecule   Two or more atoms held together by *chemical bonds* to form a stable unit. It is sometimes defined as the smallest part of a chemical compound (that is, a combination of at least two atoms) that can take part in a chemical reaction; but note that this includes the possibility that the atoms in the molecule are identical, as in the case of the hydrogen molecule, H_2. Hydrogen is, in fact, the smallest and lightest molecule, with a molecular weight of 2. The largest 'macromolecules', proteins, have molecular weights of several million. Haemoglobin, the molecule which carries oxygen around in your blood, is made up of 1,203 hydrogen atoms, 758 carbon atoms, 218 nitrogen atoms, 195 oxygen atoms, 3 atoms of sulphur and 1 atom of iron.

moment of inertia   A measure of the resistance of an object to rotational forces, in the same way that mass is a measure of the resistance of an object to being pushed around in straight lines (see *Mach's Principle*). Each minuscule part of the object contributes a share of rotational inertia equal to the mass of that bit of the object multiplied by the square of its distance from the axis of rotation; the overall moment of inertia is given by adding up all of these tiny moments, by integration. This is then used in place of mass in the rotational equivalents of the familiar equations of linear mechanics – for example, just as the kinetic energy of an object with mass m moving at velocity v is $\frac{1}{2}mv^2$, so the kinetic energy of an object with moment of inertia I rotating with an angular velocity of ω cycles per second is given by $\frac{1}{2}I\omega^2$.

momentum   The product of the mass (m) of an object and its velocity (v), usually denoted (for historical reasons) by the letter p, so that $p = mv$. Momentum does not change unless a force is applied to the object; this is one of the fundamental laws of classical mechanics (the law of conservation of momentum). Newton discovered that the rate of change of momentum is equal to the force applied. See also *angular momentum*.

monopoles   Hypothetical particles which each possess a single 'flavour' of magnetism, either north or south. Some *grand unified theories* predict the existence of monopoles, but they have never been found.

Morley, Edward Williams (1838–1923)   American chemist and physicist, best remembered for his collaboration with Albert Michelson on an experiment which famously failed to find any evidence of an 'ether' through which light is propagated.

Morley was born in Newark, New Jersey, on 29 January 1838. He was educated at home, then studied at Williams College from 1857 to 1860, and at the Andover Theological Seminary (his father was a Congregational minister) from 1861 to 1864. He started out on a career as a minister, but kept up his scientific studies as well (he was awarded his master's degree by Williams College in 1863). In 1869, after a spell as a teacher in Marlboro, Massachusetts, he became professor of natural history and chemistry at Western Reserve College in Hudson, Ohio, where he was also a regular preacher in the college chapel; from 1873 to 1888 he was, in addition, professor of chemistry and toxicology at Cleveland Medical College (after Western Reserve College had moved to Cleveland, where it eventually became part of Case-Western Reserve University). Morley retired in 1906. His speciality was the painstakingly accurate measurement of chemical abundances, including variations in the amount of oxygen in the air. This skill in precision measurements stood him in good stead in the famous collaboration with Michelson, which was carried out in the late 1880s in Cleveland. Morley died in Hartford, Connecticut, on 24 February 1923.

Moseley, Henry Gwyn Jeffreys (1887–1915)   British physicist who was a pioneer of X-ray spectroscopy, and was the first person to realize that the atomic number (a term he coined) of an *element* is a measure of the charge on the nucleus of an atom of the element. This led to a better understanding of the periodic table of the elements, and to the prediction of the existence of several 'new' elements which were soon found. It was Moseley who made the link between the place of an element in the periodic table and the internal structure of its atoms.

Born in Weymouth on 23 November 1887, Moseley graduated from the University of Oxford in 1910, and worked with Ernest Rutherford in Manchester before returning to Oxford in 1913. He volunteered for the army on the outbreak of the First World War, and served as an officer in the Royal Engineers. He was killed by a sniper at Gallipoli on 10 August 1915.

Mössbauer, Rudolf Ludwig (1929–)　German physicist who was awarded the Nobel Prize for Physics in 1961 for his discovery of the *Mössbauer effect*, which involves the absorption of gamma radiation by an atomic nucleus.

Born in Munich on 31 January 1929, Mössbauer graduated from the Munich Institute of Technology in 1952, was awarded his master's degree in 1955, and received his PhD from the same institute in 1958, for research carried out at the Max Planck Institute for Medical Research, in Heidelberg. He discovered the eponymous effect during the course of his postgraduate studies. Mössbauer then became a professor in Munich, but went to the United States in 1960 and in 1961 became professor of physics at Caltech. From 1965 onwards, he was professor of physics at the Technical University in Munich, apart from a spell (1972–7) as director of the Institut Laue-Langevin, in Grenoble.

Mössbauer effect　A process in which a nucleus is made to absorb gamma radiation without recoiling. This is achieved by locking the nucleus of interest into the structure of a crystal, so that the whole crystal absorbs the recoil and the nucleus of interest does not move. The importance of this is that, because the target nucleus does not move, it 'sees' precisely the wavelength of gamma radiation that is produced by the gamma ray source (in practice, the source of the gamma rays is also a nucleus locked into a crystal structure to avoid recoil).

The nucleus will absorb the radiation only if it is precisely fine-tuned to match a resonance in the nucleus (just as an atom can only absorb sharply defined lines in the spectrum corresponding to transitions of electrons between energy levels; see *Bohr model*). If the nucleus can recoil, this enables it to absorb gamma rays with slightly different wavelengths, but in the Mössbauer effect only very precisely defined wavelengths can be absorbed. The exact wavelength that will be absorbed can be fine-tuned by moving the absorbing material at a steady velocity, which may be as small as a few millimetres per hour. The effect is so precise that it can be used to measure the change in wavelength of gamma rays caused by the difference in the gravitational field of the Earth between the top and bottom of a tall building, providing a direct confirmation of the accuracy of some of the predictions of the general theory of relativity. It has also been used to confirm the accuracy of the predictions of the special theory of relativity.

Mott, Nevill Francis (1905–1996)　British physicist who was awarded the Nobel Prize for Physics in 1977 for his work on the electronic structure of magnetic and disordered systems (by which they mean *semiconductors)*.

Born in Leeds on 30 September 1905, Mott graduated from the University of Cambridge in 1926 and spent three years in research there, but never completed a PhD. He spent two years as a lecturer in physics at the University of Manchester (1929–30). After a spell as a Fellow of Gonville and Caius College, Cambridge, and a visit to Niels Bohr's Institute in Copenhagen, in 1933 he moved to the University of Bristol, as professor of theoretical physics. He became the director of the H. H. Wills Physical Laboratories in Bristol in 1948, and returned to Cambridge as Cavendish Professor of Physics in 1954, succeeding Lawrence Bragg and holding the post until he retired in 1971, but remaining active in research. Mott played a major role in shaping the development of the Cavendish Laboratory in the second half of the 20th century, and was knighted in 1962. He died in Milton Keynes on 8 August 1996.

In the 1930s, Mott worked on the quantum theory of atomic collisions (scattering) and wrote a series of important textbooks. He later studied the electronic structure of solid-state systems such as semiconductors.

Mottelson, Ben Roy (1926–) American-born Danish physicist who shared the Nobel Prize for Physics in 1975 with Aage Bohr and James Rainwater, for their work on the theory of the structure of the atomic nucleus.

Born in Chicago, Illinois, on 9 July 1926, Mottelson studied at Purdue University, graduating in 1947, then at Harvard (under the supervision of Julian Schwinger), where he was awarded his PhD in 1950. He then went to Niels Bohr's Institute in Copenhagen, and in 1953 took up a post with CERN as a member of a theory group based in Copenhagen. He became a professor at NORDITA in 1957, and a Danish citizen in 1971. For information about the work for which he shared the Nobel Prize, see *Bohr, Aage Niels*.

moving mirror radiation In a process analogous to the *Casimir effect*, a mirror that is moving through space in a certain way will produce real photons out of the energy of the *quantum vacuum*. The process does not work for a stationary mirror, nor one moving at constant velocity, nor even one experiencing a uniform acceleration; but it does work for a mirror moving in such a way that the rate of its acceleration is changing. Paul Davies, who (together with Stephen Fulling) first calculated the effect, has described its efficiency as 'pitifully low', saying that appreciable numbers of photons will be created from a single mirror only by accelerative forces of unimaginable violence; but other researchers have pointed out that the effect could be amplified if two moving mirrors in a configuration like that of the Casimir effect are used. This could lead to a resonance, amplifying the process of photon production in the gap between the mirrors.

There are two ways in which this might be achieved – either moving the whole cavity set-up to and fro bodily, or moving the mirrors in and out in antiphase with each other, repeatedly squeezing and expanding the cavity. It is estimated that motion gentle enough to prevent the mirrored cavity breaking apart could produce as many as ten photons per second. If the experiment is ever successfully carried out, it will provide direct evidence in favour of the quantum theory of the vacuum.

M-theory See *string theory*, *membranes*.

Müller, (Karl) Alex (1927–) Swiss physicist who shared the Nobel Prize for Physics with Georg Bednorz in 1987, for their discovery of high-temperature superconductivity.

Born in Basle on 20 April 1927, Müller studied at the Swiss Federal Institute of Technology and was awarded his PhD in 1958. He then worked at the Battelle Institute, in Geneva, until 1963, when he joined the IBM Zurich Research Laboratory. Alongside this work, he became a lecturer at the University of Zurich in 1962, and professor in 1970.

multiwire proportional chamber   A type of particle detector devised by Georges Charpak at CERN at the end of the 1960s. The basic unit of such a detector is made up of three layers of fine parallel wires rigged like a sandwich so that the central layer is at a positive electric potential of 3–5 kilovolts relative to the two outer layers. When a charged particle passes through the detector, it triggers an avalanche of electrical activity in the vicinity of a single wire in the central layer of the sandwich, so with a series of these basic units stacked together it is possible to follow the track of the charged particle as it passes through the chamber. A multiwire chamber can monitor the passage of a million particles a second, and they have become an integral component of most particle physics experiments.

muon   A heavy counterpart to the electron, a *lepton* with a mass 206.7683 times that of the electron, but otherwise identical to an electron. The muon is unstable, with a half life of 2.19709 microseconds, and decays into an electron, a muon neutrino and an electron antineutrino. See *generation*.

muon neutrino   The type of neutrino associated with the *muon*. Discovered in the early 1960s by Leon Lederman, Melvin Schwartz and Jack Steinberger. See *neutrino*.

naked charge   = *bare charge.*

naked mass   = *bare mass.*

nano   Prefix denoting 1 billionth, 10^{-9}. From the Greek nanos, meaning dwarf.

nanometre   1 billionth of a metre. $1\,nm = 10^{-9}\,m = 0.000000001\,m.$

nanosecond   1 billionth of a second.

nanotechnology   Engineering on the very small scale, literally the scale of atoms and molecules. The first person to suggest this as a real possibility was Richard Feynman, in a talk given at Caltech at the end of 1959. The idea was reinvented and made a practical reality by Eric Drexler, at MIT, in the 1970s. The dream is to make tiny machines that will, for example, be able to travel inside the arteries of someone with heart problems, cleaning the furred-up walls of the arteries; or little submicroscopic robots that could take the raw materials (basically, grass) and turn them into 'meat'. As yet, the actual achievements have been more on the scale of party tricks, such as arranging individual atoms to spell out the letters IBM or to write the first page of *A Tale of Two Cities* on a scale where it can be read only by using an electron microscope. But every new technology has to start somewhere.

   Further reading: Ed Regis, *NANO!*

natural units   Units in which the fundamental constants of physics – the speed of light, Planck's constant and Boltzmann's constant – are defined to have the value 1. This enables physicists to work with simplified versions of their equations, which are not cluttered up by symbols such as c and h. For example, in natural units Einstein's equation $E = mc^2$ becomes $E = m$; you often find physicists quoting the masses of particles in electron volts (strictly speaking, a unit of energy), with the c^2 term taken as read. If it is necessary, or you want to, it is easy to put factors like c^2 back in at the end of a calculation to give a numerical value for some quantity in more familiar units (in this case, grams). It may seem puzzling that the speed of light can be defined as 1, but this is, in fact, straightforward – it is equivalent to choosing to measure time in seconds and distances in light seconds, so that light travels at a speed of 1 light second per second; the other constants can be defined in a similar way.

Ne'eman, Yuval (1925–)   Israeli physicist who developed the *eightfold way* idea independently of Murray Gell-Mann and played a significant part in establishing the theoretical basis for the quark model.

Ne'eman was born in Tel Aviv (then in Palestine) on 14 May 1925. He studied engineering at the Institute of Technology in Haifa (the Technion), graduating in 1945, then served with the Israeli army during the war that led to the establishment of Israel as an independent state in 1948. Although still a member of the Israeli defence forces, Ne'eman was later allowed to carry out research alongside his other duties. He studied at the Ecole de Guerre in Paris in 1952, rose to the rank of colonel in 1955 and was deputy director of the Intelligence division from 1955 to 1957. He then became a military attaché at the Israeli embassy in London, and carried out research at Imperial College, a five-minute walk from the embassy, where he completed a PhD in physics (under the supervision of Abdus Salam), awarded by the University of London in 1961.

From 1955 onwards, Ne'eman was also a member of Tel Aviv University. From 1961 to 1963 he worked for the Israeli Atomic Energy Commission, and in 1963 he joined Tel Aviv University full time, where he became a full professor and vice-rector of the university in 1964. He has served as president of the university (1971–5) and has been a member of the Israeli parliament, where he was Minister of Science from 1982 to 1984 and again from 1990 to 1992, after which he was briefly Minister of Energy. He has also had a long association with the University of Texas at Austin.

Ne'eman's classification of the elementary particles, known as SU(3) from the mathematical terminology describing the symmetry properties of the group (see *group theory*), was proposed in his thesis work, developed at the end of 1960 and published in 1961, and predicted the existence of the omega-minus particle, which was discovered in 1964. Also in 1964, Ne'eman and Gell-Mann together produced a book, *The Eightfold Way*. In an autobiographical note in his book *The Particle Hunters* (co-written with Yoram Kirsch), Ne'eman describes how he was 'captivated' by group theory when introduced to it by Salam, and hoped to be able to explain the relationship between particle properties in terms of a group known as G(2), which corresponds to patterns shaped like the Star of David, but discovered instead that SU(3) gave a perfect fit, provided the omega-minus existed. Ne'eman has also made important contributions to astrophysics.

negative feedback See *feedback*.

Neumann, John von (1903–1957) Hungarian-born mathematician (christened Janos, but also known, during his time in Germany, as Johann) who became an American citizen and was an adviser on the Manhattan Project.

'Johnny' von Neumann, as he was usually known, was a brilliant mathematician who was interested in the theory of games (which, in spite of its jolly sounding name, has serious implications for conflicts such as war) and the fundamentals of the theory of computers (modern electronic computers are sometimes referred to as 'von Neumann machines'). But his main contribution to quantum theory was a mistake which, thanks to his awesome reputation, held back the development of the field for decades.

Born in Budapest on 28 December 1903, von Neumann studied at the University of Berlin from 1921 to 1923 and in Zurich (at the ETH) from 1923 to 1925, then returned to Budapest where he completed his PhD studies at the University in 1926.

He was a *Privatdozent* at Berlin University from 1927 to 1929, then joined the University of Hamburg, but left Germany for the United States in 1930. There he worked in Princeton, first at the university and from 1933 at the Institute for Advanced Study. He died from cancer in Washington, DC, on 8 February 1957.

Von Neumann's big mistake appeared in a book on quantum theory that he published in 1932. The book was regarded as a definitive *tour de force*, and included a 'proof' that no *hidden variables* theory could ever properly describe the workings of the quantum world. This inhibited generations of physicists (with the notable exception of David Bohm) from working on hidden variables versions of quantum theory. But there was an error in von Neumann's proof.

Although one mathematician, Grete Hermann, did point out the flaw in 1933, nobody believed that she could be right and the great von Neumann could be wrong. It was only in 1966, when the physicist John Bell, who had a reputation to match that of von Neumann, established why the von Neumann proof was not just flawed but silly, that the physics community began to take notice of the mistake. One reason for the delay may have been because the mistake was so elementary that, even now, it is hard to see how a mathematician of von Neumann's stature could have blundered so blatantly. In an interview with *Omni* magazine (May 1988), Bell later said 'the proof of von Neumann is not merely false but *foolish!*'

Bell was astonished that anyone had ever been taken in because the proof rests upon a basic mistake involving commutation (see *Abelian group*). Von Neumann had used the fact that a particular property of a quantum system obeys the commutative rules on average, and then applied the rule to individual components of the quantum system. This is as daft as saying that, if the average height of a class of schoolchildren is 1.2 m, therefore the height of each child is 1.2 m.

The unfortunate consequence of this mistake persisted even after Bell spoke out about it in 1966, and the claim that hidden variables theories cannot work because von Neumann said so can still be found in some textbooks today. This is why the investigation of hidden variables versions of quantum theory only really began to take off in the 1990s, and is now one of the most exciting areas in the development of models of the quantum world.

neutral current An interaction in which no charge is carried by the *boson* that mediates the interaction. Strictly speaking, the term can apply to the electromagnetic interaction, which is mediated by uncharged photons; the term is more familiar, though, in the context of the weak force, where some interactions are mediated by the Z particle. These particles were first observed in 1973, and provided important confirmation of the accuracy of the gauge theory of the weak interaction. See also *charged current*.

neutrino One of the *elementary particles*, a lepton with zero charge, spin ½ and extremely small (possibly zero) mass. Neutrinos interact with other particles only through the weak interaction and (if they do have mass) gravity. They come in three varieties, each associated with an electron-like lepton: the electron neutrino, the muon neutrino and the tau neutrino.

The need for neutrinos (strictly speaking, for electron neutrinos) was first pointed out by Wolfgang Pauli, in 1930, to explain where the 'missing' energy in *beta decay* was going (see *forces of nature*). The idea that the energy was being carried off by an otherwise undetectable particle did not catch on at first, in part because only two fundamental particles (the proton and the electron) were known at the time; the name proposed by Pauli, neutron, was soon used for another particle, similar to the proton but without any electric charge, identified by James Chadwick in 1932.

Soon after the discovery of the neutron, Pauli's idea was taken up by Enrico Fermi and developed in the context of a more comprehensive theory of the weak interaction. When asked if Pauli's and his neutral particle was the neutron discovered by Chadwick, he replied 'no, Pauli's is only a neutrino'. The name stuck.

Neutrinos are produced in large quantities in nuclear fission reactions, and the first neutrinos were identified in experiments carried out in 1956 alongside a nuclear reactor at Savannah River in the United States (strictly speaking, these were electron antineutrinos). At first, it was assumed that neutrinos had zero mass. In the 1980s, however, cosmologists became increasingly convinced that some form of *dark matter* dominates the Universe, providing more mass overall than all the bright stars and galaxies put together. Neutrinos left over from the Big Bang fill the Universe in enormous quantities, and could provide a significant proportion of the dark matter if each neutrino has even a tiny mass, a few to a few tens of electron volts. Stimulated by these cosmological speculations, particle physicists have carried out several experiments designed to detect even such a small trace of mass associated with individual neutrinos (for comparison, the mass of an electron is 500,000 eV). The results are not yet conclusive, but hint that electron neutrinos do have masses in the range required by the cosmologists (in round numbers, somewhere between 0.5 eV and 5 eV, which means between 1 millionth and 1 hundred-thousandth of the mass of an electron). The other varieties could have larger masses (still small compared with the proton mass), but do not exist in such copious quantities as electron neutrinos.

neutrino oscillations   Several different models suggest that it may be possible for neutrinos to change from one flavour to another – for example, electron neutrinos might become tau neutrinos. The particles could then change back into the original flavour (or into the third flavour), oscillating between the allowed possibilities. Such oscillations from one variety of a particle (one member of a family) to another are known to occur with other quantum entities, but they are possible only if the particles involved have mass; no matter how small that mass is, it cannot be zero. So one of the key tests for neutrino mass is to monitor a beam of one kind of neutrons produced in a collider experiment to see whether other flavours of neutrino are appearing in the beam (or, equivalently, if the original flavour disappears from the beam as the neutrinos oscillate into other varieties).

By the middle of 1996, an experiment at Los Alamos that had been running for three years had produced evidence for 22 events caused by electron neutrinos in a beam of supposedly pure muon neutrinos – some of the best experimental evidence, up to that time, that muon neutrinos can oscillate to become electron neutrinos, and suggesting that one of these two flavours of neutrino has a mass of a few tenths of an electron volt.

neutron   one of the *elementary particles*, a *baryon* with a rest mass of $1.6749542 \times 10^{-27}$ kg (= 939.5729 MeV), zero electric charge and a spin of ½, which obeys *Fermi–Dirac* statistics. See also *atomic physics*.

neutron decay = *beta decay.*

neutron diffraction Demonstration of the wave nature of a beam of neutrons, which can be made to diffract in the equivalent of the double-slit experiment and other classic demonstrations of the wave nature of light. Neutron diffraction was demonstrated particularly clearly in a series of experiments carried out at the end of the 1970s by Tony Klein and his colleagues at the University of Melbourne, confirming the dual wave–particle nature of the neutron.

neutron star A star made almost entirely of neutrons, containing roughly as much mass as our Sun but packed into a sphere only about 10 km across, with the density of an atomic nucleus. Although theory had predicted the existence of such stars, the concept was not taken entirely seriously by astronomers until they were discovered (accidentally), in the form of pulsars, in the second half of the 1960s.

The discovery was important to quantum physics, not just to astronomy, because it confirmed the accuracy of those earlier predictions, which had been made on the basis of calculations of the quantum properties of stellar material after the nuclear fusion reactions that keep a star like the Sun shining had stopped.

The theory said that, with no internal source of energy to provide a pressure holding itself up against the inward tug of gravity, a star at the end of its life would shrink down into a ball of material in which individual nuclei (chiefly of carbon) produced by nuclear fusion during its lifetime would be embedded in a sea of electrons. In this state, the electrons are as close together as they are allowed to be by the quantum rules (remember that electrons are *fermions*, and no two fermions can occupy the same quantum state), in a so-called degenerate state – the star is said to be held up by degeneracy pressure. Such a star is known as a white dwarf. A white dwarf with the same mass as our Sun would be about the same size as the Earth; many such stars are known, and their properties provide confirmation of the accuracy of the calculations based on quantum theory.

But if the star has more than a certain amount of mass, known as the Chandrasekhar limit and equal to about 1.4 times the mass of our Sun, the gravity of the star forces the electrons to combine with protons in the nuclear material, forming neutrons through inverse beta decay. Such a star will collapse further to form a neutron star, where once again it is supported by quantum degeneracy pressure, but this time operating between the neutrons. Again, the observed properties of neutron stars (in particular, their masses and the absence of any white dwarf stars with masses greater than the Chandrasekhar limit) confirm the accuracy of the quantum calculations.

There has been speculation that at the heart of a neutron star matter might be broken down still further to produce a quark core, but this has not been proven.

If any star has more than about three times the mass of our Sun (the Oppenheimer–Volkoff limit) at the end of its life, it will collapse still further to become a black hole, as gravity overwhelms all quantum effects and matter is crushed out of existence.

Newlands, John Alexander Reina (1837–1898) British chemist who developed the idea of a periodicity in the properties of the chemical elements independently of (and earlier than) Dmitri Mendeleyev, but whose ideas were not accepted at the time.

Born in Southwark, London, on 26 November 1837, Newlands was largely educated at home (by his father, a Presbyterian minister) until 1856, when he spent a year studying under August Hoffmann at the Royal College of Chemistry in London. He worked at the Royal Agricultural Society as a chemist until 1864 (taking time off to fight with Garibaldi in Italy in 1860; Newlands' mother was of Italian descent), then spent four years working independently as an analytical chemist and teaching chemistry privately. In 1868 he became chief chemist in a sugar refinery, but when the business went into decline he once again became an independent analytical chemist, this time working with his elder brother. He died in London on 29 July 1898.

Newlands' work on the periodic relationship between the properties of the chemical elements was carried out between 1863 and 1866, and produced the first periodic table of the elements in 1865. He pointed out that when the elements are arranged in order of increasing atomic weight, 'the eighth element starting from a given one is a kind of repetition of the first', which he called the Law of Octaves; but when he gave a paper on the subject to the Chemical Society in 1866 the idea was so ridiculed by establishment critics (including the professor of physics at University College, London) that he gave up working on the subject until after Mendeleyev published his more complete and better-thought-out periodic table in 1869. Newlands' key paper on the subject, rejected in its original form by the Chemical Society in the mid-1860s, was not published until 1884. His contribution was eventually recognized by the award of the Davy Medal of the Royal Society (which never made him a Fellow) in 1887.

Newton, Isaac (1642–1727) English physicist who laid the basis for the development of modern science with his three laws of mechanics and his theory of gravity. He also made important contributions to optics, and invented calculus. Newton is widely regarded as the greatest scientist who ever lived, not least because in addition to making these great discoveries he had to establish the proper technique for making scientific discoveries, by experiment and observation.

He was born at Woolsthorpe, in Lincolnshire, on Christmas Day 1642. At that time, Britain was still using the Julian calendar, but the modern Gregorian calendar had already been introduced in Catholic countries, and on that calendar (the one we use today) Newton was born on 4 January 1643. Newton's father, a modestly successful farmer, died before the baby was born; his mother remarried when the infant was three years old, and until 1658 Isaac was brought up by his grandmother. His mother was then widowed for a second time, and Isaac was taken back into her home. He was supposed to take over the farming of her land, but showed no aptitude for this (mainly because he was not interested) and was lucky enough (thanks to the intervention of an uncle) to be allowed to go back to school to prepare for a university education. He went up to Cambridge in 1661 and graduated in 1665.

Newton intended to stay on in Cambridge, but just after he graduated the university was closed because of the threat of plague, and he spent the next eighteen months back in Woolsthorpe, where he thought long and hard about many things (including gravity) and carried out experiments on light using prisms. He returned to Cambridge after the plague scare was over, became a Fellow of Trinity in 1667 (at the age of 24), and was appointed Lucasian Professor of Mathematics in 1669.

Newton retained his Cambridge appointments until 1701, but had essentially

Isaac Newton (1642–1727).

ceased his scientific work in 1696, when he was made warden of the Royal Mint and moved to London. He became master of the Mint in 1698, and pushed through a much needed reform of the currency and recoinage, a job he carried out with ruthless efficiency. It was for this work, not his scientific studies, that he was knighted in 1705, by Queen Anne. From 1703 until his death (in London, on 20 March 1727), Newton was president of the Royal Society. He was buried in Westminster Abbey.

Newton was a difficult man, secretive and concerned that his ideas might be stolen by others; he worked alone and often did not publish his findings until pressed to do so. He also spent a lot of time carrying out alchemy experiments, which were still perceived as an acceptable part of science at that time. His greatest work, *Philosophiae Naturalis Principia Mathematica* (always known as the *Principia*) was only published (in 1687) at the urging of Edmond Halley, who had discovered that Newton had proved that the orbits of the planets around the Sun could be explained only by an inverse square law of gravity. The discovery probably dated back to the plague year, when Newton was in Woolsthorpe, but he had never made it public. In his notebooks, written much later (1716), Newton told how he had worked on mathematical problems, what is now called calculus, his theory of colours *and* gravity in 1665 and 1666, 'for in those years I was in the prime of my age for invention, and minded Mathematics and Philosophy more than at any time since'.

What emerged at Halley's urging in the mid-1680s was a volume which provided not only a theory of gravity, but the laws of motion on which modern physics was built.

Publication of Newton's work on optics was also long delayed, this time because of a bitter argument he had with Robert Hooke, in the early 1670s, about the significance of his work and how much he owed (according to Hooke) to Hooke's earlier studies. It was this row which largely led to Newton withdrawing from scientific debate and hugging all his ideas to himself in Cambridge for the next fifteen years. Although he did publish the *Principia* when talked into it by Halley, Newton simply sat on the rest of his work on optics until Hooke died, in 1703, then published his book *Opticks* in 1704.

Newton's theory of light saw it as a stream of tiny particles, or corpuscles, like miniature cannon balls. The image held sway for a hundred years (in spite of the work of Christiaan Huygens), until the work of Thomas Young and Augustin Fresnel established the wave nature of light. In the 20th century, however, it became clear, in the context of quantum mechanics, that light has to be regarded as being both particle and wave.

But the single most important feature of Newton's work was not the specific form of his theory of gravity, or light, or the laws of motion. What mattered was that he showed that the laws of physics are universal – that it is the same law of gravity which makes an apple fall from a tree and holds the Moon in its orbit around the Earth and the Earth in its orbit around the Sun; that all those moving objects all obey the same laws of mechanics; and that these are the same laws that apply to balls rolling down inclined planes in a laboratory on Earth. It was Newton who made the Universe the physicists' plaything, and it is thanks to the success of the Newtonian approach, over more than 300 years, that physicists today talk confidently about what went on in the Big Bang, and about what goes on as quarks exchange gluons inside the proton.

newton   The SI unit of force, named after Isaac Newton. 1 N is the force required to give a mass of 1 kg an acceleration of 1 ms^{-2}.

Newtonian gravity   The description of the gravitational interaction developed by Isaac Newton, centred on the inverse square law, which says that the force of attraction between any two masses, anywhere in the Universe, is proportional to the product of the two masses (m_1 multiplied by m_2) divided by the square of the distance (r) between them. The constant of proportionality in the equation (the gravitational constant) is written as G, so that the equation describing the force f becomes

$$f = Gm_1m_2/r^2$$

It is implicit in this equation that the distances are measured from the centres of the objects involved, as if, in each case, all their mass were concentrated at a point. In his *Principia*, Newton proved that this is so for spherical objects; that proof is as important to the successful application of Newtonian gravity as the inverse square law itself.

Newtonian gravity is not a complete description of how things behave gravitationally where very strong gravitational *fields* are involved, and the general theory of relativity (see *relativity theory*) comes into its own under those conditions. Both Newtonian gravity and the general theory of relativity are classical theories (see *classical mechanics*), and as yet there is no fully satisfactory theory of *quantum gravity*, which would be needed to describe the behaviour of gravity when large masses are concentrated on very small scales, in particular at the birth of the Universe.

Newtonian mechanics = *classical mechanics*. See also *Newton's laws of motion*.

Newton's laws of motion The three universal rules which govern the workings of the everyday world (together with gravity and electromagnetism; see *forces of nature*). They were presented to the world by Isaac Newton in the *Principia*, published in 1687.

The first rule is that any (and every) object stays still or moves in a straight line at a constant speed unless it is acted upon by a force. This is counter-intuitive (that is, it doesn't match up with everyday common sense), because although we see that the objects around us are indeed quite happy to sit still in one place, and we have to give an object a push to start it moving (which does seem to obey Newton's first law), it soon comes to a halt again, seemingly of its own volition (which does not seem, at first sight, to obey the first law). But in the everyday environment, on the surface of the Earth, motion is always opposed by friction, a force caused by things rubbing against one another. So the slowing down of a moving object is, in fact, precisely obeying Newton's first law. Newton's first law superseded the 'commonsense' idea, formalized by Aristotle in the 4th century BC, that a force is required to keep something moving. Among other things, this removed the need to find a force (angels?) to push the planets along in their orbits; but it was still necessary to find a force to hold them in elliptical orbits, in spite of their natural tendency to move in straight lines.

Billiard balls rolling on a smooth table are slowed by friction relatively quickly; the puck on an air table used for a game of table hockey is suspended on a cushion of

Newton's laws of motion. A rocket, such as the Space Shuttle, moves forward because it is pushing material from its exhaust backward. Action and reaction are equal and opposite, so the momentum of the massive rocket moving relatively slowly is balanced by the momentum of the light exhaust gases moving very quickly in the opposite direction.

air and moves almost without friction, giving a fair example of Newton's first law (and, indeed, his second and third laws) at work. And today most people have seen pictures from manned spacecraft in orbit around the Earth, where moving objects travel through the cabin of the spaceship in straight lines and bounce off the walls in perfect agreement with Newton's laws, affected only by the tiny friction of their passage through the air.

But Newton had never played table hockey, nor seen how things move inside a spaceship in free fall. His genius was to extrapolate from the messy conditions that exist on the surface of the Earth to see the underlying simple truth.

Newton's second law says that, when a force is applied to an object, this changes the **momentum** of the object in a precise way. (Since the mass of the object can be assumed to stay constant in most simple examples, such as a collision between two billiard balls, this is equivalent to saying that the velocity of the object changes in a precise way.) The rate at which the momentum (velocity) changes is proportional to the force applied, so that twice as much force produces twice as much rate of change in velocity; and the direction in which the change in momentum occurs is in the direction of the applied force. Since the rate of change of velocity is called acceleration, this is equivalent to saying that a force F applied to an object with mass m gives it an acceleration a in the direction in which the force is applied, so that $F = ma$. This distinguishes the mass of an object (its **inertial mass**) from its weight; the mass is measured by the amount of acceleration any given force produces; the weight represents the response of the mass to one particular force, the gravitational pull of the Earth (see also **gravitational mass**).

One important point about Newton's second law is that a change of direction is a change in velocity (an acceleration), even if the speed of the object remains unchanged. So an applied force may simply change the direction in which an object is moving, without changing its speed. This can happen, for example, if you tie a stone to a piece of string and whirl it round in a circle; the stone may be travelling at a constant speed, but its velocity is constantly changing to keep it moving in a circle, because there is a force acting along the string to your hand. Another point to notice is the effect of mass on the equation. A particular force applied to a small mass will give it a larger acceleration than when the same force is applied to a large mass – turning the equation around, the acceleration is equal to the force *divided by* the mass.

Newton's third law says that, whenever a force is applied to an object, the object pushes back with an equal and opposite force. Newton actually used the term 'action', so that his third law is sometimes still quoted in the form 'for every action there is an equal and opposite reaction'; but it is better to avoid using that word in this way today, since **action** is now used by physicists (especially quantum physicists) to mean something completely different.

You can see Newton's third law at work on the billiard table (if you make allowance for the effect of friction). When one moving ball strikes another, the second ball is deflected at an angle, while the first ball is itself deflected in the opposite direction. Both balls change their direction (and therefore the velocity and momentum, irrespective of whether the speed changes) as a result of a force acting on them. And in the example of the stone being whirled around on a piece of string, there is not only a force along the string pulling the stone inward and making it move in a circle, but an equal and opposite force tugging outward along the string, which you feel trying to move your hand outward. (That force is resisted by your muscles, and transmitted through your feet to become a force trying to jiggle the entire Earth about in response; technically, the Earth does move as a result, but by far too small an amount to notice!)

The same rules apply on the squash court. You hit the ball with the racket, and the ball flies off, while the racket pushes back on you and, through you, on the Earth, moving it a tiny bit the other way. The ball hits the wall, and its velocity is reversed, while the wall and, through the wall, the entire planet recoils in the opposite direction. It is because mass enters into the equation, and the mass of the Earth is, to say the least, considerably more than the mass of a squash ball, that you do not notice the recoil; but it is still there, just as in the case of two colliding billiard balls.

But one of the neatest examples of all of Newton's laws of motion, and his law of

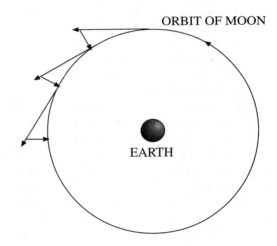

Newton's laws of motion. Newton explained how the Moon tries to move in a straight line, but is constantly being tugged towards the centre of the Earth by gravity. The repeated tugging bends the Moon's path into a closed orbit around the Earth.

ORBIT OF MOON

EARTH

gravity, is the one he used himself to explain how gravity holds the Moon in its orbit around the Earth. One of the most important features of Newton's *Principia* is that he proved mathematically that the force of gravity produced by an object like the Earth acts, if you are outside the Earth or on its surface, as if all of the mass of the planet is concentrated at its centre; so the famous inverse square law of gravity can be used with all measurements taken from the centre of the Earth and the centre of whatever other astronomical object we are interested in. Proving this is not easy, but once it is proved it makes calculations involving gravity and orbits a lot simpler. Newton also knew that the acceleration caused by gravity near the surface of the Earth will make any object (such as an apple falling off a tree) fall through a distance of 16 feet in the first second of its fall – we may as well use the same units that Newton did in this historic example, but if you want to make the conversion, 16 feet is just under 4.98 m. According to Newton's first law, the Moon should 'want' to travel in a straight line through space, at a constant speed. He reasoned that it must be being held in orbit around the Earth by the same force, gravity, that makes an apple fall to the ground.

In principle, this is easy to understand, as the diagram shows. Imagine that for a second the Moon does travel in a straight line. Then gravity gives it a tug towards the Earth. It moves in a straight line for another second, and gets another inward tug. If all the inward tugs happen with the right strength, the result will be to bend the motion of the Moon into a perfect orbit around the Earth. But did the numbers work out?

The Moon is 60 times further away from the centre of the Earth than we are, at the surface of the Earth. So, according to the inverse square law of gravity, the force acting on the Moon should be less than the force acting on an apple at the surface of the Earth by a factor of 60 squared, which is 3,600. In that case, in 1 second the effect of the Earth's gravity should be to make the Moon shift sideways in its orbit by a distance given by dividing 16 feet by 3,600. This works out at a little bit more than one-twentieth of an inch (in modern units, just under 1.4 millimetres). As Newton showed, for an object travelling at the speed of the Moon, at the distance of the Moon from the Earth, a sideways nudge of just this size, every second, is exactly what is required to make it travel in a closed orbit around the Earth, completing one circuit in just over 27 days – it travels at about 3,700 km per hour round an orbit just under 2,400,000 km long.

So we've seen the first and second laws of motion at work (as well as gravity), explaining the motion of the Moon. What about the third law? In fact, it isn't really true that the Moon is in orbit around the Earth. The Moon pulls back on the Earth, through gravity, with an equal and opposite force to the pull of the Earth on the Moon (in line with the third law), and as a result the two objects dance around each other, orbiting their mutual centre of gravity – if you like, the balance point in the Earth–Moon system. It is like trying to balance a see-saw with a child on one side and an adult on the other; the adult, being heavier (more massive), has to sit much closer to the balance point than the child. And because the Earth is much more massive than the Moon, the balance point for the Earth–Moon system actually lies below the surface of the Earth, so it is still a good approximation to say that the Moon orbits the Earth. Nevertheless, even in this famous example the effect of the third law is still there, wobbling the Earth from side to side during the course of a month.

Newton's laws really do explain everything that is going on in the world above the quantum level – and they certainly cannot be ignored even when we are dealing with quantum entities such as photons and electrons; it's just that at the quantum level there are other rules which also have to be taken into consideration.

Newton's rings   Circular pattern of concentric rings, now explained as *interference* fringes, produced when a convex lens is placed on a flat glass plate so that light is reflected from the two surfaces, which are close to each other near the point of contact. If white light is used, it produces a pattern of coloured rings; if monochromatic light is used, it produces a pattern of light and dark rings, with a dark spot in the middle.

The phenomenon is named in honour of Isaac Newton, who observed it when he placed the object lens of a telescope on to a flat glass plate, while he was working at home in Woolsthorpe in the mid–1660s, when Cambridge University was closed because of the risk of plague. He studied the phenomenon in detail in the 1670s, but did not publish a full account until 1704, in his *Opticks*, because of his row with Robert Hooke.

Although Newton favoured the idea that light was carried by a stream of tiny particles, he appreciated that the regular spacing of the rings could be explained only in terms of some sort of vibration or wave, which he suggested was a vibration of 'the aether' caused by the passage of corpuscles of light through it. The rings are now interpreted as a straightforward interference effect of light waves reflected from the curved surface of the lens and light waves reflected from the flat surface of the glass. The spacing of the rings depends on the wavelength of the light involved.

Nishijima, Kazuhiko (1926–) Japanese physicist who arrived at the idea of *strangeness* independently of Murray Gell-Mann in 1953.

Born on 4 October 1926, Nishijima studied at the University of Tokyo (BSc 1948) and at the University of Osaka (PhD 1954). He worked at the University of Osaka until 1959, when he became a professor at the University of Illinois, Urbana. In the 1950s, he also spent brief periods as a visitor at the Max Planck Institute in Göttingen and at the Institute for Advanced Study in Princeton.

In 1967 he returned to Japan as professor of physics at the University of Tokyo. As well as his discovery of strangeness, in 1957 Nishijima predicted that there must be a second kind of neutrino, to accompany the muon in the same way that the electron neutrino accompanies the electron.

NMR = *nuclear magnetic resonance.*

Noether, (Amalie) Emmy (1882–1935) German mathematician who made important contributions to the development of the concept of symmetry groups and non-commutative (non-Abelian) fields, and worked with David Hilbert. Born in Erlangen on 23 March 1882, she was the daughter of the mathematician Max Noether (1844–1921), who was then professor of mathematics at the University of Erlangen. Her brother was also a mathematician. Emmy Noether studied at Erlangen and at the University of Göttingen, and was awarded a PhD in 1907, after a long struggle to be allowed to register as a research student even though she was a woman.

Because there were no academic posts for women in Germany at that time, she worked independently for several years, but sometimes stood in for her father to give lectures at Erlangen. In 1915 she was invited by Hilbert to give a series of lectures in Göttingen. Hilbert was so impressed with her that he wanted her to stay as a professor. At the time, still, only men were allowed to hold such posts there, but she was tolerated in a semi-official capacity (this echoes the treatment of Marie Curie in Paris a little earlier). Her key work on non-commutative algebras was largely carried out in Göttingen, from 1927 onwards. As well as publishing her own work, she was at the centre of the thriving mathematics community there at that time, and made many contributions to the work of her colleagues and students, pointing the way when they were stuck with a problem. When the Nazis came to power in 1933, Noether, a Jew, fled to the United States, where she worked briefly at Bryn Mawr College and at the Institute for Advanced Study in Princeton. She died from an infection following an operation, at Bryn Mawr on 14 April 1935. Emmy Noether is generally regarded as the greatest woman mathematician of all time.

Noether's theorem See *group theory.*

noise Any kind of unwanted background interference which tends to obscure the phenomenon being studied.

non-Abelian gauge theory A *gauge theory* in which the elements of the appropriate group (see **group theory**) do not commute. Many features of the quantum world are described by this kind of non-commutative process; in simple terms, this means that if you carry out two operations A and B one after the other then the end product depends on the order in which the two separate operations are carried out.

Don't be put off by the unfamiliarity of the words; this is actually a very simple concept which is frequently encountered in everyday life. Cookery is often non-Abelian. If a recipe for making a cake tells you at one particular point in the recipe to add 250 millilitres of water and bake for 40 minutes at a certain temperature, you had better be sure that you do the two operations (adding water and baking) in the specified order; if you start out with the same mixture but bake for 40 minutes at the specified temperature and then add the specified amount of water, you will end up with something quite different. See also **Abelian group**.

non-locality Term used to describe the way in which the behaviour of a quantum entity such as an electron is affected not only by what is going on at one point (the 'locality' of the entity), but also by events that are going on at other places (other localities), which may in principle be far away across the Universe. These non-local influences occur instantaneously, as if some form of communication, which Albert Einstein called a 'spooky action at a distance', operates not just faster than the speed of light, but infinitely fast.

It is important to appreciate that the non-local nature of the quantum world has been demonstrated in experiments. In the **double-slit experiment**, for example, the behaviour of an electron passing through one of the slits depends on whether the other slit is open or closed; and the **Aspect experiment** uses **Bell's inequality** to show that non-locality cannot be avoided if we want to treat things like electrons as real objects which exist independently of whether or not they are being observed.

See also **hidden variables, transactional interpretation**.

NORDITA Acronym for the Nordic Institute for Theoretical Atomic Physics, based in Copenhagen.

normalization A standard technique used by physicists and mathematicians to scale the numbers they are working with to a convenient size – for example, to make it easy to plot the numbers on a graph. If you divide each member of a set of numbers by the smallest number in the set, they are all normalized so that the smallest number becomes 1, but the relationships between the numbers (in particular, their ratios) are unchanged; similarly, they can be normalized by dividing each number by the largest number, so that the range of numbers after normalization only goes up to 1 (we are using the word 'set' here in its everyday sense, to mean a collection of numbers, although there is no reason why the numbers could not form a set in the rigorous mathematical sense). Conceivably, you might want to normalize the set of numbers by dividing by some other number that is not in the set – for example, π – in order to get rid of this common factor and highlight some feature of the relationship between the numbers. Normalization is a straightforward, common and uncontroversial aspect of physics. But see **renormalization**.

November revolution Slightly jokey name (the punning reference is to the Russian revolution of 1917, which occurred on 7 November on the Gregorian calendar) given by particle physicists to the events of November 1974, when two separate teams, one headed by Samuel Ting and the other by Burton Richter, announced that they had independently discovered the particle now known as the J/psi (sometimes just as psi). Richter and Ting shared the Nobel Prize for the discovery just two years later, in 1976 – one of the swiftest such awards ever made, and an indication of the importance of their discovery.

Before the November revolution, the known particles could be explained in terms of different combinations of just three quarks, dubbed up, down and strange. But the unified theory of the electroweak interaction, developed at the beginning of the 1970s (see *forces of nature*), required the existence of a fourth kind of quark, dubbed charm, which had never been detected. The properties of the J/psi particle showed that it is a bound state of a charm–anticharm quark pair. This showed that quarks come in 'generations', with up/down forming one generation and charm/strange forming another. A third generation, top/bottom, has since been identified. Together with the corresponding pairs of *leptons*, these six particles and their interactions form the basis of the *standard model* of particle physics.

In his book *Inward Bound*, physicist and historian Abraham Pais describes the situation in physics following the announcement of the discovery of the J/psi as 'general pandemonium', and says that the turmoil continued for a few months until it became clear that the new discovery could be explained by charm. This 'strongly reinforced belief in QCD [*quantum chromodynamics*] *and* in electroweak theory'. It was in 1975 that the four-quark theory (as it then was) first became known as the 'standard model'.

Novosibirsk   City in Siberia which is the location of a major Russian accelerator and particle physics research centre.

nuclear energy   Energy released during the processes of *fission* and *fusion*, as a result of mass being converted into energy in line with Albert Einstein's equation $E = mc^2$. See also *nuclear power*.

nuclear fission   See *fission*.

nuclear forces   See *forces of nature*.

nuclear fusion   See *fusion*.

nuclear magnetic resonance   The selective absorption of specific frequencies of very short-wavelength radio waves by some atomic nuclei in a magnetic field. The effect occurs because in some nuclei there are unpaired protons or neutrons which act like tiny magnets spinning in the magnetic field (see *magnetic moment*). This makes the whole nucleus precess, like a wobbling spinning top. If the natural precession frequency is the same as the frequency of the radio waves fired at the material, the nuclei will absorb energy from the radio waves.

First observed in 1946, the phenomenon is used to measure magnetic moments; these measurements also give information about the electromagnetic environment that the nuclei are sitting in, and therefore about the molecular structure of the material. NMR can now be applied in medicine, to obtain 'pictures' of the inside of a person's body by observing the magnetic moments of hydrogen nuclei in the water and fats of the body. Unlike other ways of seeing inside the body (such as X-rays, or surgery) there are no known health hazards associated with NMR. NMR is also used in chemical analysis, a technique known as NMR spectroscopy.

See also *Rabi, Isidor Isaac*.

nuclear physics   The investigation of the structure and behaviour of atomic nuclei, and the way they interact with particles and with each other. This branch of physics began in 1911, when Ernest Rutherford explained the way in which alpha particles are scattered by atoms in terms of a model for the atom in which a small, dense central nucleus (he actually gave it that name the following year) is surrounded by a large, tenuous cloud of electrons. The structure of the nucleus itself was later probed by beams of particles, such as electrons. Practical applications of nuclear physics include the preparation of radioactive isotopes for use in industry and medicine, and the development of nuclear weapons and nuclear reactors.

nuclear power   Power (in the everyday sense, meaning energy) generated (usually in the form of electricity) from heat released as a result of nuclear reactions. So far (1997), all commercial nuclear power stations rely on *fission* to generate power; it is widely hoped that *fusion* may one day provide a clean and reliable source of energy for society.

nuclear reaction   A reaction which directly involves one or more atomic *nuclei*. This includes radioactive decay, when a nucleus emits particles such as electrons and alpha particles and is transformed into another kind of nucleus, *fission*, *fusion* (including the processes that keep the Sun shining), and interactions that result when nuclei are bombarded with other particles, such as neutrons.

nuclear reactor   An apparatus in which *nuclear power* is generated in a controlled way.

nuclear weapons   Devices in which *nuclear energy* is released explosively. In the misnamed 'atomic bomb' (really a nuclear bomb), the energy source is *fission*; in the so-called hydrogen bomb, although fission is used as a trigger, the main source of energy is *fusion*.

nucleon   Generic name for the two kinds of particle found in the nucleus of an atom, the proton and the neutron. Protons and neutrons may be referred to as nucleons even when they are not actually inside a nucleus, just as a European does not stop being a European if he or she travels to Brazil.

nucleon number   See *mass number*.

nucleosynthesis   The process by which heavier elements are built up from nuclei of hydrogen, initially in the Big Bang in which the Universe was born, and later inside stars.

Because the Universe is seen to be expanding today, because the Universe is filled with microwave background radiation left over from a fireball era, and because the equations of the general theory of relativity automatically provide a description of expanding spacetime, astronomers know that the Universe as we see it now emerged from a very hot, very dense state, the Big Bang. Although there is still debate about how the seed of the expanding Universe formed, the known laws of physics, tested in accelerator experiments and other laboratories on Earth, can be applied to everything that has happened in the expanding Universe since it had roughly the density of an atomic nucleus. This was less than 1 hundred-thousandth of a second after 'the beginning', calculating from the time when, if the present expansion were wound backwards, all of the present visible Universe would have been contained in a single mathematical point.

At still earlier times, conditions in the early Universe were so extreme (in particular, it was so hot) that matter and energy were interchangeable, in line with

Albert Einstein's equation $E = mc^2$, and particles such as electrons, protons and neutrons were manufactured out of pure energy. Because of a tiny imbalance in the laws of physics (see **CP conservation**), although these particles were mostly made in particle–antiparticle pairs (such as electron–positron pairs) which promptly annihilated one another, for every billion or so antiparticles that were created there were about a billion and one particles. So at the end of this era, a one in a billion residue of protons, neutrons and electrons was left embedded in a sea of hot radiation. About 1 hundred-thousandth of a second after the beginning, this was at a temperature of about 100 billion kelvin.

Over the next 3 minutes and 46 seconds, as the Universe expanded and cooled, primordial protons and neutrons were able to fuse together to make nuclei of helium (specifically, helium-4). Because neutrons themselves are unstable particles which, when outside a nucleus, decay with a half life of a few minutes, by the time this process was complete there were only enough neutrons around for about 25 per cent of the baryonic mass of the Universe to end up in helium nuclei; the rest remained as single protons, nuclei of hydrogen. There were also traces, but no more than traces, of very light elements such as deuterium manufactured in the Big Bang fireball. Less than 4 minutes after the beginning, Big Bang nucleosynthesis was over.

So when stars began to form, the raw material that went into them was about 75 per cent hydrogen, 25 per cent helium and a fraction of 1 per cent of other very light

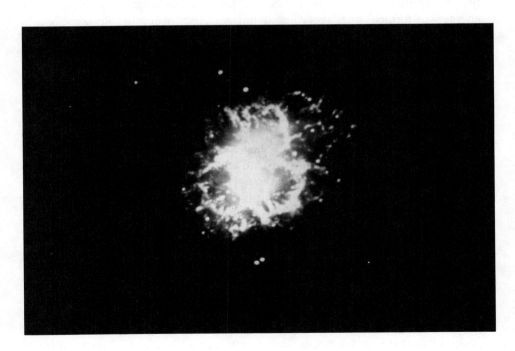

Nucleosynthesis. When a large star explodes as a supernova, elements manufactured inside the star during its lifetime, and in the supernova explosion itself, are scattered through space and help to form the raw material of later generations of stars, and planets. This is one still-expanding cloud of supernova debris, the Crab Nebula, formed in the explosion of a supernova seen from Earth in 1054 AD.

elements. This was (and still is, since new stars continue to form today) the raw material for stellar nucleosynthesis. In the conditions at the heart of a star, it is relatively easy for hydrogen nuclei to be converted into nuclei of helium; this is the process which generates energy in most stars (including the Sun) even today. Later in the life of a star, there is plenty of helium around (in the form of nuclei of helium-4, alpha particles) to provide the building blocks for further nucleosynthesis. But the nucleus obtained by sticking two alpha particles together, beryllium-8, is extremely unstable, and breaks apart within 10 millionths of a second after it forms. In a *tour de force* paper published in the 1950s, Fred Hoyle and his colleagues showed that it is possible, under the conditions which exist inside a star older than our Sun, for three alpha particles to combine almost instantaneously to make one nucleus of carbon-12, which is stable.

The trick depends on the fact – predicted by Hoyle and confirmed in experiments carried out at the Kellogg Radiation Laboratory at Caltech – that a carbon-12 nucleus can exist in a specific excited state, known as a resonance. This is analogous to the way an atom can absorb a photon and move to a higher energy state, in that case because one of its electrons has moved to a higher energy level. Just as the atom can emit a photon and fall back into the ground state, so an excited carbon-12 nucleus can emit a much more energetic photon and fall back into its ground state.

All this was well known by the 1950s; Hoyle's unique insight was to appreciate that there must be an energy state at exactly the right level in carbon-12 to match the energy state formed when a beryllium-8 nucleus 'absorbs' another alpha particle. In the so-called triple alpha process, adding a third alpha particle to an unstable beryllium-8 nucleus does not, as you might expect, blow it apart, but provides just the right amount of energy to ease the combined nucleus into the excited carbon-12 state – if you like, slotting it into place on a high step of the energy staircase appropriate for carbon-12.

You can extend the staircase analogy. If you try to throw a bouncy ball from the bottom of the staircase on to a high step, it will tend to bounce off and end up somewhere else. But if you throw the ball with exactly the right speed, it will be slowed by gravity as it moves upwards and will arrive at the chosen step with zero vertical velocity, exactly at the top of its trajectory. Then it can settle gently on to the chosen step without bouncing off. This is like the way in which the triple alpha process eases three helium-4 nuclei into an excited state of carbon-12.

Once a star contains carbon-12, further nucleosynthesis is relatively straightforward. The addition of another alpha particle produces oxygen-16, and similar fusion processes readily build up elements such as neon-20, magnesium-24 and silicon-28. At each step, the process releases energy because the mass of the end product (for example, neon-20) is slightly less than the mass of the components from which it is made (in this case, oxygen-16 and helium-4). Nuclides with intermediate masses are produced by radioactive decay processes that eject electrons or (more rarely) positrons from the nuclei, but nuclides which contain, in effect, whole numbers of alpha particles remain the most common.

The final step in this process occurs when pairs of silicon-28 nuclei fuse to form elements of the iron group, including iron-56 itself, cobalt-56 and nickel-56. Elements heavier than iron can be made only if there is an input of energy, because they store energy less effectively than iron-group elements; in effect, energy has to be put in to make some additional mass to manufacture these heavier elements. (This, of course, is why energy is released if one of the heavier nuclei, such as uranium-235, fissions; now, the mass of the components is slightly less than the mass of the single nucleus.) To some extent, the build-up of nuclides heavier than iron-56 goes on all the time inside old stars (known as red giants), where energetic neutrons produced as by-products of the various fusion reactions that we have described can be captured one at a time by nuclei to build up elements from iron-56 to bismuth-209. But if a nucleus of bismuth-209 captures a neutron, it promptly splits apart through alpha decay.

Heavier elements still, and many nuclides in the mass range between iron-56 and bismuth-209, are manufactured during supernova explosions, when the core of an old, massive star collapses, releasing a great deal of gravitational energy as it does so. This violent process breaks many nuclei apart, producing a flood of very energetic neutrons which can be captured in rapid succession by other nuclei. Many of the neutron-rich nuclei produced in this way then decay, through beta decay or alpha decay, or fission, to produce a variety of heavy elements.

All of these processes are thoroughly understood because each of the individual nuclear interactions involved can be studied in particle accelerator experiments here on Earth. Data from these experiments can then be combined with an understanding of how stars work to make predictions about how effective nucleosynthesis should be in producing particular elements (indeed, particular isotopes of those elements). The predictions of models of nucleosynthesis based on the experiments closely match the observed abundances of elements in the Universe at large, and the behaviour of supernova explosions also closely matches the predictions of astrophysical calculations which draw on particle physics data from experiments here on Earth.

There is a very close link between particle physics and astrophysics, and the implication is that astrophysicists do indeed have a very good understanding of what went on during the first few minutes of the birth of the Universe, and what goes on inside stars today. It also means that every nucleus of every atom of elements heavier than hydrogen and helium (apart from those tiny traces of deuterium and the like) has been manufactured inside stars and scattered through space in supernova explosions to form the raw material of future cosmic generations of stars, planets and even people.

Further reading: John Gribbin, *Companion to the Cosmos*.

nucleus   The central part of an atom. The nucleus contains most of the mass of the atom (in the form of protons and neutrons) and all of its positive electric charge. It is surrounded by a cloud of electrons, which carry a little mass (about 0.05 per cent of the mass for hydrogen; less than 0.025 per cent of the mass for all other atoms) and all of the negative charge associated with the atom. Nuclei have diameters in the range from 10^{-15} m to 10^{-14} m, while an atom has a diameter of about 10^{-10} m, between 10,000 and 100,000 times larger than the nucleus. If the nucleus were a few millimetres across, the diameter of the whole atom would be a few hundred metres across – roughly the size to fit on to a football pitch.

The term 'nucleus' was chosen (by Ernest Rutherford, in 1912) in conscious mimicry of the use of the word (which comes from the Latin for 'little nut') to refer to the central part of a cell in biology. Where there is scope for confusion, it is best to refer specifically to the 'atomic nucleus'; but that problem does not arise in this book.

nuclide   A specific atomic nucleus, defined by the number of protons and number of neutrons it contains (and, strictly speaking, by its energy state). For example, He-3 (two protons and one neutron) and He-4 (two protons and two neutrons) are both nuclides of helium. If the nuclides have the appropriate number of electrons associated with them to make complete atoms, they are called *isotopes*. The two terms are often used interchangeably.

Oersted, Hans Christian (1777–1851)   Danish physicist who discovered that an electric current produces a magnetic field.

Born at Rudkøbing on 14 August 1777, Oersted was the son of an apothecary and studied at the University of Copenhagen, where he was awarded a degree in pharmacy in 1797 and went on to complete a PhD in philosophy in 1799. After working for only two years in pharmacy, he toured Europe between 1801 and 1803, catching up with the scientific developments of his day, and on his return to Denmark he supported himself by giving public lectures and through journalism (a kind of Danish Carl Sagan) until he became professor of physics at the University of Copenhagen in 1806. In 1829 he became director of the Polytechnic Institute, also in Copenhagen, where he stayed until his death on 9 March 1851. A great popularizer of science, Oersted founded the Danish Society for the Promotion of Natural Science in 1824.

The work for which Oersted is now remembered was carried out between 1812, when he predicted that an electric current should produce a magnetic field, and 1820, when he first measured the effect. Unfortunately (but hardly surprisingly), his prediction was based on no real understanding of what was going on, but on the belief that all of the known forces must be interconvertible. This developed from the philosophy of Immanuel Kant and the idea that all of the forces of nature had a common origin, so Oersted was not so much wrong as about 150 years ahead of his time. Before his demonstration of the link, it had been accepted (except by Kantian philosophers) that electricity and magnetism were entirely separate phenomena.

The study of electromagnetism was then carried forward by other researchers, notably André Ampère and Michael Faraday. Oersted's own later work dealt with the compressibility of gases and liquids, and the phenomenon of thermoelectricity, where an electric current is generated by heat.

Ogawa, Hideki   See *Yukawa, Hideki*.

omega minus   A baryon with a mass of 1,672 MeV (more than 1.5 times the mass of the proton) and strangeness –3, predicted by the *eightfold way* model and first detected in experiments at the Brookhaven National Laboratory in 1964. It has a relatively long lifetime of nearly 10^{-10} sec, a charge of –1 (the same as the charge on an electron) and spin of ½ (so it is a fermion), and it is now understood to be made up of three strange quarks. The discovery of the omega minus and the confirmation that it provided of the validity of the eightfold way approach (based on group theory) was a key step in the development of the quark concept and the whole standard model of particle physics.

Onnes, Heike Kamerlingh (1853–1926)   Dutch physicist who carried out pioneering investigations of low-temperature physics.

Onnes was born at Groningen, where his father owned a tile factory, on 21 September 1853. He studied at the University of Groningen from 1870, and then (from 1871) in Heidelberg, where he was supervised by Robert Bunsen and Gustav Kirchoff. He returned to Groningen and completed the research for his PhD in 1876, but did not formally receive the degree until 1879. Between 1876 and 1882 Onnes worked at the Polytechnic in Delft, and it was during this time that he met Johannes van der Waals (at that time a professor in Amsterdam) and became interested, through discussions with van der Waals, in the possibility of investigating the behaviour of matter at very low temperatures. He became professor of experimental physics at Leiden University in 1882 and in 1894 he founded the cryogenic laboratory there, which became the leading centre for low-temperature research in the world.

Onnes was the first person to liquefy helium (in 1908), and he discovered the phenomenon of superconductivity in 1911. He was awarded the Nobel Prize for Physics in 1913. He died in Leiden on 21 February 1926.

ontological interpretation   See *Bohm, David Joseph*, and *hidden variables*.

operator theory   A version of quantum mechanics largely developed by Pascual Jordan in the late 1920s, in which the fundamental wave equation of Erwin Schrödinger's version of the theory is interpreted in terms of a mathematical expression known as an operator. A very simple example of an operator would be the familiar square root symbol ($\sqrt{}$); it indicates that the operation of taking the square root of the variable following the operator should be carried out.

Observable quantities such as position, energy and momentum are all replaced, in this model, by operators. An operator is something which changes a function that it operates on in a more subtle way than, for example, simple multiplication; the most important point in this context is that, like *matrices*, in general operators do not commute (see *Abelian group*). It turns out that the mathematics needed to describe the quantum world in this way had already been developed in the 19th century by William Hamilton (see *Hamiltonian*).

Oppenheimer, (Julius) Robert (1904–1967)   American physicist who is best remembered as the 'father of the atomic bomb', but who also made significant contributions to the development of quantum theory and astrophysics.

Born in New York City on 22 April 1904, Oppenheimer graduated from Harvard University in 1925 (having completed the four-year course in three years) and then visited the Universities of Cambridge and Göttingen (where he received his PhD in 1927) at the time when the breakthrough in developing quantum mechanics was being made; he worked alongside Werner Heisenberg, Paul Dirac and Max Born during these seminal years.

After completing his PhD, Oppenheimer spent two more years in Europe, at Leiden and Zurich, before returning home. Back in the United States, he brought the new ideas from Europe with him, and played a large part in introducing them to his American colleagues. In 1929 he took up simultaneous appointments as assistant professor at Caltech and at the University of California at Berkeley. He commuted between the two campuses for the next thirteen years, becoming a full professor (or rather, two full professors) in 1936. Oppenheimer made himself into a good teacher who influenced the development of a generation of American physicists.

Robert Oppenheimer (1904–1967).

In 1943 he was appointed as director of the new Los Alamos Scientific Laboratories in New Mexico, where much of the work leading to the development of the nuclear bomb (the Manhattan Project) was carried out. He returned briefly to California at the end of the Second World War, but became director of the Institute for Advanced Study, in Princeton, in 1947. He retired as director in 1966, but stayed on at the Institute as a professor until his death, from cancer of the throat, in Princeton on 18 February 1967.

One of Oppenheimer's key early contributions to quantum physics was to use Paul Dirac's equation of the electron to show, in 1930, that there should be a positively charged counterpart to the electron, with the same mass as the electron (crucially, Dirac himself had missed this point in his own work, and had suggested that the proton was the positively charged particle required by his equation). A few years later Oppenheimer, with his students in California, developed the mathematical foundation of the theory of black holes. But it was as a team leader and teacher that he made his most important contributions – on the West Coast in the 1930s, on the Manhattan Project in the Second World War, and then at Princeton.

Oppenheimer held left-wing political views and was a communist 'fellow traveller', although not a member of the Communist Party. As a result, he was victimized during the McCarthy era in the early 1950s, and lost the security clearance allowing him access to secret information. Partly as an open acknowledgement that Oppenheimer had been treated unjustly at that time, in 1963 President

Johnson conferred on him the Enrico Fermi Prize, the highest honour awarded by the US Atomic Energy Commission.

orbit The trajectory followed by an object moving in the gravitational field of another object – for example, the Earth moving around the Sun, or an artificial satellite moving around the Earth. The orbit need not be a closed path, as it is in the case of the Earth around the Sun; it can be a parabola or a hyperbola, as in the case of some comets passing by the Sun. By analogy with planetary orbits, the term is used in the *Bohr model* to refer to the hypothetical trajectories of electrons moving around atomic nuclei under the influence of the electric forces operating between the nuclei and the electrons. But this image was soon superseded by the idea of an *orbital*.

orbital The region around an atomic nucleus in which an electron in a particular quantum state (which means, essentially, one with a particular energy) is likely to be found. This replaces the idea from the *Bohr model* of an orbit, with the electron as a well-defined particle moving around that orbit. Because of quantum uncertainty and wave–particle duality, an electron associated with a nucleus in an atom can be regarded not as existing at any particular point, but rather as being constrained within a certain region, the orbital. For discussion of the chemical properties of atoms and molecules, the orbital can be regarded as a region in which the electric charge of the electron is distributed, as if the electron were smeared out (not necessarily evenly) over the entire orbital.

The shapes and orientations of the orbitals can be calculated using quantum mechanics, and their nature helps to explain many physical and chemical properties of atoms and molecules. The lowest energy state accessible to an electron in an atom corresponds to a spherical orbital (known as an *s* orbital) surrounding the nucleus. There is room for two electrons in such an orbital, with opposite spin to each other. At the next energy level, as well as an *s* orbital there is another kind of orbital with the same energy. The shape of an individual orbital of this second kind is like a double sphere, one on either side of the nucleus, like two tennis balls touching one another. Again, two electrons can 'fit' into each orbital, but each of the electrons occupies both of the spheres. In this case, however, there are three possible orientations for the orbital, at right angles to each other, so three of these orbitals (known as

Orbital. An *s* orbital (top left) surrounds the nucleus of an atom symmetrically, like a sphere. The *p* orbitals are like double spheres, on either side of the nucleus, and can be oriented in three different ways, at right angles to each other, which all have the same energy.

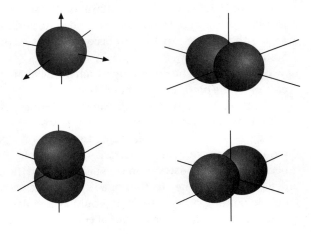

p orbitals) share essentially the same energy level, and a total of six electrons can occupy those orbitals, plus two more in the *s* orbital, making eight in all at that level in the atom.

Things rapidly become more complicated as you increase the number of electrons, but these simple examples indicate how the orbital concept relates to the behaviour of electrons in atoms determined by spectroscopy (see also *atomic physics*) and explains the structure of the *periodic table*. When electrons are shared between atoms to form *chemical bonds*, they can be thought of as occupying orbitals which are shared between the nuclei of those atoms. These shared orbitals can be more complicated, as when an *s* orbital overlaps with a *p* orbital to produce a hybrid orbital. Such *sp* hybrids are what hold the four hydrogen atoms in a methane molecule to the central carbon atom – the *s* orbital from each hydrogen atom 'overlaps' with a mixture of the *s* and *p* orbitals available in the outermost layer of electrons in the carbon atom. The *sp* hybrids form four identical bonds, symmetrically oriented with respect to the carbon nucleus, pointing towards the corners of an imaginary tetrahedron with the carbon nucleus at its centre.

order of magnitude   An estimate of the size of a quantity expressed to the nearest power of ten (the nearest order of magnitude). If two numbers are, say, roughly three orders of magnitude different from one another, one is roughly 1,000 (10^3) times bigger than the other. The nucleus of an atom has a radius of about 10^{-15} m, while an atom has a radius of about 10^{-10} m. So an atom is five orders of magnitude larger than a nucleus.

organic chemistry   See *quantum chemistry*.

Osheroff, Douglas Dean (1945–)   American physicist who shared the Nobel Prize for Physics in 1996 for his work on the *superfluidity* of liquid helium-3.

Born in Aberdeen, Washington, on 1 August 1945, Osheroff studied at the California Institute of Technology (BSc 1967) and at Cornell University (PhD 1973). It was in 1971, while he was a graduate student at Cornell, that Osheroff first noticed the peculiar behaviour of supercooled helium-3, at a temperature of a few thousandths of a kelvin; he reported the discovery to his thesis adviser, David Lee, and their colleague Robert Richardson, and over the next few months the team established that they were observing superfluidity.

From 1972 onwards Osheroff was on the staff of the AT&T Bell Laboratories, and in 1981 he was appointed head of solid-state and low-temperature physics research at Bell Labs. He left in 1987 to become professor of physics and applied physics at Stanford University.

pair production   The process by which particles are created out of energy – or, in the case of *virtual particles*, out of nothing at all.

Although Albert Einstein's equation $E = mc^2$ describes the interconvertibility of matter and energy, this is not the whole story. Even though there may be enough energy available (for example, in the form of a gamma ray) to provide the mc^2 required to make a single particle (for example, an electron), this is not allowed because various conservation laws also apply. In the case of an electron, which carries negative charge, the creation of a single particle would violate the conservation of charge (among other things); it would also violate the conservation of leptons. So whenever particles are made out of energy, they are produced in pairs, with opposite physical properties. This means that an electron, which has negative charge and a lepton number of +1, is always accompanied, in this process, by a positron, which has positive charge and a lepton number of –1.

At least, this is the case in all experiments carried out on Earth; a tiny imbalance in the laws of physics (see **CP conservation**) allowed for the production of about one extra electron for every billion positrons manufactured out of energy in the Big Bang, and a precisely equivalent excess of protons over antiprotons. This tiny excess is all that has survived from the Big Bang to make up the visible Universe today.

parallel vectors   Arrows that point in the same direction. See *antiparallel vectors*.

paramagnetism   See *permeability*.

parity   The operation which reverses the signs of the coordinates used to describe a system, so that a position described in three dimensions by the coordinates x, y and z is now described as the position $-x$, $-y$, $-z$. This is equivalent to studying the mirror image of the original system.

Individual particles can be assigned a parity, which is equivalent to saying that they are lefthanded or righthanded, although physicists usually use the terms 'even parity' and 'odd parity' (or sometimes, positive and negative parity). The two kinds of parity actually correspond to two different kinds of quantum wave function for a particle: a symmetric wave function (which means that the value at the point $-x$, $-y$, $-z$ is the same as at the point x, y, z,) or an antisymmetric wave function (which is mirrored so that the value at $-x$, $-y$, $-z$ is *minus* the value at x, y, z). There are also more complicated possibilities, which do not have a definite parity.

It was originally assumed that parity must be conserved in all particle interac-

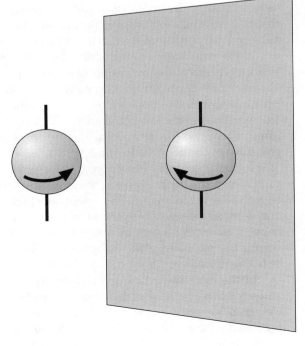

Parity. Parity reversal is equivalent to reflecting a particle, or an interaction, in a mirror. One effect of this (shown here) is to reverse the spin of the particle.

tions, but this idea had to be modified in the mid-1950s in order to explain some puzzling features of the decay of particles then known as theta and tau. These two particles seemed to be identical, except that the decay of the theta produced two pions, while the decay of the tau produced three pions. The problem was that, although a single pion has odd (negative) parity, a pair of pions has even (positive) parity, while a trio of pions has odd (negative) parity. This is because the overall parity is given by multiplying the individual parities together, just like the way negative numbers multiply together: $(-1) \times (-1)$ gives you $(+1)$, but $(-1) \times (-1) \times (-1)$ gives you (-1).

Following a suggestion made by the experimenter Martin Block, who was his room-mate at a conference in 1956 (at Rochester, in New York), Richard Feynman made the heretical proposal to the meeting that the theta and tau are different states of the same particle, which itself has no definite parity, and that parity is not always conserved. The idea was quickly taken up by Chen Ning Yang and Tsung Dao Lee, who showed that parity is not conserved in the weak interaction; their theory was immediately tested and proved correct by Chien Shiung Wu, and Yang and Lee received the Nobel Prize for their work in 1957 – one of the quickest such awards ever made. The 'theta' and 'tau' particles were renamed as the kaon, a single particle which can decay either into two pions or three pions, with violation of parity conservation. Processes in which parity is not conserved would look different in a mirror-image world – the mirror image of a weak interaction might describe a process which is impossible in the everyday world.

parity non-conservation   See *CP conservation*.

participatory universe   The idea, largely developed by John Wheeler, that the Universe exists only because we are looking at it. This is an extreme, but logical, development of the **Copenhagen interpretation** of quantum mechanics, which says that the probabilistic wave function of an entity such as an electron collapses to become a real particle only when it is measured or observed.

In his contribution to the book *Some Strangeness in the Proportion* (ed. by Harry Woolf), Wheeler gives an example of how the questions we choose to ask about the Universe may dictate the answers we get. He describes a parlour game in which one person is supposed to find out, by asking questions which can be answered 'yes' or 'no', what object the rest of the group are thinking about. In the usual form of the game, one person (the prospective questioner) leaves the room while the rest of the group agrees on the object in mind. In this case, after his series of questions led up to 'Is it a cloud?' and the answer 'yes', Wheeler was told that his friends had tried a variation on the theme, without warning him. Instead of choosing an object while he was out of the room, they had agreed only to give a consistent set of answers to his questions, so that each answer had to be logically compatible with all the previous answers, but with no definite object in mind at the start. Wheeler's repeated questioning had narrowed down the options until the answer to his final question had to be in the affirmative.

In the same way, the questions we ask about the Universe may 'create' the answers that we perceive as indicating the underlying reality about the nature of the Universe. 'We have no more right to say "what the photon is doing" – until it is registered – than we do to say "what word is in the room" – until the game of question and response is terminated', says Wheeler. Similar ideas are tackled in a different way by Andrew Pickering in *Constructing Quarks*.

History, as some philosophers have long argued, has no meaning, on this picture; the past has no existence except in the way it is recorded in the present, by experimental data or in our memories. The whole Universe can be thought of as a *delayed choice experiment*, in which the existence of observers who notice what is going on is what imparts tangible reality to the origin of everything.

particle accelerator   = *accelerator*.

particles   The fundamental building blocks of matter. Although studies of the world of the very small show that individual particles must under some circumstances be treated as waves (so that you could argue that everything is made of waves and *fields*), those same studies show that what we are used to thinking of as waves (notably electromagnetic radiation) must under some circumstances be treated as particles, so the particle concept is still fundamental to physics.

At one level, everything that we have direct experience of can in principle be explained in terms of six particles and the way they interact with one another – the matter particles (electron, neutrino, proton and neutron) and the force carriers (photon and graviton). At a deeper level, the proton and the neutron can be described as being made up of two kinds of *quark*, up and down, held together by the exchange of another kind of force carrier, the gluons; and radioactive decay is explained in terms of the weak interaction, requiring another set of force carriers, the intermediate vector bosons.

Many (more than a hundred) kinds of other particle can be manufactured in particle accelerators, but their structure can be explained by the addition of just a few more basic building blocks, the heavier *leptons* (members of the same family as the electron and the neutrino) and the heavier quarks (members of the same family as the up and down quarks), giving a total of just twelve matter particles plus the variety of force carriers.

There are both astronomical reasons and theoretical reasons from particle physics to expect the existence of other kinds of particle in the Universe (see *dark matter*), and these may be so prolific that they contribute by far the bulk of the mass of the observable Universe; but they have not yet been detected in experiments on Earth.

parton   Name coined by Richard Feynman as a generic description for any particles within the proton, neutron and other *baryons*. These particles are usually referred to today as *quarks* and *gluons*; Feynman's choice of terminology at the end of the 1960s was deliberately agnostic, an attempt to ensure that the interpretation of experiments probing the structure of matter at this level were not coloured by preconceived theoretical ideas about what might be discovered – an attempt by Feynman, as he would have phrased it, to keep the researchers honest.

parton model   Generic name for any model of the structure of particles such as protons and neutrons in terms of other particles (partons) inside them. The specific parton model that has become incorporated into the standard model of particle physics involves quarks and gluons as the partons; but the parton idea was introduced (by Richard Feynman, in the late 1960s) specifically to avoid prejudging the question of whether quarks existed, and to provide a broader theoretical framework for discussing the results of scattering experiments (then being carried out at SLAC) in which electrons were being fired into protons at high energies, and were scattering in a way reminiscent of the way the scattering of alpha particles from gold atoms revealed the existence of the atomic nucleus to Ernest Rutherford in 1911.

Feynman interpreted the results from SLAC in terms of a simple model, in which he assumed that the beam of electrons 'sees' a nucleon as a box filled with long-lived, point-like entities (the partons) from which they scatter. Each electron bounces off (scatters from) a parton in an *elastic collision*, to produce exactly the kind of scattering seen in the experiments, shaking up the contents of the box which then settle down again after the electron has gone on its way.

This model is, in fact, much more powerful and general than the idea that a nucleon is simply made up of three 'real' quarks, like three marbles in a bag. Indeed, the observed scattering (in particular, the *scaling* of the scattering process) cannot be matched up with such a simple three-quark model. You have to have those quarks (sometimes called the 'valence' quarks, by analogy with the *valency* of atoms in chemistry) embedded in a sea of virtual quark–antiquark pairs, and you also have to include the gluons holding the quarks together, in order to get the right kind of scaling. All of these particles – valence quarks, virtual quarks and gluons – are partons.

One of the key features of the parton model is that each scattering between an electron and a parton occurs as if the parton were a freely moving particle, not a tightly bound entity. This is an example of *asymptotic freedom* at work.

path integrals   A way of adding up quantum probabilities, originally devised by Richard Feynman when he was a PhD student in the early 1940s, which is based upon the use of *Lagrangian* functions, instead of the *Hamiltonian* approach previously used in quantum mechanics. This led to a completely new formalism for quantum theory (presented in Feynman's thesis) which brings out the holistic nature of the quantum world and turned out to be the most powerful and elegant version of quantum mechanics; indeed, Feynman's formalism includes the *classical mechanics* of Newton within the same mathematical framework. According to John Wheeler, Feynman's PhD supervisor, Feynman's presentation of these ideas in his thesis in 1942 marked the moment 'when quantum theory became simpler than classical theory' (but the ideas were not formally published in a journal until 1948 because of the delay caused by Feynman's involvement with the *Manhattan Project*).

The essence of the path integral approach is that, in order to calculate the probability that a particle will go from A to B, you have to take account of every possible path from A to B, not just the most direct route, or the trajectory described by Newtonian mechanics. Each path (each 'world line') has a certain probability, which is related to the *action* associated with that path – indeed, it was starting out from the idea of action that led Feynman to the path integral approach. The overall probability is calculated by adding up (integrating) the contributions from all the paths. Although this means adding up an infinite number of probabilities, most of them are very small (infinitesimally small) and this is the kind of 'infinite addition' that calculus can cope with, to give finite answers. In practice, all of the probability *amplitudes* cancel out except for the paths close to the classical trajectory, so that the classical trajectory emerges as a consequence of the quantum probabilities.

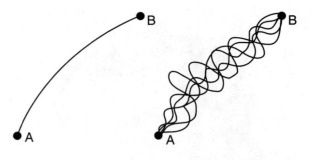

Path integral. Classical mechanics says that a particle follows a single trajectory from A to B. The path integral version of quantum mechanics says that the quantum entity takes every possible path from A to B (not just the few shown here, but literally every path). But, except under special circumstances (as in the experiment with two holes), the different quantum paths interact in such a way that they cancel each other out everywhere except near the classical trajectory.

There is a particularly neat way of imagining what is going on, which comes from Feynman himself. Start with the experiment with two holes, using light. The probability of getting a spot of light at a particular point on the detector screen is worked out by adding up (in the right way) the probabilities corresponding to light passing through each of the two holes. Now make four holes in the obstructing screen. Obviously, we have to add up four probabilities, for four possible paths. Go on to eight holes, sixteen, thirty-two, and so on. We can imagine dividing the obstructing screen into as many holes as you like and still calculating the same kind of addition of probabilities. Indeed, we can carry on making holes until there is no screen left – only an infinite number of holes and an infinite number of paths to integrate over. So, among other things, the path integral approach automatically describes how a single photon (or electron) goes through both holes at once in the experiment with two holes, and interferes with itself.

See *sum over histories*.

Pauli, Wolfgang (1900–1958) Austrian-born (later American) physicist who made many contributions to the development of quantum physics in the 1920s, notably the *exclusion principle*, for which he received the Nobel Prize for Physics in 1945.

Born in Vienna on 25 April 1900, Pauli was the son of a professor of physical chemistry at the University of Vienna, and had Ernst Mach as a godfather. His precocious talent was demonstrated (and his reputation in the world of physics established) when he published a comprehensive review of both the special and the general theories of relativity in 1920, the year *before* he completed his PhD at the University of Munich (where he studied under Arnold Sommerfeld). The review appeared first as a 200-page article in the *Mathematical Encyclopedia*, and two years later as a slim book. Albert Einstein himself commented favourably on the book, saying that it showed 'the deep physical insight, the mastery of clear systematic exposition, the familiarity with the literature, and the trustworthiness of the critical faculty' of its author.

After completing his PhD, Pauli visited Göttingen, where he worked with Max

Wolfgang Pauli (1900–1958).

Born, and Copenhagen, where he worked with Niels Bohr, before taking up a post as *Privatdozent* at the University of Hamburg in 1923. In 1928 he moved to the Swiss Federal Institute of Technology (the ETH), in Zurich, as professor of physics. He took leave from this post to spend the academic year 1935–6 as a visiting professor at the Institute for Advanced Study, in Princeton.

During the Second World War, Pauli spent the years from 1940 to 1945 in America, at the Institute for Advanced Study (but still nominally retained his post in Zurich). He had gone to the United States in 1940, fearing that the Nazis might overrun Switzerland. He was doubly at risk from them, since his mother was Jewish and Pauli himself still held an Austrian passport – Austria had been annexed by Germany, and under German law he was now a German citizen. But he returned to the ETH early in 1946 and stayed there for the rest of his career; although he now became a Swiss citizen, he continued to visit Princeton from time to time. Pauli died in Zurich on 14 December 1958.

The famous Pauli exclusion principle was first expressed in 1924 and published in detail the following year; it explained why each *orbital* in an atom (at that time, physicists still thought in terms of orbits, but the principle is the same) can be occupied by no more than two electrons. Pauli's other great insight came in 1930, when he suggested (originally in a letter to Lise Meitner) that the 'lost' energy in beta decay might be being carried off by an unseen particle. The idea was taken up by Enrico Fermi, who gave the particle the name *neutrino*.

Pauli's other contributions to quantum physics were wide ranging, but solid rather than spectacular. He had a biting wit and would soon punch holes in any flawed arguments being offered at a lecture in which he was in the audience. He was a bad lecturer himself, but expressed his ideas with energy and clarity in informal discussions and (as with the neutrino idea) letters. Pauli was such a hopeless experimenter that a legend grew up that he had only to walk through a laboratory for the apparatus to break – a phenomenon referred to by his colleagues as 'the Pauli effect'.

Pauli exclusion principle   = *exclusion principle.*

Pauling, Linus Carl (1901–1994)   American chemist (the greatest chemist of the 20th century) who applied quantum physics to establish the foundations of quantum chemistry in the late 1920s and early 1930s. He received the Nobel Prize for Chemistry for this work in 1954, and a second Nobel Prize, for Peace, in 1962, in recognition of his work in the campaign to end nuclear weapons testing. Pauling also determined the molecular structure of amino acids and proteins, and later became convinced that massive doses of vitamin C could not only protect against the common cold and other ailments, but help the body fight cancer. This controversial claim has not yet been accepted by the medical establishment, although many people (including the principal author of this book) claim from first-hand experience that vitamin C does indeed stop colds, at least.

Pauling was born in Portland, Oregon, on 28 February 1901. His father, who died when the boy was nine, was a pharmacist. Pauling studied at Oregon State Agricultural College (now Oregon State University), where he received his BSc in chemical engineering in 1922, and at Caltech (PhD in physical chemistry, 1925). He then spent two years in Europe (on a Guggenheim Fellowship), where he worked with Arnold Sommerfeld in Munich, Niels Bohr in Copenhagen, Erwin Schrödinger in Zurich and William Bragg in London, gaining first-hand insight into the new quantum theory and the technique of X-ray crystallography. He became an assistant professor at Caltech in 1927 and full professor in 1931.

Although he was director of the chemical laboratories at Caltech from 1936 to 1958, by the end of the 1950s there was a groundswell of feeling against Pauling in the Chemistry Division, where many of his colleagues regarded him as left wing and objected to his anti-nuclear activities. He was deeply hurt when the Chemistry Division took no notice of the award of the Peace Prize in 1962 (although the Biology Division gave a party in recognition of the award), and resigned his Caltech posts. He was then in his early sixties, and officially retired, but he stayed on the Caltech faculty lists as an unpaid research associate and then, after his seventieth birthday, as professor emeritus. The old wounds had healed sufficiently by 1991 for the Chemistry Division to give him a rousing ninetieth birthday party.

In 'retirement', Pauling moved to a country house that he had had built at Big Sur, where he developed his ideas about vitamin C and other non-mainstream medical views. He became a member of the Center for the Study of Democratic Institutions (described by his biographers Ted and Ben Goertzel as 'a liberal think tank') at Santa Barbara from 1963 to 1967, was appointed as a professor at the University of California, San Diego, in 1967, then took up a similar post at Stanford University in 1969.

Pauling's best-selling book *Vitamin C and the Common Cold* was published in 1970. He retired from Stanford in 1973 (under pressure from the authorities because of the publicity generated by his unconventional medical ideas, and also because of his continuing leftist sympathies) and became an emeritus professor there, as well as at Caltech. Undaunted by the rebuff from Stanford, Pauling then established the Linus Pauling Institute to continue the vitamin C research. He remained active until his death, at Big Sur, on 19 August 1994.

Pauling's great work in chemistry started with a classic paper, published in 1931, on 'The nature of the chemical bond', which used quantum mechanics to explain how two electrons (one from each of two atoms) can be shared to make a bond between the two atoms (see *quantum chemistry*). He co-authored an influential book *Introduction to Quantum Mechanics*, published in 1935, introduced the idea of hybrid **orbitals**, and summed up his work so far in another book, *The Nature of the Chemical Bond*, published in 1939 and regarded as the most influential chemistry textbook of all time. This and his other textbooks are still among the best introductions to quantum chemistry.

From the mid-1930s he turned his attention to the study of the chemistry of life, including the structure of proteins and the difference between haemoglobin in normal people and in those suffering from sickle-cell anaemia, using as one of his tools X-ray crystallography. He came close to determining the structure of DNA, but was beaten on that occasion by Francis Crick and James Watson. As the 1950s developed, Pauling's energies were increasingly concentrated on the nuclear weapons issue, and he did little more mainstream scientific research.

Further reading: Ted and Ben Goertzel, *Linus Pauling*.

Peierls, Rudolf Ernst (1907–1995)   German-born British theoretical physicist who investigated, among other things, quantum theory and the theory of nuclear reactions. He was a hard-line devotee of the *Copenhagen interpretation*.

Born in Berlin on 5 June 1907, Peierls studied at the Universities of Berlin, Munich and Leipzig, and at the ETH in Zurich. He worked briefly in Rome, Copenhagen, Manchester and Cambridge, then settled at the University of Birmingham in 1937, where he stayed until 1963. He then became professor of physics at the University of Oxford, retiring in 1974 and then spending three years as professor of physics at the University of Washington, Seattle. He returned to Oxford in 1977 and became an Honorary Fellow of New College.

Together with Otto Frisch, in 1940 Peierls wrote a paper *On the Construction of a 'Superbomb'*, which demonstrated that there was a realistic possibility of developing nuclear weapons; this top-secret memorandum was instrumental in persuading the British government to begin research on nuclear weapons, and after the United States entered the war Peierls became the leader of the British group on the **Manhattan Project**. He was knighted in 1968, and died, in Oxford, on 19 September 1995. His son, Ronald (born in Manchester in 1935), became a particle physicist at the Brookhaven National Laboratory.

PEP   Acronym for Positron–Electron Project, a collider at SLAC designed to reach energies of 36 GeV. It was completed in 1980. Together with PETRA, PEP provided direct evidence for **gluons** from 'three-jet' events.

periodic law   See *elements*.

periodic table   The classification of the *elements* according to their properties, first proposed by Dmitri Mendeleyev in 1869 (but see also *Newlands, John Alexander Reina*). The periodic relationship between the elements is explained by *quantum chemistry* (see also *Bohr model*).

Perl, Martin Lewis (1927–)   American physicist who received the Nobel Prize for Physics in 1995 for the discovery of the *tau* particle.

   Born in Brooklyn on 24 June 1927, Perl studied chemical engineering at the Polytechnic Institute in Brooklyn, but his studies were interrupted by service with the Coast Guard and the Army during the Second World War, and he only received his first degree in 1948. He then spent two years working as a chemical engineer for General Electric, in Schenectady, but took courses at Union College in his spare time, and decided that what he really wanted was a career in physics. So in 1950 he went to Columbia University, where he was awarded his PhD in 1955 (he was a student of Isidor Rabi at Columbia). Perl then worked at the University of Michigan until 1963, when he became professor of physics at Stanford and a group leader at SLAC.

   It was at SLAC that Perl and his colleagues discovered the tau particle, in a series of experiments in the mid-1970s. Because this was the third charged *lepton* to be discovered, Perl gave it the name tau from the initial letter of the Greek word for 'third'. He later became involved in experiments at the 10 GeV electron–positron collider at Cornell, but remained based at SLAC.

permeability   A measure of the way in which a substance changes the strength of a magnetic field in which it sits. The strict definition of magnetic permeability is the ratio of the magnetic flux density inside the substance to the strength of the field outside the substance. Because of the way the various units are defined, in some systems of units the permeability of free space is not 1 – although you would expect it to be, since empty space does not affect a magnetic field relative to the strength of the field in empty space. This is entirely a result of the bad choice of units, and can be regarded as a normalization problem (see *magnetic constant*). If you are working in those units, it is better to work with the relative permeability, which is the permeability divided by the permeability of free space. If the (relative) permeability is less than 1, the material is said to be diamagnetic; if it is greater than 1, the material is said to be paramagnetic.

permeability of free space   See *magnetic constant*.

permittivity   The electrical equivalent of *permeability*, permittivity is a measure of the extent to which a substance can resist the flow of electric charge. As with permeability, the permittivity of free space ought to be 1 in any sensible system of units, and where this is not the case, the numbers can be normalized by dividing by the actual value of the permittivity of free space in those units. The relative permittivity, defined in this way, is also known as the dielectric constant. See *electric constant*.

permittivity of free space   See *electric constant*.

Perrin, Jean Baptiste (1870–1942)   French physicist who demonstrated the reality of atoms by studying *Brownian motion*. He was awarded the Nobel Prize for Physics for this work, in 1926.

   Born in Lille, on 30 September 1870, Perrin studied in Lyons and in Paris, where he graduated from the Ecole Normale Supérieure (a teacher training college) in 1894, but stayed on there as a physics teacher until 1897, when he received his PhD. He was a reader in chemistry at the Sorbonne from 1897 to 1910, then became professor of physical chemistry at the same institution. He stayed there for the next 30 years, apart from service as an officer in the Engineers from 1914 to 1918, when he worked on devices for detecting submarines underwater. In 1940, after the German occupation, he had to leave France because of his anti-fascist activities, and he went to New York, where he died on 17 April 1942.

Perrin's most important early work, in 1895, was a study of cathode rays which provided a rough indication of the ratio of the charge on an electron to its mass, paving the way for the definitive work by J. J. Thomson which established that the electron behaves as a particle. But his key contribution to physics came in 1909, when his painstaking observations of Brownian motion established that this process occurs exactly in the way required if the particles involved are being buffeted by atoms, providing the definitive experimental support for the theory proposed in 1905 by Albert Einstein. This was accepted as the final proof of the reality of atoms.

Perrin was an enthusiastic popularizer of science, whose book *Les Atomes*, first published in 1913, was widely translated and sold tens of thousands of copies in France alone. His son Francis (born in 1901) and grandson Nils (born in 1927) both became physicists.

perturbation   Any relatively small secondary influence on a system that is dominated by some other influence. The perturbation modifies the behaviour of the system, so that it does not behave precisely as it would if it were affected by the main influence alone, but the modification is not so great as to destroy the effects of the main influence and make it unrecognizable.

The classic example of perturbation at work (in a scientific sense) comes from astronomy, where (ignoring everything outside the Solar System) the orbit of a single planet around the Sun under the influence of the Sun's gravity alone can be described perfectly by Newton's laws, and is an ellipse. The real orbits of the planets are perturbed by the gravitational influences of the other planets themselves, and deviate slightly from the simple orbits determined from Newton's laws using the gravitational influence of the Sun alone. Similarly, the behaviour of an electron in an atomic *orbital* is perturbed by the influence of other electrons in the atom.

perturbation series   See *perturbation theory*.

perturbation theory   Not really a theory at all, but a technique for obtaining approximate (but possibly very accurate) solutions to problems that are difficult or impossible to solve precisely. It is used both in classical physics and in quantum physics. The idea is to find a way to describe the system being investigated in terms of one component which can be calculated precisely and describes the overall behaviour of the system, plus a secondary component which acts as a perturbation and whose influence can be calculated approximately to an acceptable degree of accuracy.

In practice, the perturbation term often turns out to involve an infinite series of ever smaller corrections, so that the overall behaviour of the system can in principle be calculated to whatever precision you like by calculating enough terms in the series. But practical difficulties (such as the amount of computer time available to solve the equations) usually determine the cut-off point and limit the accuracy of the solution.

In classical mechanics, perturbation theory can be used, for example, to calculate the orbit of the planets in the Solar System (see **perturbation**). In quantum mechanics, perturbation theory has many applications, but its greatest triumph is in **quantum electrodynamics**, where the behaviour of electrically charged particles inter-acting with magnetic fields can be explained to enormous accuracy by calculating a few terms in the appropriate perturbation series.

One of the reasons why **Feynman diagrams** are such a powerful tool in quantum field theory is that they provide a clear and powerful way to represent the successive perturbations involved in such a calculation. But it should never be forgotten that perturbation theory is by definition an approximation, no matter how good an approximation. As Abraham Pais has commented in his book *Inward Bound*, in a *cri de coeur* that speaks for all physicists: 'Even though this method has served us well we would like to be free of it. We don't know how.'

PETRA   Acronym for Positron–Electron Tandem Ring Accelerator, a collider built at DESY which reached its design energy of 19 GeV per beam in 1980. Even before reaching full power, in 1979 PETRA produced evidence for **gluons**, in the form of **three-jet events**; it was later upgraded to reach energies of 23 GeV per beam.

phase   In astronomy, a term used to refer to the changing appearance of the Moon (or other celestial objects), in this case from new Moon through first quarter to full Moon and last quarter.

In chemistry, a reference to the physical state of a substance – liquid, solid or gas.

In classical physics, a reference to the stage reached in an oscillatory (wave) motion; usually denoted by an angle, between $0°$ and $360°$, representing the point the wave has reached in its cycle (how far away it is from a peak or a trough). Two waves are said to be 'in phase' if the sets of peaks and troughs from one wave precisely coincide with those of the other wave, and the two waves are in antiphase if the peaks from one wave coincide with the troughs of the other wave. It is the difference in phase of the waves involved which produces interference effects such as the pattern of light and shade seen in the **double-slit experiment**.

phase (quantum)   A property which determines the most probable trajectories that quantum entities, such as electrons and photons, follow. The best way to think of this is in terms of little arrows, following the presentation used by Richard Feynman in his book *QED: The strange theory of light and matter*.

Imagine that the quantum entity has associated with it a circular wheel, on which there is an arrow drawn from the centre of the wheel to a point on its rim. If we want to work out the most likely way for the entity to travel from A to B, we have to take account of every possible path, and add them together in the right way, to find the most probable path (see **path integrals**, **sum over histories**). Consider just one of the paths from A to B, and imagine rolling the little wheel along that path, starting out with its arrow pointing straight up. At the end of the journey, after turning round many times, the arrow will be pointing in a certain direction relative to the vertical. That is the phase associated with that particular path. You can also, if you like, think of this as representing the stage of its cycle that a wave following that path will have got to when it arrives at B.

You can do the same thing for every possible route from A to B, including routes that, say, detour via Mars on the way across the room. In each case, you end up with the arrow pointing in a definite direction associated with that particular path.

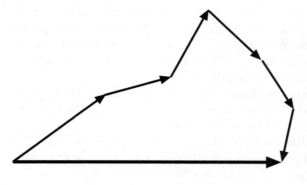

Phase. Adding up the arrows for path integrals. The direction of each arrow represents phase; the length of each arrow represents amplitude. The arrows are added by joining them nose to tail, then drawing a total arrow from the tail of the first little arrow to the nose of the last little arrow. This total arrow gives you the overall amplitude, and therefore the probability. It is because paths with very different phases have arrows pointing in very different directions that they tend to cancel each other out. Paths with the same phase have arrows which point in the same direction and add together.

The final position and size of the arrow, in each case, represents the amplitude for that particular path, and the probability is given by the square of the amplitude. The size of each of these particular arrows is the same (because we are dealing with the same 'particle' going on the same journey, from A to B), and the overall effect is found simply by adding together all of the little arrows representing the different paths.

The total probability is the square of the sum of the amplitudes, not the sum of the squares of the amplitudes. The way to add up arrows (***vectors***) is to join them nose to tail, preserving their relative orientations (the phase differences), but in any order you like, and to draw a line from the end of the first arrow to the tip of the last arrow. It is this final arrow that determines the overall probability, and what matters is the difference in phase for all the different paths (the way the different little arrows point). The fact that the addition can be carried out in any order is crucial, since it means that none of the quantum paths has any special status.

It turns out that the phase is only nearly the same along trajectories that are very close to the classical paths from A to B (in the case of light, straight-line paths) and that the phases are wildly different even for neighbouring paths for more exotic trajectories, such as a detour to Mars and back on the way from the light bulb in my room to my eyes. This is why the contributions from all the other paths cancel out, leaving the illusion that only the classical paths are followed in everyday life.

This geometrical description is good for getting a picture of what is going on; in algebraic terms, the little arrows (vectors) can be represented in terms of ***complex numbers***, and this is the way calculations are usually carried through.

phase difference   See *phase.*

phase transition   The change of a physical system from one state (one phase) to another at the same temperature, but without any change in its chemical composition. Phase transitions are familiar in everyday life, when liquid water changes to ice at $0°C$ or to steam at $100°C$. They also occur with other rearrangements of the constituents of a system: for example, at the Curie point, a characteristic temperature where there is a change in the magnetic properties of a material.

In the very early Universe, the change from a state in which free *quarks* existed (at very high density and temperature) to a state in which quarks became bound up inside *hadrons* was a phase change, so that in a sense hadrons are frozen quarks. Earlier still, there were phase changes associated with the way the four *forces of nature* known today split off from the original force described by *grand unified theories*.

Phase transitions are associated with the absorption or release of energy. When steam condenses to form liquid water by condensation, it gives up latent heat. In an analogous way, the phase changes that occurred in the very early Universe provided the energy which powered the early expansion of the Universe, and which provided the energy which made the fireball of the Big Bang very hot during the first few minutes, when Big Bang *nucleosynthesis* took place.

phonon   A 'particle' of sound travelling through a crystal lattice. The idea of a sound wave can be replaced by the idea of phonons in an analogous way to the description of light in terms of *photons*. This turns out to be useful in describing the thermal conductivity of some solids, and the way electrons in the crystal structure interact with sound waves.

phosphorescence   Long-lived emission of light from a substance after it has been stimulated by some outside influence other than a rise in temperature. The influence could, for example, be the effect of *X-rays*, or the impact of electrons. The definition of 'long lived' is arbitrarily set as any *luminescence* lasting more than 10 nanoseconds (10^{-8} sec) after the stimulation has stopped.

photino   A particle predicted by the theory of *supersymmetry*, but not yet detected in any experiment. The photino is the supersymmetric counterpart of the photon, and would probably be the lightest supersymmetric particle. It would have spin ½, making it a fermion.

photodiode   A device that absorbs light and uses it to change the electrical properties of a semiconductor junction, allowing an electric current to flow. This makes direct use of the *photoelectric effect*, which provided evidence for the existence of *photons*.

photoelectric effect   The release of electrons from a substance under the influence of light or other electromagnetic radiation. The effect has a key place in the history of the development of quantum physics because it was explained, by Albert Einstein in 1905, by treating light in terms of a stream of particles (now called *photons*) rather than as a wave. Although the implications were not fully appreciated at the time (not even by Einstein) this was the first step towards the concept of *wave–particle duality*. It was, incidentally, for this work, not for either of his theories of relativity, that Einstein received the Nobel Prize.

Einstein was the first person to take seriously the idea of light quanta (introduced by Max Planck a few years earlier), and to treat them as more than a mere mathematical trick used to explain the spectrum of *black body radiation*. He also knew of the experimental work on the photoelectric effect carried out separately by J. J. Thomson and Phillip Lenard at the end of the 19th century and beginning of the 20th century.

The experiments used beams of light of a single colour, which means that the waves in the light all have the same frequency, shone on to metal surfaces. Using a bright light, there is more energy shining on each square centimetre of the surface of the metal than with a dim light (or with the same light further away from the metal). You might expect that with more energy available, the electrons knocked out of the metal surface by the light would be more energetic and travel faster. In fact, to the surprise of the turn-of-the-century experimenters, it turned out that as long as the frequency (or wavelength, which is just the frequency multiplied by the speed) of the light stays the same, the energy of each electron liberated is the same; but when the light is brighter, more electrons are liberated. The electrons do, though, move faster if light with a higher frequency (shorter wavelength) is shone on the metal – blue light produces faster-moving electrons than red light.

Einstein explained this by taking the equation at the heart of Planck's description of black body radiation, $E = hf$, and applying it not to the little oscillators inside atoms but to the electromagnetic radiation itself. He said that the photoelectric effect could be explained if light itself came in definite packets, or quanta, each with an energy hf, where h is Planck's constant and f is the frequency of the light.

It takes one light quantum to knock one electron out of the metal, and for a particular frequency all the light quanta have the same energy, so all the liberated electrons have the same energy. In a brighter light of the same colour (frequency), there are more light quanta, but each quantum still carries the same energy; so more electrons, still with the same energy, are liberated. And when the frequency is increased (for example, from red to blue), f is bigger, so each light quantum carries more energy, and the liberated electrons move faster, even if only a few electrons are liberated.

In fact, it is possible to explain the photoelectric effect in terms of light as a continuous wave which carries energy that can be accepted by the atoms in the metal surface only in definite lumps. This was pointed out by David Bohm in the 1950s, leading to the tongue-in-cheek suggestion that perhaps Einstein didn't really deserve his Nobel Prize. But experiments in the 1920s (see **Compton effect**) had established once and for all the reality of photons, so there is not really any doubt that Einstein was right after all. Nor is there any doubt of the historical importance of the establishment of the reality of photons; although this took many years, it was triggered by Einstein's work on the photoelectric effect.

photomultiplier   General name for any of several different kinds of detector that absorb low-intensity electromagnetic radiation (usually light) and amplify the signal produced by the incoming radiation. One technique involves electrons being released from a cathode by the triggering effect of the arrival of the photons (an example of the *photoelectric effect* at work), with these electrons in turn triggering the release of more electrons from other electrodes, in a repeated cascade process. The final signal may be 100 million times stronger than the original. These devices are used in astronomy, in image intensifiers (TV cameras that 'see in the dark') and in *scintillation counters*.

photon   A particle of light. The idea of the photon stems directly from Albert Einstein's work on the *photoelectric effect* in 1905, but it took many years to establish the physical reality of photons, and they were only given this name in 1926, by the American physical chemist Gilbert Lewis.

The main reason for the delay (apart from the startling novelty of Einstein's proposal to physicists who had been brought up on the idea of light as a wave) was the lack of direct experimental confirmation of Einstein's ideas. In 1905 the experimental data were not sufficiently accurate to exclude the possibility of other theoretical explanations of the photoelectric effect, and there was no other evidence in favour of Einstein's idea at all. Einstein himself, as late as 1911, said that 'I insist on the provisional nature of this concept which does not seem reconcilable with the experimentally verified consequences of the wave theory.'

The person who provided the accurate experiments needed to prove that Einstein's idea was correct was, ironically, motivated by a strong desire to prove him wrong. Robert Millikan disliked the idea of light quanta intensely, and spent ten years testing Einstein's explanation of the photoelectric effect to greater and greater accuracy. He found that the rival theories failed to hold up, and that Einstein's interpretation was correct. Much later, Millikan commented: 'I spent ten years of my life testing that 1905 equation of Einstein's and contrary to all my expectations, I was compelled in 1915 to assert its unambiguous verification in spite of its unreasonableness.' It was no doubt some consolation when Millikan received the Nobel Prize for this work (and for his measurement of the size of the charge on the electron) in 1923. As a by-product of his attempt to prove Einstein wrong, he had also obtained a very accurate measurement of the value of h, Planck's constant.

Theoretical progress was also slow. Although Einstein talked of light in terms of energy quanta in 1905, it was not until 1909 that he specifically referred to 'point-like quanta' – echoing the image of a particle such as an electron, and a far cry from any description in terms of waves, or even wave packets – and made the prescient observation 'that the next phase in the development of theoretical physics will bring us a theory of light that can be interpreted as a kind of fusion of the wave and the emission [particle] theory'. Only in 1916 (after Millikan's work), though, did he introduce the idea of momentum associated with the quantum of light. And it was in 1923 that Arthur Compton's investigation of the effect that now bears his name showed electromagnetic quanta behaving exactly like particles, exchanging both energy and momentum in collisions with electrons.

In modern terminology, the photon is a **boson** with zero mass and spin 1. It obeys **Bose–Einstein** statistics. About 1,000 billion (10^{12}) photons of sunlight fall on a pinhead each second; when you look at a faint star, your eye receives a few hundred photons from that star each second.

Further reading: Abraham Pais, *Subtle is the Lord*.

picosecond   1 thousandth of a billionth (10^{-12}) of a second.

pilot wave   The wave that fills the Universe and guides a real particle (such as a photon or an electron) around in the version of quantum theory developed originally by Louis de Broglie and extended by David Bohm and his colleagues. See *hidden variables, non-locality, undivided whole.*

pion (pi-meson)   One of the particles which carry the strong force between protons and neutrons (and other *hadrons*), binding an atomic nucleus together. They were predicted by Hideki Yukawa in 1935, and were discovered in cosmic ray experiments on the Pic du Midi, carried out by a group of researchers from Bristol University, headed by Cecil Powell, in 1947. The pion comes in three varieties: the pi-zero (zero electric charge, mass of 135 MeV) and the pi-plus and pi-minus (respectively with one unit of positive or negative charge, each with mass 140 MeV). They are *bosons*, with zero spin, and obey *Bose–Einstein statistics*. The pi-zero is now regarded as being composed of either an up/antiup quark pair or a down/antidown quark pair (that is, it can be made in both ways); the pi-plus is an up/antidown pair and the pi-minus is a down/antiup pair.

Pippard, Alfred Brian (1920–) British physicist who used microwave radiation (similar to the radiation in a microwave oven) to study the behaviour of the surface layer of *superconductors*. He introduced the idea of 'coherence', in which many electrons in a superconductor behave like a single quantum entity, paving the way for the theory of superconductivity developed by John Bardeen, Leon Cooper and John Schrieffer.

Born in London on 7 September 1920, Pippard studied at the University of Cambridge, completing his first degree in 1941, but his education was then interrupted by wartime work on radar. He returned to Cambridge in 1945 and completed his PhD in 1947; he stayed in Cambridge throughout his career, becoming Plummer Professor of Physics in 1960 and then Cavendish Professor of Physics from 1971 until his retirement in 1982. He was knighted in 1975.

Planck, Max Karl Ernst Ludwig (1858–1947) German physicist who realized, just at the end of the 19th century, that the spectrum of *black body radiation* could be explained if electromagnetic energy is absorbed and emitted by atoms only in discrete packets, called quanta. Planck did not suggest that light (and other electromagnetic radiation) actually exists as quanta (now called photons), but thought of this process (which involved what he called the 'elementary quantum of action') as something to do with the way the oscillating charges within atoms worked. Nevertheless, his insight represented a breakthrough which revolutionized physics, led Albert Einstein to suggest the physical reality of light quanta, and gained Planck himself the Nobel Prize for Physics in 1918.

Planck was born in Kiel on 23 April 1858; his father was professor of civil law at the University of Kiel. When Planck was nine, he moved with his family to Munich, where he attended the Maximilian Gymnasium (high school); he was an outstanding student who could have chosen a career in linguistics or music (he had perfect pitch, and played the piano throughout his life, sometimes accompanied by Einstein on violin), but settled on physics as the area in which there was a better chance of doing original work. He entered the University of Munich in 1874 and spent one of his student years at the University of Berlin, where he was taught by Gustav Kirchoff and Hermann Helmholtz (it was usual at that time for students in Germany to visit another university in the course of their studies, on a *Wanderjahr*). He received his PhD from Munich University in 1879, for a thesis on thermodynamics, and stayed on there as a *Privatdozent*.

In 1885 Planck became a professor of physics in Kiel, and in 1888 he moved to Berlin to become the first director of the new Institute for Theoretical Physics there, and an assistant professor of physics. He became professor of physics at the University of Berlin in 1892, and stayed in the post until 1926, when he retired and was succeeded by Erwin Schrödinger.

In 1930, when he was already 72 years old, Planck became president of the Kaiser

Max Planck (1858–1947).

Wilhelm Institute in Berlin. He resigned from this post in 1937 as a protest against the Nazi regime's treatment of Jewish scientists. But he survived the Second World War and in 1945, when the Institute became the Max Planck Institute and was relocated to Göttingen, he was once again appointed as its president, a post he held until his death, in Göttingen, on 4 October 1947.

Planck was a German scientist of the old school: hard working, thorough and conservative. His great interest was thermodynamics, and when he became interested in the puzzle of black body radiation in the second half of the 1890s his hope was to resolve the **ultraviolet catastrophe** by the application of thermodynamic rules. He became increasingly frustrated by his failure to find the correct description of this spectrum, even though he published several key papers establishing a connection between thermodynamics and electrodynamics. The breakthrough that he achieved in 1900 came not through a cool, calm and logical process of scientific insight, but almost as an act of desperation, mixing guesswork with his physical intuition to join together two incomplete (and at first sight incompatible) descriptions of electromagnetic radiation (the **Rayleigh–Jeans Law** and **Wien's Law**) with some mathematical juggling to make them fit together. Planck pulled the right mathematical curve out of the hat (he later referred to it as a 'lucky intuition'), but nobody, not even Planck, realized in 1900 what physical properties of matter and radiation gave rise to that curve.

After 1900 Planck made contributions to optics, thermodynamics and other fields, and was one of the first people to realize the importance of the special theory of relativity. But his most important contributions to science after 1900 were as an administrator. These were considerable, and were achieved even though in his personal life he suffered a series of blows from 1909 onwards, when his first wife died. His elder son was killed in action in 1916, and in 1917 one of his twin daughters died in childbirth. Her sister later married the widower, Planck's son-in-law, and died the same way. Planck's home in Berlin was destroyed by bombs in 1944, and his younger son was brutally killed by the Gestapo for his part in the plot to assassinate Hitler later that year.

Planck density   The density of matter where 1 *Planck mass* occupies a volume 1 *Planck length* across. Roughly 10^{94} grams per cubic centimetre, 10^{60} times the density of an atomic nucleus.

Planck energy   The mass-energy associated with a particle which has the *Planck mass*, in line with Albert Einstein's equation $E = mc^2$. Roughly 10^{19} GeV.

Planck length   The length scale at which classical ideas about gravity and spacetime cease to be valid, and quantum effects dominate. This is the 'quantum of length', the smallest measurement of length that has any meaning. It is determined by the relative sizes of the constant of gravity, the speed of light and Planck's constant, and is roughly equal to 10^{-33} cm, about 10^{-20} times the size of a proton.

Planck mass   The mass of a hypothetical particle which would have an equivalent wavelength, according to quantum theory, equal to 1 Planck length. The Planck mass is about 10^{-5} g; this is small by everyday standards, but 10^{19} times the mass of a proton, and would be contained within a volume roughly 10^{-60} times that of a proton, representing an enormous density that has not occurred naturally since the birth of the Universe in the Big Bang. In order to create conditions in which such particles could be formed, present-day particle accelerators would have to be made 10,000 trillion times more powerful.

Planck radiation law   The mathematical equation that describes the nature of *black body radiation*. Discovered by Max Planck in 1900.

Planck scale   The scale on which events occur over a few multiples of the *Planck length* and the *Planck time*.

Planck's constant   A fundamental constant, denoted by h, which relates the energy of a quantum of electromagnetic radiation (a *photon*) to its frequency, through the equation $E = hf$. In other words, it relates the particle nature of a quantum entity to its wave nature. The value of h is 6.626 x 10^{-34} joule seconds (6.626 x 10^{-27} erg seconds) and it appears in many calculations involving quantum mechanics – for example, in the **uncertainty principle** discovered by Werner Heisenberg. In many of these equations, it is more convenient to use as the fundamental constant h divided by twice the vale of pi ($h/2\pi$). This is written as $\hbar$, and is pronounced 'h-cross'; slightly confusingly, h-cross is also referred to as Planck's constant, since to physicists it is usually obvious where the factor of 2π goes. To non-physicists, it doesn't matter anyway; all that matters is to notice how incredibly small both h and $\hbar$ are, which explains why we are not aware of quantum effects such as wave–particle duality in the everyday world. Named in honour of Max Planck; see also *black body radiation*.

Planck temperature   The temperature associated with the *Planck energy*. About 10^{32} kelvin.

Planck time   The time it would take light, moving at a speed of a few times 10^{10} cm per second, to cross a distance equal to the *Planck length*. This is the 'quantum of time', the smallest measurement of time that has any meaning, and is equal to 10^{-43} sec. No smaller division of time has any meaning. So, for example, within the framework of the laws of physics as understood today we can say only that the Universe came into existence when it already had an age of 10^{-43} sec, and was at the *Planck density*.

Planck units   See under the name of the specific unit – for example, *Planck length*, *Planck time*, *Planck mass*. These are the natural units to use in discussions of *quantum gravity*.

planetary model   See *Bohr model*.

plasma   A hot state of matter in which *electrons* have been stripped from *atoms* to leave positively charged *ions*, which mingle freely with the electrons. Plasma can be regarded as a fourth *phase* of matter, together with the familiar solid, liquid and gas phases.

Poincaré, (Jules) Henri (1854–1912)   French mathematician who made many important contributions to science, including (in 1906) an independent derivation of many of the results of Albert Einstein's special theory of relativity, and who showed that the transformation equations discovered by Hendrik Lorentz form a *group*.

Born in Nancy on 29 April 1854, Poincaré studied at the Ecole Polytechnique and at the School of Mines, intending to become an engineer. But he became diverted into mathematics and received his PhD, in 1879, for a thesis dealing with the properties of functions defined by partial differential equations. He taught briefly at the University of Caen, then became a professor at the University of Paris, from 1881 until his death, in Paris, on 17 July 1912.

Poisson, Simeon-Denis (1781–1840)   French mathematician and physicist, best known for his work on probability. He discovered the Poisson distribution. This applies to, among other things, the probability of radioactive decay.

Born in Pithiviers on 21 June 1781; died in Sceaux on 25 April 1840. Poisson studied at the Ecole Polytechnique in Paris, where his teachers included Pierre Laplace and Joseph Lagrange. He later taught at the Ecole and at the Sorbonne. He investigated the theory of heat, elasticity, electricity, magnetism and light, as well as undertaking his famous work on probability.

See also *Fresnel, Augustin Jean*.

Poisson spot   See *Fresnel, Augustin Jean*.

polarization   A property of *light* (or other *electromagnetic radiation*) that is best understood in terms of the wave nature of the radiation.

In James Clerk Maxwell's description of light, it is made up of two sets of transverse waves, at right angles to each other, representing the electric and magnetic components of the radiation. For simplicity, think only of one of these (the electric component, say). In ordinary unpolarized light, there is a jumble of waves with all different orientations around the direction in which the light is moving. Each wave can be represented by an arrow at right angles to the direction of motion, but some arrows point up, some sideways and some at in-between orientations. But in plane polarized light, all of the arrows are oriented in the same direction, as if they are spears being carried stiffly upright (or at some other angle) by a column of marching soldiers.

In so-called circularly polarized or elliptically polarized light, there is still an 'arrow' indicating the direction of orientation of the wave, but this arrow rotates, like the second hand of a clock, as the wave advances.

Polarization is produced naturally when light passes through certain crystals, when it is reflected, or when it passes through artificial material such as the Polaroid used in sunglasses (see *Malus, Etienne Louis*). The behaviour of polarized light highlights some of the key features of the behaviour of the quantum world (see *probability*).

polarized light   See *polarization*.

positive feedback   See *feedback*.

positron   The positively charged, *antimatter* counterpart to the *electron*.

positronium   A kind of pseudo-atom in which a negatively charged electron and a positively charged positron orbit around one another, with the positron in a sense playing the role of the proton in a hydrogen atom. Such systems have been made artificially, but have a very short lifetime (less than 1 ten-millionth of a second) before the electron and the positron annihilate one another in a burst of gamma radiation. See *antimatter*.

potential energy   Energy that an object possesses as a result of its position – for example, in a gravitational or an electric *field*. An object at the top of a flight of stairs has more gravitational potential energy than it has when it is at the bottom of the stairs. In order to carry the object upstairs, work has to be done, and this provides the extra potential energy; if the object falls down the stairs, the gravitational potential energy is converted initially into *kinetic energy*, and then the kinetic energy is converted into heat (thermal energy) as the object comes to a halt.

Potential energy always has to be measured relative to some standard state – in this case, we might choose to set the zero point of our measurements as the state with the object at the bottom of the staircase. There may, however, be a minimum value of the potential associated with a particular field, below which it is impossible to drop.

Another example of potential energy is the energy stored in a spring (associated with electromagnetic forces, and therefore ultimately a quantum phenomenon described by the theory of *quantum electrodynamics*) when it is compressed. One of the strange features of *gravity* is that the zero point of energy for a mass m made up of many particles is with the particles dispersed to infinity. If the particles then fall together, under the influence of gravity, energy is released (just like the example of an object falling downstairs), so that the remaining potential energy is now negative. If the particles merged at a point, the total gravitational potential energy associated with them would then be reduced to $-mc^2$, exactly balancing the mc^2 of their total Einsteinian mass-energy. This raises the intriguing possibility that all of the matter in the Universe could have appeared out of nothing at all, with a total energy of zero, at a point in spacetime.

Powell, Cecil Frank (1903–1969)   British physicist who developed the technique of studying elementary particles from the tracks they leave in photographic emulsion. He was awarded the Nobel Prize for Physics in 1950, for his discovery of the *pion*.

Born in Tonbridge, in Kent, on 5 December 1903, Powell graduated from the University of Cambridge in 1925, then worked at the Cavendish Laboratory under the supervision of Charles Wilson (inventor of the *cloud chamber*) on improving techniques for photographing particle tracks in cloud chambers. This led to the award of his PhD in 1928.

Powell then moved to the University of Bristol, where he stayed for the rest of his career, becoming professor of physics in 1948 and director of the Wills Physics Laboratory in 1964. He was deeply concerned about arms control and the social responsibilities of scientists, took part in the Pugwash conferences on science and world affairs, and was one of the scientists involved in the creation of CERN. Powell died in Bellano, in Italy, on 9 August 1969.

The technique which led to the discovery of the pion was developed from 1938 onwards, when Powell used a stack of photographic plates, instead of a cloud chamber, to trace the tracks of particles from cosmic rays, providing a direct, permanent record. The technique was boosted by the development of sensitive photographic emulsions during the Second World War for military purposes; the pion was discovered in 1947.

Poynting, John Henry (1852–1914)   British physicist who determined the density of the Earth and made an accurate measurement of the gravitational constant in the 1890s. In 1903 he suggested that radiation from the Sun can have the seemingly paradoxical effect of making small particles in orbit around the Sun spiral inwards until they are consumed by the Sun's heat. This idea was taken up and developed in the context of *relativity theory* by the American physicist Howard Robertson, in the 1930s; it is now known as the Poynting–Robertson effect.

Poynting also carried out important work on the theory of electromagnetism, and showed that the flow of electromagnetic energy across a surface can be expressed in terms of a *vector* quantity, now known as the Poynting vector. And in 1884 he published a statistical analysis of the fluctuations in prices on the London Stock Exchange.

Born in Monton, near Manchester, on 9 September 1852, Poynting studied at Owens College in Manchester (later Manchester University) and received his degree (like all students at Owens College) as an external student of the University of London (BSc 1872). He then studied at the University of Cambridge from 1872 to 1876 (although he never formally received a PhD, he was awarded a Doctor of Science degree by Cambridge in 1887). From 1876 to 1878, Poynting worked at Owens College. He then spent two years as a research fellow in Cambridge (working under James Clerk Maxwell at the Cavendish Laboratory), before he was appointed professor of physics at Mason Science College (which later became the University of Birmingham) in 1880, a post that he held until he died (from diabetes), in Birmingham, on 30 March 1914.

Poynting–Robertson effect   See *Poynting, John Henry*.

Poynting vector   See *Poynting, John Henry*.

principle of equivalence   See *relativity theory*.

principle of least action   See *action*.

principle of least time   = Fermat's Principle; see *variational principle*.

principle of terrestrial mediocrity   See *anthropic principle*.

Privatdozent   An unpaid university lecturer. Such a post provided the way in which young academics used to get their foot on the first rung of the ladder of a university career, particularly in Germany in the 19th century. A Privatdozent was allowed to give lectures in the university, but received no salary and obtained an income, such as it was, by charging an admission fee for his (it was almost always an exclusively male prerogative) lectures. The story that some British universities are thinking of introducing the system for the 21st century are reliably reported to be scurrilous rumours.

probabilistic interpretation   See *ensemble interpretation*.

probability   In physics, probability means exactly what it means in everyday life, or in the casino. It is the chance, or likelihood, that a particular event will happen. If a perfectly balanced coin is tossed fairly, there is an exact fifty-fifty chance that it will come down heads, and a fifty-fifty chance that it will come down tails. The probability for each outcome is therefore ½. As this simple example shows, if every possible outcome of a process is considered, the total probability is exactly 1; probabilities add up for different outcomes to the same event.

A slightly more complicated example shows how probabilities can be calculated. If two dice are rolled (from now on, we take it as read that everything is perfectly balanced and there is no cheating), what is the chance of a particular total number, between 2 and 12, coming up? For each die, there are six equally likely outcomes, so the chance of getting any particular number, from 1 to 6, is ⅙. But for the combinations possible for two dice, the probabilities are not all equal. To get the total 2, for

UNE MAISON DE JEU A L'ÉPOQUE DE LA RENAISSANCE

Probability. At the quantum level, the Universe operates in accordance with the same laws of probability as an honest dice game.

example, there is only one possible combination, with each die showing the number 1. The probability of each die coming up 1 is ⅙; the probability of both together coming up 1 is found by multiplying these two probabilities together, so it is 1/36.

But the probability of getting a combined total of 3 is different. You can get this in two ways: if the first die comes up 1 and the second die comes up 2, or the other way around. Each different outcome has a probability of 1/36, so the total probability of getting 3 is 1/36 *plus* 1/36, which is 1/18. The most likely outcome of throwing two dice is to get the total 7, which can be made in six different ways (1 + 6, 2 + 5, 3 + 4, 4 + 3, 5 + 2, 6 + 1) and has an overall probability of ⅙.

With larger numbers of things interacting with one another, and many possible combinations of outcomes, calculating the probabilities in this way would be very tedious, but the overall pattern of probabilities can be described statistically, and is represented by a distribution in which there are many ways for more likely outcomes (equivalent to throwing a 7 with two dice) to happen, and only a few ways for less likely outcomes (equivalent to throwing a 2 with two dice) to happen. Rare events are said to lie in the 'tail' of the distribution. But even though we can't take the time to add up all the probabilities in the right way, nature does so, through the workings of 'blind chance'.

In his book *The Cosmic Code*, Heinz Pagels gives an example of chance working on a large scale by analysing the number of dog bites reported to doctors or hospitals in a big city over the years. Over an interval of five years, the numbers of reported bites were 68, 70, 64, 66 and 71, an average of just under 68 bites per year. 'Because the events are random and independent,' says Pagels, 'the distribution is stable.' It is *possible* that there will be a year in which there are just 5 reported bites, or as many as 500; but these probabilities lie way out in the tail of the distribution.

One important feature of this kind of study, though, is that with data from enough years (enough independent outcomes of the 'experiment'), statisticians can work out these probabilities and get a good idea of the overall distribution, so they can calculate the likelihood of rare events that have never actually been observed. It is the same kind of statistical analysis that leads meteorologists to refer to a 'once in a century' storm, or flood, even where records have not been kept for 100 years. They mean it is the kind of event that has a 1 in 100 probability of occurring in any year, just as you would expect to get 2 once in every 36 rolls of a pair of dice. But just as you might get 2 on two consecutive rolls, you might get two 'once in a century' floods in successive years.

Probability lies at the heart of the mystery of quantum reality, because the quantum world obeys strict probabilistic rules. It is *not* the case that by having a large number of quantum events happening at the same time you can predict what will happen in a statistical way, like the prediction that in the city Pagels mentions there will be about 68 reported dog bites next year (see **ensemble interpretation**). Quantum probability can be seen to be at work at the level of individual atoms, photons and electrons. Worse, the probability associated with a quantum entity confronted with a choice of outcomes at one location is affected – instantaneously – by what is going on at another location (see **Aspect experiment**, **double-slit experiment**). It is as if the probability of my throwing a 6 on a die in my study depends on the number you have thrown on a die in your room.

One clear example of probability at work in the quantum world comes from radioactive *decay*. Each nucleus in a sample decays spontaneously, at random, in accordance with the blind workings of chance. Yet the combined effect is that exactly half of the nuclei in a sample (no matter what number you start with) decay within a precisely determined time, the *half life*. It is this kind of quantum behaviour that led Albert Einstein to make his famous remark, 'I cannot believe that God plays dice.' But all the evidence is that the quantum world really is ruled by chance.

Quantum probability is not always, though, like everyday probability. An example involving *polarization* makes this clear.

The Polaroid in an ordinary pair of sunglasses is made so that it transmits only vertically polarized light. You can think of it as like a picket fence, with vertical slots in it, and the light falling on the lens as carrying spears, or arrows, oriented in all directions across the line of flight of the photons. Photons carrying vertical arrows will slip through the slots in the fence, but photons carrying arrows oriented in any other direction will not get through.

The choice of vertical polarization for sunglasses is because light reflected from horizontal surfaces (such as the road, or the sea) tends to be horizontally polarized, so the glasses cut out reflected glare – as well, of course, as cutting out all incoming sunlight except the vertically polarized waves. So if you take two pairs of Polaroid sunglasses, and turn one so that the earpieces are at top and bottom, instead of on either side, you will have two sets of polarizing filters, one that transmits only vertically polarized light and the other that transmits only horizontally polarized light. Put the glasses together so that one lens from the second pair is directly behind one lens from the first pair, and look through the lined-up lenses. As you would expect, no light is transmitted. The view is black.

But what happens if you hold the second lens so that its polarization is oriented at 45 degrees to that of the first lens? Common sense says that if the first lens is vertically polarized, only vertically polarized light, with upward pointing arrows, gets through it, and that upward-pointing arrows cannot get through a slot oriented at 45 degrees. Alas for common sense, what actually happens is that half of the light gets through the second polarizing filter. And what does get through comes out on the other side with diagonal polarization.

You have probably guessed what's coming. If a third, horizontally polarized filter is now put in line with the other two, half of the diagonally polarized light (a quarter of the original light) gets through. And now, it is horizontally polarized! Taking the three-lens system as a whole, you have seemingly passed horizontally polarized light through a system which includes a vertical polarizing filter – like passing a horizontal spear through a vertical slot in a picket fence.

You can, if you like, start with two polarizing filters, crossed at right angles so that no light gets through. You then add a third filter, between the two of them, oriented in such a way that it ought, according to common sense, to block light coming through either of these original filters. And now, 25 per cent of the light is transmitted.

The experiment can be carried out with two polarizers in various orientations, and the number of photons that gets through varies in a smooth and regular way, from 100 per cent when the polarizers are oriented in the same way, to zero when

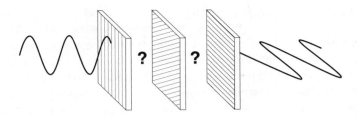

Probability. The bizarre behaviour of polarizing filters. Without the middle filter in place, the other two would be 'crossed' (at right angles to each other) and no light would get through, because there is zero probability that light passing through the vertical filter will end up horizontally polarized. With the 45 degree polarizing filter in place, 25 per cent of the light gets through the whole set-up. This is because there is a 50 per cent probability that the light getting through the vertical filter is polarized at 45 degrees, and a 50 per cent chance that the fraction of the light that passes through the 45 degree polarizer is oriented horizontally and can get through the last filter.

they are crossed at right angles. And the same effect happens if individual photons are fired, one at a time, through the experiment. In the basic three-filter set-up with crossed and diagonal filters, one photon in four gets through the triple system – but there is no way to predict in advance which photons will pass and which will get stopped at any of the barriers they encounter.

It seems as if each 'vertically polarized' photon (that is, one which has passed through a vertically polarizing filter) actually has a well-defined probability of being oriented with a different polarization – zero chance of being horizontally polarized, a 50 per cent chance of being diagonally polarized, and so on. The photon has to be regarded, on the standard interpretation of quantum physics (see *Copenhagen interpretation*) as being in a *superposition of states* until it reaches a polarizing filter. Then it 'decides' whether or not it is polarized in the right way to get through (choosing at random, just as if it were throwing dice). If it does get through, it is now in a new superposition of states, with different probabilities that will govern the outcome of its next encounter with a polarizing filter. Its behaviour is always governed by the strict rules of probability, but the probabilities are changed by events. As Paul Davies has put it, in his book *Other Worlds*:

> It must be emphasised that the quantum indeterminacy does not merely mean that we cannot know which polarization direction the photon really possesses, it means that the notion of a photon with a definite polarization direction does not exist. There is an inherent uncertainty in the *identity* of the photon itself, not just in our knowledge of it.

That 'inherent uncertainty' means that we can describe the outcome of quantum events *only* in terms of probability.

See also *wave–particle duality*.

probability wave   The interpretation of the *wave function* of a quantum entity such as an electron (see *Broglie, Louis de* and *Schrödinger, Erwin*) as a wave of *probability* passing through the Universe and specifying the probability of finding the electron (or whatever) at a particular point in space. This idea, developed by Max Born, is one of the main planks in the platform of the *Copenhagen interpretation*. The square of the *amplitude* of the probability wave at any point gives the probability of the particle being found at that point. Probability waves can interfere with one another, like everyday waves, changing the appropriate probabilities. The probability wave can even, as in the *double-slit experiment*, pass through the Universe in a complicated fashion so that it interferes with itself to produce a pattern of probabilities which depends on the nature of its surroundings – or on the nature of the experiment we choose to carry out.

An analogy which some people find useful is with a crime wave. A so-called crime wave is really a change in the probability of a crime, or a particular kind of crime, being committed. The crime wave may move about, in both time and space, as the probabilities change (for example, from day to night, or as the population of a city moves from the centre to the suburbs).

The whole idea of probability waves is seriously flawed, though, because it separates the waviness of the quantum world from its particle facet, as if things like electrons and photons 'really are' point-like particles. You might find that easy to accept for electrons, but it would be interesting to go back in time and ask James Clerk Maxwell, let alone any modern physicist, if he thinks of light waves as merely probability waves. The best that can be said of the probability wave idea, as with the whole Copenhagen interpretation, is that it works in the sense that it provides a way to calculate the probabilities associated with quantum events. But what it *means* is very hard to fathom. In any case, there is now direct experimental evidence which shows individual quantum entities behaving as both wave and particle at the same time. See *wave–particle duality*.

Project Poltergeist   See *neutrino*.

Prokhorov, Alexander Mikhailovich (1916–)   Russian physicist who shared the Nobel Prize for Physics in 1964 with Nikolai Basov and Charles Townes, for his contribution to the development of *masers* and *lasers*.

Prokhorov's parents were Russian émigrés living in Australia when he was born (on 11 July 1916), but the family returned to Russia after the 1917 revolution. He graduated from Leningrad University (now the University of St Petersburg) in 1939, served with the Red Army during the Second World War, completed his PhD in Leningrad in 1946, and then spent his scientific career at the Lebedev Physical Institute, in Moscow, where he worked with Basov. Alongside this post, he was appointed professor of physics at the University of Moscow in 1954.

proportional counter   A type of particle detector in which the size of the output pulse of electricity is proportional to the number of ions produced when an ionizing particle passes through the chamber of the counter. One of the first particle counters, developed by Ernest Rutherford and Hans Geiger, worked in this way, using a single wire. Some of the most effective modern counters, developed by Georges Charpak, extend the technique using many wires. See *multiwire proportional chamber*.

proton   One of the *elementary particles*, a *baryon* with a rest mass of $1.6726231 \times 10^{-27}$ kg (= 938.2796 MeV), an electric charge of $+1.60218925 \times 10^{-19}$ coulombs and a spin of ½, which obeys *Fermi–Dirac statistics*. See also *atomic physics*.

proton decay   A prediction of the *grand unified theories*, which attempt to unite the electroweak and the strong forces (see *forces of nature*) and relate *quarks* to *leptons*. These theories require that quarks, which are heavier than leptons, should ultimately decay into leptons. Protons are made of quarks, so they should decay as a result (neutrons, of course, decay into protons).

The typical half life for the proton predicted by these theories is in excess of 10^{32} years, so there is little point in watching an individual proton and waiting for it to decay. But the rules of *probability* which govern the quantum world imply that, if you have 10^{32} protons in one lump of matter, one of them will decay every year (if the half life is indeed 10^{32} years). One kilogram of matter (such as a lump of iron or a tank of water) contains about 1.7×10^{27} nucleons (just under half of them protons in most substances; all of them protons in water). So 1 million kilograms (1,000 tonnes) of water contain 1.7×10^{33} protons.

This is a manageable amount of material to work with. Several experiments are now under way which monitor huge tanks of water, or lumps of iron or steel, to search for effects attributable to proton decay. The characteristic nature of this kind of decay is predicted by theory (typically, one quark inside a proton might turn into a positron, leaving the other two quarks paired up as a pion with zero charge) and observations of such events should provide a characteristic 'signature' of proton decay. Alas, none has yet been seen. This rules out the simplest versions of the grand unified theories, but leaves open the possibility that protons decay on even longer timescales, as predicted by other GUTs.

psi meson   See *J/psi particle*.

QCD   = *quantum chromodynamics*.

QED   = *quantum electrodynamics*.

q numbers   See *quantum algebra*; note that this is *not* a shorthand for *quantum numbers*.

qualitative calculation   A rough calculation intended to give a feel for the sort of size of the variables involved. When a physicist says, for example, that a neutron star is 'about 10 km' across, it should be no surprise if a detailed *quantitative calculation* comes up with a diameter of 8 km, or 16 km (or even 20 km).

quantitative calculation   An accurate calculation giving a precise value for the size of the variables involved, strictly speaking to a specified level of accuracy (a specified number of decimal places). The calculation of the *magnetic moment* of the electron in *quantum electrodynamics* is the supreme example of a quantitative calculation in quantum physics.

quantization   The division of physical parameters such as *energy* (the archetypal example), length and time into discrete units. This means that in the world of the very small it is not possible to have continuous, infinitesimal changes in the properties associated with entities such as photons and electrons, and changes must occur in discrete jumps (quantum leaps).

A helpful analogy can be made with the operation of an automated cash dispenser at a bank. The machine at my own bank is set so that it can deliver cash only in multiples of £10. I can draw, say, £60 or £70, but not £65. This does not mean that in-between amounts (like £65) do not exist, but that the machine is incapable of handling them. When Max Planck developed the first quantum theory of electro-magnetic radiation (see *black body radiation*), it was assumed that something like this was going on, with atoms able to 'dispense' electromagnetic energy only in lumps of a certain size, even though in-between energies could exist. But this is not the case. As quantum theory was developed, it became clear that there are no in-between energies allowed for photons, as if money came only in multiples of £10.

In fact, the analogy can be extended further, since in a sense money really is quantized – in the example of sterling currency, the smallest unit is the penny, and it is impossible to have a physical cash sum which does not add up to a whole number of pennies. Because pennies are a very small unit of currency, they add up to give almost any value you like, giving the impression of a continuously variable system of pricing things; because each photon carries only a tiny amount of energy, they add up to give the impression of a continuous spectrum of light. But don't try to push the analogy too far, since, unlike photons and other quanta, a penny could be physically cut into smaller pieces, even though they would have no monetary value.

The term 'quantization' is also used to refer to the way a classical field can be described in terms of particles: for example, the description of the electromagnetic field in terms of the exchange of photons, or the description of gravity in terms of gravitons. See also *second quantization*.

quantum   Used as a noun, the smallest amount of something that it is possible to have. Electric charge, for example, comes only in fixed multiples of a basic unit, the quantum of charge. For everyday purposes, this quantum of charge is equal in magnitude to the charge on the electron; quarks have charges which are a fraction of the charge on the electron, which means that (even though we never detect a free quark) the basic quark charge, equal in magnitude to one-third of the charge on the electron, can be regarded as the fundamental quantum of charge. Plural, quanta. See *quantization*.

Used as an adjective, a reference to the world of the very small, where the rules of quantum physics have to be applied.

quantum algebra   Version of quantum mechanics developed by Paul Dirac in 1925, after learning of Werner Heisenberg's breakthrough realization that quantum entities can be described in terms of variables which do not commute (see *Abelian group*). Dirac realized, independently of the Göttingen group, that this is of fundamental importance to an understanding of the quantum world, and that such entities could be described using matrices. Dirac's version of quantum mechanics was developed using a mathematical formalism that is different from both *matrix mechanics* and *wave mechanics*, but which was later shown to include both of these variations on the theme within itself, and to do more besides. Like matrix mechanics, Dirac's version includes *classical mechanics* within quantum mechanics as a special case, corresponding to setting *Planck's constant* equal to zero or to large values of the *quantum numbers*.

Quantum algebra involves the manipulation of entities which Dirac called 'q numbers', which have bizarre properties by the standards of the everyday world – for example, it is impossible to say which of two numbers a and b is bigger; the concept of one number being bigger or smaller than another has no place in quantum algebra. But, as with other variations on the quantum theme, the rules of this mathematical system exactly fitted the observations of the behaviour of atomic processes. By including the effects of the special theory of relativity, Dirac found among other things that the equations predicted the recoil of an atom when it emits light (see *Compton effect*). The theory also included, from the outset, a formulation for half-integral *spin* quantum numbers; and it was using quantum algebra that Dirac later developed the idea of *antimatter*. Heisenberg, who was sent a copy of the key quantum algebra paper by Dirac before the end of 1925, was generous in his praise:

> I have read your extraordinarily beautiful paper on quantum mechanics with the greatest interest, and there can be no doubt that all your results are correct ... [The paper is] really better written and more concentrated than our attempts here.

Unfortunately, in the years following 1926 most physicists were seduced away from the quantum algebra (or even the matrix mechanics) approach by the seemingly comfortable familiarity of Erwin Schrödinger's wave mechanics, which actually obscures the weirdness of the quantum world and held back any proper understanding of what was going on for decades.

quantum chemistry The description of the way *atoms* combine to form *molecules*, and the way molecules interact with one another, using the rules of *quantum physics*. One of the key insights in quantum chemistry is that, because an electron is not a classical particle located at a definite point in space, even a single electron can 'surround' the nucleus of an atom, filling a volume roughly as big as the whole atom. The region in which a particular electron is contained is known as an *orbital*. But instead of thinking of the electrons as in different shells, neatly nested inside one another like onion rings, it is better to visualize them all interpenetrating, like a lot of ripples on a pond. Each individual electron cloud extends down to 'touch' the nucleus, and all the electrons in an atom come under the direct influence of the nucleus, although some are influenced more strongly than others. The most important distinction is that some orbitals are more concentrated further out from the nucleus than others. This corresponds to electrons being further up the rungs of the energy ladder associated with that atom, and therefore more easily detached from the atom than electrons lower down the ladder.

Some electrons fit symmetrically around the nucleus, but very many electron orbitals have a definite shape and a definite orientation with respect to one another. In very many cases, the orbitals stick out from atoms in clearly defined directions, which are determined by the laws of quantum physics. Since the electron cloud of an atom is what another atom 'sees' and interacts with, this affects the chemical behaviour of the atom. And each orbital has room for two electrons, one with spin up and one with spin down.

The spacing of orbitals on the energy ladder occurs because there is not enough room next to the nucleus for all of the orbitals to fit. Only two electrons can fit into the closest, most tightly bound orbital. At the next energy level (in a sense, a little further out from the nucleus) there is room for four orbitals, containing a total of eight electrons when they are all full. Such a set of orbitals with essentially the same energy as each other corresponds to the older idea of an electron shell. Full shells are particularly stable, and as a result atoms 'like' to arrange themselves to get filled outer shells. This means, for example, that helium (with just two electrons, filling its only occupied shell) and neon (with a total of ten electrons, exactly filling its two occupied shells) are both stable, unreactive elements – members of the family of inert gases. It also means that an atom of hydrogen, with a lone electron in its single occupied shell, is eager to combine with another atom of hydrogen (among other things), so that the two electrons shared between the two nuclei give each an illusion of having a filled shell (see *chemical bond*).

This is the basis of quantum chemistry. Atoms exchange, or share, electrons in such a way that they get as close as possible to the desirable state of having a fully occupied outer shell. The quantum rules which describe how this happens were largely developed by Linus Pauling, soon after quantum physics itself was established. The idea of sharing electrons between atoms came originally from the American Gilbert Lewis, as early as 1916. But at that time even the concept of electron spin had not yet been dreamed of, and there was no explanation as to why paired electrons, and filled shells, are important in chemical bonding. That had to await the development of quantum mechanics, after 1925. The quantum rules do explain why the energy levels are where they are, and why filled shells are particularly stable. And they provide other insights. For example, at a very simple level, the two electrons shared between two hydrogen nuclei in a hydrogen molecule should not be thought of as sitting in between the two nuclei forming a link between them, but as forming a single elongated cloud (a molecular orbital) surrounding both nuclei.

Within a year of Erwin Schrödinger publishing his quantum mechanical wave equation, two German physicists, Walter Heitler and Fritz London, had used this mathematical formalism to calculate the change in energy when two hydrogen atoms, each with its own single electron, combine to form one molecule with a pair of shared electrons. The change in energy depends on the rearrangement of the electrons within the electric field produced by the two nuclei, reminiscent of the change in *potential energy* of an object moving to a different position in a gravitational field. When Heitler and London calculated the difference in energy between two hydrogen atoms and one hydrogen molecule, they came up with a number very close to the amount of energy which chemists already knew, from experiment, is required to break the bonding between atoms in a hydrogen molecule. Later calculations, including refinements made by Pauling, gave even better agreement between theory and experiment (and the agreement is just as good, of course, whichever version of quantum mechanics you work with).

This was a really startling development in 1927 and did much to establish quantum mechanics in general, and wave mechanics in particular, as a practical tool. Instead of just saying that, for reasons unknown, electrons 'like' to pair up and atoms 'like' to have filled shells, physicists were able to calculate the change in energy when electrons paired up and shells were filled. The calculations showed, once and for all, that the way atoms react with one another is not an arbitrary process, but one in which the most stable arrangements of atoms in molecules are always the arrangements with least energy. It was in 1927 that theoretical chemistry truly became a quantitative science, rather than mere qualitative description, and became indisputably a part of physics.

In principle, the same kind of calculations can be applied to any molecule. In practice, with larger numbers of electrons involved this becomes extremely difficult for anything much more complicated than a hydrogen molecule, and various approximation techniques (which we won't go into here) are used to make the calculations easier. But we can't resist going into a little detail about the way quantum chemistry gives an insight into the chemistry of more complicated molecules, including those based on carbon (organic chemistry), the basis of life itself.

A carbon atom has six electrons, two in a full inner shell and four in the next shell, one in each of the four orbitals available at that level. These orbitals are identical, each made up of a mixture of the one spherical (*s*) orbital and three elongated (*p*) orbitals (at right angles to each other) that simple theory says can exist on this rung of the energy ladder. In an early triumph of the quantum theory of chemistry, in 1931 Pauling showed how these four simple orbitals interact to form the four hybrid (*sp*) orbitals, which point towards the corners of an imaginary tetrahedron centred on the carbon nucleus. Each of these orbitals can form a chemical bond by capturing, or sharing, an electron from another atom. In the simplest example, each orbital latches on to the single electron from a different hydrogen atom, forming a molecule of methane, CH_4.

The quantum mechanical calculation explained why, in molecules such as methane, carbon does indeed have four bonds, all identical in strength to one another, aligned as if pointing towards the corners of a regular tetrahedron. The electrons in each of these bonds, though, are in a hybrid state, a mixture of *s* and *p* orbitals, reminiscent of the way an electron has to be thought of as being both wave and particle at the same time.

This idea led Pauling on to develop the idea of resonance, which in this context asserts that, if a molecule can be described in two (or more) equally acceptable ways (that is, arrangements which have equal energies), then the molecule has to be thought of as existing in all of those states simultaneously. The example of ozone, a relatively simple molecule, gives a picture of what is going on.

Oxygen has six electrons in its outermost occupied shell. It therefore has a *valency* of two, and would 'like' to capture two more electrons to fill the shell. This is why it forms the stable compound water, with a chemical formula H_2O. Each of the two hydrogen atoms in a water molecule shares its lone electron with the oxygen atom, giving the oxygen atom the illusion of eight electrons in the outermost occupied shell; and each of the hydrogen atoms gets a share in one of the electrons from the oxygen atom, giving it the illusion of a full shell. In a slightly more complicated variation on the theme, an oxygen molecule, O_2, forms because a pair of oxygen atoms can each contribute two electrons to the bonding process, forming a double bond. But how can we explain the existence of a tri-atomic form of molecular oxygen, ozone, O_3?

Pauling's insight, backed up by quantum mechanical calculations which correctly give the energies associated with the bonds in ozone, is that this is an example of resonance. One possibility that we have not yet mentioned is that one oxygen atom could give up an electron completely to the other oxygen atom. The first oxygen atom would be left with a net positive charge, and the second oxygen atom would have a net negative charge, so they would be attracted to one another (this is called an ionic bond). The first atom is also left with an outer occupied shell containing just five electrons, so it could now make three covalent bonds, the kind we have already discussed. The second oxygen atom, though, now has seven

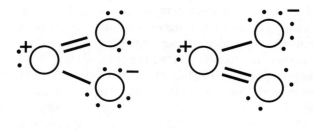

Quantum chemistry. The two ways in which three oxygen atoms can get together to make an ozone molecule, according to classical ideas. In fact, quantum processes smear out the contributions of the electrons to give each of the two bonds in an ozone molecule the strength of one and a half ordinary bonds, intermediate between a single bond and a double bond.

electrons in its outer occupied shell, with room for just one more, so it could now make just one covalent bond. So the first atom (strictly speaking, it is now a positive ion) could form a double bond with a third oxygen atom, and still have the capacity to form a single bond with the second atom (really, a negative ion).

This provides two ways to make a molecule of ozone, indicated in the diagram. A double bond is stronger than a single bond, and provides a tighter link between the atoms involved. But spectroscopic studies show that there is no difference in the bond strengths of the two bonds in an ozone molecule. Each of the bonds acts as if it were a 'one and a half' strength bond. The real structure is a resonance between the two possible quantum states we have shown here, a molecule with two equally strong bonds and a rearrangement of electric charge such that one end of the molecule is slightly positive and the other end is slightly negative. And, we stress, the strength of the bonds, determined by spectroscopy, exactly matches the predictions of the quantum calculations.

Something very similar happens when the carbonate ion, CO_3, forms. This structure has a double negative charge because it has taken up two electrons from other atoms, with which it forms an ionic bond (for example, in carbonate rocks such as limestone). But the CO_3 structure itself consists of a carbon atom held to three oxygen atoms by three hybrid bonds, a resonance in which each has a strength equivalent to 1.333 normal bonds; we'll leave you to work out the details.

One particularly important example of resonance occurs in the benzene ring, a structure made of six carbon atoms joined in a loop, with each bond between adjacent pairs of carbon atoms equivalent to 1.5 ordinary bonds. So each carbon atom uses three bonds (1.5 on each side) to hold itself in place in the ring, leaving one bond free to latch on to something else, outside the ring. This structure is at the heart of many complex organic molecules, including some amino acids, sugars and proteins; variations on the ring shape, involving other atoms as well as carbon but still with hybrid bonds, are found in DNA, RNA and the bases which are attached to DNA and RNA and carry the genetic code. And *hydrogen bonds*, themselves a quantum phenomenon, hold the two strands of a DNA double helix together (see *molecular biology*). The chemistry of life is very much a quantum phenomenon, which can be fully understood only in terms of quantum physics.

Further reading: John Gribbin, *In Search of the Double Helix.*

quantum chromodynamics   Theory that describes the way *quarks* interact with one another by the exchange of *gluons*. The name was deliberately chosen to echo the name of *quantum electrodynamics*, the theory that describes the way charged particles interact with one another by the exchange of *photons*, and on which quantum chromodynamics (often referred to as QCD) is based. The 'chromo' part of the name comes from the way in which a property of quarks and gluons that is analogous to electric charge is given the names of colours, and referred to as colour charge. This is purely a whimsical convention; it does not mean that quarks and gluons are coloured in the everyday sense of the term.

In the mid-1960s, two families of *leptons* were known, each made up of an electron-like particle and a neutrino. These are the electron and its neutrino, and the muon and its neutrino. But at that time, when the idea of quarks was introduced, only three different kinds of quark were needed to explain all the known *hadrons*. The up and down quarks formed a pair, but the strange quark was an oddity, a seemingly truly fundamental particle (if the theory were correct) without a partner. Although Murray Gell-Mann did suggest, in the paper in which he introduced his version of the idea of quarks, that there ought to be a fourth quark to partner the strange quark, the idea was not taken seriously at first because there was no evidence of any particle which incorporated the fourth quark.

Indeed, the whole idea of quarks was not taken seriously at first, until the experiments at SLAC at the end of the 1960s and in the early 1970s, which probed the structure within the proton (see *parton model*). But that didn't stop a few theorists from toying with the idea and developing it further. Those theorists were much more concerned by the puzzle that a particle such as the omega minus could be made up of three quarks (in this case, three strange quarks), all seemingly in the same quantum state. Since quarks are *fermions*, and obey the rules of *Fermi–Dirac* statistics, this ought to be impossible – no two identical fermions, let alone three, can be in the same quantum state.

The situation echoed the problem some 40 years earlier, when physicists had to explain how two electrons in an atom could occupy the same state by sitting on the same energy level. The answer then was that electrons possess a property called *spin*, and that the two electrons sharing an energy level in an atom are not in identical quantum states because one has spin up and the other has spin down. The obvious suggestion (obvious, that is, with hindsight) was that quarks must have some additional property which distinguished between the otherwise identical triplets in the omega minus and some other particles. In fact, quarks also possess spin, so two strange quarks could, in principle, occupy the same state inside an omega minus, one with spin down and the other with spin up; but there would be no way for another strange quark, which must have one or other of these two spin orientations, to fit in. So the argument that an additional property is needed to distinguish between quarks still holds, and in any case it turns out that all the three strange quarks in the omega minus do, indeed, have the same spin.

As very often happens in particle physics, it turned out that the mathematics needed to describe such triplets already existed (this power of mathematics has led at least one philosopher of science, Bruno Augenstein, to suggest that 'every version of mathematical concepts has a physical model somewhere, and the clever physicist should be advised to deliberately and routinely seek out, as part of his activity, physical models of already discovered mathematical structures'; see *Schrödinger's Kittens*). Oscar ('Wally') Greenberg, a theorist working at the University of Maryland, had been developing his exotic variations on the field theory approach for several years by 1964, when the idea of quarks was introduced. Greenberg had been playing with the mathematics for its own sake, with no thought of practical applications. But he quickly realized, and pointed out, that what he called 'parastatistics' could be applied to the problem of distinguishing the members of triplets of otherwise identical quarks from one another.

The idea was taken up by two Japanese theorists, Yoichiro Nambu (of the University of Chicago) and M. Y. Han (of Syracuse University), who collaborated, in 1965, in developing a version of Greenberg's highly mathematical theory that could be understood by physicists. It was at this point that the idea of labelling the property that distinguishes otherwise identical quarks from one another with the names of colours was introduced. But QCD was not established in all its glory until after the discovery of the fourth quark, charm, in the early 1970s (see *J/psi particle*).

We stress that the terminology is no more than a convenient semantic device, used to label the quarks – just as, indeed, the use of the words 'up' and 'down' to describe some property of particles that we choose to label 'spin' is a convenient semantic device. But it enables us to understand that there is a difference between a red up quark and a blue up quark, just as there is a difference between a red up quark and a red down quark.

The three 'colours' used to label quarks are often named red, blue and green; some physicists prefer to use the label 'yellow' instead of 'green', but it makes no difference to the argument. We shall skip the sometimes muddled development of the concept in the 1960s, leave out the elegant mathematics which explains the relationships between coloured quarks, and just give you the simple physical version of the full theory.

The key idea is that only 'colourless' combinations of quarks can exist. There are two ways in which this can be achieved. A combination of three quarks, each with a different colour, is colourless, in the same way that a combination of an electron and a proton, one with positive charge and the other with negative charge, has zero overall charge. And a quark–antiquark pair is colourless because the colour of the quark is cancelled out by the anticolour of the antiquark. This applies to all particles, not just to those which contain three otherwise identical quarks. So the proton, for example, should be thought of as made up not simply of two up quarks and one down quark, but of one red/up quark, one green/down quark and one blue/up quark. Mesons are made of quark–antiquark pairs – the pion, for example, can be thought of as a red/up quark paired with an antired/antidown quark. But single quarks, or groups of four, say, would carry a net colour, which seems to be forbidden.

The idea of a fourth quark, charm, was revived by Sheldon Glashow and two of his colleagues at Harvard (John Iliopoulos and Luciano Maiani) at the beginning of the 1970s, and Murray Gell-Mann and Harald Fritzsch took up the idea and developed a field theory approach to describe the various interactions of these particles. This is a Yang–Mills type of *gauge theory*, in which the coloured quarks interact with one another through the exchange of gluons. Colour takes on the role of electric charge in QED, with the complication that there are three kinds of colour charge but only one kind of electric charge (positive charge, in this sense, being 'anti-negative', in the same way that colour charge can be red or antired, and so on). Like colours repel one another, and unlike colours attract – so two red quarks, say, repel one another and are not found in the same triplet, but a red and an antired attract one another and form a meson. Quarks of different colours – for example, red and blue – attract one another, but less strongly than the attraction between a colour/anticolour pair.

But whereas only one kind of particle, the photon, is needed to mediate the electromagnetic force, and just three kinds of particle (the W^+, W^- and the Z^0) are needed to mediate the weak interaction, a total of eight kinds of particle are needed to mediate what became known as the colour force, which operates between quarks.

These mediating particles are now known as gluons. Things are made even more complicated because quarks can change colour. This means that the gluons themselves must be coloured, in order to be able to carry colour from one quark to another. And that in turn means that gluons themselves are affected by the colour force. Because they carry colour, gluons could get together to make a kind of particle consisting entirely of glue, held together by the colour force, with no quarks; they would also be forced to travel in clusters with no overall colour visible. These clusters are called *glueballs* (or sometimes, gluonium); there are hints (but, as yet, no more than hints) that glueballs may have been produced in some particle accelerator experiments. The self-interaction between gluons results from the fact that QCD is a *non-Abelian gauge theory*. It is also responsible for the *asymptotic freedom* associated with the colour force.

The basic ideas can be understood in terms of *symmetry* and *symmetry breaking*. Imagine that each quark has within itself a means of altering its colour – a kind of internal pointer which can point in any one of three directions, at 120 degrees to one another (see diagram; this is like the idea of *isotopic spin* as a way of distinguishing nucleons). A symmetric global gauge transformation would be one that turned every pointer clockwise (say) by 120 degrees, changing the colour of every quark in the Universe, but leaving the laws of physics unchanged. A local symmetric gauge transformation might change the pointer setting (that is, the colour) of just one quark inside one hadron, but leave the rest of the world unchanged. It is the need to restore symmetry under local transformations that requires the existence of new fields, corresponding to the eight massless gluons each with zero electric charge.

Quantum chromodynamics 1. The 'colour' associated with a quark can be thought of as a property like isotopic spin. The visible 'label' on an individual quark may be red, blue or green (indicated by the uppermost colour in this diagram); but we can imagine rotating the quark to reveal one of the other colours. This is an example of a symmetry operation.

Because QCD (like QED) is a gauge-invariant theory, its basic equations work in the same way and give the same results at all points in space and time. The specific form of the symmetry is called SU(3), which means special unitary group in three dimensions; however, the three 'dimensions' here correspond not to the physical dimensions of space, but to the three colours (or pointer settings) available to quarks.

In physical terms, any quark inside a hadron is free to change its colour, independently of all other quarks, but can do so only by emitting a gluon, which is promptly absorbed by another quark. Most individual gluons carry a mixture of a colour and an anticolour *of a different kind*; this is why there are so many kinds of gluon. You might think that there ought to be nine kinds of gluon – red/antiblue, red/antigreen, blue/antigreen, blue/antired, green/antiblue, green/antired, plus three kinds of gluon which do not cause colour changes, red/antired, green/antigreen and blue/antiblue. In fact, there is a redundancy built into the group-theoretical description of gluons which says that, in effect, antiblue is equivalent to a red/green combination, antired is equivalent to a green/blue combination, and antigreen is equivalent to a red/blue combination. So one of the three non-colour-changing gluons doesn't count – it doesn't matter which one, but for the sake of argument take red/antired, which is equivalent to red/blue/green, which is equivalent to antigreen/green, which we already have. (This is a gross oversimplification of a sophisticated concept which is discussed fully in, for example, *The New Physics*, ed. by Paul Davies.)

When a colour-changing gluon is absorbed by another quark, the anticolour cancels out the colour of that quark, giving the quark the 'new' colour carried by the gluon. The second quark therefore undergoes a colour change which is of exactly the kind required to cancel out the change in the first quark and keep the hadron colourless. In other words, gluons are exchanged in order to keep the particle world white. For example, a blue up quark may emit a blue/antired gluon, becoming a red up quark; the gluon is absorbed by a red down quark, say, which becomes a blue

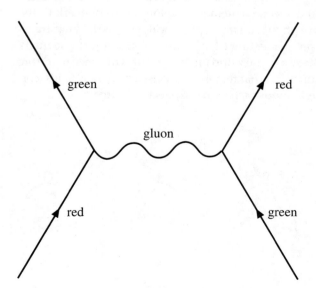

green red

gluon

red green

Quantum chromodynamics 2. Quarks interact with one another by the exchange of gluons. Gluons themselves carry colour, so they can change the colour state of the quarks they interact with. In this example, the gluon has rotated the state of one quark from red to green, and the other one from green to red.

down quark as a result. So all hadrons are colourless, even though the quarks within them may be undergoing kaleidoscopic changes of colour every instant.

Although free gluons, like free quarks, cannot exist and have therefore never been 'seen' directly in particle experiments, the way in which electrons scatter from protons shows that the three quarks inside the proton carry only about half of its momentum. The rest is carried by electrically neutral, massless particles which have no influence on the electrons that scatter from the internal constituents of the protons. They can only be the gluons which are constantly being exchanged between the quarks inside the proton. See also *parton model*. It is this constant exchange of gluons that binds quarks together inside hadrons; the resulting colour force is so powerful that it has been estimated that, in order to separate two quarks by 1 m, you would need as much energy as is required to lift 1 tonne of matter by 1 m from the surface of the Earth. In other words, the gravitational force of the whole Earth acting on a tonne of matter is roughly the same as the colour force between two individual quarks. On this picture, the strong interaction (see *forces of nature*) is a residual effect of the colour force, leaking out from within hadrons to affect the hadrons next door. The 'real' strong interaction is the force described by QCD.

The discovery of a third generation of quarks (associated with the third generation of leptons) was comfortably accommodated within this framework (see *standard model*).

quantum computers Proposed computers that would operate on quantum principles, and which in the most extreme incarnation would exist as a vast (potentially infinite) number of computers in the alternative realities of the *many worlds interpretation*, all working on a problem together and solving it in a fraction of the time required for a conventional computer.

This science-fiction-like scenario stems partly from the proven success of so-called parallel processing, in which many computer processors are linked to work on different aspects of a problem simultaneously. For example, if you wanted to calculate the orbits of 1,000 stars in a cluster under the influence of their mutual gravitational interaction, one way to do it would be to hold 999 stars 'still' in the computer simulation, calculate the forces on the other star, and allow it to move a tiny bit in response to those forces. Then hold that star still, and allow one of the others to move a tiny bit in the new pattern of forces, and so on. Alternatively, you could have 1,000 processors working in parallel, each calculating the orbit of a single star and making allowance for the continuing changes in positions of the other 999 stars.

Parallel processing works – it was developed by Danny Hillis and his colleagues at Thinking Machines Corporation in Cambridge, Massachusetts, in the 1980s, and has since begun to be applied to real problems in physics. As the name suggests, Thinking Machines was originally interested in artificial intelligence (and still is). It was, in fact, Richard Feynman who first appreciated the power of parallel processing at solving physics problems; his son, Carl, worked for Thinking Machines, and Richard Feynman spent some time there as a consultant in the mid-1980s. It was also Richard Feynman, in 1982, who pointed out the possibility of building a computer that operated on quantum principles. He further conjectured that a quantum computer could act as a 'universal quantum simulator', which would simulate the behaviour of any quantum system, following the rules of the quantum world rather than the logic of everyday life; it was only in 1996 that Seth Lloyd, of MIT, proved that this conjecture is correct, and that 'a variety of quantum systems, including quantum computers, can be "programmed" to simulate the behavior of arbitrary quantum systems whose dynamics are determined by local interactions' (*Science*, vol. 273, p. 1073, 1996).

The proof is technical, and the possibility of building such a machine is remote at present, but the effort may be worth while because, as Lloyd explains, a quantum computer with a few tens of quantum bits could carry out in a few tens of steps simulations that would require **Avogadro's number** of memory sites and operations in a classical computer. So such systems could offer our best hope of understanding what really goes on in the quantum world.

As usual, the best analogy to use to get an insight into what quantum computing is all about (especially quantum parallelism, an idea promoted vigorously by David Deutsch of Oxford University) is the **double-slit experiment**. Deutsch has suggested that, in the same way that a single photon goes by both routes through the experiment with two holes, so a single quantum computer will carry out calculations along all possible paths in parallel universes (see also **path integral**; although the enthusiasts for quantum computers tend to use the language of the many worlds interpretation, everything can be described equally conveniently in the path integral formulation, where the close similarities between what happens in a quantum computer and what happens in the real quantum world of light rays and the like become obvious).

The key to all modern computers is the use of binary switches of one kind or another to store and manipulate information in a string of zeros and ones ('on' or 'off' states of a switch). In a quantum computer, this might be done by using a single electron in an atom to represent 0 (the electron in its ground state) or 1 (the same electron in an excited state). Such an ultimate binary switch is sometimes referred to as a quantum dot. The way to excite the electron would be to shine a pulse of laser light on it; but if you shine the light for exactly the right amount of time, you will create a **superposition of states** in which there is a fifty-fifty chance of the quantum dot represented by the electron indicating 1 or 0 (see also **Schrödinger's cat**). In the many worlds interpretation, this means that in one world the quantum dot is set at 1, and in the other world it is set at 0. Extending this simple idea (vastly!) provides the image of a single quantum computer which is calculating all possibilities at once (for example, a superposition of a thousand states each calculating the orbit of a single star). Of course, all the different solutions would exist in different realities, and would become visible to us only when they interfered with one another, just as what we actually observe in the experiment with two holes is the **interference** pattern on the final screen. What you would see on the screen of the quantum computer in our world is the result of that kind of interference, a **sum over histories** representing the fruits of all the labours of the computer and its ghosts in the parallel realities.

Such a device would be practicable (if at all) only because of a key feature discovered by Peter Shor of the AT&T Bell Laboratories, in New Jersey. Shor was interested in whether it might be possible, in principle, to use a quantum parallel computer to crack certain kinds of allegedly uncrackable code. These codes depend on multiplying very large prime numbers together, and they could be cracked only by factorizing the resulting (much larger) numbers. This can be done in principle, but requires enormously powerful conventional computers and even then takes months or years to achieve. In 1994 Shor showed that a quantum computer no more powerful than a mid-1990s PC could solve such problems in a few seconds. He proved that if quantum parallelism works, then as all the ghost computers in alternate realities work together on the problem, in effect each guessing a number and trying it out to see if it is a factor of the big number, all the incorrect guesses cancel out through destructive interference. Only the routes through the quantum realities that lead to the right answer reinforce one another, combining as a single solution to the problem in what we regard as the real world.

For this particular problem, it doesn't even matter if the machine makes mistakes (perhaps because of background noise caused by imperfections in the crystals where the quantum dot electrons are trapped). Even if it only got the right answer once in a thousand times, it is so fast that you could run it a thousand times and find the right answer by simply multiplying up the claimed answers to see which one gave the correct large number!

With such obvious military/political/industrial espionage applications, it is certain that research into quantum computers will continue. These are real possibilities that could soon become fact. The Japanese computer industry, in particular, is already working on the possibilities of using quantum effects (such as *tunnelling*) in more effective conventional computers, and it would be a relatively small step from there to a true quantum computer.

quantum cookery   The use of the quantum equations, such as Erwin Schrödinger's *wave equation*, as recipes with which to solve problems involving quantum entities, without bothering to try to understand what is going on in the quantum realm and what the equations really mean.

quantum cosmology   The application of *quantum physics*, including the ideas of the *grand unified theories,* to describe the very early stages in the life of the Universe and, in some models, the origin of the Universe out of nothing at all. It deals with the era when what is now the entire visible Universe was confined within a volume no bigger across than an atom. This startling development, which brought within the bounds of respectable physics questions and ideas which had previously seemed to belong in the realms of philosophy and metaphysics (or even religion), developed in the 1980s largely following the introduction of the idea of cosmological inflation.

The key piece of information in all cosmology is that the Universe is expanding. If we imagine winding that expansion backwards, it means that some 15 billion years ago the Universe was in a superhot, superdense state (the Big Bang); taken to the ultimate conclusion, the expansion implies that the Big Bang itself emerged from a point of zero volume (a singularity) at 'time zero'. Standard cosmology (if you like, classical cosmology) developed in detail in the 1960s and 1970s, but went back to the discovery of the expansion of the Universe at the end of the 1920s. It used observations of the expansion (notably from the famous redshift in the light of distant galaxies) and the cosmic microwave background radiation (electromagnetic radiation left over from the Big Bang in which the Universe was born), plus the general theory of relativity (which describes, among other things, expanding spacetime) to paint an impressively detailed picture of everything that went on in the Universe after the time it had roughly the density of an atomic nucleus (about 10^{14} times the density of water).

This was about 0.0001 sec after time zero, when the temperature of the Universe was 1,000 billion kelvin. We have a thorough understanding of what went on under such extreme conditions (and under the less extreme conditions that followed as the Universe expanded and cooled) because experiments involving nuclear matter can be carried out in accelerators here on Earth, so we know how things like protons and neutrons behave. This standard model of cosmology shows how, in the first few minutes of the existence of the Universe, primordial protons, neutrons and electrons were 'cooked' into a mixture of roughly 75 per cent hydrogen and 25 per cent helium, with tiny traces of very light elements such as deuterium. This was the raw material which went into the first generation of stars, and eventually produced the galaxies that we see scattered across the Universe today.

Of course, this cooking of hydrogen and helium in the early Universe involved quantum processes, including *pair production* and annihilation, *fusion*, inverse beta decay and *beta decay* itself. But these processes are so routine, and so thoroughly understood, that they are regarded not as being part of quantum cosmology in particular, but just as part of the standard toolkit of physics. The question addressed by quantum cosmology is how the rapidly expanding fireball with the density of nuclear matter (the most extreme density known to occur naturally today) got to be in that state 0.0001 sec after time zero.

One of the key ideas in quantum cosmology (perhaps *the* key idea) is that there never was a time zero. The Universe as we know it must have been born with an age equal to the *Planck time*, the smallest unit of time that can exist. It would then have had the *Planck density*. Such a seed of the Universe could have arisen as a *quantum fluctuation* out of nothing at all, just as a pair of *virtual particles* can be manufactured out of nothing at all, using energy borrowed from quantum *uncertainty*. Although the mass-energy of such a quantum seed of the Universe would be enormous, this would be exactly balanced by the equally enormous, but negative, *potential energy* of its gravitational field. So the seed would have no net energy, and could be created as a virtual 'particle' with an infinitely long lifetime.

You might think that the enormous gravitational field associated with such a superdense entity would promptly crush it out of existence, whatever quantum uncertainty might say. But this is where cosmic inflation – itself a pure *quantum field theory* phenomenon – comes in. The grand unified theories tell us that, under the extreme conditions existing when the age of the Universe was the same as the Planck time, all of the *forces of nature* were on an equal footing, so that there was just one universal force. Gravity itself would have split off from the others at the Planck time (10^{-43} sec after time zero), and there is still no completely satisfactory theory of how *quantum gravity* works. As the Universe began to cool, the strong force would have split off by about 10^{-35} sec, and this process is much better understood, at least in outline.

The key point is that the splitting apart of the components of the original grand unified force would have been associated with *scalar fields* that acted as a kind of antigravity, pouring energy (essentially, some of the original mass-energy) into the expansion of the Universe. Today, the Universe is expanding at a more or less steady rate, in a linear fashion (actually, it is slowing down slightly under the influence of gravity, but we can ignore that for this discussion). It is like taking a walk down the road, in which each step takes you the same distance as the previous step. But in the very early Universe, the scalar fields that we have mentioned caused exponential expansion. This is like taking a walk in which the second step takes you twice as far as the first step, the third step four times as far, the fourth step eight times as far, and so on. When it was just over 10^{-35} sec old, the Universe would have been doubling in size at least once every 10^{-34} sec (some versions of inflation theory suggest even more rapid expansion).

This sounds modest, but it means that in just 10^{-32} sec the seed of the Universe would have doubled in size 100 times. That is enough to take a quantum fluctuation 10^{20} times smaller than a proton and inflate it into a sphere about 10 cm across in a time of just 1.5×10^{-32} sec. At that point (well, roughly; all these numbers are qualitative rather than quantitative), the scalar fields have done their work. As they fade away, they leave a hot fireball of energy, a grapefruit-sized lump containing everything that will become the visible Universe, expanding so rapidly that even though gravity does immediately start slowing the expansion down, it will take hundreds of billions of years first to halt the expansion and then to reverse it, sending everything (eventually) hurtling back together in a Big Crunch, the mirror image of the Big Bang.

Quantum cosmology makes many successful predictions about the nature of the Universe we live in. First, the era of rapid inflation would have made spacetime flat, in much the same way that, if you inflated a child's balloon to be the size of the Earth, its surface would look flat on a human scale. In the context of the general theory of relativity, this is equivalent to saying that the Universe sits very close to the dividing line between eternal expansion and the kind of recollapse into a Big Crunch that we have just described. Before inflation theory was invented, it was one of the great mysteries of cosmology that the Universe does indeed exist in this state, too close to precise flatness for any observational test yet devised to tell us which side of the line it sits.

Second, because the entire visible Universe expanded out from a seed much smaller than a proton, in which there was little room for any variations in, for example, density, it should be very smooth and uniform. This is not quite so obviously a feature of the visible Universe to the lay person, because we see bright stuff concentrated in stars and galaxies. But it is clear to astronomers from the way stars and galaxies move that they are embedded in a background sea of at least ten and perhaps a hundred times as much dark matter, which is much more smoothly distributed. The true smoothness of the Universe is shown by the incredible uniformity of the cosmic microwave background radiation, which last interacted with matter about 300,000 years after time zero, and comes to us from all parts of the sky with almost exactly the same temperature.

But not *precisely* the same temperature. In the early 1990s, data from the COBE satellite showed that there are tiny differences in the temperature of the background radiation from one part of the sky to another, changes in temperature from place to place on the sky of about 1 hundred-thousandth of a kelvin. These temperature dif-

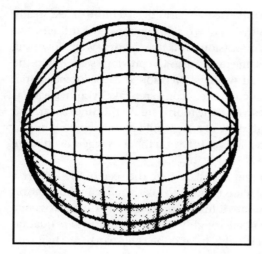

Quantum cosmology. Inflation flattens the Universe in the same way that the surface of a sphere gets flatter as the sphere expands.

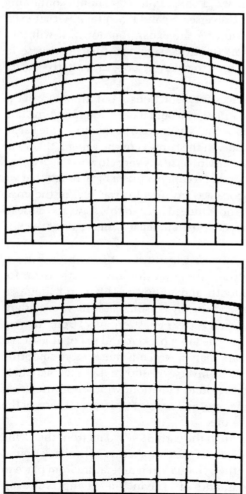

ferences indicate that there were differences in the density of matter in the Universe from place to place 300,000 years after time zero of just the right size to have acted as the seeds from which the structures we see today, clusters and superclusters of galaxies, would have grown under the influence of gravity. Even better, the pattern of these ripples in the background radiation (the 'spectrum' of the fluctuations) exactly matches the kind of pattern that should have been imprinted by quantum fluctuations during the era of inflation, when what is now the entire visible Universe was smaller than an atom. Indeed, the ripples we see today were probably imprinted when the Universe was only about 10^{-25} cm across – 100 million times bigger than the Planck length, but 1,000 billion times smaller than a proton. The very stars and galaxies we can see today – and, indeed, ourselves – exist because of quantum processes that occurred in the first split second after time zero.

And as icing on the cake, quantum physics predicts the existence of just the kind of almost uniform, dark matter sea required by cosmologists to explain the way stars and galaxies move. Apart from the possibility that some of this matter may be in the form of *neutrinos* with mass, it could be in the form of *axions*, or a sea of *SUSY particles*, predicted by some of the same grand unified theories that predict the occurrence of inflation itself.

Because of all this, particle physicists are now extremely interested in cosmology, and in particular in what observations of the Universe today can tell us about the conditions that existed before the Big Bang. Theories such as *supersymmetry* cannot be easily tested in accelerator experiments on Earth because such tests require experiments to be carried out at energies greater than any that could be obtained in practicable accelerators – indeed, to test some of the predictions of some of the grand unified theories you would need an accelerator as big as the Solar System. But those energies occurred naturally in the first split second after time zero, and the resulting quantum processes seem to have left visible imprints on the Universe which can be used, like fossils are used in geology, to tell us what conditions were like long ago. Studying these fossils from the dawn of time is the best way to test the predictions of the grand unified theories.

This marriage between quantum physics and cosmology is the most profound development in science since Isaac Newton realized that the laws of physics are strictly *universal* laws. Cosmologists and particle physicists had worked essentially independently of one another for 60 years, but it turned out that the theory of the very early Universe based on observational cosmology (the study of the very large) closely matched the theory of how light and matter ought to behave under extreme conditions (based on the study of the very small), and that ideas from quantum physics (notably, but not exclusively, inflation) could resolve puzzles in cosmology. This agreement suggests that our physical understanding of the Universe is a good one, and really does tell us what is happening on all scales from the Planck length to the size of the visible Universe itself – from 10^{-33} cm to 10^9 light years (10^{27} cm). This spans a range of 10^{60} powers of 10. There are enough details to be filled in to keep generations of physicists happily occupied, but it does seem that they are on the right track.

Further reading: John Gribbin, *In the Beginning*; and *Companion to the Cosmos*.

quantum cryptography The technique of using quantum physics to create an uncrackable code. It uses code systems that depend on a 'key' of random numbers. A message to be sent is turned into numbers (perhaps simply by assigning 1 to letter A, 2 to letter B, and so on). Then random numbers taken from a particular, predetermined page in a book of random numbers are added to the numbers corresponding to the message before it is transmitted. At the other end, anyone equipped with the same book of random numbers who knows which page has been used can subtract the random numbers to reveal the message.

The snag is that *anyone* with the pad of random numbers can do this, and there has to be a way to tell the recipient which page of which pad you are using. In quantum cryptography, the random numbers themselves are transmitted in a quantum form, perhaps as a stream of polarized photons. The only way a third party can eavesdrop is to read the *polarization* of the photons – but this changes the polarization and reveals that the eavesdropper is at work. It is even possible to ensure that the act of eavesdropping scrambles the quantum information so that the message cannot be read. The technique works, but has so far been used only on the laboratory scale.

quantum dot See *quantum computers*.

quantum electrodynamics Theory that describes the way electrically charged particles interact with one another and with magnetic fields through the exchange of photons. Usually referred to as QED, this is the jewel in the crown of quantum physics, a theory that has been tested experimentally to a very large number of decimal places, and has passed every test.

QED describes all interactions involving photons (light) and charged particles – in particular, all interactions involving photons and electrons. Because the interactions between atoms and molecules depend on the arrangement of electrons in clouds around the nuclei, and the forces between atoms and molecules are electrical forces, this means that QED is all you need to explain all of chemistry. It explains how a spring stretches, how dynamite explodes, why the sky is blue and how your eye works. It is, as Richard Feynman called it in the subtitle of his book on the subject, the theory of light and matter, in the everyday use of the word 'matter' – given, that is, that you do need gravity as well to explain how large objects interact with one another. Outside the nucleus, on the scale of atoms and above, all that matters is QED and gravity.

QED had its roots in the 1930s, when Hans Bethe and Enrico Fermi suggested (in 1932) that interactions between charged particles could be described as being mediated by photons. *Pair production*, in which a photon is physically transformed into a pair of charged particles, also indicated the close links between light and matter, and led Paul Dirac (and others) to try to develop a version of QED based on the idea of *positrons* as 'holes' in a sea of negative-energy electrons. This proved horribly complicated and was plagued by infinities; and progress with these ideas was stopped in its tracks by the Second World War. It is, though, interesting to note that Hideki Yukawa's suggestion that the strong nuclear force is mediated by the exchange of bosons was already, in 1935, based on an analogy with QED, just as, three decades later, *quantum chromodynamics* would be based on an analogy with modern QED.

What is now known as QED was developed independently by three great physicists in the mid to late 1940s. Two of them, Julian Schwinger in America and Sin-itiro Tomonaga in Japan, used what might be called the traditional mathematical formalism of quantum mechanics at that time, building on the approach used in the 1930s; the third, Richard Feynman, used a completely different approach, which involves a much better physical insight into what is going on and which, with the aid of *Feynman diagrams*, provides a more powerful computational tool for solving problems in QED. (He actually reinvented quantum mechanics from scratch, making the whole business much clearer and showing how it relates to classical mechanics.) We shall concentrate on Feynman's approach, but without the heavy mathematics.

The clearest way to see the power and importance of QED as it developed in the 1940s is to look at how it resolved experimental puzzles. Early in 1947, Willis Lamb and his colleague Robert Retherford, working at Columbia University, had detected the *Lamb shift* in the spectrum of hydrogen. They were measuring the spacing between *energy levels* for the various quantum states of the electron in the hydrogen atom. According to Dirac's theory of the electron, in a hydrogen atom the electron could sit in either of two states, with identical energies; but Lamb found that one of these quantum states has slightly more energy than the other, so that there is a tiny difference between the two energy levels. At about the same time, Isidor Rabi made an accurate measurement of the *magnetic moment* of the electron, which was also to play a key role in the development of QED.

The discovery of the Lamb shift showed that the Dirac theory was incomplete. But this was no real surprise because physicists already knew that in Dirac theory (and, indeed, in classical theory) the *self-interaction* of an electron would be infinite. When the electron was embedded in another electromagnetic field (such as the field associated with the proton in a hydrogen atom), that would correspond to an infinite shift of energy levels – an infinite Lamb shift. Ignoring the self-interaction (which was the standard procedure until Lamb and Retherford actually measured the Lamb shift) would give zero Lamb shift, while including the self-interaction in the most obvious way would give infinite Lamb shift. So the measurement of a small Lamb shift was actually a step in the right direction. The measurement of the Lamb shift told physicists that what they had to find in a better theory of QED was not zero or infinity, but a small, sensible and precisely known number. It was this discovery that prompted Feynman (with some urging from Hans Bethe) to apply his new ideas based on *path integrals* and Feynman diagrams to the problem (see also *phase*). Everything worked out beautifully, with the aid of *renormalization* to remove the last of the infinities that were still causing difficulties. Schwinger obtained equivalent results using his technique, and this gave both of them confidence that they were on the right track; it was Freeman Dyson who showed that the two approaches (and Tomonaga's approach, which was similar to Schwinger's) were mathematically equivalent to one another, and who publicized the breakthrough in a paper published in 1949.

The classic example of the accuracy of QED is its prediction of the size of the magnetic moment of the electron (there are many other examples, but one is enough to make the point). Using Dirac's theory and an appropriate choice of units, this has a value of exactly 1. But the experimental determination gives a value of

1.00115965221, with an uncertainty of ±4 in the last digit. QED predicts a value of 1.00115965246, with an uncertainty of ±20 in the last two digits. Theory and experiment agree to an accuracy of one part in ten decimal places, or 0.00000001 per cent. As Feynman points out in his book *QED*, this is equivalent to measuring the distance from Los Angeles to New York to the thickness of a human hair; it is also the most precisely determined agreement between theory and experiment for any theory and any experiment ever carried out on Earth. We can give you a feel for how this impressive precision is achieved, and how QED works, using Feynman diagrams.

In an interview in 1988 (see Jagdish Mehra, *The Beat of a Different Drum*), Feynman stressed that:

> The diagrams were intended to represent physical processes and the mathematical expressions used to describe them. Each diagram signified a mathematical expression. Mathematical quantities were associated with points in space and time. I would see electrons going along, being scattered at one point, then going over to another point and being scattered there, emitting a photon and the photon goes over there. I would make little pictures of all that was going on; these were physical pictures involving the mathematical terms. These pictures evolved only gradually in my mind … they became a shorthand for the processes I was trying to describe physically and mathematically … I was conscious of the thought that it would be amusing to see these funny-looking pictures in the *Physical Review*.

One important feature of these diagrams is that they treat particles and antiparticles on an equal footing. This automatically ensures that the theory meets the requirements of the special theory of relativity (it is said to be Lorentz invariant), and this also clarifies the nature of the singularities that plague QED. It was Dyson who proved that the infinities that arise in interactions described by Feynman diagrams can always be renormalized; this dramatic discovery was instrumental in persuading other physicists to follow Feynman's lead. Today, one of the main criteria used to decide whether a new idea in particle physics is worth pursuing is whether or not it is renormalizable – that is, whether it can be described using Feynman diagrams. If not, it is rejected (except, that is, in the case of **quantum gravity**, where the prize on offer is so great that physicists persist in trying to find a theory in spite of difficulties with renormalization!).

Leaving out the mathematics and using the diagrams in a qualitative way, we can describe the interaction between an electron and a magnetic field in a diagram like the first figure here. A photon from the magnetic field is absorbed by the

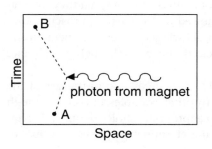

Quantum electrodynamics 1. An electron travelling from A to B interacts with a photon from a magnet.

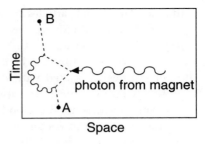

Quantum electrodynamics 2. At a more subtle level, we have to take account of the possibility that the electron interacts with itself en route from A to B, by emitting a virtual photon, which it promptly re-absorbs.

electron. In fact, this simple process is the Feynman version of Dirac theory, and if nature really were that simple, the magnetic moment of the electron would be precisely 1. Such simple Feynman diagrams, with no loops in them, exactly describe the classical behaviour of particles and fields.

But the electron also has a self-interaction, in which it emits a ***virtual photon*** and reabsorbs it, making a loop in the diagram. The second figure shows how this might happen while the electron is interacting with the magnetic field – and, using QED, it is possible to calculate the effect of this on the magnetic moment (remember that the single Feynman diagram represents all possible interactions of this kind). It makes the calculated magnetic moment a little bigger than 1, closer to the measured value. It was this simple version of the calculation that gave physicists confidence in the new theory in the 1940s.

The next step is to consider the possibility that the electron emits two photons, one after the other, and reabsorbs them. The calculation is harder (and took two years to be completed), but it gives a prediction even closer to the measured value of the magnetic moment. It wasn't until the middle of the 1980s that the even more complicated calculations for the case involving three virtual photons at a time were completed, giving the value for the magnetic moment quoted above. And we know why we still don't have precise agreement between theory and experiment – we have not yet included the effects of processes involving five, six or more virtual photons. But the technique – an example of ***perturbation theory*** at work – works so well because at each stage (each time more virtual photons are included) the correction is smaller. The calculation is said to converge on the experimentally determined value for higher 'order' in the calculation. (One of the problems with trying to develop a theory of ***quantum gravity*** using this kind of approach is that the corrections get *bigger* for higher-order terms – the numbers diverge, and it is impossible to get a clear prediction of whatever it is you are trying to calculate, because going up one more level of complexity gives a completely different answer to the one you had before.)

There are further complications, which cannot be calculated at all as yet, but which (in an example of the power of Feynman's approach) can be pictured with the aid of the diagrams. Thanks to quantum ***uncertainty***, even a single virtual photon involved in the self-interaction of an electron can turn itself into an electron–positron pair (or any other pair of ***virtual particles***) for a very brief time; and the electron and positron in that pair can be involved in other self-interactions, and so on *ad infinitum*. Each 'real' electron is actually surrounded by a frothing cloud of virtual photons and other entities, popping in and out of existence all the time. It is this cloud of virtual particles around an electron which prevents it from behaving as a *bare charge*, and reduces the self-interaction from infinity to the small amount responsible for the Lamb shift.

This whole picture can be carried over into QCD, with clouds of gluons and other entities associated with quarks, and therefore with hadrons, such as the proton. (Of course, the proton also has charge, and so have virtual photons and their associated entities in the cloud as well.) The best calculation of the magnetic moment of the proton obtained using QCD gives a value of 2.7 ±0.3, compared with an experimental value of 2.79275. Not bad, but nowhere near as good as QED.

The big remaining conceptual problem with QED is the business of renormalization, which works only because you know what answer you are trying to get. Dirac was never happy with the idea at all, and Feynman said that 'it is what I would call a dippy process!'. Nevertheless, it works. In Feynman's words, 'nearly all the vast apparent variety in Nature results from the monotony of repeatedly changing just these three basic actions': the movement of a photon from one place to another, the movement of an electron from one place to another, and the interaction of an electron with a photon.

It is worth emphasizing that the key word here is 'interaction'. When two electrons interact by exchanging a photon, or when an electron interacts with itself by emitting and absorbing a photon, it is natural to use everyday language in which the photon is emitted at an earlier time and absorbed at a later time. But there is nothing in QED to say which way round this happens; we are just as entitled to say that the photon is emitted 'in the future' and absorbed 'in the past'. Even an electron 'moving forwards in time' is exactly the same as a positron 'moving backwards in time'. The whole of QED is entirely symmetrical in time – as, indeed, is the whole of quantum mechanics. See *transactional interpretation*.

In the language of *quantum field theory*, QED is a *gauge theory*, with a symmetry represented by the one-dimensional unitary group, U(1) (see *group theory*). The most obvious example of this *gauge symmetry* is that the electric potential itself is not a meaningful quantity; all that can be measured is the potential difference. It is the simplicity of this *symmetry group* that makes it so easy (relatively speaking) to carry out calculations in QED.

Further Reading: Richard Feynman, *QED: The strange theory of light and matter*; John and Mary Gribbin, *Richard Feynman: A life in science*; Silvan Schweber, *QED and the Men Who Made It*.

quantum field theory　This means just what it says – it is the *quantum* theory of *fields*. The technical terminology may be intimidating when you first encounter it, but conceptually quantum field theory is very simple (the mathematics needed to carry out calculations in quantum field theory is not always so simple, but we won't be dealing with that here).

In classical physics, a field is something which stretches out from an object and conveys a force (there are really only two forces in classical physics, gravity and electromagnetism). The force can be described in terms of ripples in the field, or waves. But in quantum physics, we know that waves can be described in terms of particles (see *wave–particle duality*). So the concept of a field in the classical sense is replaced by the concept of particles which carry forces as they are exchanged between other quantum entities. The classic example is the **photon**, which mediates the electro-magnetic force; the three **intermediate vector bosons** are associated with the weak force; and the eight **gluons** mediate the colour force between **quarks**. **Gravitons** ought to be associated with the gravitational force, but they have not yet been detected directly.

None of these force carriers can be detected while they are doing their job of carrying a force – they are all **virtual particles** and do not, for example, leave tracks in a **bubble chamber**. If they were detected, they would not be doing their job of carrying a force between two other quantum entities. But both photons and the intermediate vector bosons can be manufactured out of energy and made visible as independent particles (members of the **boson** family); gluons are a special case and cannot exist in isolation (see **quantum chromodynamics**).

Extending this idea, entities that we are used to thinking of as matter particles (such as the electron) can be described in terms of waves. These waves can be described as ripples in another kind of field (one field for each type of particle), and the particles themselves can be described as field quanta, just as the force carriers are described as field quanta. (Since the idea of a quantum field theory of force carriers came first, this idea of matter particles as field quanta is sometimes referred to as 'second quantization'.) The key difference is that these particles are **fermions**, and so they are described by a different kind of field; but the principle is the same. In quantum field theory, there are *only* fields to worry about, and all particles (whether matter particles like the electron, or force carriers like the photon) are regarded as excited states of the appropriate fields – field quanta.

quantum fluctuation The temporary appearance of energetic particles (either **bosons** or **fermions**) out of nothing, as allowed by **quantum uncertainty**. The classic example is **pair production**. See **vacuum fluctuation**. Note that 'temporary' here need not necessarily mean short lived by human standards (although it does usually); it is possible that the entire Universe appeared out of nothing as a quantum fluctuation, and will disappear again in a few hundred billion years. See **quantum cosmology**.

quantum gravity General name for any of several attempts to describe gravity in terms of **quantum field theory**, by combining the general theory of relativity (see **relativity theory**) with quantum mechanics. None of these has proved fully successful, although partial theories of quantum gravity do suggest that it may eventually be possible to describe gravity in the same way as the other **forces of nature**, and then to combine all four of these forces in one mathematical framework, as a **theory of everything**. The key feature of any theory of quantum gravity is that the gravitational force between two massive objects is mediated by the exchange of particles known as gravitons – the quanta of the **gravitational field** – in an analogous way to the way that exchange of **photons** between electrically charged particles mediates the electromag-netic interaction.

The main reason why it is difficult to develop a full theory of quantum gravity is the equivalence of **inertial mass** and **gravitational mass** (see **Mach's Principle**). This means that both the size of the force experienced by a small mass in the gravitational field of a large mass *and* the acceleration caused by that force depend on the mass of the small object, in such a way that it cancels out of the calculation. So the accelera-tion produced is independent of the mass of the small object (in other words, heavy objects and light objects fall at the same rate, if air resistance can be ignored). This is quite unlike, say, the workings of the electric force, where a particle with more charge will experience a stronger acceleration than a particle with less charge if they are both in the same electric field. Although the reasons for this equivalence are not fully understood, it is related to the description of gravity in terms of the curvature of spacetime that is a feature of the general theory of relativity; this gives spacetime a dynamic role in any theory of gravity, unlike the passive role it plays in all other quantum field theories, where it is merely the stage on which the quantum action takes place.

Another problem is that gravity is so weak. This means that the effects of quantum gravity should only be noticeable at about the *Planck scale*, over distances that are twenty orders of magnitude (twenty powers of 10) smaller than a proton. This makes it impossible to carry out experiments to test the predictions of any quantum theory of gravity (this would require an accelerator operating at an energy of about 10^{18} GeV, compared with the 10^3 GeV or so of the most powerful modern machines); the theories can be 'tested' only by the accuracy with which they explain already known phenomena.

But even though the Planck scale is small, it is bigger than zero, and one of the bugbears of the general theory of relativity (indeed, its only really embarrassing feature) is that it predicts the ultimate collapse of matter (inside black holes) into a mathematical point of zero volume, a singularity. The only way to prevent this happening seems to be if quantum effects operating at the Planck scale can do the trick, and this is an additional powerful motivation (alongside the hope of finding a theory of everything) for research into quantum gravity.

What might now be called the traditional approach to quantum gravity, using the archetypal example of *quantum electrodynamics* as the template, was pioneered by Richard Feynman in the early 1960s (interestingly, *before* the development of *quantum chromodynamics*). The major idea that Feynman introduced to thinking about gravity at that time (about 1962) was the perturbation technique (see *perturbation theory*) that had been developed in the context of QED. He showed, in a series of influential postgraduate lectures that were widely circulated in note form, but have only recently been published in book form (*Feynman Lectures on Gravitation*, ed. Brian Hatfield) that the entire classical theory of gravity, including the equations of the general theory of relativity (see *relativity theory*), can be derived from a quantum description of interactions between particles that have mass, involving the exchange of gravitons, which are massless particles that each have two units of quantum spin.

This impressive achievement actually echoes Feynman's own approach to QED, developed two decades earlier, in which he started out from basics and showed that the entire classical theory of electromagnetism, including *Maxwell's equations*, can be derived from a quantum description of interactions between particles that have charge, involving the exchange of photons, which are massless particles that each have one unit of quantum spin.

The situation is more complicated in quantum gravity than in QED because the gravitons can interact with one another, as well as with massive particles. This is because all energy produces a gravitational field, including the energy of the gravitational field itself. It is this interaction between gravitons that makes it impossible to renormalize quantum gravity. The other important difference between quantum gravity and QED is that, whereas in the electromagnetic case like charges repel one another and opposite charges attract, in the gravitational case there is only one kind of 'charge', and like gravitational 'charges' (that is, masses) attract one another.

It can be shown that a spin 1 particle carries a repulsive force between like particles (matching the known spin and behaviour of the photon), while a messenger particle with either spin 0 or spin 2 could in principle carry an attractive force between like particles. A spin 0 graviton would be the quantum associated with the simplest possible gravitational field, but this could not provide all the subtlety of the general theory of relativity; however, a spin 2 graviton exactly matches the properties required for a force of gravity operating within the framework of four-dimensional spacetime (including the inverse square law of gravity).

But the philosophical basis of the two theories, QED and quantum gravity, is the same. Just as in QED, in quantum gravity *Feynman diagrams* without any loops in them describe interactions that follow the rules of the classical theory (in this case, the general theory of relativity), which is why it is possible to derive the classical theory from the quantum theory, in spite of the difficulties with renormalization; those difficulties arise only in the full quantum theory, where there are loops in the diagrams. (This means that simple Feynman quantum gravity, using diagrams without loops, does provide a perfectly good description of everything that is going on above the Planck scale; but so does the classical general theory of relativity, so don't get too excited).

Feynman actually got as far as making the single loop correction (or perturbation calculation) for quantum gravity, in the summer of 1962. In order to get this approach to work at all, he had to include an allowance for the influence of 'ghost' fields, responsible for the presence of particles which exist only as self-contained loops in the Feynman diagrams and have no 'real' existence. This insight has enabled other researchers to develop the technique to higher order, but with great difficulty and without any success in finding a way to get rid of the infinities that plague the theory.

Feynman himself was not greatly bothered by the difficulties with this standard approach to quantum gravity, or the infinities that arise and cannot be renormalized away. He thought that the *Lagrangian* which describes what is going on in the context of the general theory of relativity must be considered as only an 'effective Lagrangian' that describes the (relatively) low-energy behaviour of a more fundamental theory. This would echo the way in which Newtonian gravity is an effective theory that describes very well the behaviour of objects under even less extreme conditions, where we do not have to take account of the full implications of the complete equations of the general theory. Feynman suggested that there must be another layer to gravity, a more fundamental theory underpinning both the general theory and Newtonian gravity, and operating on the tiniest scales, close to the *Planck length*. The most exciting developments in the theory of gravity in the 1990s concern just such a possibility, the suggestion that gravity and the other forces of nature arise naturally from the behaviour of so-called string with dimensions comparable to the Planck length. *String theory* automatically produces (among other things) massless particles with two units of quantum spin – gravitons. And, to the delight of the theorists, it is not riddled with infinities. See also *supergravity*.

quantum interpretations See *models*, or under the particular interpretation of interest. The fact that there are many different ways of looking at the quantum world, and that these interpretations all make the same predictions about the outcomes of experiments, is seen by some people as unreasonable democracy. A good interpretation is one that makes you feel happy with your own personal picture of the quantum world, and what you think is a good interpretation may not be the same as what I think is a good interpretation. Indeed, what looks like a good interpretation in one context may not look so good in another context. But this does not matter at all, since we both agree on how experiments, like the *double-slit experiment*, will turn out. You can even favour one interpretation on weekdays and a different one at the weekend. But the one thing you must not do is believe that *any* quantum interpretation is The Truth. They are all simply crutches for our limited human imaginations, ways for us to come to grips with the weirdness of the quantum world, which never goes away and is outside the scope of everyday experience.

For what it is worth, we like the *sum over histories* approach and the *transactional interpretation* best. See also *consistent histories interpretation*, *Copenhagen interpretation*, *decoherence*, *ensemble interpretation*, *hidden variables*, *many worlds interpretation*, *quantum logic*.

Further reading: John Gribbin, *Schrödinger's Kittens*.

quantum jump = *quantum leap*.

quantum leap The important feature of a quantum leap is that it is a discontinuous transition between quantum states. For example, an electron 'in' one *energy level* in an *atom* jumps instantly into another energy level, emitting or absorbing energy as it does so. There is no in-between state, and it does not take any time at all for the leap to occur.

Because *Planck's constant* is so small, these transitions essentially occur only on a very small scale, which is why the world seems smooth and continuous to us (see *black body radiation*). They also occur at random, selecting from the options available to the quantum entity in accordance with the strict rules of probability. So a quantum leap is a sudden change in a system that occurs on a very small scale and is made at random. This puts the claims of some advertising which uses the term in a different perspective from that intended by the advertisers.

quantum logic A desperate attempt, pioneered by John von Neumann in the 1930s, to find an interpretation of quantum mechanics that would solve the quantum mysteries. The quantum logic interpretation says that everyday logic cannot be applied to the quantum world. Everyday logic is called Boolean logic, after George Boole, an Irish mathematician who lived from 1815 to 1864 and was the first person to use a symbolic language and notation to describe purely logical processes. In the mathematical logic that develops from these ideas, terms like 'and' and 'either' are represented by mathematical symbols, and logical arguments can be written out as mathematical equations. The 'quantum logic' approach to solving the quantum mysteries says that terms like 'and' and 'either' do not have the same meaning in the quantum world as they do in the everyday world, so that offering a photon a choice of either one of two slits to go through takes on a different logical significance. We cannot improve on the comment made by Heinz Pagels, in his book *The Cosmic Code*, describing the response of a person whose brain is wired up to operate on quantum logic to the puzzles of the quantum world:

> If we tell them about the two-hole experiment they just smile – they have no idea what the problem is. Now we see what the trouble with quantum logic is – it is more restrictive than ordinary Boolean logic. You cannot prove as much with quantum logic, and that is the reason you do not have any sense of weirdness in the physical world. Adopting quantum logic would be like inventing a new logic to maintain the earth was flat if confronted with the evidence that it is round.

See *quantum interpretations*.

quantum mechanics   The laws of mechanics that apply to the world of the very small. Named in contrast to *classical mechanics*; in fact, though, any proper formulation of quantum mechanics includes classical mechanics within itself, as a special case for very large values of *quantum numbers*, or equivalent to setting *Planck's constant* equal to zero. The term 'quantum mechanics' is essentially synonymous with *quantum physics*.

quantum numbers   Initially, sets of numbers that were assigned to quantum entities (such as electrons) as labels, relating to their observed properties, derived from techniques such as *spectroscopy*. They were assigned ad hoc, with no underpinning of a secure theoretical foundation to explain (let alone predict) why the numbers were required. But as a proper theory of *quantum physics* developed in the 1920s, it became possible to understand these numbers in terms of physical properties of the entities they were attached to.

The best example is electron *spin*. One of the spectroscopic puzzles that could not be explained by the *Bohr model* of the *atom* is that some spectral lines that should, according to the simplest version of that model, occur as single lines are split into closely spaced multiplets. In order to explain this *energy level* splitting, the electron had to be assigned four separate quantum numbers, corresponding to the way it could occupy slightly different energy levels. Three of these numbers could be explained in physical terms as describing: first, the *angular momentum* of the electron in its orbit (this was in the days when physicists still thought in terms of orbits, but even an *orbital* has angular momentum associated with it); second, the shape of the orbit; and third its orientation.

At first, the fourth quantum number was a mystery. It simply represented some extra physical property of the electron, which came in two varieties, to account for the observed splitting of the spectral lines. It was Wolfgang Pauli, in 1924, who came up with the idea that this fourth quantum number described the electron's 'spin', which could only point up or down, giving the required double-valued quantum number.

Although it is nice to be able to think of the behaviour of the quantum world in terms of cosy pictures, like spin, familiar from everyday life, it is arguable that this actually conceals the mysterious nature of the quantum world, and that we would be better off keeping the quantum numbers as mysterious labels for unknown properties of quantum entities. The term is, though, now largely obsolete and only of historical interest. It would be possible, for example, to describe the labels attached to *quarks* as quantum numbers, rather than *colours*; but nobody ever does.

A related use of the term (also pretty much obsolete, but important historically) is to refer to the actual number of quanta of a particular kind associated with a quantum state. For example, the angular momentum of an electron in an atom is quantized in units of $\hbar$ (see *Planck's constant*). If the angular momentum quantum number of a particular state of the electron is L, that would mean that the angular momentum of that state is L multiplied by $\hbar$. In this sense, quantum states with large quantum numbers correspond to classical behaviour, because the difference between the L state and the $L + 1$ state is tiny ($\hbar$ compared with L times $\hbar$), and the quantum entity will seem to be behaving in a smooth, continuous manner as it jumps from one state to the next.

quantum phase   See *phase (quantum)*.

quantum physics   The physics that describes the behaviour of the world on very small scales, the scale of *atoms* and *molecules* and below. The feature of the quantum world that gives quantum physics its name is that at this level physical processes are discontinuous and occur in *quantum leaps*. But these discontinuous jumps are very small because *Planck's constant* is very small, so the enormous number of quantum leaps going on at this sub-microscopic level add up to give us the illusion of a world in which change is smooth and continuous.

There are other peculiar features of the quantum world. It works in accordance with the laws of chance, so that a quantum entity selects from the options available to it at random – this discovery was much to the disgust of Albert Einstein, who made his famous comment that he could not believe that 'God plays dice', but the randomness of the quantum world has been borne out in countless experiments. The quantum world is also affected by *non-locality*. Two quantum entities that have once interacted with one another (or been part of the same system, such as two photons ejected simultaneously from an atom) remain somehow entangled with one another for ever, and 'aware' of each other's state and any change in the partner's state (see *Aspect experiment*). And there is nothing in the equations of quantum physics to distinguish the past from the future and indicate an *arrow of time* in the Universe – but this last feature it shares with *classical mechanics*.

This whole book is about quantum physics and I do not intend to try to summarize the entire book in a couple of thousand words here. But this does seem like a good opportunity to emphasize that, although quantum physics may be weird, it works. It isn't just a rather abstract and abstruse collection of theories and hypotheses that physicists can play with to keep themselves amused, but a highly successful, practical package which underpins many aspects of modern society that we take for granted. Here are a few examples.

One of the most common examples of quantum physics at work in a practical application in the home is the *laser*. Many people own a CD player, which works by scanning the disc on which information is stored (words, music, pictures or whatever) with a laser beam. Like its cousin, the *maser*, the laser operates on fundamental quantum principles, involving the *stimulated emission* of radiation from atoms. Albert Einstein, investigating *spontaneous emission* and laying the statistical ground rules for quantum theory in 1916 (the same statistical rules that he was later to find so abhorrent) was the first person to appreciate that an excited atom might be

Quantum physics. Quantum physics in everyday life – the bar code reader in a library uses a laser which operates on quantum principles first explained by Albert Einstein.

triggered into releasing a quantum of energy (a photon) and falling back into its ground state by receiving a nudge from another photon with the same wavelength as the one the atom is primed to release. Rather like the cascade of neutrons that is involved in a *chain reaction*, the laser process produces a cascade of photons from an array of excited atoms.

There are many variations on the theme, but they all use the same principle. Incoherent energy (radiation in which all the photons are jumbled up, so that the wave from one photon is partially cancelled and partially reinforced by the waves from the photons next door) is used to excite the atoms, and the laser (or maser; the same thing at longer wavelengths) mechanism releases that energy in a coherent beam in which all of the photons march in step, so that the energy from each wave reinforces the energy from all the other waves in constructive *interference*. Some lasers provide the ultimate 'straight edge' used in surveying work; others produce short-lived, powerful pulses of energy that can be used to drill holes in hard objects. Laser cutting tools are used in applications as diverse as microsurgery and bulk cutting of cloth in the garment industry. And you also see lasers at work in the super-market, reading the bar codes on the products. It is all quantum physics.

Nuclear energy, in the form of both nuclear weapons and power generation, is also a practical application of quantum physics. Whether we are dealing with *fission* or *fusion*, quantum processes such as *tunnelling* are crucially important, along with (in the case, for example, of fission-powered electricity-generating stations) the statistics of how nuclei *decay*, either spontaneously or under the influence of an impact from outside.

Then there are computers. Leaving aside the exotic possibility of developing true *quantum computers*, all modern computers, from the chip in your washing machine to a home PC to the biggest electronic superbrain in the world, depend on the properties of *semiconductors*, in which the properties of the conduction electrons (as in all conductors) depend on *Fermi–Dirac statistics*. A promising possibility for future generations of computers is to make use of another quantum phenomenon, *super-conductivity*; see also *Josephson junction*.

And what about life? *Quantum chemistry* is the branch of quantum physics that deals with interactions between atoms and molecules, and that is what life is all about. It is no surprise that in the middle of the 20th century many of the great advances in the understanding of the life molecule, DNA, and the genetic code were made by scientists who had training in physics. Erwin Schrödinger wrote an influential book called *What is Life?*; Francis Crick, who (with James Watson) discovered the structure of DNA, was a physicist who turned to molecular biology largely under the influence of that book; Linus Pauling was a quantum chemist who also turned to molecular biology; and George Gamow, who made many contributions to quantum physics, also played a part in the cracking of the genetic code. Such later developments as genetic engineering and cloning have also been achieved because of the understanding of how DNA works that has been achieved directly through the application of quantum physics to biology. But I know of no biologist who switched to quantum physics and made a mark in the field. This is not just because biology is easier than physics (although, according to biologist John Maynard Smith, who trained as an aeronautical engineer, it is), but because quantum physics is more *fundamental* than biology. Indeed, quantum physics is as fundamental as science gets.

But all of this, it is worth emphasizing, has been achieved through *quantum cookery*. Computer technology, lasers, nuclear power, genetic engineering and more besides use quantum physics, to be sure; but they use it in recipe-book fashion, without any need to worry about what it all means. The average quantum mechanic is no more worried about the philosophical implications and the interpretations of quantum physics than the average car driver worries about what is going on inside the engine of the car. As long as it works, their attitude seems to be, why worry about how it works? But if you do want to know what is going on inside the engine, take a look at the entry on the *double-slit experiment* and follow your nose from there.

quantum statistics   See *Bose–Einstein statistics* and *Fermi–Dirac statistics*; all quantum entities obey one or the other of these two systems.

quantum teleportation   The (as yet hypothetical) idea of sending perfect copies of quantum systems across space (perhaps just to the other side of the lab) in the style of *Star Trek's* transporter system (as in 'Beam me up, Scotty'). Doing this for macroscopic objects will not be possible, if at all, for a very long time. But it ought to be possible for entities on the scale of atomic particles. If you want to transport an electron, in a particular quantum state, across the Universe, you first need to prepare a pair of quantum entities which have interacted with one another and are in an entangled state (see *EPR experiment*). Take one of these entangled entities off in a box, without looking at it, leaving your partner at home with the other entangled entity (in its own box). Much later, when you are far away, your partner allows an electron to interact with the entangled entity back home, and measures the outcome of the interaction (which has destroyed the original quantum state of the electron). The entity in your box will be changed, instantaneously, as a result; but you don't know it yet. The results of the measurement made back home can be sent to you by any conventional means. Armed with this information, you can 'subtract out' the entangled entity's influence and be left with an exact copy of the original electron, in its original quantum state – for all practical purposes, it *is* the original electron.

The process would be tedious, and it still requires information to be sent slower than the speed of light, in order for you to carry out the subtraction. But it is not simply like sending information (like a fax) on how to make a duplicate, because the original is destroyed in the process. It is, as has sometimes been remarked, 'teleportation, Jim, but not as we know it'.

quantum theory   See *quantum physics*.

quantum uncertainty   See *uncertainty*.

quantum vacuum   In quantum physics, the vacuum is not 'nothing at all', but a seething mass of *virtual particles* produced by *quantum fluctuations* in accordance with the rules of *uncertainty*. See *Casimir effect*.

quark   General name for one kind of *elementary particle*, the fundamental building blocks from which all *hadrons* are constructed. Quarks feel the colour force (see *quantum chromodynamics*) and form a level of matter below that of *neutrons* and *protons*. All quarks have spin ½; some have ⅔ units of electric charge, and some have –⅓ units of electric charge (where the *electron* has –1 units). They come in six varieties of *flavour* (up, down, strange, charm, bottom and top) and three varieties of *colour* charge (red, green and blue).

quark confinement   See *confinement, asymptotic freedom*.

quark–gluon plasma   An extreme state of matter, thought to have existed in the first split second of the birth of the Universe, when conditions are so hot and dense that individual *hadrons* cannot exist, but are ripped apart to form a soup of *quarks* and

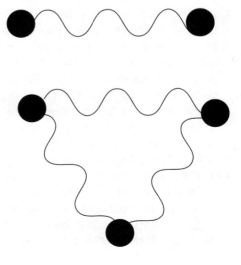

Quark. Isolated quarks are never seen, because of the way they are held together by the exchange of gluons (see asymptotic freedom). A quark/antiquark pair make up a meson; a triplet of quarks form a baryon. Mesons and baryons are collectively known as hadrons.

Quark-gluon plasma. Attempts to create a quark-gluon plasma will soon be made at the LHC, in CERN. This computer graphic shows how the new tubes for the LHC will be positioned inside the existing LEP tunnel.

gluons. The quark–gluon 'era' ended about 1 hundred-thousandth of a second (10^{-5} sec) after the Universe began expanding from the *Planck scale*. At that time, it had cooled sufficiently for a *phase transition* to take place (the quark–hadron phase transition), allowing hadrons to 'freeze out' from the plasma.

In round terms, the mass-energy of a proton is 1 GeV, and the volume of a proton is about 1 cubic femtometre (1 fm = 10^{-5} m). This represents the ultimate density of everyday matter, found in an atomic nucleus – 1 GeV per fm³. The quark–hadron phase transition is calculated to occur at a density of about 3 GeV per fm³ – when there is enough pure energy contained in the volume of a single proton to make three protons. Experiments at CERN and at the Brookhaven National Laboratory are attempting to achieve such densities, and to recreate the quark–gluon plasma that existed in the Big Bang, using techniques in which nuclei of heavy elements (such as gold, which has 118 neutrons and 79 protons in each nucleus) are accelerated to 99.9957 per cent of the speed of light and smashed into each other head on. The required energies and densities should be achieved before the end of the 20th century (perhaps before you read these words).

quark–hadron phase transition   See *quark–gluon plasma.*

quark nuggets   Hypothetical form of matter containing roughly equal numbers of up, down and strange *quarks*. Some calculations suggest that such *strange matter* may have been produced in the Big Bang in which the Universe was born, and that leftover nuggets of the stuff might still exist in the Universe today. I wouldn't bet on it.

Rabi, Isidor Isaac (1898–1988)   Austrian-born American physicist who won the

Nobel Prize for Physics in 1944 for his development of the resonance method, which led to the measurement of the *magnetic moment* of the electron. This was a key step in the development of *quantum electrodynamics*.

Rabi was born in Rymanow, then part of the Austro-Hungarian Empire (now part of Poland), on 29 July 1898. He emigrated to the United States with his family as a baby (in 1899), and later studied at Cornell University, where he received his first degree (in chemistry) in 1919. Rabi then worked for three years in a non-scientific job, before returning to scientific studies, first at Cornell and then at Columbia University, where he received his PhD in 1927. He received fellowships which enabled him to spend most of the next two years in Europe, working with (among others) Max Born, Wolfgang Pauli, Werner Heisenberg and Otto Stern. Stern, in particular, greatly influenced the direction of Rabi's career by introducing him to molecular beam techniques (see *Stern–Gerlach experiment*). Rabi returned to the USA in 1929 to take up a post at Columbia, where he spent the rest of his career, becoming a full professor in 1937. It was during the 1930s that he carried out his most important work, developing techniques for studying the magnetic properties of atoms, nuclei and molecules from measurements of the *magnetic moments* of nuclei spinning in magnetic fields.

During the Second World War, Rabi worked on radar at MIT. He turned down an invitation to become deputy director of the *Manhattan Project* team in Los Alamos, and later became a strong advocate of the peaceful uses of atomic energy. Rabi also helped to establish both the *Brookhaven National Laboratory* and (during a spell as a member of UNESCO), *CERN*. The technique of *nuclear magnetic resonance*, which stemmed from Rabi's work, has widespread applications. Among other things, Rabi's work on the behaviour of atoms and nuclei forms part of the basis of practical applications such as *lasers* and *atomic clocks*. The measurement of the *Lamb shift* also used a development of techniques devised by Rabi.

He died, in Manhattan, on 11 January 1988.

radiation   Term widely used to describe any process which carries energy through space. The most familiar form is *electromagnetic radiation* (light and radiant heat), but the word is also used to describe streams of energetic particles, such as *alpha radiation* or *beta radiation*, and is even applied to *gravitational radiation*.

radiation pressure   Pressure exerted on a surface by radiation, but specifically by *electromagnetic radiation*, such as light. This can be thought of as a result of *photons* hitting the surface and conveying *momentum*, like a stream of tiny cannonballs. Except under extreme conditions (such as at the heart of a star), radiation pressure is very small. But it can have a noticeable influence on very light particles, and helps to push the tails of comets outward from the Sun (see p.328).

radiation resistance   A kind of electromagnetic *inertia*, an extra resistance of charged particles to being accelerated, over and above the resistance to acceleration associated with their mass. It is called radiation resistance because an accelerated charged particle radiates electromagnetic energy, so the resistance seems to be associated with the emission of radiation. When, for example, electrons are pushed through a wire to make a changing electric current, they radiate away energy in the form of radio waves – but not as much energy as is going into pushing the electrons through the wire.

Radiation pressure. The pressure of light from the Sun makes the tenuous tail of a comet point away from the Sun and out into space. This engraving illustrates a comet seen in 1680.

Radiation resistance is a sign that the electron (or whatever charged particle is being pushed around) is interacting with something. In the same way that gravitational inertia can be understood in terms of an interaction between the mass that is being pushed around and all the other mass in the Universe (see *Mach's Principle*), radiation resistance can be understood in terms of an interaction between the charge that is being pushed around and all the other charges in the Universe. This idea was developed by Richard Feynman and John Wheeler in the 1940s (see *Wheeler–Feynman absorber theory*). In the 1980s, John Cramer used a similar formalism as the basis for a new interpretation of quantum theory (see *transactional interpretation*).

radioactive chain = *radioactive series.*

radioactive dating General term used for any technique for estimating the age of a sample of material from a process involving radioactive decay. The archetypal example is ***carbon dating***, but the term is also applied to dating techniques involving the measurement of the proportions of potassium and argon *isotopes*, rubidium and strontium isotopes, or uranium and lead isotopes in a sample.

radioactive decay Process whereby an unstable ***nucleus*** or particle spits out one or more particles and transforms into a stable nucleus or particle. The extreme case where a massive nucleus splits into two roughly equal parts is generally referred to as nuclear ***fission*** rather than decay, but also involves the ejection of other particles as well as the two 'half nuclei'.

The two main kinds of radioactive decay associated with atomic nuclei are *alpha decay* and *beta decay*. They have the effect of transforming a radioactive original nucleus (called the parent) into a nucleus of another *element* (called the daughter), which may or may not be radioactive itself. Decay happens on a characteristic timescale known as the *half life*. This kind of decay may occur in a chain several steps long before ending up with a stable nucleus. Decay may also involve the release of *energy* in the form of *electromagnetic radiation*.

Unstable particles decay in similar ways – the classic example is the beta decay of a lone *neutron* to produce a *proton*, an *electron* and a *neutrino*. According to the accepted models of particle physics (including the *grand unified theories*, or GUTs) the only really stable particles are the lightest family of *quarks* and the electron and its associated neutrino. Everything else – more massive particles created out of energy in particle accelerators on Earth, or in violent events in the Universe – will decay, ultimately into either quarks or electrons. The quarks form protons and neutrons, and the neutrons decay into protons and electrons. Even the proton ought to be unstable, changing itself into a *positron* and a *pion*, which itself decays into two *gamma rays*; the positrons produced in this way would annihilate with electrons to make more gamma rays, mirroring the way in which matter is thought to have been created out of energy in the *Big Bang*. The *supersymmetry* versions of the GUTs suggest that the lifetime of the proton is, however, about 10^{45} years, so there is no immediate prospect of the Universe disappearing in a puff of gamma rays.

See also *Becquerel, Henri.*

radioactive series   Any of four known chains of *radioactive decay*. There are three naturally occurring radioactive series ('natural' under conditions that exist on Earth today, that is). All three involve repeated *alpha decay* and end with an isotope of lead. One (the thorium series) starts with naturally occurring thorium-232 and ends with lead-204; the second (the actinium series) starts with uranium-235 and decays via actinium-227, ending up with lead-207; the third (the uranium series) starts with uranium-238 and ends with lead-206. The fourth series (the neptunium series) starts with artificially produced plutonium-241, which decays (again, by repeated alpha decay) via neptunium-237 to bismuth-209.

radioactivity   See *radioactive decay.*

radiocarbon   Another name for the radioactive *isotope* carbon-14. See *carbon dating.*

radioisotopes   Shorthand for radioactive *isotopes*. Some radioisotopes are used as labels to trace the path of some substances through a mechanical system, a chemical system or a living system. In this process, some of the atoms in a compound are replaced by the radioactive versions and their movements are monitored, perhaps using a *Geiger counter.*

radiology (radio therapy)   The use of *X-rays* (in particular) or other forms of radiation in medicine – for example, in the treatment of cancer. Radiation reduces the activity of dividing cells, and even kills them, so that if it can be applied directly to a malignant tumour, it can kill the cancer (or at least stop its growth). The problem is that any radiation which interacts with the healthy part of the patient's body (or the bodies of the medical staff) may do harm, so extreme care is necessary in such procedures.

Radioactive cobalt is widely used as a source of highly penetrating *gamma rays* in this kind of work, which is known as radio therapy. Radiology is also used in diagnostic medicine, such as in the familiar X-ray pictures of patients (diagnostic radiology).

radio waves *Electromagnetic radiation* with wavelengths in the range from a few millimetres (just longer than *infrared radiation*) up to hundreds of kilometres – in principle, as long as you like.

The existence of what are now called radio waves was predicted by James Clerk Maxwell in the 1860s, after he discovered the equations that describe the propagation of all forms of electromagnetic radiation including *light*. Radio waves were first knowingly produced artificially, confirming Maxwell's prediction, by Heinrich Hertz in 1888.

Rainwater, (Leo) James (1917–1986) American physicist who received the Nobel Prize in 1975 for his work on the theory of the structure of the *nucleus*.

Born on 9 December 1917 at Council, Idaho, Rainwater took his first degree, in physics, at Caltech (1939), then moved to Columbia University, where he received his master's degree in 1941 and his PhD in 1946. His PhD studies were interrupted by work on the *Manhattan Project*, but after completing them he stayed on at Columbia, becoming a full professor in 1952. Apart from his Nobel Prize-winning work (see *Bohr, Aage Niels*), his research interests included pion scattering studies and measurements of neutron *cross-sections*.

He died at Yonkers, New York, on 31 May 1986.

Raman, (Chandrasekhara) Venkata (1888–1970) Indian physicist who discovered the *Raman effect* in 1928 and was awarded the Nobel Prize as a result, in 1930. He was the first Asian to receive a Nobel Prize, and this achievement inspired a generation of young researchers.

Born at Trichinopoly, Madras, on 7 November 1888, Raman studied at the AVN College in Vizagapatam, where his father was professor of mathematics and physics. He then went on to Presidency College of the University of Madras, where he received his first degree in 1904 and his master's degree in 1907, at the age of nineteen. In both cases, he achieved the highest possible grades in the examinations. Because there were no openings for a career in scientific research in India at that time, he became an accountant with the civil service in Calcutta, where he stayed for the next ten years; but he carried out research into the nature of sound and the theory of musical instruments in his spare time (using the facilities of the Indian Association for the Cultivation of Science), making a reputation which led to him being appointed professor of physics at the University of Calcutta, in 1917.

Raman stayed in this post until 1933, carrying out his most important work and being knighted in 1929. He then moved to become head of the physics department of the Indian Institute of Science (and also, for a time, director of the Institute – the first Indian to hold the post) until 1948, when the government of newly independent India established the Raman Research Institute in Bangalore, with Raman himself as the first director. He stayed in the post until he died, in Bangalore, on 21 November 1970.

Raman also studied the way light is scattered by sound waves, the vibrations of atoms in crystals, the optical properties of diamonds and other jewels, and colour vision.

Raman effect The scattering of monochromatic light (light of a single wavelength or colour) as it passes through a transparent medium. The scattering is caused by the interaction of individual *photons* with the molecules in the medium. Because of the random motions of the molecules, some photons are given a boost in energy, scattering them to shorter wavelengths, and some lose energy, being scattered to longer wavelengths. So the scattered light is no longer purely monochromatic, and the way it has been scattered provides information about the molecules. This is an example of *inelastic scattering*. Like the Compton effect, it demonstrates the reality of photons.

The effect is named after Chandrasekhara Raman, who first studied it in detail in 1928. He had become interested in the scattering of light in 1921, while travelling by ship through the Mediterranean back to India from a conference in England. He realized that the blue colour of the sea could not be explained by scattering caused by particles suspended in the water (the standard explanation at the time, following a suggestion made by Lord Rayleigh), and showed that it is caused by the scattering of white light (which contains all the colours of the rainbow) by water molecules. (In a similar way, more than a decade earlier, Albert Einstein had shown that the blueness of the sky is caused not by scattering from dust in the air, but by scattering from the molecules of the air; in fact, the blueness of the sea is partly a reflection of the blueness of the sky). Short-wavelength blue light is more easily scattered than other colours, and gets bounced around to come at the observer from all directions.

After Arthur Compton discovered the **Compton effect** in 1923 (inelastic scattering of X-rays), in 1925 Werner Heisenberg predicted that a similar effect would occur with visible light. Raman had already thought of this, and had carried out some preliminary experiments; spurred on by Heisenberg's work, he improved on those experiments and showed the effect at work in completely clean air and pure liquids in 1928.

Raman scattering   = *Raman effect*.

Ramsay, William (1852–1916) Scottish chemist who discovered the inert gases argon, helium, neon, krypton, xenon and radon – making him the only person to discover an entire group of elements in the *periodic table*.

Ramsay was born in Glasgow on 2 October 1852. His father was an engineer, his grandfather founded the Glasgow Chemical Society, and one uncle was a professor of geology. Although his family hoped that he would have a career in the Church, and he received a classical education at Glasgow University between 1866 and 1869, Ramsay was proud of his scientific heritage, and learned chemistry by working in the laboratory of the city analyst in Glasgow for the next two years. He spent a year at the University of Heidelberg with Robert Bunsen before working with Rudolf Fittig (1835–1910) in Tübingen, where he received his PhD in chemistry in 1873. He returned to Glasgow to work as an assistant in the chemistry department of Anderson's College, soon moving to a similar post at the university. In 1880 he became professor of chemistry at University College, Bristol, and in 1887 he moved to University College, London, again as professor of chemistry.

Although he had previously worked in organic chemistry, in 1892 Ramsay learned of Lord Rayleigh's work which showed that nitrogen derived from the air is denser than nitrogen derived chemically. Rayleigh thought that this meant that synthetic nitrogen is contaminated with a lighter gas, but Ramsay suggested that it was because the nitrogen from the atmosphere contains a previously unknown, denser gas. In 1894 he was able to extract a small bubble of this gas from a sample of air, and it was identified spectroscopically by William Crookes. Ramsay and Rayleigh jointly named the 'new' element argon, from the Greek word for 'inert', because of its extreme reluctance to interact chemically with anything (see *quantum chemistry*; argon has *atomic number* 18).

This led Ramsay on to discover the other inert gases mentioned above. The discovery of helium in turn led him to work with Frederick Soddy which showed, in 1903, that helium is produced by the *radioactive decay* of radium. This was a key discovery in the development of *nuclear physics*; we now know that the *alpha particles* emitted during radioactive decay are helium nuclei.

Ramsay was a popular and friendly man, who did all his great work in collaboration with other people. He was knighted in 1902 and received the Nobel Prize for Chemistry in 1904 (the year Rayleigh received the Physics Prize). He retired in 1912 and died at High Wycombe on 23 July 1916.

random walk See *Brownian motion*.

Rayleigh, Lord (John William Strutt, Third Baron Rayleigh of Terling Place) (1842–1919) British physicist who received the Nobel Prize for Physics in 1904 for the discovery of the element argon; William Ramsay received the Nobel Prize for Chemistry for the discovery in the same year.

Born at Langford Grove, Essex, on 12 November 1842, he started life as John William Strutt, and became Lord Rayleigh on the death of his father, the second Baron, in 1873. For consistency, we shall simply refer to him as Rayleigh.

Rayleigh was a sickly child and youth, who attended both Eton and Harrow briefly, each time having to leave because of ill health. From 1857 to 1861 he was taught privately, then he went up to Cambridge, where he graduated in 1865. He then became a Fellow of Trinity College, but travelled in the United States until 1868, when he returned to England and set up a private laboratory in the family home at Terling Place. He had to resign his Fellowship when he married in 1871, and an attack of rheumatic fever soon after the marriage brought him near to death. He travelled up the Nile in a houseboat to recuperate, accompanied by his wife Evelyn, the sister of Arthur Balfour, a British statesman who served as Prime Minister from 1902 to 1905.

Very much a gentleman amateur, but a serious scientist, Rayleigh continued his research and was elected a Fellow of the Royal Society in 1873, the year his father died. He studied electromagnetism, sound and the behaviour of light, worked with *diffraction gratings* and developed John Tyndall's suggestion that the blue colour of the sky is caused by the scattering of light from dust molecules in the air; Albert Einstein later showed that this is, in fact, a result of scattering of light by the molecules of gas in the air, but Rayleigh had the right idea.

By 1879 Rayleigh's reputation was so sound that he succeeded James Clerk Maxwell as Cavendish Professor of Physics in Cambridge. He held the post for five years, then returned to private research; although he was also professor of natural philosophy at the Royal Institution in London from 1887 to 1905 (succeeding Tyndall), the post carried minimal responsibilities. When he was awarded the Nobel

Prize, he gave the money to provide an extension to the **Cavendish Laboratory**. Rayleigh helped to establish the National Physical Laboratories at Teddington in 1900, and in 1908 he became chancellor of Cambridge University, holding the post until he died, at Terling Place, Witham, on 30 June 1919.

Rayleigh was an able experimenter and wrote an epic book on *The Theory of Sound*, published in two volumes in 1877 and 1878. He made solid contributions across a wide range of the physical sciences, but is best remembered for the discovery of argon. He had found that the density of nitrogen extracted from the air is 0.5 per cent greater than that of nitrogen manufactured chemically, and in 1892 he published a letter in the journal *Nature* drawing attention to the puzzle. This led William Ramsay to the discovery of argon, essentially simultaneously with Rayleigh, who had continued his own research into the problem. They announced the discovery jointly in 1895.

But to physicists Rayleigh's most important contribution was the **Rayleigh–Jeans Law**, published in 1900 – an equation which describes the distribution of energy at different wavelengths in **black body radiation**. The Rayleigh–Jeans equation describes accurately longer wavelengths; Wilhelm Wien found an equation accurately describing the shorter-wavelength end of the black body spectrum; and Max Planck found a way to reproduce the combined black body curve by introducing the idea of **quanta** into physics.

In spite of his early health problems, Rayleigh remained active and alert in old age, working on scientific papers until a few days before his death, at the age of 76.

Rayleigh–Jeans Law An equation, originally developed by Lord Rayleigh (in 1900) and later refined by Sir James Jeans (in 1905) that accurately describes the long-wavelength end of the spectrum of **black body radiation**. Earlier, approximate versions of the law had been known, but it was Rayleigh who put it in precise terms. The equation treats light as a classical wave, and leads to the **ultraviolet catastrophe** at shorter wavelengths; the shape of the shorter-wavelength end of the black body curve is more accurately indicated by **Wien's Law**. It was the need to reconcile early versions of these two descriptions of black body radiation and smooth over the gap between the region where the Rayleigh–Jeans Law works and the region where Wien's Law works that led Max Planck to introduce the idea of **quanta** into physics.

reactor A device for obtaining energy steadily from nuclear **fission** using a self-sustaining **chain reaction**. There are hopes that reactors based on **fusion** may be developed, but so far this has not been achieved.

The basis of a fission reactor is that a naturally radioactive substance, such as uranium-235, uranium-233 or plutonium-239, is used as the fuel. If a large enough quantity of such material is put together in one place, the **decay** of some nuclei in the material will trigger the decay of other nuclei. In order to prevent this process running away explosively, the reactor also contains material which acts as a moderator, absorbing some of the neutrons produced by the decay. The moderator may be something as simple as water or graphite, or something more exotic, such as a heavy water, rich in **deuterium**. The amount of moderator is adjusted so that the self-sustaining chain reaction in the fuel produces a steady output of energy, as heat which can be used to drive steam turbines to generate electricity (or to provide propulsion in some ships and submarines).

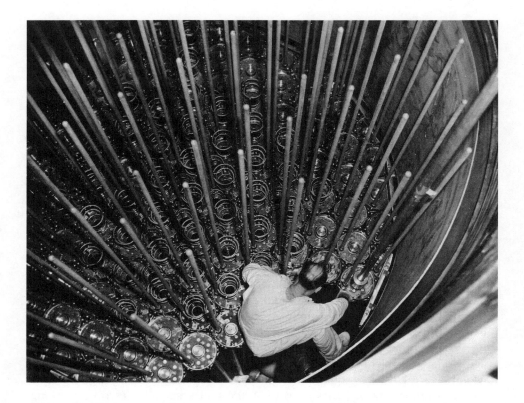

Reactor. Inside a fission reactor. When the reactor is running, the rods absorb neutrons and prevent a runaway chain reaction.

As a by-product, reactors produce particles such as neutrinos, *radioisotopes* and other *nuclides*.

reality   See *local reality*.

reflection   The way in which light or sound is bounced back by a sufficiently smooth surface (smooth compared with the wavelength of the waves involved). The term is also used to describe similar processes, such as the way a quantum wave may be reflected when it encounters an energy barrier, or the way a stream of particles bounces off an unyielding surface (you could say that a squash ball is reflected from the walls of the court, but this is not a common usage of the term).

In classical physics, light is regarded as travelling in straight lines, and reflects from a mirror in such a way that the angle between the incoming ray and a line perpendicular to the surface of the mirror is the same as the angle between the outgoing ray and the line perpendicular to the surface of the mirror (the angle of incidence is equal to the angle of reflection). In *quantum electrodynamics* the situation is more complicated, and the apparent simplicity of reflection in classical physics is regarded as being a result of light travelling by many different paths, most of which cancel each other out (see *path integral*, *phase*).

refraction   The way in which the direction of travel of light or sound is bent when it crosses the boundary between two media that have different optical or acoustic properties. The most familiar example is the way light bends when it moves across the boundary between air and a liquid, which explains why a stick pushed into a pond, or a swizzle stick in a drink, looks bent.

Refraction is explained in classical physics as a result of waves moving at different speeds in different media. If light moves from a medium in which it has a higher speed (such as air) to one in which it has a lower speed (such as glass), the angle between the incoming ray and a line perpendicular to the surface is greater than the angle between the refracted ray and a line perpendicular to the surface (the angle of refraction is, in this case, less than the angle of incidence). The opposite holds when the light emerges from the glass. In *quantum electrodynamics*, the situation is more complicated, and the apparent simplicity of refraction in classical physics is regarded as being a result of light travelling by many different paths, most of which cancel each other out (see *path integral*, *phase*). See also *action*.

Reines, Frederick (1918–) American physicist who received the Nobel Prize for Physics in 1995 for his experimental detection of the *neutrino*.

Born at Paterson, New Jersey, on 16 March 1918, Reines studied at the Stevens Institute of Technology, in Hoboken, New Jersey (MSc 1941) and at New York University, where he was awarded his PhD in 1944. He had actually left New York to work on the *Manhattan Project* before finishing writing up his thesis (on the *liquid drop model*) and for the next fifteen years he was a group leader at the Los Alamos laboratory, studying the physics and effects of nuclear explosions. That is where he was based when (with Clyde Cowan) he carried out the experiments which detected *neutrinos* (strictly speaking, antineutrinos) produced by the Savannah River *reactor*. Reines and Cowan had originally planned to search for neutrinos in the blast of radiation produced by a nuclear bomb, but J. M. B. Kellogg, the head of the physics division at Los Alamos in 1951, suggested that it might be more convenient if they used a reactor instead. After their first unsuccessful attempts to find neutrinos at a reactor near Hanford, in Washington, it was John Wheeler, in 1955, who suggested the Savannah River site.

From 1959 Reines worked at the Case Institute of Technology in Cleveland, as professor of physics. In 1966 he became a professor of physics at the University of California, Irvine. Apart from his Nobel Prize-winning work, Reines developed improved techniques for detecting elementary particles using *scintillation counters*, and studied cosmic rays. His group at Irvine was part of a team that detected neutrinos from a supernova explosion in 1987, and has also been involved in the search for *proton decay*.

relative atomic mass The mass of an atom in units where the mass of an atom of the *isotope* carbon-12 is defined as 12. This replaces the term *atomic weight*, although that term is still used.

relative permeability See *permeability*.

relativistic mass The *mass* of an object measured by an observer who is moving relative to the object (or, if you prefer, the mass of an object moving relative to an observer, who is measuring the mass, entitled to regard themselves as stationary). See *rest mass*.

relativistic mechanics The version of *classical mechanics* that describes the behaviour of objects under conditions where either or both of the special theory of relativity and the general theory of relativity (see *relativity theory*) has to be taken into account. Relativistic mechanics is still classical, in the sense that it does not involve quantum effects, but it is not in accordance with common sense. This is because the deviations of relativistic mechanics from *Newtonian mechanics* become significant only under conditions very different from those we experience in everyday life – for example, for objects travelling at a large fraction of the speed of light, or in intense *gravitational fields*. For this reason, the term *classical mechanics* is often taken to refer only to non-relativistic Newtonian mechanics.

Among other things, the equations of relativistic mechanics describe the correct way to add up velocities, so that no matter how you try to increase the speed of an object, it can never be accelerated from below the speed of light to above the speed of light. In this case, the appropriate equations are those of the special theory of relativity. But the special theory also describes perfectly the behaviour of objects moving at slower speeds, like cars moving down the road. If the speeds involved are much less than the speed of light, the special theory and Newtonian mechanics give exactly the same answers to the calculations. Newtonian mechanics is contained within relativistic mechanics as a special case, which actually makes it more special (in this sense of the term, meaning 'restricted') than the special theory.

relativistic quantum mechanics   The version of *quantum mechanics* that also takes account of the requirements of the special theory of relativity (see *relativity theory*). As yet, there is no satisfactory theory that includes quantum mechanics and the general theory of relativity (see *theory of everything*), but this does not matter when dealing with practical problems such as the way atomic spectra are produced, because *quantum gravity* is likely to be important only at the *Planck scale*.

But the requirements of the special theory are extremely important in practical applications such as *spectroscopy*. This is most easily understood by thinking of the electrons in an atom as tiny particles moving very fast in their orbits; any system in which high speeds are involved (that is, speeds that are a sizeable fraction of the speed of light) has to be described in terms of the special theory if we want to predict its behaviour accurately.

The first person to achieve this marriage of quantum theory and the special theory was Paul Dirac, in 1928. Among other things, his development of relativistic quantum mechanics led to the prediction of the existence of *antimatter*.

relativity theory   Overall name for two theories developed by Albert Einstein in the first two decades of the 20th century. The special theory of relativity describes the behaviour of objects moving at constant velocities ('special' is used to mean 'restricted', as in an example being a 'special case' of a more general process). The general theory of relativity deals with accelerations (which is why it is more general), but, crucially, is also a theory of gravity. Indeed, one of the cornerstones of the general theory of relativity is the principle of equivalence (or the equivalence principle), which says that acceleration and gravity are precisely equivalent to one another. Both theories are classical, in the sense that they do not take account of quantum effects. It is, however, essential to include the special theory in quantum mechanics to obtain good agreement between theory and observations (see *relativistic quantum mechanics*).

The special theory was first published in 1905 and deals with the relationships between moving objects. By the end of the 19th century, there was a clear conflict between the Newtonian description of such relationships (see *classical mechanics*) and *Maxwell's equations* of electromagnetism. Classical mechanics is what we think of as common sense, so that if you are riding in a car at a certain speed and throw a ball out ahead of you, the speed of the ball relative to the road is equal to the speed of the car plus the speed of the throw. But Maxwell's equations include the *speed of light* as an absolute constant, c, which makes no mention of the relative motion of the person measuring that speed. Einstein created the special theory by starting out from this feature of Maxwell's equations and accepting that Newtonian mechanics must be wrong. He built up a system of mechanics (*relativistic mechanics*) in which if you ride in a car and send out a beam of light ahead of you at speed c relative to the car, the speed of the light beam will also be measured as c (*not* as c plus the speed of the car) by an observer standing by the roadside, or by any observer anywhere in the Universe, however they are moving. Because the speed of light is so large, though, the differences between the special theory and classical mechanics become noticeable only for objects moving at high speeds.

Although Einstein developed his theory in mathematical (algebraic) terms, in 1908 Herman Minkowski (1864–1909) showed that it could be described geometrically, by treating time as a fourth dimension. This geometrization of the special theory, although not immediately to Einstein's taste, made it easier to understand and eventually helped to lead him to the general theory.

You can get a feel for how the special theory 'works' by looking at a pencil and the shadow it casts on a table. The pencil exists in three dimensions and has a definite length. But its shadow exists in two dimensions, on the surface of the table, and the length of the shadow depends on how the pencil is oriented in three-dimensional space. In Minkowski's version of the special theory, every object exists in four-dimensional spacetime and has a four-dimensional 'length', which is called its extension. Depending on how the object is oriented in four-dimensional spacetime, it can have a different length in three dimensions. 'Twisting' an object in spacetime means changing its speed. The faster an object moves, the shorter its three-dimensional shadow (its length) gets; the shadow disappears entirely if the object is moving at the speed of light.

Similarly, time (or rather, the duration of time, the interval between successive ticks of a clock) is also a shadow from four-dimensional spacetime, but this shadow must be thought of as being cast on a wall, perpendicular to the table, not on the table, so that when the length of the object (the shadow on the table) shrinks, the time taken for one tick of the moving clock (the shadow on the wall) lengthens. But the appropriate combination of the time 'length' and the space 'length', given by the equations developed by Einstein in 1905, stays the same – the mathematically inclined may have already realized that the extension of an object in spacetime is given from its shadows in space and time by the four-dimensional equivalent of Pythagoras' famous theorem for determining the length of the hypotenuse of a right-angled triangle; the mathematics of the special theory really is very straightforward when expressed in terms of geometry.

The key features of the special theory are that, from the point of view of an observer who is entitled to regard him or herself as stationary (that is, in an inertial frame, which means they are not experiencing accelerations), time recorded on a moving clock will run slow, a moving object will shrink in the direction of motion, and a moving object will gain mass. The speed of light is the same for all observers in all inertial frames, no matter how the source of the light is moving, and no object that starts off moving more slowly than the speed of light can ever be accelerated to above light speed (but see **tachyon**). The famous equation $E = mc^2$ is a consequence of the special theory. It is worth emphasizing that all of the implications of the special theory have been tested many times, and that it has passed every test with flying colours. Particle **accelerators**, for example, routinely deal with particles moving at sizeable fractions of the speed of light, where all these effects are important. The accelerators have to be built to take account of such things as the change in mass of a particle as it moves faster, and – given that they are indeed designed in accordance with relativistic mechanics, not classical mechanics – they simply would not work if the special theory was a bad description of reality. The special theory also describes perfectly the behaviour of objects moving at steady speeds much less than the speed of light; it includes Newtonian mechanics within itself, as a 'special case' of the 'special theory'.

Like the special theory, the general theory of relativity can be most easily understood in terms of the geometry of four-dimensional spacetime. The general theory was first presented by Einstein in a series of lectures to the Prussian Academy of Sciences at the end of 1915, and was published the following year.

One of the key steps towards the development of the general theory was the equivalence principle, which Einstein first formulated in 1907. This idea, that the effects of acceleration are indistinguishable from the effects of a uniform **gravitational field**, results from the equivalence between **gravitational mass** and **inertial mass**. Einstein realized that a person falling from a roof would not feel the effects of gravity – the acceleration of their fall would exactly cancel out the feeling of weight.

In modern language, the equivalence is best described in terms of a spaceship being accelerated through space by constant firing of its rocket motors. When the motors are not firing, everything inside the spaceship floats about in free fall, just as weightless as the person falling from a roof. In principle, when the motors are firing, the acceleration of the rocket could be adjusted so that everything inside felt a force exactly as strong as the force of gravity on Earth (or any other strength you chose), pushing things to the back of the vehicle as it moved forward through space. Any scientific experiments carried out in this accelerating frame of reference (for example, studies of the swing of a pendulum, or the way balls roll across the floor) would give exactly the same results as if the spaceship were standing on its launch pad on Earth, and not accelerating at all.

There is one important caveat. Acceleration is equivalent to a *uniform* gravitational field. Strictly speaking, the Earth's gravitational field is not uniform, because it spreads out from a point at the centre of the Earth. If you were sealed in a lift in free fall down a shaft drilled to the centre of the Earth, you would be able to tell that you were not falling freely in space by careful observations of any two objects sharing the lift with you – perhaps a pair of oranges. As you moved closer to the centre of the Earth, the oranges would move closer together, on converging paths. But at the

Relativity theory. Albert Einstein's insight. He realized that if there is no way for a person in a freely falling lift to tell whether they are falling or floating in space, then the path of a light ray must be bent by gravity. Then, for the people in the falling lift (bottom) the light ray seems to follow a straight path horizontally across the lift from A to B, just as it does when they are floating in space (top). This insight led to the general theory of relativity.

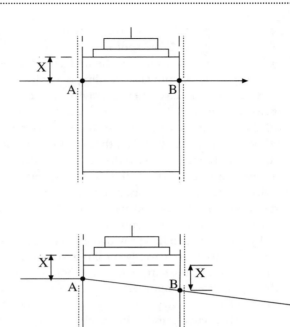

surface of the Earth, the equivalence between gravity and acceleration is very nearly perfect.

Einstein also used the example of a falling lift to come up with the most famous example of the effect of gravity as described by the general theory: the way it bends light. Einstein imagined a windowless elevator falling freely inside its shaft, the cable having snapped and all the safety devices having failed. According to the equivalence principle, there would be no way for a physicist inside the elevator, equipped with all the usual instruments of a physics lab, to tell whether the elevator was accelerating towards an unpleasant collision with the ground or was floating freely in the depths of space.

But what would happen to a beam of light shone across the falling elevator from one side to the other? In the weightless 'room', **Newton's laws of motion** must apply, and the light beam must travel in a straight line from one side of the elevator to the other. But now imagine how things will look to somebody outside the elevator. Suppose that the elevator has walls of glass, and that the path of the light beam is measured by sensitive instruments on each floor that the elevator falls past. Because the 'weightless' elevator and everything in it is actually being accelerated by gravity, in the time it takes the light beam to cross the elevator the falling laboratory has increased its speed. The only way in which the beam can hit the spot on the other wall exactly opposite the spot it started from is if it has followed a curved path, bending downwards to match the increase in speed of the elevator. And the only thing that could be doing the bending is gravity.

So Einstein inferred that if gravity and acceleration are precisely equivalent, gravity must bend light, by a precise amount that could be calculated. This was not entirely a startling suggestion; Newton's theory, based on the idea of light as a stream of tiny particles, also suggested that a light beam would be deflected by gravity. But in Einstein's theory, the deflection of light is predicted to be exactly twice as great as it is according to Newton's theory. When the bending of starlight caused by the gravity of the Sun was measured during a solar eclipse in 1919, and was found to match Einstein's prediction rather than Newton's, the general theory of relativity was hailed as a scientific triumph.

By then, Einstein had provided a physical picture of what goes on to bend light. Imagine empty space (strictly speaking, spacetime) to be represented by a stretched rubber sheet, like the surface of a trampoline. On such a surface, you can represent light rays by rolling marbles across the surface; they travel in straight lines. Now imagine placing a heavy object, such as a bowling ball, on the sheet, to represent the Sun. The sheet curves under the weight, and if you roll a marble across the sheet, near to the 'Sun', its trajectory will be bent as it follows the curve around the heavy weight. The presence of matter, said Einstein, causes four-dimensional spacetime to curve in an equivalent way. The curvature of spacetime then affects the motion of everything passing through the region of curved spacetime, including light beams and planets. The situation is summed up in a neat aphorism: 'matter tells space how to bend; space tells matter how to move'.

There is one important point about this picture that sometimes causes confusion. We are dealing not just with curved space (in spite of the aphorism!), but with curved spacetime. The orbit of the Earth around the Sun, for example, forms a closed loop in space, and the Earth is held in its orbit by the Sun's gravity. If you imagine that this closed orbit represents the curvature of space caused by the Sun, you might leap to the conclusion that space itself is closed around the Sun – which is obviously not true, because the Sun is not a black hole and light (and other things) can escape from the Solar System.

In fact, the Sun and the Earth are both following their own so-called world lines through four-dimensional spacetime, first described in those terms by Minkowski in 1908. In that description, time and space are geometrically equivalent, but are related by the speed of light, which is 300 million metres per second. So each second of time is equivalent to 300 million metres in the time direction. The Earth and the Sun are moving through spacetime in very nearly the same direction, from the past into the future, and instead of the orbit of the Earth around the Sun being a closed loop, in four dimensions it is a stretched-out helix, twisting around the world line of the Sun.

The general theory of relativity, like the special theory, makes many predictions which have been tested by experiment many times. It has passed every test with flying colours. There is no doubt that the general theory is a good and accurate description of the behaviour of matter in the Universe, and of the relationship between space, time and matter. If it is ever improved upon, any better theory will have to incorporate the general theory within itself, just as the general theory incorporates Newton's theory of gravity within itself (see *quantum gravity*). One of the most important features of the general theory, although outside the scope of the present book, is that it provides a complete description of the Universe, and of the way that the Universe has expanded away from a singularity at the beginning of time, via the Big Bang, to reach its present state. See also *quantum cosmology*.

Further reading: John Gribbin, *Companion to the Cosmos* and *In Search of the Edge of Time*.

renormalization   Trick used by physicists to get rid of unwelcome infinities that plague *quantum electrodynamics* and other aspects of *quantum physics*. Many of these infinities are associated with *self-interaction*, such as the interaction of an electron with its own electromagnetic field.

Because the electric force, for example, goes as 1 divided by the distance squared, and the distance of an electron (the source of the field) from itself is zero, the interaction with itself ought to be infinite (1 divided by zero is infinity). Among other things, because *energy* and *mass* are equivalent, this means that the electron should have infinite mass, which it clearly does not. The way renormalization works is, in effect, by subtracting one infinity from another to leave a finite answer. And it only works because you know the finite answer you are trying to get – for example, the mass of an electron. But once a theory such as QED has been renormalized in this way, it can then be used to make further calculations and predictions without intrusive infinities (for example, the calculation of the *magnetic moment* of the electron).

This problem arises even in classical theory. Renormalization was first used in the context of classical theory, by the Dutch physicist Hendrik Kramers (1894–1952), who started developing the idea at the end of the 1930s and completed it in 1947, after the Second World War. Following a suggestion made by his former teacher, Hendrik Lorentz, he introduced the idea that the mass of a charged particle such as an electron should be regarded as being made up of two contributions, the *bare mass* plus the (infinite) mass resulting from the electromagnetic self-interaction. But *Maxwell's equations* can be solved to give the total mass-energy of the field associated with the electron, which is another infinity. If we subtract this infinity from the first infinity, the two infinities disappear, leaving the bare mass behind. This is renormalization.

You might think that infinity minus infinity, if it means anything at all, ought to be zero. But infinity is a funny thing. Imagine making infinity by adding up all of the integer numbers $(1 + 2 + 3 + 4 \ldots)$. You could also make infinity by doubling every integer and then adding up all the resulting numbers. You might think that the second infinity is twice as big as the first one. But it isn't, because it includes only all the even numbers $(2 + 4 + 6 + 8 \ldots)$, while the first infinity includes both odd numbers and even numbers! If you subtract the second infinity from the first one, in this case you will be left with infinity once again – the sum of all the odd numbers $(1 + 3 + 5 + 7 \ldots)$.

What is needed in QED is to be able to subtract one infinity, made up of (infinity), from another infinity, made up of (infinity plus a little bit), to leave behind (a little bit). Kramers presented his ideas in detail to a conference held in 1947 (the Shelter Island Conference) – the same conference where Willis Lamb first presented details of the **Lamb shift**. Hans Bethe almost immediately found a way to use renormalization to provide an approximate calculation of the Lamb shift. It was only approximate because it did not take account of the requirements of the special theory of relativity (see **relativistic quantum mechanics**). Nevertheless, Bethe found that the infinities involved could indeed be cancelled out of the equations in the right way to produce renormalization. This was an important discovery because renormalization cannot be made to work with any old set of equations and any old infinities; they have to be the right kind of infinities for the trick to work, or else you would indeed end up with either zero or infinity when you subtracted one of them from another. Only certain kinds of theory, notably the **gauge theories**, are renormalizable.

To some people, Bethe's success in calculating even a non-relativistic version of the Lamb shift using renormalization seemed like a miracle; to others, it seemed like a fraud. To most physicists, it suggested that he had made a fundamental discovery about the way the world works, even though they weren't at all sure just what that discovery was (this latter position is still the one held by most physicists today).

It was Richard Feynman who, jumping off from Bethe's example, showed that renormalization works in a fully relativistic version of QED, and he presented his version of quantum electrodynamics, complete with renormalization, to another major conference held in April 1948. The versions of QED developed by Julian Schwinger and Sin-itiro Tomonaga were introduced at about the same time, and are also renormalizable in this way. Renormalization is a key feature of QED, and without renormalization the theory would not work. Even so, this tinkering with infinities is deeply unsatisfactory and always troubled Paul Dirac (among others). Feynman himself said, in his book *QED*, 'having to resort to such hocus-pocus has prevented us from proving that the theory of quantum electrodynamics is mathematically self-consistent … [renormalization] is what I would call a dippy process!' And yet, it works.

See also **bare charge**.

resonance   In classical physics, the oscillation of a system (for example, a guitar string) at its natural frequency of vibration, triggered by an outside stimulus with the appropriate frequency. If you play a note on a piano with exactly the frequency of one of the open strings of a guitar, that string of a guitar lying near the piano will resonate, even though it has not been plucked. This kind of resonance can occur in mechanical systems (which is why some cars vibrate a lot when the engine is turning over at a certain speed, and why a pure note sung by a trained singer can shatter a wine glass), in electrical circuits, and in atoms and molecules.

In chemistry, the term 'resonance' is sometimes used to refer to the mixture of molecular states that occurs when a compound can be represented by two or more arrangements of the same atoms. Ozone is a good example of this kind of resonance (see **quantum chemistry**).

In particle physics, a resonance is a short-lived particle, a member of the **baryon** or **meson** families, that can be regarded as an excited state of a more stable particle. Such resonances decay into more stable forms through the influence of the strong interaction (see *forces of nature*). They have lifetimes typically in the region of 10^{-23} sec. To put this in perspective, the ratio of 10^{-23} sec to 1 millionth of a second is the same as the ratio of 1 millionth of a second to 1,000 years. They cannot be detected directly, but influence the rate at which some particle interactions occur, and evidence for their existence was first found (to the astonishment of physicists) in the early 1950s, by Enrico Fermi's group in Chicago. They were studied in detail only in the 1960s.

Resonances exist because particles such as protons are made up of quarks. An atom can be put into an excited state if energy is provided to boost one or more of its electrons into a more energetic *orbital*; the electron soon falls back, releasing electromagnetic energy as it does so. Resonances are excited states of *hadrons*, with some of their constituent quarks boosted into higher energy levels. They decay by emitting particles such as pions and kaons, which carry away the excess energy. Very many of these short-lived particles are now known, and the term 'resonance' is becoming obsolete in this context.

rest energy The energy associated with the *rest mass* of a body, in line with the famous equation $E = mc^2$. This is the minimum amount of energy that can be associated with the object.

rest mass The *mass* of a body (or particle) measured by an observer who is not moving relative to the body (particle). The measured mass is significantly different from the rest mass only for objects moving at a sizeable fraction of the speed of light relative to the observer. Particles which travel at the speed of light, including *photons*, have zero rest mass, even though they are never at rest.

retarded wave A wave that travels outwards from its source and forwards in time. See *transactional interpretation*, *Wheeler–Feynman absorber theory*.

Richardson, Owen Williams (1879–1959) British physicist who investigated the emission of electrons from hot surfaces (thermionics, a name he invented), a key step in the development of electron tubes (electronic valves, TV tubes and the like). He received the Nobel Prize for Physics for this work in 1929 (it was actually the 1928 Prize, held over for a year).

Born in Dewsbury on 26 April 1879, Richardson studied at the University of Cambridge, graduating in 1900, and simultaneously received a BSc from the University of London. He stayed on to do research at the *Cavendish Laboratory* (picking up a DSc from the University of London in 1903), then taught at Princeton from 1906 to 1913. He returned to England to become professor of physics at King's College, London – a post he held until he retired in 1944. He worked on telecommunications for the military during the First World War, was knighted in 1939, and worked on radar during the Second World War. After he retired, he bought and ran a dairy farm in Hampshire.

Richardson died at Alton, in Hampshire, on 15 February 1959.

Richardson, Robert Coleman (1937–) American physicist who shared the Nobel Prize for Physics in 1996 for the discovery of *superfluidity* in liquid helium-3.

Born in Washington, DC, on 26 June 1937, Richardson studied at the Virginia Polytechnic Institute, where he received his BSc in 1958 and MSc in 1960, and at Duke University (PhD 1966). He then joined the staff of Cornell University, where he became a full professor in 1975 and was appointed director of the Laboratory of

Atomic and Solid State Physics in 1990. In 1984 he was a visiting professor at the Bell Laboratories research centre in Murray Hill, New Jersey. The work for which Richardson received the Nobel Prize was carried out in 1971 and 1972 at Cornell, after Douglas Osheroff (then a research student) noticed the peculiar behaviour of helium-3 at very low temperatures. Together with David Lee, Richardson and Osheroff established that this was a result of superfluidity.

Richter, Burton (1931–)   American physicist who received the Nobel Prize for Physics in 1976 for the discovery of the *J/psi particle*.

Born in Brooklyn on 22 March 1931, Richter studied at MIT (BSc 1952; PhD 1956), then moved to Stanford, rising to become a full professor in 1967. Richter was a driving force behind the construction of the SPEAR accelerator, which led to the discovery of the J/psi by his team in 1974, essentially simultaneously with the discovery made by a team led by Samuel Ting, with whom Richter shared the Nobel Prize. He became the director of SLAC in 1984.

Röhrer, Heinrich (1933–)   Swiss physicist who received the Nobel Prize for Physics in 1986 for his part in the development of the tunnelling electron microscope.

Born in Buchs, St Gallen, on 6 June 1933, Röhrer studied at the Federal Institute of Technology (ETH) in Zurich, receiving his first degree in 1955 and his PhD (for research on *superconductivity*) in 1960. He worked at Rutgers University in New Jersey from 1961 to 1963, then returned to Zurich to work at the IBM Research Laboratory, where he carried out the work with Gerd Binnig that led to the Nobel Prize.

Röntgen, Wilhelm Conrad (1845–1923)   German physicist who discovered *X-rays* and, as a result, was awarded the first-ever Nobel Prize for Physics, in 1901.

Born in Lennep on 27 March 1845, Röntgen studied in the Netherlands (his mother was Dutch), and then at the Federal Institute of Technology (ETH) in Zurich, where he received his PhD in 1869. He was influenced by August Kundt (1839–94) while at the ETH, and followed Kundt first to the University of Würzburg and then, from 1872 to 1874, to the University of Strasbourg, working as Kundt's assistant. He then worked as a *Privatdozent* in Strasbourg in 1874–5, and after a year in Hohenheim returned to Strasbourg as professor of mathematics and physics until 1879. He then became director of the Physics Institute in Giessen, where he stayed until 1888, moving on to become professor of physics at Würzburg University. In 1900 Röntgen was appointed professor at the Ludwig-Maximilian University in Munich, where he stayed until he retired, in 1920. He died, in Munich, on 19 February 1923.

Although Röntgen worked on a wide variety of problems in physics, he is best remembered for the discovery of X-rays, made in November 1895 while he was in Würzburg. This stemmed from his investigations of *fluorescence*, which he had begun the year before using a *Crookes tube*. Although there was an element of luck concerning the discovery, Röntgen was a first-rate experimental physicist, who made full use of that luck and carried out an immediate and thorough investigation of the phenomenon he had discovered.

Röntgen rays   Old name for *X-rays*.

Rubbia, Carlo (1934–)   Italian physicist who received the Nobel Prize for Physics in 1984, in recognition of the efforts of a large team of researchers at CERN, which he headed, that had led to the discovery of the W and Z particles in 1983.

Born in Gorizia on 31 March 1934, Rubbia studied at the Scuola Normale Superiore in Pisa and at the University of Pisa, where he received his PhD in 1957. He taught in Pisa for two years before working at Columbia University in the United States as a research fellow. He then (in 1960) joined the faculty of the University of Rome, but stayed only until 1962, when he moved to CERN, where he has stayed, holding his post there jointly with a post as professor of physics at Harvard University from 1972 to 1988. He was director-general of CERN from 1989 to 1993, and pushed through the LEP and LHC projects.

See also *forces of nature*.

rubidium–strontium dating = *strontium dating*.

Rumford, Count (Benjamin Thompson) (1753–1814) American-born physicist who proved that heat is a form of motion.

Born at North Woburn, Massachusetts, on 26 March 1753, Benjamin Thompson was the son of a farmer. He became a schoolteacher, moved to Rumford (now Concord), New Hampshire, and married a wealthy widow much older than himself in 1772, at the age of nineteen. He served as a major in the 2nd New Hampshire Regiment, which led to problems with his fellow Americans, by then itching for independence; at first, this led Thompson to disassociate himself from the British, but in the end he sided with them and acted as a secret agent for the Crown in the run-up to the American War of Independence. He had to flee to London in 1776 (he had left his wife and baby daughter the previous year).

Thompson stayed in England for six years, where he began a series of experiments involving gunpowder. His reputation as a scientist was made by the investigations of different kinds of gunpowder and the force they applied to a cannon ball fired from different guns, which led to him being elected a Fellow of the Royal Society in 1779. In 1780, at the age of 27, he became Undersecretary of State in the Colonial Office. He came back to America in 1782 as a lieutenant-colonel in a regiment in New York, but at the end of the war he returned to Europe. He was promoted to colonel in 1783 and knighted by the English in 1784, but in the same year he moved to Bavaria. For several years he was a minister in the government of Bavaria, where he carried out an enormous amount of work, including reforms of the army, draining of the marshes around Mannheim, the introduction of the potato to the local agriculture, and the introduction of a Poor Law system.

In Bavaria, Thompson (as well as his government work) investigated the nature of heat and carried out his own experiments in social engineering. In his capacity as Minister of War, he employed beggars to manufacture military uniforms, devised recipes for cheap but nutritious soups to feed the hungry workforce, and used soldiers to tend the gardens where the vegetables to make the soup were grown. And in an effort to promote the drinking of coffee (as an alternative to alcohol), he invented the first percolator. The interest in heat led to the design of better uniforms with more effective insulation; the interest in gardening led to the establishment of the famous Englischer Garten in Munich.

In 1791 he was made a Count of the Holy Roman Empire in recognition of his government service in Bavaria, and it is as Count Rumford that he is now usually known, and that he carried out the work for which he is best remembered. This was when he was involved in the manufacture of cannon in Bavaria. The cannon were made by being bored out from blocks of metal by drills, and the metal became very hot in the process. Rumford's studies, reported to the Royal Society in 1798, showed that there was no change in the weight of the material (adding in the bored-out stuff), and that the amount of heat produced was related to the amount of work done in turning the drill. This showed that heat is a form of energy, not a fluid ('caloric') as others had thought. This was not fully accepted, though, until well into the 19th century.

Rumford was keenly interested in promoting science and spent the later part of his life in this work. In 1796 he established prizes, known as the Rumford Medals, in the Royal Society and in the Boston Academy of Arts and Sciences; these prizes are still awarded for important scientific achievements in the field of light and heat today. Rumford left Bavaria in 1798 and returned to London, where he founded the Royal Institution in 1800 and personally chose Humphrey Davy to lecture there. He also invented the kitchen range (a 'fire in a box'). Two years later, after a year of travel on the continent, he moved to Paris, where in 1805 he married the widow of Antoine Lavoisier (Rumford's first wife had died in 1792); but like his first marriage this was not a success, and the couple separated in 1809. Rumford died at Auteuil, near Paris, on 21 August 1814. His will provided an endowment for the Rumford Chair at Harvard, a post which still exists.

Ruska, Ernst August Friedrich (1906–1988) German electrical engineer who received the Nobel Prize for Physics in 1986 for his work on the development of the electron microscope.

Born in Heidelberg on 25 December 1906, Ruska studied at the Technical University in Munich (1925–7) and then in Berlin, graduating as an engineer from the Technical University of Berlin in 1931 and receiving his PhD in 1934. Together with Max Knoll, he built the first experimental instrument that might be called an electron microscope in 1931, while still a graduate student, hot on the heels of the experimental confirmation of the wave nature of the *electron*. It had a magnification of just 17x. It was only in 1933 that Ruska (now working on his own) succeeded in building an electron microscope with better resolution than an optical microscope.

Ruska worked on the development of television in the 1930s, and was with the Siemens & Halske company in Berlin from 1937 to 1955. He then held the post of director of the Institute for Electron Microscopy at the Fritz Haber Institute, from 1955 until he retired in 1972. Alongside this work, he became a professor at the Free University of Berlin in 1949, and was later appointed a professor at the Technical University of Berlin, in 1959. He spent much of his career developing the electron microscope further.

He died in Berlin on 27 May 1988.

Rutherford, Ernest (Baron Rutherford of Nelson) (1871–1937) New Zealand-born British physicist who made many pioneering investigations of the structure of the *atom* and *radioactivity*, and was awarded the Nobel Prize for Chemistry in 1908 – at that time, the Nobel Committee thought that this was the appropriate branch of science for categorizing Rutherford's work, but the award caused much mirth because Rutherford himself always regarded chemistry as an inferior field of study, and once said that 'all of science is either physics or stamp collecting'.

Born at Spring Grove (later renamed Brightwater), near Nelson, New Zealand, on

Ernest Rutherford (1871–1937).

30 August 1871, Rutherford graduated from Canterbury College, Christchurch, in 1892, with a BA degree. He then switched to science and was awarded an MA in 1893, with first-class honours in mathematics and physics. He stayed on for another year to do research at Christchurch (financing himself by teaching part time), and along the way received his BSc in physics in 1894. At this time, he was interested in the magnetic properties of iron, which he probed using what would now be called high-frequency radio waves (this was just six years after Heinrich Hertz discovered radio waves); in the course of this work, Rutherford built a sensitive detector for these waves, one of the first radio receivers. He produced two papers about this work, which won him a scholarship providing funds to work in England.

In 1895 Rutherford moved to England and began to carry out research at the Cavendish Laboratory, where he made what are now accepted as the first long-range (that is, outside the laboratory) wireless transmissions, over a distance of some 2 miles. But under the influence of J. J. Thomson, he soon gave up this work and concentrated on **atomic physics**. He moved on to become professor of physics at McGill University, in Montreal, in 1898, where he carried out his early pioneering work on radioactivity. In 1907 he came back to England to become professor of physics in Manchester, where he was responsible for the discovery of the atomic **nucleus** and the **transmutation of elements**.

Rutherford worked for the Admiralty during the First World War (on secondment from Manchester), investigating techniques for detecting submerged submarines. In 1919 he succeeded Thomson as head of the Cavendish Laboratory, and he held the post until he died, in Cambridge, on 19 October 1937. He was knighted in 1914, was granted a peerage (Baron Rutherford of Nelson) in 1931, and is buried in Westminster Abbey.

Rutherford's early work in Cambridge, under the supervision of Thomson, involved the effect of X-rays, which had just been discovered by Wilhelm Röntgen, on gases. He moved on to investigate radioactivity in 1897, just after it had been discovered by Henri Becquerel. He found that there are two kinds of radiation, which he named **alpha radiation** and **beta radiation**; in 1900 he identified a third type of radiation, which he named **gamma radiation**. In Montreal, working with Frederick Soddy, Rutherford found that radioactive decay involves an atom being spontaneously broken down to make atoms of a different **element**, and found the relationship between the intensity of the radiation and the **half life** of the radioactive substance. With Bertam Boltwood (1870–1927), he pioneered the investigation of **radioactive series** and invented **radioactive dating**, providing the first evidence that some rocks in the Earth's crust are more than 1,000 million years old. A long investigation of alpha radiation culminated in 1908, after Rutherford's return to England, when he showed that alpha particles are identical to helium atoms carrying two units of positive charge – that is, a helium atom which has *lost* two negatively charged electrons.

It was in 1909 that Rutherford suggested the experiment, actually carried out by Hans Geiger and Ernest Marden, that led to the nuclear model of the atom, which Rutherford announced in 1911 (he first used the term 'nucleus' to describe the kernel of an atom in 1912). In 1912 and 1913 Niels Bohr visited Rutherford's group at Manchester, where he developed the **Bohr model**.

After his war work Rutherford made his last great direct contribution to physics in 1919, when he showed that the impact of an alpha particle with a nitrogen nucleus could cause the nitrogen nucleus to break up, forming a hydrogen nucleus and an oxygen nucleus – the first artificial transmutation of one element into another. As head of the Cavendish, Rutherford was instrumental in the construction of the first particle **accelerator**, built by John Cockroft and Ernest Walton in 1932, the same year that James Chadwick, also at the Cavendish, discovered the **neutron** (a particle Rutherford had predicted in 1920). In 1934 a team under Rutherford bombarded **deuterium** with **deuterons** and produced **tritium**; this is regarded as the first achievement of an artificial nuclear **fusion** reaction, but a case can also be made that fusion occurred as an essential intermediate step in the pioneering 'atom splitting' experiments of Cockroft and Walton.

For all his achievements, both as a hands-on scientist and as an administrator, Rutherford is regarded as the father of modern nuclear physics. He had a forceful, outgoing personality, but was well liked (as well as respected) and was a superb team leader. He died in Cambridge on 19 October 1937.

Rutherford model   See *atomic physics*.

Rydberg, Johannes Robert (1854–1919)   Swedish physicist remembered for discovering a mathematical formula which gives the frequencies of the lines in the spectrum from an **element**. The formula includes a number named the Rydberg constant, in his honour.

Born in Halmstad on 8 November 1854, Rydberg studied at the University of Lund, where he received a mathematics degree in 1875 and a PhD in 1879. He stayed there for the rest of his career, rising to become professor of physics in 1897 (the post was made permanent in 1901). He retired from the post just a month before he died, in Lund, on 28 December 1919.

Rydberg's work stemmed from his interest in the periodic table and his conviction that the patterns found by Dmitri Mendeleyev must be related to the structure of the atom. His great achievement was to organize a mass of spectroscopic data, largely obtained by other researchers, into order. This provided a more general framework for the *Balmer series*. Although this was a purely empirical discovery, and Rydberg had no idea what the physical basis of the patterns he had found might be, Niels Bohr later showed that the *Bohr model* of the atom exactly explained these relationships between spectral lines. Rydberg was also an early proponent of the idea of *atomic number*.

See *spectroscopy*.

Rydberg constant   See *Rydberg, Johannes Robert*.

Sakharov, Andrei Dimitrievich (1921–1989)   Russian physicist who became well known as a dissident under the Soviet regime in the 1970s and 1980s (he was awarded the Nobel Peace Prize in 1975 for his efforts to obtain a nuclear test-ban treaty), but had earlier been the leading scientist involved in the development of the Soviet hydrogen bomb. Among his many other contributions to physics, in the 1960s he proposed a mechanism for the formation of matter (in preference to *antimatter*) in the Big Bang. This suggestion went largely unnoticed at the time, but became a cornerstone of the standard model of the Big Bang in the 1980s.

Sakharov was born in Moscow on 21 May 1921, the son of a physics teacher. He started studying physics at Moscow State University in 1938, but after the invasion of Russia by Hitler's armies the department was evacuated to Ashkebad, where he graduated in 1942. After graduation, he worked for three years as an engineer at an arms factory in Ulyanovsk, on the Volga.

In 1945 Sakharov joined the Lebedev Institute, in Leningrad, where he studied under Igor Tamm, receiving a PhD for work on *cosmic rays* in 1947, and going on to work with Tamm on the problem of achieving nuclear fusion for both civil and military use, developing the hydrogen bomb and proposing the use of a 'magnetic bottle' to trap *plasma* in a fusion reactor. This has been developed into one of the most promising designs for a fusion reactor, called Tokamak. Although in 1953 he had become the youngest person ever elected to the Soviet Academy of Sciences, by the 1960s he was campaigning both for the test-ban treaty and for civil rights, and became increasingly distanced from the establishment. He produced no scientific papers at all between 1958 and 1965, but was instrumental in obtaining a reform of the Soviet educational system which encouraged the early entry of physicists and mathematicians (who are most creative when young) into university, and also played a major part in breaking the stranglehold of the bizarre biological ideas of Trofim Lysenko (1898–1976), who had been a favourite of both Josef Stalin and Nikita Khrushchev, on Soviet work in the life sciences.

Partly as part of a deliberate rejection of his earlier work, Sakharov became increasingly interested in the apolitical science of cosmology; he also worked on the

Andrei Sakharov (1921–1989).

theory of *quarks*, and on attempts to develop a theory of *quantum gravity*. Sakharov puzzled over the question of what came 'before' the Big Bang, and was one of the first people to suggest (in 1969) that there might be a large amount of dark matter in the Universe.

But Sakharov's most important contribution to cosmology was undoubtedly his investigation of baryon asymmetry – the fact that the matter in our Universe is in the form of *baryons*, not an equal mixture of matter and antimatter. The puzzle is that when matter is formed out of *energy* on Earth (in experiments using particle *accelerators*), each particle is accompanied by its antiparticle counterpart. The only way to 'make' an *electron*, for example, is by making a *positron* as well. But if matter and antimatter had been made in equal quantities in the Big Bang, all the particles and antiparticles would have met up and annihilated one another, leaving nothing but *electromagnetic radiation* behind. We are made of baryons, and we exist only because the Big Bang produced an excess of baryons over antibaryons, so that a little matter was left over to form stars, galaxies, planets and people after all the annihilation had finished.

In a paper published in 1967, Sakharov showed that there is a tiny asymmetry in the laws of physics, which actually says that under the conditions operating in the Big Bang roughly a billion and one baryons would have been produced for every billion antibaryons. We are made of some of the one-in-a-billion particles that didn't get annihilated; the rest (along with all the antiparticles) got turned into the cosmic microwave background radiation.

But this could only have happened in a hot Big Bang which is in the process of cooling (and therefore has an inbuilt *arrow of time*). It is connected with the way the four fundamental *forces of nature* split apart from one another as the energy density of the Universe decreases (see *grand unified theories*). Sakharov was way ahead of his time, and his work received the attention it deserved only after a similar model was developed independently by a Japanese physicist, Motohiko Yoshimura, in 1978. Sakharov's insight, explaining the requirements that had to be met in order for matter to exist in the Universe today, made before the grand unified theories were developed and providing not only a powerful theoretical argument for the existence of a hot Big Bang, but also an actual (correct!) prediction of the strength of the background radiation (roughly a billion photons for every baryon) was itself worthy of a Nobel Prize, and is one of the most perceptive insights in all of cosmology. But, although he repeatedly returned to the study of the baryon asymmetry throughout the rest of his life, Sakharov was distracted from promoting the idea as he increasingly turned his attention to the more pressing political problems of the 1970s.

In 1980, as part of a clampdown on dissidents, Sakharov was sent into internal exile in the city of Gorky, where he went on repeated hunger strikes to try to obtain permission for his wife, Yelena Bonner, to travel outside the Soviet Union for medical treatment. In spite of this, and his isolation from other scientists, he continued to work and publish important papers in cosmology, investigating (among other things) the properties of evaporating *black holes*. He was released by Mikhail Gorbachev in December 1986, and renewed his campaign for civil rights in the USSR, while still working on cosmology and in particular on the origin of the baryon asymmetry. He was elected to the Congress of People's Deputies in 1989, shortly before he died, in Moscow, on 14 December that year.

Further reading: John Gribbin, *In Search of the Big Bang*.

Salam, Abdus (1926–1996) Pakistani physicist who received the Nobel Prize for Physics in 1979 for his contribution to the development of the theory of the electroweak interaction (see *forces of nature*).

Salam was born in Jhang Maghiana (then part of India; now in Pakistan) on 29 January 1926. He studied at Government College in Lahore, graduating in 1946 with an MA awarded by Panjab University, then at the University of Cambridge, receiving a double first degree in mathematics and physics in 1949 and going on to research at the *Cavendish Laboratory*. Salam started work in experimental physics, but found that he was more suited to theoretical work and turned to *quantum field theory*, working in Paul Dirac's department. He returned to Pakistan as head of the mathematics department at Panjab University in Lahore in 1951, a year before the formal award of his PhD by Cambridge (although he had completed the work for his PhD, the rules said it could not be awarded until three years after he began the research!). Simultaneously he held the post of head of mathematics at Government College. He went back to Cambridge as a lecturer in mathematics in 1954, and was appointed professor of theoretical physics at Imperial College, London, in 1957, a post he held until 1993, even though he was also linked with other research centres.

Salam retained close links with Pakistan, and was scientific adviser to the President from 1961 to 1974; he was also deeply committed to encouraging science and technology in developing countries, and it was largely thanks to his efforts that the International Centre for Theoretical Physics was set up in Trieste in 1964 to foster the growth of science in the Third World. He was the first director of the ICTP, from 1964 to 1993, alongside his post at Imperial College, and in 1994 became its president.

He died in Oxford on 21 November 1996, after a long illness, and was buried in Rabweh, near his birthplace.

scalar   A quantity that has magnitude, but no sense of direction.

scalar field   A field which does not have an inbuilt sense of direction. A commonly used example is a field that represents the temperature of the air at every point in a room. Obviously, the field 'fills' the room – there is a number corresponding to the temperature at every point. But although a thermometer placed in the field will record a low temperature near the door, where there is a draught coming in, and a high temperature by the radiator on the other side of the room, there is no force which always pushes thermometers towards (or away from) hot spots. Differences in temperature do set up convection currents, which flow because of unevenness in the field, but there is no equivalent of the way a tiny charged particle is forced to move in a certain direction along electric field lines (see *vector field*, *field*).

One of the peculiarities of a scalar field is that it may be very difficult to detect. A completely uniform scalar field will have no influence on its surroundings (for example, if every point in the room is at the same temperature, there will be no convection currents, no matter how high that temperature is). An imperfect (but insightful) analogy can be made with two physics laboratories, one in the basement of a tall building and the other on the top floor. The strength of *gravity* is not *quite* the same in the two laboratories, but it would be very hard to tell this from any experiments that involve collisions between balls moving about on a smooth pool table. In either laboratory, you would find *Newton's laws of motion* describing the behaviour of the pool balls in the same way. But if the pool table in the top lab were pushed out of the window, it would plummet to the ground and be smashed, releasing gravitational *energy* in the form of heat.

scalar particle   Field particle associated with a *scalar field*. In *quantum field theory*, all fields can be represented by particles. Scalar particles are the simplest of these quantum particles and have zero spin.

scaling   Term coined by James Bjorken to describe the way in which the pattern of behaviour observed in *deep inelastic scattering* (specifically, when energetic electrons were being fired into protons at SLAC in the 1960s) is the same at different energies. Only the scale has changed. A fair analogy might be to compare a normal tree with its bonsai equivalent – they are essentially the same thing but on different scales. It was when Richard Feynman explained scaling in terms of *partons* that the *quark–gluon* model of *hadrons* became accepted.

scanning tunnelling microscope   A microscope which uses quantum *tunnelling* to probe the surfaces of materials. The tip of the probe of the microscope floats across the surface being scanned, held a few atomic diameters above it by superconducting levitation. Electrons tunnel out of the surface being scanned and into the probe, producing a changing electrical signal which reveals details of the surface.

scattering   General term used to refer to the way waves or particles bounce off an obstruction. It was scattering of *alpha particles* by gold *nuclei* that revealed the

Scattering. A beam of light passing through water is only visible because particles in the water scatter some of the light sideways. If the water were perfectly clean, the light beam would be invisible from the side.

nuclear structure of the atom in the first decade of the 20th century, and *inelastic scattering* of light that confirmed the reality of *photons* some two decades later. Particle scattering has remained a prime tool of physics in *accelerator* experiments, including the work that led to the discovery of structure within *hadrons* (see *quarks*).

In a slightly different use of the term, light is also regarded as being scattered, rather than reflected, from a rough surface because incoming photons are reflected at different angles with no coherent overall pattern. In this case, each individual photon might be regarded as being reflected, but the overall result is a more or less random scattering of the incident light. It is this kind of scattering of light that produces the blueness of the sky, as blue photons are more easily bounced around the atmosphere than red photons.

Schawlow, Arthur Leonard (1921–) American physicist who received the Nobel Prize for Physics in 1981 for his development (independently of Nicolaas Bloembergen) of the use of *lasers* in *spectroscopy*.

Born at Mount Vernon, New York, on 5 May 1921, Schawlow studied at the University of Toronto, receiving his first degree in 1941 and his PhD in 1949. He then joined the group led by Charles Townes at Columbia University (and later married Townes' sister). His work with Townes laid the foundations for the development of the laser, but Theodore Maiman is credited with building the first working laser. In 1951 Schawlow joined the staff of the Bell Telephone Laboratories, where he stayed for ten years, although also maintaining contact with Columbia as a visiting professor, before moving on to Stanford University as professor of physics in 1961. It was there, in the 1970s, that he developed the techniques of laser spectroscopy which led to the award of his Nobel Prize.

Schrieffer, John Albert (1931–)   American physicist who shared the Nobel Prize for Physics in 1972 with John Bardeen and Leon Cooper for their theory of *superconductivity*.

Born in Oak Park, Illinois, on 31 May 1931, Schrieffer studied at MIT and at the University of Illinois (his PhD, awarded in 1957, was for work on superconductivity under the supervision of Bardeen, which formed his part of the BCS theory). Schrieffer then spent two years in Europe as a National Science Foundation Fellow, working at the University of Birmingham and at the Niels Bohr Institute in Copenhagen, before returning to Illinois for three years. In 1962 he took up a post at the University of Pennsylvania, Philadelphia, becoming professor of physics there in 1964. He later worked at Cornell and at the University of California, and from 1992 held the post of university professor at Florida State University.

Schrödinger, Erwin (1887–1961)   Austrian physicist who developed the *wave mechanics* formulation of quantum physics. He received the Nobel Prize for Physics as a result, in 1933.

Schrödinger was born in Vienna, on 12 August 1887. The year is particularly important because Schrödinger was a physicist of the old school, by some way the oldest of the quantum pioneers of the 1920s; he developed wave mechanics in the hope and expectation that he was bringing sanity back into quantum theory, and after he found that his wave equation did not, after all, remove the need for puzzling quantum processes (see, for example, *quantum leap*) and the role of the observer (see *Copenhagen interpretation*), he said of the theory he had helped to father, 'I don't like it, and I wish I'd never had anything to do with it.'

Schrödinger was largely educated by a private tutor until 1898, when he entered the Gymnasium (high school) in Vienna. He went on to the University of Vienna, entering in 1906. At that time, the first degree in physics at Austrian universities was a doctorate, awarded after four years' work, involving both taught courses and a little research; it was about equivalent to a modern master's degree. Schrödinger received his Dr Phil degree in 1910. After a year of military service, in 1911 he became an assistant in the university. In the First World War, Schrödinger served as an officer in the artillery, then returned to his university post. Because of the difficulty of life in postwar Austria, in 1920 Schrödinger moved to Germany, where he worked briefly in Jena (1920), Stuttgart (1920–1) and Breslau (1921), then on to Switzerland, where he settled as professor of physics in Zurich in 1921. It was in Zurich that he carried out the work on quantum theory for which he is now remembered (jumping off from the suggestion by Louis de Broglie that the electron could be treated as a wave), and the publication of this work in 1926 (when he was 39), using the *Hamiltonian* formulation, led to him being offered the opportunity to succeed Max Planck as professor of theoretical physics in Berlin, in 1927. Schrödinger's Hamiltonian version of quantum mechanics became the standard methodology, and is still the way most students are introduced to the subject today. Richard Feynman's *Lagrangian* approach (developed in the 1940s) is, however, more complete and satisfying, and widely used by the experts.

Erwin Schrödinger (1887–1961).

When the Nazis came to power in Germany, in 1933, Schrödinger left for Oxford, where he became a fellow of Magdalen College. He did not fit into the Oxford scene (where his fellow academics were, to say the least, surprised that he turned up with both a wife and a mistress), and in 1936 he returned to Austria, to work at the University of Graz. But two years later the German annexation of Austria led him to move once again (initially to Italy and then briefly to the United States), and in 1939 a post was created for him at the Dublin Institute for Advanced Studies (indeed, the whole Institute was created in order to provide a base for his work). During his time in Dublin, Schrödinger wrote an influential book, *What is Life?*, which encouraged a generation of physicists (including Francis Crick) to turn to molecular biology after the Second World War, with extremely beneficial results for biology. Schrödinger stayed at the Dublin Institute until 1956, when he returned to Austria as professor of physics in Vienna. The following year he suffered a severe illness, from which he never fully recovered; he officially retired in 1958, becoming professor emeritus. He died in Vienna on 4 January 1961.

Schrödinger's disdain for convention extended beyond his sex life. According to Paul Dirac (quoted in *Pioneers of Science*, by Robert Weber), when Schrödinger went to a conference 'he would walk from the station to the hotel where the delegates stayed, carrying all his luggage in a rucksack and looking so like a tramp that it needed a great deal of argument at the reception desk before he could claim a room'. It need hardly be added that he was popular with students and highly regarded as a teacher.

Further reading: Walter Moore, *Schrödinger: life and thought.*

Schrödinger equation The *wave equation* used in one version of quantum mechanics to describe the behaviour of a quantum entity, such as an electron. It was discovered by Erwin Schrödinger in the mid-1920s. See *wave mechanics*.

Schrödinger's cat Shorthand name for a hypothetical 'experiment' dreamed up by Erwin Schrödinger in 1935 to demonstrate the absurdity of the *Copenhagen interpretation* of quantum mechanics.

Schrödinger was particularly concerned about two things – the idea that a quantum system could be in a *superposition of states*, and the requirement (in the Copenhagen interpretation) of an intelligent observer to 'collapse the wave function' and force a quantum system to take up a unique *state*. The puzzle depends on setting up a system where there is a precise fifty-fifty chance of a particular quantum event – such as the *decay* of a radioactive nucleus – occurring.

The conventional wisdom in quantum mechanics says that the nucleus exists in a superposition of states, half decayed and half not decayed, unless its state is measured. Only at that point does it decide which state it is in. Schrödinger pointed out that the radioactive substance could be sealed in a windowless steel chamber (usually translated as a 'box', but we prefer to think of this as a large room), with a detector (perhaps a simple *Geiger counter*) to monitor it. The detector is wired up to release a cloud of poison gas into the room if the radioactive material decays, and living in the room there is the famous cat. If the chamber is sealed and nobody looks into it, then when the radioactive nucleus is in a fifty-fifty superposition of states, according to the strict Copenhagen interpretation the Geiger counter, the poison gas and the cat are all in a superposition of states. The radioactive material both has and has not decayed, the poison gas both has and has not been released, and the cat both has and has not been killed.

This is only a thought experiment, and no cat has ever been put through such indignities. But if the Copenhagen interpretation is correct, everything remains in limbo until an intelligent observer looks into the chamber. At that point, the superposition collapses and the cat becomes either dead or alive. Until somebody looks, the situation inside the chamber is described in Schrödinger's own words as 'having in it the living and the dead cat (pardon the expression) mixed or smeared out in equal parts'.

This bizarre picture encouraged some physicists to seek alternative interpretations of quantum mechanics, eventually with considerable success (see, for example, *transactional interpretation*). But in spite of such nonsense as the Schrödinger's cat 'paradox', the Copenhagen interpretation is still enshrined in most textbooks and taught to most students of quantum mechanics as the standard interpretation.

See also *Wigner's friend*.

Further reading: John Gribbin, *In Search of Schrödinger's Cat*.

Schwartz, John (1941–) American physicist who pioneered the development of *string theory* in its modern particle physics context.

Born in North Adams, Massachusetts, on 22 November 1941, Schwartz took his first degree at Harvard (1962) and his PhD at the University of California, Berkeley (1966). He worked at Princeton University from 1966 to 1972, then moved to Caltech, where he stayed, becoming a full professor in 1985 and being appointed Harold Brown Professor of Theoretical Physics in 1989. He has spent spells as a visitor at the Ecole Normale Supérieure in Paris and Queen Mary College in London, as well as at American institutions.

Schwartz, Melvin (1932–) American physicist who shared the Nobel Prize for Physics in 1988 with Leon Lederman and Jack Steinberger for their joint discovery of the muon neutrino.

Born in New York City on 2 November 1932, Schwartz studied at Columbia University, in New York, receiving his PhD in 1958. He stayed at Columbia until 1966, then became professor of physics at Stanford University, where he stayed until 1983. Alongside this post, in the 1970s he founded a computer company, Digital Pathways, and in 1983 Schwartz left the academic world to concentrate on his business interests. In 1991 he came back into the academic fold as associate director of high energy and nuclear physics at the Brookhaven National Laboratory, where his Nobel Prize-winning work had been carried out in the early 1960s.

Schwinger, Julian Seymour (1918–1994) American physicist who received the Nobel Prize for Physics in 1965 for his part in the development of the theory of *quantum electrodynamics*.

Born in New York City on 12 February 1918 (an almost exact contemporary of Richard Feynman), Schwinger was a prodigy whose talent for mathematics was apparent from an early age, but his erratic attendance at college (City College of New York, which he entered in 1934) and disdain for other subjects threatened his career. Luckily, one of his tutors, Lloyd Motz, was a PhD student at nearby Columbia University. He realized that Schwinger had a talent that should be nurtured, and drew him to the attention of Isidor Rabi at Columbia. Thanks to Rabi, Schwinger transferred to Columbia, where he blossomed. He graduated in 1936, at the age of eighteen, and received his PhD (also from Columbia) in 1939. The work on which his PhD thesis was based, which dealt with the way slow neutrons are scattered by atoms, had largely been carried out during his final year as an undergraduate, and while he was a graduate student Schwinger spent a lot of time with the experimentalists, helping to analyse their results. He also spent the academic year 1937/8 at the University of Wisconsin, where he got into the habit (which he kept) of working all night and sleeping by day, so that he did not get disturbed while working.

Schwinger then moved to the University of California, Berkeley, where he worked under Robert Oppenheimer. In 1941 he moved on to Purdue University, but like most physicists of his generation his career was then interrupted by war work. In spite of his talent and connections, Schwinger did not join the Los Alamos team working on the **Manhattan Project** during the Second World War (he was offered a role but turned it down), but instead spent the summer of 1943 at the Metallurgical Laboratory of the University of Chicago, where the first nuclear reactor was being built, then worked at the Radiation Laboratory of MIT from 1943 to 1945, mainly being involved with radar research (he had been a consultant on this work while still based at Purdue). He then joined the faculty at Harvard, and in 1947 was made a full professor at the age of 29 – one of the youngest people ever appointed to such a post at Harvard. In 1972 he moved to the University of California, Los Angeles, where he spent the rest of his career.

Schwinger developed his version of QED independently of Feynman (whose approach was unique) and of Sin-itiro Tomonaga, who used essentially the same

Julian Schwinger (1918–1994).

approach as Schwinger. But Schwinger was a mathematician who revelled in the esoteric finery of his trade, and presented his results in a complicated formalism that gave all the right answers but could be understood only by expert mathematicians. He once sneeringly remarked that Feynman's approach 'brought computation to the masses', the implication being that the hoi polloi should not be given access to such delights as QED. After his work on QED, which was complete by 1948, Schwinger made no more major contributions to physics. He worked on *synchrotron radiation*, and in 1957 he predicted that there must be two different kinds of *neutrino* (associated with the electron and the muon respectively; the tau particle was not known at that time). In the light of his early success, though, these were modest achievements. Schwinger's greatest achievement after 1948 was to act as the adviser and inspiration for a succession of first-class graduate students.

He died, in Los Angeles, on 16 July 1994.

Further reading: Silvan Schweber, *QED and the Men Who Made It.*

scientific notation   The trick of writing numbers as 'powers of ten' instead of writing long strings of zeros. So 100 can be written as 10^2 (meaning 'a 1 followed by two zeros') and 0.001 can be written as 10^{-3} ('a decimal point followed by two zeros and a 1'). The convention comes into its own when we are dealing with very large numbers or very small numbers. A number such as 137 could, for example, be written as 1.37 x 10^2 (= 1.37 x 100), but there is little point in doing so; it is, however, obviously a good idea to give the mass of the electron as 9.1 x 10^{-31} kg, rather than as 0.00000000000000000000000000000091 kg.

scintillation counter   A form of particle detector in which a flash of light is produced when an electrically charged particle passes through the material in the counter. This kind of detector was used by Ernest Rutherford and his colleagues in the classic experiments which revealed the nuclear structure of the atom. In those days, scientists had to watch the experiment and monitor the flashes by eye; today, each burst of light is turned into an electrical impulse which is amplified and fed into a computer. The computer monitors the scintillations, and if two or more counters are used in the same detector, it is possible to reconstruct the trajectory of the particle(s) passing through the experiment.

The flashes of light are produced because atoms in the detector material are excited by interactions with the passing particles and raised into a high energy state. They almost immediately fall back into their ground state, emitting a flash of light as they do so. The detector material used is often a plastic, but sometimes liquids are used.

second law of thermodynamics   The law of nature which says that things wear out. One expression of the second law of thermodynamics is that heat cannot flow from a cold object to a hotter object of its own volition. Place an ice cube in a cup of warm water, and the ice melts as heat flows into it from the water, ending up with a cup of slightly cooler water than you had before. You never see ice cubes spontaneously forming in cups of water, as heat drains out of the cold ice into the hotter liquid. Ice cubes can only be made (for example, in a domestic freezer) by using *energy* to pump heat out. Another facet of the second law is the way in which a house left unattended for a long time will crumble away under the influence of wind and weather, whereas a pile of bricks left unattended will never spontaneously form itself into a house (see *arrow of time*).

In his book *The Nature of the Physical World*, Arthur *Eddington* said that:

> The second law of thermodynamics holds, I think, the supreme position among the laws of Nature. If someone points out to you that your pet theory of the universe is in disagreement with Maxwell's equations – then so much the worse for Maxwell's equations. If it is found to be contradicted by observation – well, these experimentalists do bungle things sometimes. But if your theory is found to be against the second law of thermodynamics I can give you no hope; there is nothing for it but to collapse in deepest humiliation.

It may seem strange that such an important law is the 'second' law of anything; but the first law of thermodynamics is merely a kind of throat-clearing statement that work and heat are equivalent to one another.

second quantization   The description of *fields* in terms of *quanta*. The first quantization (although almost never referred to in that way) dealt with the properties of particles, such as electrons jumping from one energy level to another in an *atom*. The second quantization dealt with fields, and showed that they could be described in terms of the exchange of particles. The name is really only of historical interest, and not something worth worrying about.

Segre, Emilio Gino (1905–1989)   Italian-born American physicist who received the Nobel Prize for Physics in 1959 for the discovery of the antiproton.

Born in Tivoli, near Rome, on 1 February 1905, Segre studied at the University of

Rome, initially as an engineer, but switching to physics under the influence of Enrico Fermi. He became Fermi's first graduate student and received his PhD in 1928. After a year's compulsory military service he returned to the University of Rome to pick up his career. He spent brief periods working with Otto Stern in Hamburg and Pieter Zeeman in Amsterdam, but in 1932 he was appointed as an assistant professor in Rome, to work with Fermi on the investigation of nuclear physics, bombarding nuclei with neutrons and studying artificial radioactivity. In 1936 Segre became a professor at the University of Palermo, where he discovered, in 1937, a previously unknown element (which he called technetium; it has **atomic number** 43) that had been produced in the heavily irradiated parts of an old accelerator at Berkeley. Segre started down the road that led to the discovery of technetium using pieces of radioactive scrap that he had picked up on a visit to Berkeley and brought home with him; but with what now seems breathtaking disregard for safety, various items of radioactive material were regularly sent to Segre from the USA by ordinary mail, and it was in one of those samples, a piece of molybdenum foil that had once been part of a **cyclotron**, that he identified technetium. It was the first identification of a previously unknown man-made chemical element.

Segre often visited the USA, and in 1938 he was on a working summer visit to Berkeley when the deteriorating political situation in Europe made it undesirable to return to Italy (Segre was Jewish, and new anti-semitic laws were passed in Italy in 1938). His family managed to join him in California. He spent the rest of his career at Berkeley (apart from wartime work at Los Alamos as a group leader on the **Manhattan Project**), and became an American citizen in 1944. He became a full professor at the University of California, Berkeley, in 1947. In 1940 Segre and his colleagues found another new element, now known as astatine (atomic number 85); he was also involved in the discovery of plutonium (atomic number 94). After the war, he studied proton–proton and proton–neutron interactions, using first the then-new cyclotron at Berkeley and later the **Bevatron**. The antiproton was discovered during the Bevatron experiments. It was only the second **antiparticle** to be discovered (after the positron).

As well as his important contributions to science, Segre made his mark as an editor of scientific journals and as a writer, producing a biography of Fermi and a volume of autobiography and memoirs (published posthumously) which gives a rare insight into what it was like to work in particle physics in the middle decades of the 20th century. He died in Berkeley on 22 April 1989.

Further reading: Emilio Segre, *A Mind Always in Motion*.

selectron   The **supersymmetric partner** of the **electron**.

self-energy   See *self-interaction*.

self-interaction   An interaction between a particle that produces a **field** and the field produced by the particle. For example, an electron has an electric charge, so there is an electric field associated with the electron. The electron interacts with that field. Because the strength of the interaction is inversely proportional to the square of the distance from the source of the field, and the electron is at zero distance from itself, the self-interaction ought, on this naive picture, to be infinite, giving the electron an infinite self-energy and infinite mass. Exactly equivalent problems arise with other fields, such as gravity. This crucial difficulty with field theory (both classical and quantum versions) is resolved by **renormalization**. See also **quantum electrodynamics**.

semiconductor   A crystalline substance with an electrical conductivity intermediate between that of conductors (like copper and other metals) and insulators (like rubber and plastics). The most familiar semiconductors are silicon and germanium. Their valuable semiconducting properties are a quantum phenomenon associated with the behaviour of the electrons in the crystals. Individual **orbitals** associated with **atoms** in the crystal overlap, creating energy bands in which electrons can move through the crystal.

Germanium and silicon atoms have a **valency** of four. By adding a trace of impurities such as arsenic or phosphorus (which each have five valency electrons) to the crystal, it is possible to provide a small surplus of electrons which move through the conduction bands in this way; this is called an n-type semiconductor. By adding other impurities, such as boron or indium (which each have only three valency electrons), it is possible to produce a deficit of electrons, creating 'holes' in the conduction bands which act just like positively charged electrons, being repelled by positive charges and attracted to negative charges. This is called a p-type semiconductor. Suitable combinations of p- and n-type semiconductors make up transistors and diodes.

Serpukhov   Russian particle accelerator laboratory located south of Moscow.

set   In mathematics, a set is any well-defined group of objects. This could be as simple as the set of all the vowels in the alphabet, or the set of all even numbers. In **quantum field theory**, particular sets and the relationships between them describe the behaviour of entities in the quantum world. See also **group theory**.

shadow matter   A hypothetical form of matter that may have been formed in the very early Universe at the moment (just after the **Planck time**) when gravity split off from the other fundamental interactions. According to some versions of **supersymmetry** theory, when that happened some of the **energy** in the Universe ended up in the form of the particles that we know today (including the **baryons** that our bodies are made of), while the rest ended up as a completely separate set of particles, which shares nothing in common with our particles except that they also feel the force of gravity. You could walk right through a shadow person and neither of you would ever notice. For our kind of matter, gravity is mediated by a particle with two units of spin, the **graviton**; this has (according to theory) a supersymmetric partner called the gravitino, with a spin of ½. Shadow matter would contain equivalents to *both* these particles, a shadow graviton and a shadow gravitino, with spin zero and spin ½, respectively. There would also be shadow equivalents of other forces and particles, but all of the shadow forces except gravity would be self-contained, affecting only the shadow world. It is because gravity is a property of spacetime, and both universes share the same spacetime, that the shadow universe and our Universe can interact gravitationally.

There is no reason to think that the two kinds of matter would have formed in equal quantities, and shadow matter could account for all of the dark matter required by some cosmological models of the Universe, which greatly exceeds the amount of visible matter. Alternatively, there could be a mirror image relationship between the two worlds, with shadow **electrons**, shadow **protons** and shadow **neutrons** making up a universe of shadow stars and shadow galaxies which even includes its own shadow dark matter (perhaps in the form of shadow *axions*). Unfortunately, the theory of shadow matter is so vague that just about any science-fiction-like possibility you can dream up is allowed (or at least, not forbidden).

Nevertheless, there is a key point which makes this more than idle speculation. Vague though the details are, it seems that there may be only one solution to the equations involving supersymmetry and **superstrings**, and that this solution inevitably produces both the real Universe and a shadow universe. Many particle physicists think that that would be a small price to pay for a theory which, if the solution to the equations can be found, would exactly predict all of the properties of everything in the visible Universe.

shadow universe   See *shadow matter*.

shell   See *electron shell*.

shell model of the nucleus   A model of the *nucleus* developed by analogy with the shell model of electrons in atoms (see *electron shell*). The model was largely developed by Marya Goeppert-Mayer and Hans Jensen. They suggested that the *magic numbers* which characterize nuclei might be explained if protons and neutrons are arranged around the nucleus in a series of nucleon shells. The magic numbers then correspond to 'complete' shells, in the same way that chemically unreactive elements correspond to atoms with full outer electron shells. Although useful for some purposes, the shell model of the nucleus is not the whole story (see, for example, *liquid drop model*).

Shockley, William Bradford (1910–1989)   British-born American physicist who shared the Nobel Prize for Physics in 1956 with John Bardeen and Walter Brattain for their invention of the transistor.

Born in London on 13 February, 1910, Shockley was the son of two American mining engineers. He studied at Caltech and at Harvard University, where he obtained his PhD in 1936. He then worked at the Bell Laboratories until 1956 (apart from wartime work on anti-submarine measures), when he set up his own company in Palo Alto, California. He became a professor of engineering at Stanford in 1963, and worked for Bell as a consultant from 1965 onwards. He retired in 1975 and died in Palo Alto on 12 August 1989.

shower   A collection of high-energy particles moving roughly together towards the surface of the Earth, produced by the impact of a primary *cosmic ray* with an atomic nucleus high in the atmosphere. The secondary cosmic rays in the shower fan out in a narrow cone from the impact site.

The term is also used to describe the spray of particles produced in an *accelerator* when the beam of high-energy particles hits its target.

Shull, Clifford Glenwood (1915–)   American physicist who received the Nobel Prize for Physics in 1994 for his contribution to the development of neutron-*scattering* techniques for studying the structure of matter. This involves the *diffraction* of neutrons by crystals, making use of the *wave–particle duality* of neutrons.

Born in Pittsburgh on 23 September 1915, Shull studied at the Carnegie Institute of Technology (BSc 1937; the Institute is now the Carnegie Mellon University) and at New York University (PhD 1941). He then worked in industry for five years, with the Texas Company in Beacon, New York, before joining the Oak Ridge National Laboratory in 1946. In 1955 he became professor of physics at MIT, where he spent the rest of his career, retiring in 1986. The work for which he received the Nobel Prize was carried out during his time at Oak Ridge, and laid the groundwork for probing materials using *elastic scattering* of neutrinos. See also *Brockhouse, Bertram Neville*.

Siegbahn, Kai Manne Börje (1918–)   Swedish physicist who received the Nobel Prize for Physics in 1981 for his pioneering work in electron spectroscopy.

Siegbahn was born in Lund, where his father was professor of physics, on 20 April 1918. He studied at the University of Stockholm, receiving his doctorate in 1944, and staying on there to begin the research for which he is now remembered. This developed the 'family business' started by his father, Manne Siegbahn, first using X-ray *spectroscopy* and then developing this to analyse the kinetic energy of electrons knocked out of material by a beam of monoenergetic X-rays or ultraviolet radiation – so-called electron spectroscopy. This has important applications in industry in the analysis of high-technology materials. In 1951 Siegbahn was appointed professor at the Royal Institute of Technology in Stockholm, and in 1955 he became professor of physics at Uppsala University.

The family tradition of physics continues into a third generation, with Kai's sons Per and Hans; a third son, Nils, is a chemist.

Siegbahn, (Karl) Manne Georg (1886–1978)   Swedish physicist who received the Nobel Prize for Physics in 1924 for his pioneering work in X-ray *spectroscopy*.

Siegbahn was born in Örebro, where his father was a railway station master, on 3 December 1886. He studied at the University of Lund, receiving his doctorate in 1911 and staying on there, initially as an assistant to Johannes Rydberg, to carry out the research for which he is now remembered. This provided important confirmation of the accuracy of the *Bohr model* of the *atom*. In 1920 Siegbahn became a full professor of physics at the University of Lund, moving in 1923 to the University of Uppsala. The year after he was awarded the Nobel Prize, he published a book, *Spectroscopy of X-rays*, which became a standard text. Also in 1925, Siegbahn and his colleagues showed that X-rays are affected by *refraction* in the same way that light is.

In 1937 Siegbahn became the first director of the new Nobel Institute of Physics, in Stockholm, simultaneously becoming professor of physics at the University of Stockholm. He retired from the university post in 1964 (becoming a professor emeritus), but stayed on as director of the Nobel Institute until 1975. He died in Stockholm on 26 September 1978.

sigma   Overall name for a family of three particles (one with positive charge, one negative, one neutral) discovered between 1953 and 1956. These were early examples of *strange particles*. The sigmas are *baryons*, about 10 per cent more massive than the *proton*, each with ½ unit of *spin* (so they obey *Fermi–Dirac statistics*). The sigma-plus is made up of two up *quarks* and one strange quark; the sigma-minus of two down quarks and one strange quark; and the sigma-zero of one up, one down and one strange quark.

SI units   The initials stand for Système International, and the international system is supposed to be the one used by scientists, although they have a predilection for the odd unit which doesn't fit the system, such as the *barn*. SI basically derives from the metre, kilogram and second (MKS) system of units, with a few extra basic units such as the *kelvin*.

SLAC   Acronym for Stanford Linear Accelerator Center, a research institute based

SLAC. The linear accelerator at Stanford is two miles (3.2 km) long. Electrons are injected into the accelerator at the end nearest us in the picture, and reach their targets in the experimental facility in the middle distance, across the freeway, which passes over the accelerator.

around a particle *accelerator* 2 miles (some 3 km) long built in California in the 1960s. The evacuated accelerator tube is aligned to an accuracy of 0.5 mm throughout its length, and electrons are accelerated along it by electromagnetic fields (in effect, surfing the radio waves) to reach energies of tens of GeV (initially designed to run at 20 GeV, the accelerator was developed first to 30 GeV and later to 50 GeV). The linear accelerator at Stanford, the largest accelerator of its kind in the world, was a key instrument in the development of the *parton* model of nuclear particles, which led to the acceptance of the *quark* model. The *SPEAR* ring was later built at the same site, and the original linear accelerator developed to become part of the *SLC*.

SLC Shorthand for the Stanford Linear Collider, a particle *accelerator* at *SLAC*. The collider was developed from the original linear accelerator at SLAC in the 1980s. Bunches of *electrons* and *positrons* can now be sent down the linear accelerator and then guided away in two curving arcs, one to the left and one to the right, before being brought together in a head-on collision on the other side of the circle. Because each pulse of particles has an energy of 50 GeV, in the head-on collisions the maximum energy of the colliding particles is 100 GeV.

slow neutrons *Neutrons* moving at such a slow speed that they each have a kinetic energy of only about 0.025 eV (essentially the energy associated with their thermal motion, as heat). Neutrons produced naturally by *radioactive decay* move much faster than this, but can be slowed down by passing the beam of neutrons through a substance such as water or paraffin. In the mid-1930s, Enrico Fermi realized that slow neutrons would be more easily absorbed by *nuclei* than fast neutrons, in the same way that a fast-rolling golf ball may whiz past the cup while a more gently struck putt eases into the hole. Using this technique, Fermi and his colleagues in Rome manufactured more than 40 new artificial radioactive *isotopes* in a few months.

Soddy, Frederick (1877–1956) British chemist who received the Nobel Prize for Chemistry in 1921 for his investigations of *radioactive decay* and the study of *isotopes*.

Born in Eastbourne on 2 September 1877, Soddy was a student at Eastbourne College, where he was inspired to become a chemist by his teacher, R. E. Hughes; together, they published a scientific paper in 1894, before Soddy went on to study at the University College of Wales, Aberystwyth. From there, he moved on to Oxford University in 1895, graduating top of his year in 1898. He stayed on at Oxford for two more years, and in 1900 he applied for the job of professor of chemistry at the University of Toronto. In spite of visiting Toronto in person, he failed to get the job; but on the way home he stopped off in Montreal, where he was offered a more junior post at McGill University, working with Ernest Rutherford.

After two fruitful years with Rutherford, in 1902 Soddy moved back to England to work with William Ramsay at University College, London. In 1904 he visited Australia as a representative of the University of London, but on his return to Britain he moved to the University of Glasgow, where he developed his theory of isotopes. He became professor of chemistry at the University of Aberdeen in 1914. After war work, he moved to Oxford in 1919, as Lees Professor of Chemistry, and played a part in modernizing the laboratories there, but did little further research. He took early retirement in 1936, after the death of his wife, and travelled in Asia for a time. He became deeply concerned about the implications of the release of atomic energy, and devoted much of his energies in the last two decades of his life to trying to encourage a sense of social responsibility among scientists. He died in Brighton on 22 September 1956.

With Rutherford, Soddy made important contributions to the theory of radioactive decay, and identified the two *radioactive series* that begin with uranium and thorium, respectively, and end with lead. With Ramsay, he confirmed the prediction (made with Rutherford) that helium is produced by the decay of radium. In 1913 Soddy showed that lead comes in different varieties; Theodore Richards independently made the same discovery, and other researchers were thinking along similar lines, but only Soddy went on to explain this, and the existence of different forms of other elements, in terms of isotopes, which he gave that name. He also discovered that when an *atom* (strictly speaking, a *nucleus*) emits an *alpha particle*, it is transformed into an atom of an element two places lower in the *periodic table* (reducing its *atomic number* by 2), but that when it emits a *beta particle*, it moves one place higher (increasing its atomic number by 1).

soft scattering = *elastic scattering*.

solid-state devices Electronic devices such as transistors and diodes which are entirely solid, being made out of crystalline *semiconductor* material and taking advantage of quantum phenomena such as tunnelling for their operation. The distinction is with the older form of electronics using thermionic valves (tubes). In the extreme version of this, we have electronic devices in which the working parts

Participants in the second **Solvay Congress**, held in 1913. The group includes Albert Einstein (standing in the centre), Marie Curie (standing, centre left) and Ernest Rutherford (seated, second from left).

(resistors, transistors and so on) are made together on one piece of material, usually silicon, known as a chip. This makes them small, light and fast compared with devices made by the older technique of joining different components together by wires or printed circuits.

Solvay Congress   Overall name for a series of scientific meetings sponsored by Ernest Solvay (1838–1922), a Belgian chemist who made a fortune from his method for manufacturing sodium carbonate (the 'soda' used in glass, soap and porcelain manufacture). Because of his interest in more abstract science, Solvay provided funds for these meetings, at which the leading physicists of the day were able to meet and exchange views. The first was held in 1911, the fifth in 1927, and they provided a strong catalysing influence on the development of physics, particularly quantum physics, in the second two decades of the 20th century. The congresses continued after the 1920s (the twelfth was held in 1961), but with less impact than in the heyday of the development of quantum mechanics.

Sommerfeld, Arnold Johannes Wilhelm (1868–1951)   German physicist who made major contributions to the quantum theory of the structure of the *atom*, and was an influential teacher.

Born in Königsberg (now the Russian enclave of Kaliningrad), on 5 December 1868, Sommerfeld studied at the local Gymnasium (high school) and went on to the University of Königsberg in 1886, studying mathematics and receiving his PhD in 1891. He then worked at the Mineralogical Institute in Göttingen, before becoming a *Privatdozent* at the University of Göttingen in 1895. In 1897 he became professor of mathematics at the Mining Academy in Clausthal, moving on in 1900 to become professor of technical mechanics at the Technical Institute in Aachen, and in 1906 to become director of the new Institute of Theoretical Physics at Munich University. Sommerfeld made the Institute (which had been created especially for him) one of the leading centres of theoretical physics at one of the most exciting times in the history of physics, numbering Peter Debye, Wolfgang Pauli, Walter Heitler, Werner Heisenberg and Hans Bethe among its students. He also spread the news about quantum physics to the United States on extended visits to the University of Wisconsin (1922–3), Caltech (1928) and the University of Michigan (1931).

Sommerfeld retired from this post in 1940, when he was in his seventies, having made a brave effort, in spite of his age, to defend Jewish scientists from the institutionalized anti-semitism of Germany in the 1930s. He took up the post again at the end of the Second World War to help in the reconstruction of German science, and died in Munich on 26 April 1951, after being hit by a car.

Although he made many contributions to physics (including the theory of the gyroscope), the most important were in quantum theory, where Sommerfeld encouraged Niels Bohr to develop his model of the atom and then developed the **Bohr model** further by introducing the idea of elliptical orbits. He later wrote two influential books on wave mechanics, and laid the foundations of the quantum theory of metals when in his late fifties.

space Traditionally, the void between the stars and planets. In the context of **relativity theory**, however, even 'empty space' has to be thought of as having a well-defined structure and properties – it is the stage on which material events take place. The properties of space, and in particular the way it is curved, determine the way objects move. In relativity theory, three-dimensional space is united with time to make a four-dimensional continuum.

This picture breaks down, according to **quantum physics**, on the smallest scales, over distances comparable to the **Planck length**. There, both space and time lose their identity in a 'quantum foam'. The American physicist John Wheeler has suggested that the presence of what we regard as a real particle in space is no more significant in the context of the activity of the quantum foam than the presence of a cloud is to the dynamics of the atmosphere. We see the cloud, or the particle, but it is only a minor disturbance in a sea of activity.

spark chamber A form of particle detector, developed in the 1960s, which consists of parallel sheets of metal separated by a distance of a few millimetres and immersed in an inert gas, such as neon. When an electrically charged particle passes through the chamber, it leaves an ionized trail in the gas, in the same way that such a particle produces a trail in a **cloud chamber**. If a high voltage is applied to every other plate in the chamber soon after the particle has passed through, sparks like miniature lightning bolts flash along these ionized trails, revealing the paths of the particles. **Scintillation counters** just outside the chamber note the passage of charged particles, and trigger the electric discharges.

sparticles The bosonic partners to everyday fermionic particles required by the theory of **supersymmetry** are given names by adding an 's' to the front of the name of their everyday matter counterparts – **selectron**, **squark**, sneutrino and so on.

SPEAR Acronym for the Stanford Positron–Electron Asymmetric Rings. The linear accelerator at **SLAC** feeds SPEAR with bunches of *electrons* and *positrons* which circulate in opposite directions around the oval ring until the experimenters bring them together in a head-on collision. SPEAR was built on a former parking lot at SLAC, and was completed in 1972; it reaches energies of a few (up to 9) GeV. In November 1974 it provided evidence for the existence of the *J/psi particle* (see also *November revolution*).

special theory of relativity   See *relativity theory*.

special unitary group (SU)   The group which defines the *local symmetry* for all known quantum forces. See *group theory*.

specific heat   Short for specific heat capacity, a number which indicates the increase in temperature of a given mass of a particular substance for a specific input of heat. It is usually measured in joules per kelvin per kilogram. The important point is that the *same* amount of heat applied to the *same* mass of different substances (such as water or lead) will give a *different* rise in temperature. The way in which this happens cannot be explained by classical theory, and was a puzzle until Albert Einstein used quantum concepts to tackle the problem in 1906. This was the first indication that quantum ideas had a wide application and were not relevant only to the study of *black body radiation*.

The specific latent heat of a substance, a related property, is the heat absorbed or released by a specific mass of the substance as it changes from solid to liquid or liquid to gas (or, in each case, vice versa) at the same temperature (see also *phase transition*).

spectral lines   Sharply defined lines seen in the *spectrum* of light. The lines are associated with the transitions of electrons from one well-defined energy level to another well-defined energy level (see *atom*, *Bohr model*) with the absorption or emission of a precise amount of energy, corresponding to a precise wavelength of light. If the transition is from a higher energy to a lower one, energy is emitted,

Spectral lines. The distinctive patterns of lines in a spectrum show up clearly even in this black and white reproduction. Red light (corresponding to longer wavelengths) is to the right here.

producing a bright line in the spectrum; if the transition is from a lower energy to a higher one, energy is absorbed, producing a dark line in the spectrum. The pattern of lines produced by atoms of a particular element (their number and the spacing between them) is unique, and can be used to identify the presence of that element. 'Light' is used here in its broad sense, to mean 'electromagnetic radiation'; the same effects are at work outside the visible region of the spectrum, and studies of 'lines' (peaks or troughs in the spectrum) at radio wavelengths have, for example, revealed the presence of complex molecules in clouds of gas and dust in space.

spectroscopy The study of the *spectrum* of electromagnetic radiation, and in particular the study of *spectral lines*. At the beginning of the 20th century, such studies were crucially important in the development of the first quantum theory of the *atom*; it was one of the triumphs of the *Bohr model* that it explained the pattern of lines seen in the spectrum of hydrogen. Spectroscopy is now used in chemical analysis (including in astrophysics, to identify the compositions of the stars), in determining molecular structures, and even in cosmology, thanks to the way features in the spectrum of a distant object are moved to longer wavelengths by the expansion of the Universe (the famous redshift).

The science of spectroscopy can be said to have begun with the discovery by William Wollaston, early in the 19th century, of dark lines in the spectrum of light from the Sun; but it took a hundred years of fumbling more or less in the dark before the origin of these lines was properly understood. Even without knowing how the lines were produced, researchers such as Joseph Fraunhofer, Robert Bunsen and Gustav Kirchoff were able to determine from experiment that each element produces its own set of spectral lines. It is no coincidence that this is the same Robert Bunsen whose name is now attached to the famous burner used in the laboratory (although he did not, in fact, invent it); the bunsen burner was a key tool in this work, which found that when an element is heated in the clear flame of the burner, it produces light with a characteristic colour, caused by emission at sharply defined lines in the spectrum. In an example familiar from everyday life, when sodium is heated it produces a bright yellow-orange light caused by emission lines in the yellow-orange part of the spectrum; you can see this by heating sodium in the flame of a bunsen burner (a practice we do *not* recommend), or (more safely!) by the yellow colour of common street lights, where traces of sodium in the gas in the tubes of the lights are excited into emission by the passage of an electric current. As in the street lights, the distinctive colour is emitted even if the atoms (in this case, of sodium) are locked up in a compound. So the characteristic sodium yellow is seen if, for example, ordinary salt (sodium chloride) is heated in the flame of a bunsen burner, or even if the salt is thrown on a fire.

Every element has its own pattern of spectral lines, and in every case the pattern stays the same (although the intensity of the lines will change) even if the temperature changes. Indeed, if a particular element is present in a cold gas and light from a hot object passes through the gas, the same pattern of lines shows up as dark features in the spectrum, where energy from the light is being absorbed. By comparing their laboratory studies of spectra with the lines seen in the light from the Sun and stars, spectroscopists were able to account for most of these lines in terms of the presence in the Sun and stars of elements known on Earth. (This kind of study began in 1859, when Kirchoff found that there is sodium in the atmosphere of the Sun.) In a famous reversal of this procedure, Norman Lockyer explained lines in the solar spectrum that did not correspond to any known element on Earth by attributing them to an unknown element, which he dubbed helium. Helium was later discovered on Earth and found to have exactly the spectrum required to fit these particular solar lines. But here we are more interested in how spectroscopy was used to probe the structure of the atom.

The spectrum of hydrogen is particularly simple – we now know that this is because the hydrogen atom consists of a single proton associated with a single electron. The lines in the spectrum that provide the unique fingerprint of hydrogen are named after Johann Balmer, a Swiss schoolteacher who worked out a mathematical formula describing the pattern in 1884, and published it in 1885 (see *Balmer series*). Balmer's formula was so simple that it clearly contained some deep truth about the structure of the hydrogen atom. But nobody knew what – until Bohr came on the scene. Bohr was not a spectroscopist and, although he had done some spectroscopy as an undergraduate, when he began working on the puzzle of the structure of the hydrogen atom the Balmer series did not immediately occur to him as the key with which to unlock the mystery. It was only when a colleague pointed out to him just how simple the Balmer formula really is that he appreciated its importance. This was in 1913 and it led directly to Bohr's model of the hydrogen atom, with the single electron able to jump from one *energy level* to another. The spacing between the energy levels depends on the size of *Planck's constant*, and this in turn determines the spacing between the lines in the Balmer series. So by using the observed spacing between the lines in the series, Bohr was able to calculate the spacing of the energy levels, and Balmer's formula could be rewritten, in a very natural way, to include Planck's constant. It was this work by Bohr, drawing on 19th-century physics, which showed definitively that the hydrogen atom contains exactly one electron.

Molecules also produce characteristic spectral features, often in the region of the spectrum corresponding to radio waves with millimetre wavelengths; very energetic sources produce characteristic spectral signatures at *X-ray* and *gamma ray* wavelengths, and the study of these spectra are important in astrophysics. The term 'spectroscopy' is also sometimes used in particle physics to describe the study of the distribution pattern of the number of particles with different amounts of energy: for example, in *cosmic rays* or in the *shower* of particles produced in an *accelerator* experiment. Even the study of *gravitational radiation*, if and when it is discovered, will be called spectroscopy.

spectrum Any representation of how the strength of the *electromagnetic radiation* from a source depends on its *wavelength*. The most familiar example is the rainbow spectrum of visible *light*, which can be displayed using a prism, or seen in a rainbow itself. White light is made up of a mixture of wavelengths. The spectrum of colours seen by the human eye covers the range from red (with longest wavelength) through orange, yellow, green, blue and indigo to violet (with shortest wavelength). The brightness of each colour in a spectrum shows how strongly that component of white light contributes to the overall brightness of the source. The properties of the spectrum can be determined accurately either photographically or using electronic detectors.

The spectrum extends beyond the visible range at both ends, into the *ultraviolet* and beyond, and into the *infrared* and beyond. The strength of radiation at different wavelengths outside the range of human eyes can be recorded by suitable instruments and displayed either as a set of numbers or as a graph. In both cases, there are places where there is a strong peak in the *energy* (corresponding to emission lines in the optical spectrum) and places where the energy dips (corresponding to absorption lines in the optical spectrum). The entire electromagnetic spectrum is broken up for convenience into separate bands, just as the optical spectrum is divided naturally into different colours. The full spectrum goes from *radio waves* (with longest wavelength) through *microwaves*, infrared, visible light, ultraviolet and *X-rays* to *gamma rays* (with shortest wavelength).

The term 'spectrum' is also used, by analogy, to refer to other phenomena, such as the pattern of energies of different particles in a *shower*. See *spectroscopy*.

speed of light   A universal constant, denoted by the symbol c, which is the same (in a vacuum) for all observers, no matter how they are moving relative to the source of the light (see *relativity theory*). The speed is 299.792458 x 10^6 metres per second. This is now (since 1983) used as the definition of the metre; the opportunity was missed then to make the minor adjustment of the metre so that it would be defined in such a way that light travels at 300 million metres per second, but this convenient round number (3 x 10^{10} cm/sec) is the one used in all but the most sophisticated calculations involving c. Since 1983, any experiments described as 'measuring the speed of light' have really been measuring the length of the standard metre against the light standard.

spin   A property of quantum entities which is related to the concept of rotation in *classical physics* – like the spin of the Earth in space – but which, as is usually the case in the quantum world, has no exact counterpart in the classical world.

Like other properties of quantum entities, spin is quantized and always comes in multiples of a basic unit of spin, which is equal to half of (*Planck's constant* divided by 2π), or $\frac{1}{2}\hbar$. For convenience, the $\hbar$ bit is usually taken as read, and physicists refer to a particle as having spin ½, spin 1, spin ¾, and so on. It turns out that the kind of spin a particle has is crucially important in determining its place in the quantum world. Particles which have an odd number of multiples of the basic unit of spin (and therefore have 'half-integer' spin overall) are *fermions* – particles, such as electrons and protons, that are what we think of as material particles. Particles which have zero spin or an even number of multiples of the basic unit of spin (and therefore have 'integer' spin overall) are *bosons* – particles, such as photons and gluons, that are what we think of as force carriers.

One of the strangest features of quantum spin is shown by the behaviour of fermions, also known as 'spin ½ particles'. If an object like the Earth turns in space through 360 degrees, it returns to where it started. But if a spin ½ particle rotates through 360 degrees, it arrives at a quantum state which is measurably different from its starting state. In order to get back to where it started, it has to rotate through another 360 degrees, making 720 degrees, a double rotation, in all. One way of picturing this is that the quantum particle 'sees' the Universe differently from how we see it. What we see if we turn through 360 degrees twice are two identical copies of the Universe, but the quantum particle is able to discern a difference between the two copies of the Universe.

The orientation of a spinning quantum particle such as an electron is also quantized, and this is why an electron can exist in either of precisely two states (with spin up or with spin down) for each *energy level* available to it in an *atom*.

See *Uhlenbeck, George Eugene*, and *Goudsmit, Samuel Abraham*.

splitting of the atom   See *nuclear fission*.

spontaneous emission   The emission of energy (as when an electron falls from a higher *energy level* to a lower energy level in an *atom*) or particles (as in radioactive *decay*) without any external stimulus triggering the emission. Spontaneous emission obeys the laws of chance and probability, a feature of the quantum world that led to Albert Einstein's famous remark, 'I cannot believe that God plays dice'; but Einstein was wrong.

spontaneous symmetry breaking   A situation that occurs in the physical world whenever the lowest energy state of a system (the *ground state*) has less symmetry than the most symmetrical state of the system. At high energies, the system is symmetric; but at lower energies, it falls into its ground state and the symmetry is broken spontaneously.

The usual example of spontaneous symmetry breaking is the way in which a hot piece of magnetic material such as iron has no overall magnetism because its internal components are jostling about in all orientations. So the iron is completely symmetric as far as magnetism is concerned, and will not line up with, for example, the Earth's magnetic field. But when the iron cools, it 'sets' into a preferred magnetism, with its internal components aligned, and will tend to line up with the Earth's magnetic field (see *ferromagnetism*).

It is thought that, under the extreme conditions that existed at the birth of the Universe, there was a complete symmetry between the *forces of nature*, which were equally strong and indistinguishable from one another; but as the Universe cooled, this symmetry was spontaneously broken, in the same sort of way as the symmetry breaking in ferromagnetism, giving the four forces known today their distinctive appearance.

SPS   See *Super Proton Synchrotron*.

squark   The bosonic counterpart to the *quark* in the theory of *supersymmetry*.

SQUID   Acronym for superconducting quantum interference device. SQUIDs behave like single quantum particles in some ways, but have macroscopic size – typically, 0.5 cm or so across, big enough to see and handle. The archetypal SQUID, developed by Terry Clark and his colleagues at the University of Sussex, is a ring of superconducting material (about 0.5 cm in diameter), just like the kind of ring you might wear on a (very small!) finger, but in which there is a constriction at one point, a narrowing of the ring's material down to a cross-section of 1 ten-millionth of a square centimetre. This weak link acts as a *Josephson junction*, and the system behaves in some ways as if it were an open-ended cylinder, like an organ pipe or a tin can with both ends removed. Just as sound waves can be made to resonate in an organ pipe, electrons in the superconducting ring form standing waves, described by the wave equation developed by Erwin Schrödinger, and can be 'tuned' by applying a varying electromagnetic field at radio frequencies. The electron wave behaves like a single quantum particle 0.5 cm across – for example, the whole of the quantum entity responds instantly to a stimulus applied to one side of the ring, without waiting for

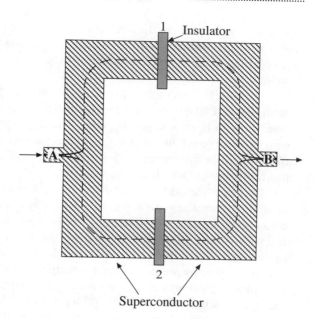

SQUID. Schematic representation of a variation on the SQUID theme. In this case, electrons travelling through the superconducting material from A to B cross two Josephson junctions ('Insulator') and can be made to interfere like light in the double slit experiment.

a signal to travel across the ring at the speed of light. This is an example of quantum *non-locality*.

SQUIDS have many practical applications, notably as sensitive detectors for measuring voltages and electric currents and as high-speed (in a sense, faster than light!) switching components in computers. But their most exciting use is in direct tests of the implications of quantum physics. One of the ambitions of Clark's team is to develop a 'macroatom' in the form of a cylinder several metres long which will behave in this non-local way, so that a stimulus applied to one end of the cylinder will cause an instantaneous quantum transition throughout the entire length of the cylinder (a cylinder 1 metre long would, after all, be only 200 times bigger than the original SQUIDS, which are themselves about 100 *million* times bigger than atoms). The Sussex team is also looking at the background *noise* in their existing systems to see if there is any evidence of advanced quantum waves arriving at their detectors from the future (see *transactional interpretation*).

standard model   In particle physics, the description of the world in terms of fundamental particles and their interactions, which are mediated by the exchange of *gauge bosons*.

According to the standard model, all of the material world is composed of *leptons* and *quarks*, which are truly fundamental particles; all other particles are composites made up from these building blocks. There are four forces which operate between particles – the colour force between quarks (which also shows up as the strong nuclear force), the weak nuclear force, electromagnetism and gravity (see *forces of nature*).

The archetypal theory that describes the behaviour of such a force is *quantum electrodynamics* (QED), which describes the electromagnetic interaction to very high precision, and involves the exchange of *photons* between charged particles. This has been extended to take in the weak interaction (which operates through the exchange of *intermediate vector bosons*) as well, giving a single unified description of the electroweak interaction (so that, arguably, there are now only three interactions to worry about).

Using QED as a template, physicists have developed the theory of *quantum chromodynamics* (QCD) to describe the interaction that operates between quarks – in this case, through the exchange of *gluons*. This is an extremely successful theory in its own right, although not as precise as QED. It is hoped that QCD and the electroweak theory may be combined in a single *grand unified theory*, and there are several promising lines of attack on this problem.

Gravity is the odd one out. Although it is described by a highly successful and accurate theory, the general theory of relativity (see *relativity theory*), this is conceptually different from the description of the other three forces of nature and cannot, strictly speaking, be regarded as part of the standard model. There are, however, promising indications that it may be possible to develop a theory of *quantum gravity* which would incorporate the same principles as QED and QCD (and involving the exchange of *gravitons*), and reproduce all of the successful predictions of the general theory of relativity. That would make it possible to describe all of the forces of nature in one mathematical framework, a *theory of everything*. See also *supergravity*, *supersymmetry*.

Some hints of physics 'beyond the standard model' emerged from experiments with the HERA accelerator at DESY, in Hamburg, at the beginning of 1997. In these experiments, the results of very high-energy collisions between protons and electrons suggest that there may be a previously unseen kind of interaction between quarks and electrons at very short distances. If these results are borne out by further experiments, it will mean not that the standard model is overturned, but that there is a deeper level of activity. (In a similar way, the general theory of relativity does not overturn Newtonian gravity, but goes beyond Newtonian theory to a deeper level of understanding.)

The term 'standard model' is also used in cosmology, but that is another story.

standard temperature and pressure   See *STP*.

Stanford Linear Accelerator Center   See *SLAC*.

Stanford Positron–Electron Asymmetric Rings   See *SPEAR*.

Stark, Johannes (1874–1957)   German physicist who made two major contributions to science, and received the Nobel Prize for Physics in 1919 in recognition of his achievements.

Born in Schickenhof, Bavaria, on 15 April 1874, Stark studied at the University of Munich, graduating with a PhD in 1898, and then worked for two years as a teacher at the Physical Institute in Munich. From 1900 to 1906 Stark was a *Privatdozent* at the University of Göttingen, then he moved to the Technische Hochschule in Hanover as a professor of physics. He held the post of professor at the equivalent institution in Aachen from 1909 to 1917, and was then appointed professor of physics at the University of Greifswald; from 1920 to 1922 he taught at the University of Würzburg.

In 1922 Stark left academic life and tried to set up a business manufacturing porcelain, but this failed and, now in his late forties, he was unable to get another academic post. This was largely because he had burnt his boats by being unpleasant to almost everybody who was now in a position to offer him a post, and by distancing himself from the scientific developments of the time. Strongly anti-semitic (the only one of his colleagues Stark ever really approved of was Philipp Lenard), Stark became a bitter opponent of both relativity theory and quantum theory in the 1920s, partly because these ideas had, as he saw it, largely been developed by Jews.

Further embittered by his repeated failure to get an academic post (he was rejected by six German universities between 1922 and 1928), Stark became an early supporter of the Nazis, and a member of the Nazi Party in 1930.

When Hitler came to power, Stark became president of the Reich Physical-Technical Institute and President of the German Research Association. From these positions he got his own back on the academics who had rejected him in the 1920s, and intensified the opposition to the 'Jewish science' of researchers such as Albert Einstein and Max von Laue, but he had to resign in 1936 because his abrasive personality led to further clashes with authority (even the Nazis found Stark to be too unstable and disruptive to make a good administrator). Because of his role as an active Nazi, Stark was sentenced to four years in a labour camp in 1947. He died in Traunstein, in West Germany (as it then was), on 21 June 1957.

Stark's first important piece of physics was carried out during his time in Göttingen, where he first predicted, and then confirmed experimentally, that the light emitted by streams of what were then known as canal rays (positively charged ions), produced in a cathode-ray tube, should exhibit the *Doppler effect* because of the rapid motion of the 'rays'. His second, and best known, major piece of work came in 1913, when he showed that when canal rays pass through a strong electric field there is a splitting of the lines in the spectrum of light from the 'rays' – the *Stark effect*.

Stark effect   Splitting of spectral lines in the light from an *atom* when it is placed in a strong electric field. The effect is caused by the distortion in the shape of the electron *orbitals* caused by the electric field, which changes the spacing of *energy levels* in the atom (the Stark shift). See also *Zeeman effect*.

Stark shift   See *Stark effect*.

state   A list of numbers which spells out the set of quantum properties of a quantum entity in a particular quantum state (for example, the *spin* orientation, *orbital* and other properties of an *electron* in an *atom*). It is also known as the state vector. The state can also be thought of in terms of the Schrödinger wave function, which also represents the state of the system. The *ground state* of a system is the quantum state with the lowest energy; states with higher energy are *excited states*. See also *superposition of states*.

state vector   See *state*.

statistical mechanics   The mathematical description of (in particular) the behaviour of gases in terms of statistics. During the 1860s and 1870s, pioneers such as James Clerk Maxwell and Ludwig Boltzmann developed the idea that a gas is made up of very many atoms or molecules (see *Avogadro's number*), which can be thought of as tiny, hard spheres which bounce around, colliding with one another and with the walls of the container in which they are confined. When the gas is heated, the molecules move faster and collide harder with the walls, increasing the pressure on the walls and, if the walls are not fixed in place, causing the container to expand. The key feature of these ideas was that the behaviour of gases could be explained by applying the known laws of physics – Newton's laws – in a statistical sense to a very large number of atoms or molecules. Any one molecule might be moving in any direction in the gas, and it might have more or less than the average speed of the molecules in the gas; but the combined effect of many molecules colliding with the walls each second produces a steady pressure. The mathematics of classical statistical mechanics was summed up by the American physicist Willard Gibbs in his book *Elementary Principles of Statistical Mechanics*, published in 1902; it was Gibbs who gave this discipline its name.

The same basic ideas carry over into quantum mechanics, but the statistical rules obeyed by the particles are different – see **Bose–Einstein statistics**, **Fermi–Dirac statistics**. **Black body radiation**, for example, can be described entirely in terms of a gas of particles (**photons**) obeying Bose–Einstein statistics. Statistical mechanics can be used to describe the behaviour of a wide range of properties of both fluids and solids (with or without using quantum mechanics, as appropriate), including electrical conductivity and **specific heat**.

Stefan, Josef (1835–1893)   Austrian physicist who was the first person to identify the relationship between the rate at which an object radiates energy and its temperature (see **Stefan–Boltzmann Law**).

Born at St Peter, near Klagenfurt, on 24 March 1835, Stefan studied at the University of Vienna, graduating in 1856, and became a teacher at the university, but carried out research alongside his lecturing duties. He was awarded a PhD by the University of Vienna in 1863, and became professor of higher mathematics and physics there, holding the post for the rest of his life. He discovered the radiation law for which he is remembered in 1879, following up experiments by John Tyndall. Using this law, Stefan made the first accurate estimate of the temperature at the surface of the Sun (about 6,000 °C). He died in Vienna on 7 January 1893.

Stefan–Boltzmann Law   A mathematical expression for the amount of heat radiated by a hot body, first derived by Josef Stefan in 1879 and refined by his former student Ludwig Boltzmann in 1884. The law says that the amount of energy radiated per unit area per second is proportional to the temperature of the object raised to the fourth power (T^4); it was Boltzmann who showed that, strictly speaking, the law applies only to radiation from a **black body**.

Stefan's Law   = *Stefan–Boltzmann Law*.

Steinberger, Jack (1921–)   German-born American physicist who shared the Nobel Prize for Physics in 1988 with Leon Lederman and Melvin Schwartz, for their joint work (carried out in the early 1960s) which revealed the existence of the muon neutrino.

Born in Bad Kissingen on 25 May 1921, Steinberger emigrated to the United States with his family in 1934, shortly after the Nazis came to power in Germany. He studied at the University of Chicago, where he received his PhD in 1948, and moved on to Columbia University (in New York), where he became a professor in 1950 and stayed until 1972. Overlapping with this post, starting in 1968 Steinberger also worked at CERN, in Geneva, where he stayed until 1986 (when he retired), including a spell as director.

Stern, Otto (1888–1969)   German-born American physicist who was a member of the team which carried out the experiment which revealed the existence of electron spin (the Stern–Gerlach experiment), and who received the Nobel Prize in 1943 for his development of the technique used in this experiment and for his measurement of the *magnetic moment* of the proton.

Stern was born in Sohrau (then in Germany; now Zory, in Poland) on 17 February 1888. He studied at several German universities, ending up at Breslau University, where he was awarded his PhD in 1912, and then worked briefly in Zurich (so that he could be near Albert Einstein). During the First World War, Stern served on the Russian front, in the meteorological service. In 1918 he took up a junior position at the University of Frankfurt, and he was made associate professor of theoretical physics at the University of Rostock in 1921. In 1923 he moved to the University of Hamburg as professor of physical chemistry, and over the next ten years he collaborated with quantum pioneers such as Niels Bohr, Wolfgang Pauli and Paul Ehrenfest.

His work with Walther Gerlach was carried out at Hamburg in 1920 and 1921. In order to carry out that experiment, Stern refined the technique of firing beams of molecules through magnetic (and later electric) fields, to investigate the properties of the molecules by looking at how they were deflected by the fields; he used a variation of this technique to measure the magnetic moment of the proton in 1933. He had already, in 1931, shown that the 'particles' in molecular beams also behaved as 'waves'. In 1933 Stern (who was Jewish) was forced to leave Germany when the Nazis came to power, and he moved to the USA, where he became professor of physics at the Carnegie Institute of Technology in Pittsburgh, and became a US citizen in 1939. He retired in 1945 and moved to Berkeley, California, where he kept in touch with scientific developments until he died, of a heart attack while at the cinema, on 17 August 1969.

Stern–Gerlach experiment Experiment carried out by Otto Stern and Walther Gerlach in the early 1920s. It involved firing a beam of silver atoms through a magnetic field, which split the beam into two streams, and was suggested by a theoretical prediction made by Arnold Sommerfeld. This was explained by Samuel Goudsmit and George Uhlenbeck in terms of electron *spin* – a previously unsuspected quantum property.

stimulated emission The emission of energy (as when an electron falls from a higher *energy level* to a lower energy level in an *atom*) as a result of some external stimulus triggering the emission. Stimulated emission has important practical applications, notably in *lasers* and *masers*.

storage ring System used at some *accelerators* to store particles until the right time to use them in collisions. The particles are accelerated to high energies and then kept circulating in the rings ('ticking over', as it were), guided by magnetic fields. The classic storage ring system uses two rings in which a beam of *electrons* and a beam of *positrons* circulate in opposite directions until they are brought into a head-on collision.

STP Short for Standard Temperature and Pressure, defined as $0\,°C$ and 1 atmosphere (101,325 pascal). It is used by chemists as the standard conditions under which experiments are carried out, so that results from one set of experiments can be compared with those from another set of experiments without having to correct for effects due to temperature and pressure.

strange matter Matter composed of a roughly equal mixture of up, down and strange *quarks*. Everyday matter (*protons* and *neutrons*) is made up only from up and down quarks. The most stable (lowest-energy) form of this baryonic matter is in *elements* of the iron group. Theoretical calculations suggest, however, that strange matter would be even more stable than nuclei of iron – if it could ever be formed. It is possible that lumps of strange matter (also known as *quark nuggets*) could have been produced in the Big Bang in which the Universe was born, and could still exist today (in which case they might be found in *cosmic rays*).

Some astrophysicists have suggested that, under the extreme conditions of pressure that exist at the heart of a neutron star (a star containing as much mass as our Sun squeezed into a ball some 10 km across), everyday matter might be transformed into strange matter. A variation on this theme suggests that what we think are neutron stars are actually stars made almost entirely of strange matter, produced by the explosion of a star as a supernova, with a thin crust of neutron material.

Bizarre though the idea is, it could be tested. A neutron star would break apart if it tried to spin faster than once every millisecond. Spinning neutron stars are known (they are detected as pulsars), and the fastest of these spin once every few milliseconds. Because a strange star is more compact (a result of its more efficient storage of mass-energy), it could spin as fast as twice every millisecond without breaking apart. So if a 'half millisecond' pulsar is ever discovered (and this is a big if), it would be direct evidence, from astronomy, for the existence of strange quarks.

strangeness One of the six *flavours* that distinguish different types of *quark*.

strange particles Particles that contain at least one strange *quark* (see *strangeness*). They got their name because when these particles, which are relatively massive, were first identified in cosmic ray *showers* in the late 1940s and then manufactured out of energy in *accelerator* experiments in the 1950s, they turned out to have unusually long lifetimes – strange behaviour, compared with other comparably massive particles.

But this does not mean that the particles are long lived by human standards. In the archetypal example, the *kaon* decays in about 1 hundred-millionth of a second (10^{-8} sec), which is strange only because according to the understanding physicists had at the end of the 1940s it 'ought' to be able to decay, to produce *pions*, in only 10^{-23} sec – a million billion times more quickly.

It turned out that strange particles are always produced in pairs, and that their behaviour could begin to be explained if one member of such a pair were assigned a strangeness of +1 and the other member of the pair had a strangeness of –1, reminiscent of the way that electrons and positrons are produced in pairs out of energy, one with negative charge and the other with positive charge. But, unlike electric charge, strangeness can leak away from the Universe – the lightest strange particles, the kaon and the *lambda*, decay via the weak force into non-strange particles.

This is not quite so extreme as it sounds because a kaon and a lambda are produced together as a strange pair. A neutral kaon with strangeness +1 is accompanied by a lambda with strangeness –1. Their combined strangeness is zero, and they each decay independently into non-strange particles, so after they have both decayed the overall strangeness is still zero and the balance is restored, just as if they had annihilated one another. But since the kaon takes 10^{-8} sec to decay, while the lambda does so in only 10^{-10} sec, there is a very brief interval during which the strangeness of the kaon is not balanced by the equal and opposite strangeness of the lambda, which has already disappeared. Strange behaviour, indeed! This is still not fully understood.

As the family of known strange particles grew, Murray Gell-Mann (who coined the term 'strangeness' in this context) and Yuval Ne'eman independently developed a classification system (the *eightfold way*) which arranged their properties in a pattern. This led directly to the idea of quarks, with strangeness seen as a property carried by a particular kind of quark, echoing the way that electrons carry electric charge. On this picture, the positive kaon (K^+, a member of the *meson* family), for example, is seen as composed of an up quark and an antistrange quark and has a strangeness of +1; the *omega minus* particle (a *baryon*) is composed of three strange quarks and has a strangeness of –3.

Strassman, Fritz (1902–1980)   German chemist who collaborated with Otto Hahn on the first experiments that clearly revealed the phenomenon of *fission* of atomic *nuclei*, and which were interpreted in those terms by Lise Meitner and Otto Frisch.

Born in Boppard on 22 February 1902, Strassman lectured at the Hanover Institute of Technology and at the Kaiser Wilhelm Institute in Berlin, before being appointed professor of inorganic and nuclear chemistry at the University of Mainz in 1946. In 1953 he became the director of the chemistry division of the Max Planck Institute. The work for which Strassman is remembered was carried out with Hahn at the end of the 1930s.

string   See *string theory*.

string theory   Any of a class of theories in physics that describe the fundamental particles and their interactions in terms of tiny one-dimensional entities – strings. These strings form loops which are much smaller than particles such as *protons*, but the important point is that they are not mathematical points – even the *electron*, previously regarded as a point-like entity, can be described in terms of string. This has a profound effect on the equations, and leads naturally to many of the observed features of the particle world. The most exciting aspect of string theory is that it seems to include gravity automatically within the same framework as the other *forces of nature*. See *supergravity*, *supersymmetry*.

Although string theory became a major interest of physicists only in the 1980s, its origins go back to the late 1960s, when many 'new' kinds of *hadron* were being manufactured in *accelerators*, but the *quark* model had not yet been fully developed and various rival models were put forward in an attempt to explain the proliferation of particles and the way they behaved. One idea, proposed by Gabrielle Veneziano, modelled the behaviour of these particles in a mathematical way which worked (up to a point), but which initially lacked any physical picture to go with it. It soon became clear, though, that the mathematical picture developed by Veneziano corresponded to interactions between little one-dimensional entities. When the quark model was developed in the 1970s, it became clear why this unusual mathematical treatment had been relatively successful. The quarks inside hadrons are held together by the exchange of *gluons*, and the effect is as if two quarks are joined by a piece of elastic. The force between quarks (the *colour* force, which also indirectly gives rise to the *strong interaction*) is so strong that the energy in the 'elastic' is comparable to the mass-energy in the quarks themselves. Under these conditions, a pair of quarks joined by the colour force behave in many ways like a stretched piece of string. An appropriate image would be the chain shot used in sea battles in the days of sail – a pair of cannon balls joined by a chain, whirling around one another as they strike the rigging of a man-of-war, and doing far more damage than two single balls passing through the sails.

At first, string theory, although intriguing to mathematicians, really worked only as a description of **bosons** (including gluons), and in any case it was soon superseded, in the original context in which it had been developed, by the quark model. But in 1970 John Schwarz and his colleague André Neveu found a way to describe **fermions** using string theory; and in 1974, just at the time **quantum chromo-dynamics** was being developed as a satisfactory theory of hadrons, Schwarz and another colleague, Joel Scherk, discovered the link between string theory and gravity.

One of the baffling features of the mathematical description of fermions in terms of strings had been that the package provided more particles than the physicists had ordered. The mathematics automatically provided a description of a particle with zero mass and a **spin** of 2. There was no hadron matching that description, and for a while the theorists tried to find a way to get rid of the unwanted particle. Then they realized that this is just the description of the **graviton**, the messenger particle of the gravitational interaction required by any theory of **quantum gravity**. Gravity is obligatory in string theory.

Very few people took the idea seriously. For the next ten years or so, apart from Schwarz the only leading proponent of string theory was Michael Green. Between them, and working at various times with a small and changing group of pioneers, they tackled many of the problems involved in the mathematics of string theory, including the perennial problem of particle physics, the occurrence of unwelcome infinities in the equations. But string theory took off in the middle of the 1980s, when it turned out that by combining the idea of **supersymmetry** with strings to make a new and improved version, superstring theory, a seemingly powerful and complete description of everything could be produced.

But there is a peculiarity about the way physicists came up with the idea of super-symmetry. It all started in 1970, when Yoichiro Nambu of the University of Chicago came up with the idea of treating fundamental particles not as points, but as tiny one-dimensional entities, called strings. This was at about the time that the quark model was beginning to be taken seriously, and in the early 1970s Nambu's idea was overshadowed by the rapid acceptance of the quark model – it was seen as a rival to quark theory, not as a complementary idea. The fundamental entities that he was trying to model were not quarks, but the hadrons (particles, such as the neutron and proton, which feel the strong force, and which we would now describe as being composed of quarks). The success of the quark model seemed to leave this kind of string theory out in the cold; but a few mathematically inclined physicists played with it anyway.

Nambu's string theory involved spinning and vibrating lengths of string only about 10^{-13} cm long. The properties of the particles he was trying to model in this way (their masses, electric charge and so on) were thought of as corresponding to different states of vibration, like different notes played on a guitar string, or to be attached in some way to the whirling ends of the strings. And these vibrations also involved oscillations in more dimensions than the three of space plus one of time that we are used to.

Embarrassingly, though, when the appropriate calculations were first carried through, they said that the entities described by these strings would all have integer spin, in the usual quantum-mechanical sense. That is, they would all be **bosons** (force carriers, such as photons). And yet, the whole point of the model had been to describe hadrons, which are **fermions** and have half-integer spin! Then Pierre Ramond of the University of Florida found a way around the problem. He found a way of adapting Nambu's equations to include strings with half-integer spin, describing fermions. But those fermionic strings were also allowed by the equations to join together in pairs, making strings with integer spin – bosons. John Schwarz at Princeton, Joel Scherk at Caltech and the French physicist André Neveu developed this idea into a consistent mathematical theory of spinning strings which included both bosons and fermions, but required the strings to be vibrating in ten dimensions. It was Scherk, in particular, who established, by 1976, that fermions and bosons emerged from this string theory on an exactly equal footing, with every kind of boson having a fermionic partner, and every kind of fermion having a bosonic partner. Supersymmetry had been born.

There is a valuable way of looking at all this, which is often emphasized by Ed Witten, one of the main players in the supersymmetry game in the 1990s. Bosons are entities whose properties can be described by ordinary commuting relationships, familiar everyday rules such as *A* times *B* is equal to *B* times *A*. Fermions, though, have properties which do not always obey these relationships – they do not commute. (In fact, they do not commute in a special way; they are said to anticommute.) The appropriate mathematics that describes this behaviour is quantum mechanics, not classical (Newtonian) mechanics. The concept of fermions is based entirely on the principles of quantum physics, while bosons are essentially classical in nature. Supersymmetry updates our understanding of spacetime to include fermions as well as bosons; it therefore updates the special theory of relativity, Einstein's first theory of space and time, by making it quantum mechanical.

This deep insight was appreciated in 1976, and the next step was seen as being to seek out a way to bring gravity into the fold, updating the general theory of relativity, Einstein's second theory of space and time, in the same sort of way. That might have speeded the development of string theory by a decade. But it was not to be – even though the gravity problem was in many people's minds at the end of the 1970s, at that time they saw the next step in terms of an extension of supersymmetry to include gravity, in a theoretical package dubbed supergravity, without using the idea of strings at all.

Almost as soon as supersymmetry had burst upon the scene, the string theory that had given it birth had been forgotten. Never seen as more than a byway of physics by most researchers, it had by 1976 been totally eclipsed by the quark model. Once the idea of supersymmetry had been placed in the minds of physicists, it was easy to incorporate it into the then standard model of the particle world, as we have outlined below (p.392). Indeed, that is the way generations of students after 1976 were introduced to supersymmetry, without any mention of strings at all. Physics moved on and left string theory behind. Just about the only people who carried on working in the field were John Schwarz and, over in London, Michael Green (Scherk died young and made no further contributions to the idea).

But while string theory languished, its offspring, supersymmetry, flourished. A band of enthusiasts soon took up the ideas of SUSY, developing various lines of attack. One describes grand unified theories (GUTs) in terms of SUSY – the theories are known as SUSY GUTs. Another focuses on gravity – supergravity, which itself comes in various forms with family resemblances but different detailed constructions. One great thing about all the supergravity models is that they each specify a different specific number of possible types of particle in the real world – so many leptons, so many photinos, so many quarks, and so on – instead of the endless proliferation of families allowed by the older GUTs. Nobody has yet succeeded in matching up the specific numbers allowed in any of these supergravity theories with the particles of the real world, but that is seen as a relatively minor problem compared with the previous one of a potentially infinite number of types of particle to worry about. A favoured version of these theories is called '$N = 8$ supergravity', and its enthusiasts claim that it could explain everything – forces, matter, particles and the geometry of spacetime – in one package. But the best thing about $N = 8$ supergravity is that it seems not merely to be renormalizable but in a sense to renormalize itself – the infinities that have plagued field theory for half a century cancel out of $N = 8$ theory all by themselves, without anyone having to lift a finger to encourage them. $N = 8$ *always* comes up with finite answers to the questions that physicists ask of it.

The search for a unified theory of physics can be thought of in terms of the two great theories of 20th-century physics. The first, the general theory of relativity, relates gravity to the structure of space and time. It tells us that they should be treated as a unified whole, spacetime, and that distortions in the geometry of spacetime are responsible for what we perceive as the force of gravity. The second, quantum mechanics, describes the behaviour of the atomic and subatomic world; there are quantum theories which describe each of the other three forces of nature, apart from gravity. A fully unified description of the Universe and all it contains (a 'theory of everything', or TOE) would have to take gravity and spacetime into the quantum fold. That implies that spacetime itself must be, on an appropriate very short-range scale, quantized into discrete lumps, not smoothly continuous. String theory, in an extended form known as superstring theory, *naturally* produces a description of gravity, out of a package initially set up in quantum terms; but it took several years for gravity to fall out of the equations.

String theory took off only in the middle of the 1980s, after a new variation on the theme was developed by John Schwarz and Michael Green. They started working together at the end of the 1970s, after meeting at a conference in CERN, and discovering that, unlike everybody else studying particle physics at the time, they were both interested in string. They began to produce results almost immediately. The key first step that they took was to realize that what was needed was indeed a theory of *everything* – all the particles and fields – not just hadrons. In such a theory, strings would have to be very small indeed – much smaller than Nambu's strings, which were only ever designed to describe hadrons. Even without knowing how the theory would develop, Schwarz and Green could predict what scale the strings would operate on, because they wanted to include gravity in the package. Gravity becomes seriously affected by quantum effects at a scale of around 10^{-33} cm (that is, 10^{-35} m), the distance scale at which the very structure of spacetime itself becomes affected by quantum uncertainty. And, of course, from the outset the new version of string theory had SUSY built into it.

The first string model to be developed by Schwarz and Green, in 1980, dealt with open-ended strings vibrating in ten dimensions, able to link up with one another and break apart. Superficially, it doesn't look like anything more than a shrunken version of Nambu's string theory. In fact, it went far further, by including (in principle; actually carrying the calculations through was another matter) string states corresponding to all the known particles and fields, and all the known symmetries affecting fermions and bosons, plus supersymmetry.

This early version set the scene for what was to follow. The central idea of all subsequent string theories is that the conventional picture of fundamental particles (leptons and quarks) as points with no extension in any direction is replaced by the idea of particles as objects which have extension in one dimension, like a line drawn on a piece of paper, or the thinnest of strings. The extension is very small – about 10^{-35} m. It would take 10^{20} such strings, laid end to end, to stretch across the diameter of a proton.

The next big step towards a true theory of everything came in 1981, when Schwarz and Green introduced a new twist (literally) to the story. The open string theory of 1980 became known as the Type I theory, and the new, Type II theory introduced a key variation on the theme – closed loops of string. Type I theory only had open-ended strings; Type II theory only had closed loops of string. In a particularly neat piece of packaging, in closed loops fermionic states correspond to ripples running around the loop one way, while bosonic states correspond to ripples running around the loop the other way, demonstrating the power and influence of supersymmetry. The closed loop version had some advantages over the open model, not least because it proved easier to deal with those infinities that plague particle physicists in the closed loop model. But the Type II theory also had its difficulties and did not seem (at the time) capable of predicting, or encompassing, all the variety of the known particle world.

There was one other cloud on the horizon. In 1982 Ed Witten and Luis Alvarez-Gaumé discovered that the Kaluza–Klein compactification trick will work to make the forces of nature in the way required only if you start out with an odd number of dimensions before compactification. This made eleven-dimensional supergravity look more attractive than ever before, but gave the ten-dimensional string theories real problems. This didn't stop people working on those theories, but it gave them something extra to think about.

The next step forward was actually a step back. Dissatisfied with Type II theory, Schwarz and Green went back to Type I theory, and tried to remove the infinities which plagued it. The problem was that there were many possible variations on the theme, and that all of them seemed to be beset not just by infinities but by what were called anomalies – behaviour which did not match the behaviour of the everyday world, especially its conservation laws. For example, in more than one version of the theory, electric charge is not conserved, so charge can appear out of nothing at all, and disappear as well.

But in 1984 Schwarz and Green found that there is one, and only one, form of symmetry (technically, SO(32)) which, when applied to the Type I string theory, removes all of the anomalies and all of the infinities. They had a unique theory, free from anomalies and infinities, that was a real candidate for the theory of everything. It was at this point that other physicists started to sit up and take notice of strings once again.

One of the teams that was fired up by the success achieved by Green and Schwarz in 1984 was based at Princeton University. David Gross and three colleagues (together sometimes known as 'the Princeton String Quartet') took a second look at the closed loop idea, writing it down using a different mathematical approach. There was plenty for them to write down because the theory is a little more complicated than I have let on so far. The kind of vibrations associated with fermions do indeed require the ten dimensions that I have mentioned. But the bosonic vibrations described (at first, unintentionally) by Nambu's first version of string theory actually take place in 26 dimensions. Gross and his colleagues found a way to incorporate both kinds of vibration into a single closed loop of string, with the ten-dimensional vibrations running one way round the loop and the 26-dimensional vibrations running the other way around the loop. This version of the idea is called the heterotic string theory ('heterotic' from the same Greek root as in 'heterosexual', implying a combination of at least two different things).

The heterotic strings neatly tidy up a loose end in the Type II theory. In theories involving open strings, some of the properties we associate with particles (the properties which physicists call 'charges') are tied to the end points of the whirling strings (this may be electric charge, if we are dealing with electromagnetism, or the 'colour charge' of quarks, or something else). But closed loops do not have end points, so where are these properties located? In heterotic strings, these properties are still properly described, but have to be thought of as somehow smeared out around the strings. This is the main physical distinction between heterotic strings and the kind of closed strings that Green and Schwarz investigated at the beginning of the 1980s; you can picture heterotic string theory as a kind of hybrid combination of the oldest kind of string theory and the first superstring theory.

How can two different sets of dimensions apply to vibrations of the same string? Because for the bosonic vibrations sixteen of the 26 dimensions have been compact-ified as a set, leaving ten more which are the same as for the ten-dimensional fermionic vibrations, with six of those ten dimensions being compactified in a different way to leave us with the familiar four dimensions of spacetime. It is the extra richness provided by the sixteen extra dimensions that makes for the richness in the variety of bosons, from photons to W and Z particles and gluons, compared with the relative simplicity of the fermionic world, built up from a few quarks and leptons. The sixteen extra dimensions in the heterotic string theory are responsible for a pair of underlying symmetries, either of which can be used to investigate the physical implications of the theory (any other choice of symmetry groups leads to infinities). One of these is the SO(32) symmetry group, which had already turned up in the investigation of open strings (32, of course, being twice 16); the other is a symmetry group known as $E_8 \times E_8$, which actually describes two complete worlds, living alongside each other (8 plus 8 also being 16). Each of the E_8 symmetries can be naturally broken down into just the kind of symmetries used by particle physicists to describe our world. When six of the ten dimensions involved curl up, they provide a symmetry group known as E_6, which is itself broken down into SU(3) x SU(2) x U(1). But SU(3) is the symmetry group associated with the standard model of quarks and gluons, while SU(2) x U(1) is the symmetry group associated with the electroweak interaction. Everything in particle physics is included within one of the E_8 parts of the overall $E_8 \times E_8$ symmetry group.

Since only one of the E_8 components is needed to describe everything in our Universe, that leaves a complete duplicate set of possibilities. The symmetry between the two halves of the group would have been broken at the birth of the Universe, when gravity split apart from the other forces of nature. The result would, some theorists believe, be the development of two universes, interpenetrating one another, but interacting only through gravity – our world and a 'shadow' universe. There would be shadow photons, shadow atoms, perhaps even shadow stars and shadow planets, inhabited by shadow people, co-existing in the same spacetime that we inhabit, but for ever invisible (Indeed, there would be shadow sleptons, shadow squarks and shadow bosinos, as SUSY itself would be duplicated in the shadow world; see *shadow matter*).

There are great physical differences between the kind of strings that emerged from Veneziano's idea and the kind of superstrings people study today, although there is a family resemblance in mathematical terms. If you are trying to use strings to describe the behaviour of hadrons, the typical size of the string involved is 10^{-13} cm; but when you are using strings to describe gravity, the typical length scale is the *Planck length*, which is 10^{-33} cm – 10^{20} times *smaller* than an atomic nucleus. The scale on which these strings operate is as much smaller than an atom as an atom is smaller than the Solar System.

Another peculiarity of the strings is that they look one-dimensional only from a distance – that is, from further away than a few Planck lengths. In order for the theory to work, each string has to be a multidimensional entity, with all but one of the dimensions it exists in 'rolled up', or compactified, in the same way that the unwelcome fifth dimension is tucked out of sight in the *Kaluza–Klein theory*. Different versions of superstring theory, as we have seen, need different numbers of dimensions, but one successful variation on the theme requires ten dimensions (nine of space and one of time). There are versions of these theories in which both open strings and closed loops can exist, and versions which have only closed loops; the latter look more promising as of the mid-1990s.

In superstring theory, each 'point' of space can be thought of as a little six-dimensional ball, about 10^{-33} cm across, and particles are associated with the vibrations of little loops of multidimensional string, scarcely any bigger than these balls of space. The loops vibrate, in a process analogous to the vibrations of a guitar string, in different ways – with different modes of vibration, equivalent to different notes on the guitar string. The electron corresponds to one mode of vibration (one 'note'), a quark corresponds to another mode of vibration, and so on. On scales much

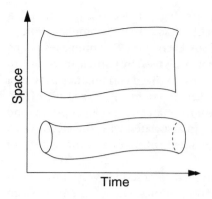

String theory 1. For a string, the equivalent of a world line for a particle is either a sheet, traced out by an open string, or a tube, traced out by a loop of string, as time passes. (For a change, time is represented across the page in this diagram.)

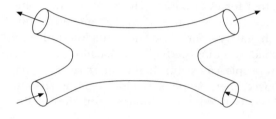

String theory 2. The equivalent of a Feynman diagram for a scattering interaction involving two loops of string sees the strings merge and then divide in a new configuration.

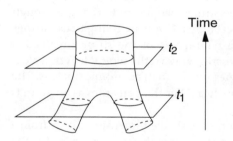

String theory 3. Another possibility is that two loops of string may merge to form a different kind of loop. For obvious reasons, this is known as the 'space-time trousers' diagram.

greater than the Planck length, however, the strings will behave like point particles, and the extra dimensions will be invisible, so that what we see is an effective field theory of particles in four dimensions (three of space and one of time); superstring theory reduces to conventional *quantum field theory* (which is contained within superstring theory) on appropriate scales, in much the same way that the general theory of relativity (see *relativity theory*) reduces to Isaac Newton's theory of gravity for most everyday purposes.

The snag is that there is no way to test any of these ideas by experiment. It would be impossible to build a particle accelerator powerful enough to probe the structure of matter on length scales comparable to the Planck length. String theory has been described as part of the physics of the 21st century that fell by accident into the 20th century; a series of lucky accidents, starting with the Veneziano model, that led to the discovery of a way of describing the world that goes far beyond our ability to probe the structure of matter – unlike all of the rest of particle physics, where theory and experiment developed alongside each other, as in the classic examples of Ernest Rutherford probing the structure of the atom, and the development of the **parton model.** There is just one prediction of string theory which might be susceptible to experiment – most of the variations on the theme predict a second kind of **Z particle**, as well as the one already known. Unfortunately the theories do not predict its mass, and even if it does exist, it may be too heavy to be manufactured in even the most powerful accelerators we have today. As Michael Green said in an article in *Scientific American* in September 1986:

> Details have come first; we are still groping for a unifying insight into the logic of the theory. For example, the occurrence of the massless graviton and gauge particles that emerge from superstring theories appears accidental and somewhat mysterious; one would like them to emerge naturally in a theory after the unifying principles are well established.

As far as we do understand strings, though, their nature is intimately bound up with the nature of space and time themselves, and this means that it may make more sense to describe spacetime in terms of string than the other way around. This approach works particularly naturally in the context of the **sum over histories** version of quantum theory, which automatically includes the interaction of the string with the background spacetime in which it moves. Superstring theory also predicts the existence of shadow matter. Some of the latest developments have seen researchers such as Green develop a new picture of black holes, the most extreme distortions of space and time, in terms of superstrings.

Abdus Salam was one leading physicist who regarded superstring theory as 'a real substitute for field theory of point particles', removing the 'intractable difficulties in quantum gravity theory' and offering exciting prospects for the future. His enthusiasm is shared by many. It should be mentioned, though, that some eminent theorists, including Sheldon Glashow, regard the whole business as a bandwagon which is about to lose a wheel, heading up a blind alley. We do not share that view.

See **membranes**.

Further reading: Paul Davies and Julian Brown (eds), *Superstrings*; F. David Peat, *Superstrings*.

strong force   = *strong interaction.*

strong interaction   The force which operates within the **nucleus** of an **atom** and holds the nucleus together, in spite of the tendency of all the positive electric charge on the **protons** in the nucleus to blow it apart. This is now known to be a manifestation of the deeper **colour** force which operates between **quarks** within the protons and **neutrons** that make up the nucleus. The strong force is about 100 times stronger than electromagnetism (over the size of a nucleus), which is why there are roughly 100 protons in the largest stable nuclei.

Physicists in the 1930s were well aware that there must be something holding the nucleus together in this way, and that this must be a short-range force, extending only over about the diameter of a nucleus (some 10^{-13} cm), otherwise it would pull all the nuclei in everything together, crushing atoms out of existence. Building on the idea that the electromagnetic force is mediated by the exchange of **photons**, in 1935 the Japanese physicist Hideki Yukawa suggested that the strong nuclear force must be mediated by the exchange of another kind of force-carrying particle, which became known as *pions*.

Yukawa had a neat explanation for the short range of the force. Electromagnetism has a very long (in principle, infinite) range because photons have no mass, and so virtual photons (see *virtual particles*) can be created in profusion out of nothing and can travel very long distances without having to disappear. The *uncertainty principle* allows even particles which have mass to appear out of nothing at all, but only for a short time that is related to their mass – the heavier the virtual particle, the shorter its lifetime. If pions had a certain mass, predicted by Yukawa, then they could live just long enough to carry the strong force between adjacent *nucleons* in an individual atom, but not long enough to reach out and influence the nucleus of the atom next door.

Yukawa predicted that his hypothetical particle would have a mass of about 150 MeV, 300 times the mass of an electron. Confusion was caused in 1937, when a previously unknown type of particle with a mass of 106 MeV was found in *cosmic rays*. At first, it was thought that this was Yukawa's pion, but it was soon established that it did not fit the bill; it was named the *muon* and is now known to be a heavy version of the electron. Real pions, both with positive and with negative charge, were identified (again, initially in cosmic ray *showers*) only in 1947; they have a mass of 140 MeV, very close to Yukawa's prediction, and there is also a neutral pion, which is slightly less massive.

The term 'strong force' is also sometimes used to refer to the colour force that operates directly between quarks; but this is misleading and strictly speaking incorrect. Pions are made of quark–antiquark pairs, specifically up–antidown, down–antiup and down–antidown or up–antiup. A proton is made up of two up quarks and one down (uud), so if it emits a positive pion (up–antidown), it loses one up quark and gains a down (to balance the 'new' antidown) and is transformed into a neutron (udd); the pion is absorbed by the neutron next door, which is transformed into a proton because it gains an up quark and one of its down quarks is cancelled out by the antidown. Similar (reverse) changes occur when negative pions are exchanged, but the exchange of neutral pions leaves a neutron as a neutron and a proton as a proton, although the individual quarks within them may be rearranged. So individual protons and neutrons lose their identity in the ever-changing web of interactions going on to hold the nucleus together, although the total number of protons and the total number of neutrons stays the same.

Just as *baryons* are composite *fermions*, so pions are composite *bosons*, which means that the colour force shows up in disguise at the level of nuclear interactions, as the strong force; but the strong force is not a truly fundamental gauge force (see *gauge theory*) because it involves the exchange of composite particles, not individual bosons – or as Vincent Icke put it, in his book *The Force of Symmetry*, the strong force involves the exchange of lumps of quarks by bigger lumps of quarks.

See also *forces of nature*.

strong nuclear force   See *strong interaction*.

strontium dating   A *radioactive dating* technique that uses measurements of the amount of strontium-87 in a sample (produced by the *decay* of rubidium-87).

Strutt, John William (1842–1919)   See *Rayleigh, Lord*.

SU(3) Shorthand way of referring to the group of three-dimensional special unitary matrices, or *special unitary group* 3 ('dimensions' here refers to the structure of the group, not to the familiar dimensions of space). This particular mathematical group (see *group theory*) underpins the pattern of relationships between elementary particles shown in the *eightfold way*, and became an integral part of the theory of *quantum chromodynamics*. It is, for example, because SU(3) is three-dimensional that there are three kinds of *colour* charge associated with *quarks*. SU(3) is an example of a *non-Abelian gauge theory*.

sum over histories A way of picturing what goes on in the quantum world (a *quantum interpretation*), originally developed by Richard Feynman in the 1940s. It is essentially synonymous with the name *path integral*, but this interpretation is so important that it is worth taking stock once again of just what it is all about.

The sum over histories interpretation gets its name because Feynman envisaged a quantum entity (such as an *electron*) travelling from A to B (perhaps through the two holes in the *double-slit experiment*) by every possible route at once. Each possible route corresponds to a 'history'. Each history has associated with it a number, called the *amplitude*, which defines the probability of that particular path being followed. This number involves the *action* associated with the path, and is a *complex number* (see *phase (quantum)*).

The probabilities of all the paths, or histories, are not 'in step' with one another, and like the amplitudes of ripples on a pond they can interfere with one another to reinforce the strength of one path while cancelling out the amplitudes of others. The probability of a particular path is just the square of the amplitude for that path; but the overall probability of going from A to B is given *not* by adding up the probabilities for each path, but by adding up the amplitudes for each path first and then squaring the total amplitude. This is a key feature of the quantum world. It is because the sum of the squares is not equal to the square of the sums that on a microlevel quantum entities behave in non-classical, seemingly weird ways. But it is a key feature of the sum over histories approach that when you do all the arithmetic, with all the adding and subtracting included in the right way, the weird routes in which an electron goes from A to B with a detour via, say, the star Arcturus all cancel out, and the history that gets reinforced is the one close to the classical idea of a trajectory. For every weird path (like the detour to Arcturus) there is a neighbouring path with opposite phase, so the two histories cancel out. Near the classical path, the phases add up, increasing the amplitude, and therefore the probability is large. Feynman's approach to quantum mechanics actually includes all of classical mechanics within the same straightforward framework. But you do have to accept that the electron in the experiment with two holes really does go through both holes at once.

superconductivity The ability of some materials to allow an electric current to flow through them with no detectable resistance. The phenomenon was first discovered at very low temperatures (close to zero on the *Kelvin temperature scale*), by Kamerlingh Onnes in 1911. It was eventually explained as a quantum phenomenon, by John Bardeen, Leon Cooper and Robert Schrieffer in 1957. The BCS theory, as it is known, explains that in a superconductor the interaction of electrons with the crystal lattice that they are passing through makes them associate in pairs, and these coupled electrons (sometimes known as Cooper pairs) behave as *bosons* because the two half-integer *spins* of each electron in the pair (which makes it a *fermion*) add up to a spin of 1, making the pair a boson. As bosons, the electrons all occupy the same energy state in the superconductor, and can be described in terms of a single wave function moving effortlessly through the conductor. The conducting electrons in the superconductor form a kind of boson gas (a *Bose condensate*).

All of this works only at low temperature because at higher temperatures (above about 20 K) the thermal motion of the electrons and the atoms in the crystal lattice breaks up the weak association of electrons in Cooper pairs. But in 1986 physicists were surprised by the discovery of superconductivity in some materials at relatively high temperatures, above 100 K. Although still operating at temperatures much lower than room temperatures (which are about 295 K), these high-temperature superconductors are important because they operate at temperatures which can be reached relatively easily, using liquid nitrogen. This raises the prospect of practical applications – for example, in communications and computing. In these superconductors, the loss of resistance is not due to the BCS mechanism and is not yet fully understood, but it is related to the way the materials involved form planes or chains of atoms in the crystal lattice, giving the conducting electrons paths of least resistance to follow.

superconductor   See *superconductivity*.

superfluidity   The way liquid helium flows without friction at very low temperatures. This is purely a quantum phenomenon. It happens because at very low temperatures the atoms of helium in the superfluid behave like a *boson* gas (or *Bose condensate*). They all occupy the same energy level, and can be described in terms of a single wave function moving effortlessly as a single unit. It is relatively easy for this to happen with helium-4, which becomes superfluid at 2.2 K, because each atom has an overall quantum *spin* of zero (the half-integer spins of each of the four *nucleons* are aligned so that they cancel out) and the behaviour of a collection of such entities is described by *Bose–Einstein statistics*. The superfluidity of helium-4 was first noticed in the 1930s in experiments at the University of Toronto; it was studied in detail by Piotr Kapitza and explained by Lev Landau.

But helium-3 atoms have an overall spin of ½, because they have three particles in each *nucleus*; so they ought to behave as *fermions*. Even so, they become superfluid at temperatures of a few thousandths of a kelvin (below about 2.6 milliK), where they can form pairs with an overall integer spin (sometimes known as Cooper pairs; see *superconductivity*) and effectively behave as bosons. Because these paired particles have an effective spin of 1, not zero, the properties of superfluid helium-3 are very different from those of superfluid helium-4 – for example, sound travels through it at different speeds in different directions. At higher temperatures, the weak association between the members of a Cooper pair is broken up by thermal motion. The superfluidity of helium-3 was first demonstrated by David Lee, Robert Richardson and Douglas Osheroff in experiments at Cornell University in 1971.

supergravity   A development from the theory of *supersymmetry* that includes gravity in a natural way. If supersymmetry is a *local symmetry*, it must be associated with a *field* carried by gauge bosons which turn out to be massless particles with spin 2 – in other words, *gravitons*. Several people made this discovery independently in the mid-1970s, and developed different variations on the theme. The preferred version at that time was called '$N = 8$ supergravity', for reasons we shall explain.

As in all versions of supersymmetry, every kind of **boson** has to be accompanied, mathematically speaking, by a 'new' kind of **fermion**. The counterpart to the graviton is called the gravitino and has spin ⅔. Gravitinos would interact very weakly with everyday matter and be extremely hard to detect; and the different versions of supergravity allow for the existence of anything from one to eight different types of gravitino (in a way roughly analogous to the way in which **quantum chromodynamics** allows for the existence of a variety of **quarks**).

The version of supergravity with eight gravitinos (the $N = 8$ supergravity) turned out to be the easiest to work with, especially when it was set in a framework of eleven spacetime dimensions (see **Kaluza–Klein theory**). The excitement felt by some physicists at the end of the 1970s was summed up by Stephen Hawking, who claimed, in his inaugural lecture as Lucasian Professor in Cambridge in 1979, that $N = 8$ supergravity offered such promise that the end of theoretical physics was in sight. But the seductive appeal of the eleven-dimensional theory at the end of the 1970s turned out to be an illusion. We know that **parity** is violated in the **weak interaction**, and it turns out that this is possible only if space has an odd number of dimensions. Since time must be included in spacetime as well, this means that spacetime must have an even number of dimensions.

As soon as supersymmetry was adapted to **string theory** to make superstring theory, however, it became clear (in the mid-1980s) that supergravity is an integral part of the new theory, even when versions involving even numbers of spacetime dimensions are considered. This is an $N = 1$ version of supergravity (that is, it has room for only one kind of gravitino), and leads to the prediction of the existence of **shadow matter**.

superposition of states   A mixture of quantum **states** for which it is impossible to specify the physical characteristics of a quantum entity. For example, a particle which has the quantum property of **spin** can be thought of in classical terms as having either spin up or spin down; on the classical picture, there are only these two possibilities. But in the quantum world the particle can exist in a mixed state, a superposition in which it has, say, a 50 per cent component of spin up and a 50 per cent component of spin down (or some other mixture of the two states). It is only when the spin of the particle is measured that it settles into a definite state and has either spin up or spin down. But as soon as we stop monitoring its behaviour, the particle dissolves into a superposition again (perhaps with a different mixture of states to before).

This applies to all quantum properties, so that superpositions can be very complicated. In the kind of simple example we have just used, however, the probability that the entity is detected in one or other state is 1, since it must have a measured spin of either up or down. The proportions of spin up and spin down in the superposition are each less than 1, but must add up to 1. This is analogous to the behaviour of sine and cosine functions for an angle – for any chosen angle, the sine and cosine are each less than 1, but add up to 1. So the extent to which the two quantum states are mixed can be described by a **mixing angle**; in the fifty-fifty mixture, the mixing angle is 45 degrees.

See **Copenhagen interpretation**, **Schrödinger's cat**.

Super Proton Synchrotron   An *accelerator* built at **CERN** in the 1970s, originally to accelerate **protons** to energies of around 400 GeV. In the 1980s, it was adapted to work as a proton–antiproton collider, reaching energies of 900 GeV.

superspace   See *supersymmetry*.

superstrings   See *string theory*.

supersymmetric partners   See *supersymmetry*, *SUSY particles*.

supersymmetry   The success of the **gauge symmetry** approach to understanding the forces and particles of nature encouraged theorists in the 1970s to attempt to find a geometrical description of everything in terms of one great **symmetry**. The last major asymmetry in the world of particle physics at that time was the distinction between 'particles' (**fermions**) and 'forces' (**bosons**). This is more than a cosmetic difference. In geometric terms, a key difference between the two kinds of entity is that a fermion has to rotate twice (that is, through 720 degrees) in order to get back to where it started, while a boson has to rotate only once (through 360 degrees) to get back to its original state (see **spin**). The breakthrough in the 1970s came with the discovery of a kind of symmetry that unites these two different patterns of behaviour in one geometrical framework – supersymmetry, or SUSY for short.

Supersymmetry works by attaching another four dimensions to the four dimensions of ordinary spacetime. The resulting eight-dimensional geometry is known as superspace, and it provides room, as it were, for the 'extra' rotation that a fermion has to make to get back to its starting configuration. But these extra dimensions are not space or time dimensions, and have nothing to do with the extra space dimensions that have to be compactified in **Kaluza–Klein theories** or **string theory**. In the mathematics of supersymmetry, there is an operation which is equivalent to rotation in the everyday world. But instead of rotating an object in three-dimensional space, this operation rotates an object from the usual four-dimensional spacetime into the eight-dimensional geometry inhabited by fermions. And, of course, there is an equivalent operation which rotates an object out of the eight-dimensional geometry inhabited by fermions and into the everyday geometry of four-dimensional spacetime. In other words, it is possible to transform bosons into fermions, and vice versa – or rather, there is no longer any distinction between bosons and fermions, and what we see as two different kinds of particle is an illusion caused by geometry.

But it is clear that rotating an electron in this way would not produce any of the known bosons. Indeed, none of the known bosons corresponds to the rotated versions of known fermions, and none of the known fermions corresponds to rotated versions of the known bosons. If supersymmetry applies in the real world, there must be a **supersymmetric partner** for every type of known boson, and one for every type of known fermion, doubling the number of varieties of particle in the world. In that case, we would be dealing with a **broken symmetry**, since the supersymmetric partners are nowhere to be seen today.

As with other broken symmetries in particle physics, the implication would be that there was a complete symmetry at higher energies – in this case, with each type of boson accompanied by a fermionic SUSY particle, and vice versa. The reason why the symmetry is broken is assumed to be that the supersymmetric partners are much more massive than the counterparts we know today, and could be manufactured only at very high energies. But there would have to be a stable lightest supersymmetric partner, or LSP, into which the other SUSY particles would decay, and this particle (perhaps the *photino*) might be detectable. Searches for such particles are under way; until they are found, however, supersymmetry, attractive though it is as a theoretical package, must remain unproven as a description of the world we live in (but see *string theory*).

See also *shadow matter*.

supersynchrotron   General name for a type of accelerator developed in the 1960s as an improvement on the original *synchrotron* design. A key feature of the design is that it uses two sets of magnets, one to bend the beam of particles and the other to focus it.

surface tension   The property that makes the surface of a liquid behave as if it were covered with a thin sheet of elastic material. It arises because the molecules in the surface layer of the liquid are attracted to their neighbours in the surface around them, and by the molecules just below the surface of the liquid, but there are (by definition!) no molecules of liquid above the surface to attract them in that direction. This restricts the motion of the molecules in the surface layer, making the surface of water, for example, rigid enough to support the weight of an ordinary paperclip carefully placed on the surface. It is because of surface tension, acting like the skin of a balloon, that droplets of liquid try to form spherical shapes.

In the body of the liquid, molecules are, on average, closer together than in the surface layer, and are repelled by each other, creating a pressure. The attractive force is experienced only when the molecules are slightly further apart, which happens in the surface layer because there are no molecules of liquid above pushing them down.

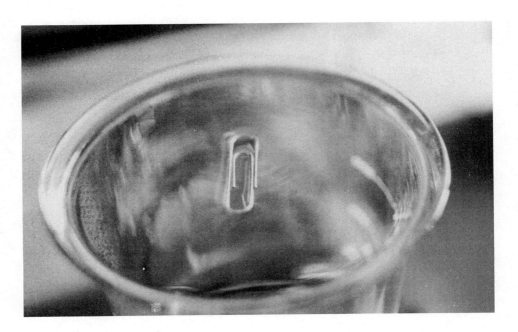

The **surface tension** of water can be demonstrated (if you are careful) by 'floating' a paperclip on the surface.

SUSY   Abbreviation for *supersymmetry*.

SUSY particles   The counterparts to everyday particles predicted by the theory of *supersymmetry* (also known as *supersymmetric partners*). For every known type of *fermion*, supersymmetry requires a 'new' kind of *boson*; for every known kind of boson, there is a fermionic counterpart. The new bosons are named by modifying the name of the equivalent fermion, adding an 's' at the front – thus, the counterpart to the electron is the selectron. The new fermions are named by modifying the name of the equivalent boson to make it end in 'ino' – thus, the counterpart to the photon is the photino.

symmetry   We all know what symmetry is in the context of a geometrical pattern, and this idea is carried over into the quantum world to describe the relationships between forces and particles. This enables scientists to describe physics in terms of geometry – if necessary, invoking more dimensions than the familiar three of space plus one of time.

A familiar example of symmetry is the reflection symmetry of some patterns, in which the right-hand side of the pattern is a mirror image of the left-hand side. Another symmetry is possessed by a perfect sphere – it looks the same no matter how it is rotated, and is said to possess spherical symmetry, or to be rotationally invariant (see *invariance*).

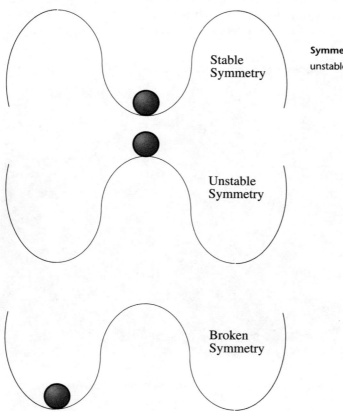

Stable
Symmetry

Unstable
Symmetry

Broken
Symmetry

Symmetry. Examples of stable, unstable and broken symmetry.

Symmetry is built into the laws of nature in a very deep way. The symmetry which says that the laws of nature are the same at every place in the Universe (translational invariance), for example, corresponds to the law of conservation of linear *momentum*; the symmetry which says that the laws of physics are the same at all times is equivalent to the law of conservation of *energy*; and the rotational invariance of the laws of physics is equivalent to the law of conservation of *angular momentum*.

Many of the symmetries in quantum physics are *broken symmetries*, where a situation that is intrinsically symmetrical has become asymmetrical – the classic example is of a ball balanced on a perfectly cone-shaped hill, which is a symmetrical situation; when the ball rolls off down one side of the hill, the symmetry is broken, but the situation you end up with still carries some imprint of the underlying symmetry. Using these ideas, physicists have discovered symmetries between the *forces of nature*, between *quarks* and *leptons*, and even between *fermions* and *bosons* (see *supersymmetry*). See *group theory*.

The term 'symmetry' is also used in quantum mechanics with a specific meaning, to refer to the kind of wave function that describes a particular kind of quantum entity. The wave function describing a boson is said to be symmetric, and the wave function describing a fermion is said to be antisymmetric.

symmetry breaking See *broken symmetry, spontaneous symmetry breaking*.

symmetry group See *group theory*.

synchrocyclotron General name for a type of *accelerator* developed from the *cyclotron* principle after the Second World War. In a synchrocyclotron, bunches of particles are accelerated together, with the field that is causing the acceleration steadily altered (synchronized) to allow for the increase in mass of the particles as they move faster. Such machines are designed in accordance with the special theory of relativity (see *relativity theory*), and would not work if this theory were incorrect.

synchrotron The basis of all large modern *accelerators* (except *linear accelerators*). A synchrotron accelerates particles not in a spiral (like a *cyclotron* or a *synchrocyclotron*), but in a toroidal (doughnut-shaped) ring, keeping them in the same 'orbit' by increasing the magnetic field holding the particles in the ring as the particles gain energy.

synchrotron radiation Characteristic radiation produced by *electrons* or other charged particles moving at high speed (a sizeable fraction of the *speed of light*) in a magnetic field. The faster the electrons move, the shorter the wavelength of the radiation. The radiation is produced by particles accelerated in a *synchrotron*, hence its name, but it also occurs naturally in the Universe wherever high-speed electrons spiral along magnetic *lines of force*, and is an important means by which astronomers can probe conditions in objects such as pulsars.

Szilard, Leo (1898–1964) Hungarian-born American physicist who played a major part in the development of the atomic bomb (see *Manhattan Project*) because of concern that a similar weapon might be developed by Nazi Germany, but who opposed the idea of actually using it on the Japanese. He also solved the puzzle of *Maxwell's demon*.

Born in Budapest on 11 February 1898, Szilard studied electrical engineering in Budapest, but was drafted into the Austro-Hungarian army during the First World

Synchrocyclotron. One of the first CERN accelerators, a 600 MeV synchrocyclotron (see p.395).

War, and after the war moved to the University of Berlin and took up physics, obtaining his PhD in 1922. He stayed at the University of Berlin until 1933, when the Nazis came to power, then moved (via Vienna) to England, where he worked as a member of the physics staff at St Bartholomew's Hospital, in London, and at the Clarendon Laboratory, in Oxford. He emigrated to the United States in 1937, taking up a post at Columbia University in New York, and became a US citizen in 1943. After the Second World War, he left physics to work in molecular biology, and was appointed professor of biophysics at the University of Chicago in 1946. Ten years later he moved to the Salk Institute of Biological Sciences at La Jolla, in California. He worked there until he died, in La Jolla, on 30 May 1964.

As early as 1934, Szilard realized that a ***chain reaction*** could be triggered in a suitable radioactive substance, and he took out a patent on the process, which he assigned to the Admiralty in Britain as a security measure. When he learned of the work of Otto Hahn and Lise Meitner on uranium fission, at the end of the 1930s, he became concerned that nuclear weapons might be developed by the Nazis, and (together with Edward Teller and Eugene Wigner) approached Albert Einstein, persuading him to write a letter to President Roosevelt drawing attention to the possibility. This initiative led to the Manhattan Project. Szilard's main role in this, during 1942, was working with Enrico Fermi, in Chicago, on the construction of the first nuclear reactor (then called an atomic pile); he then worked at Los Alamos for the rest of the war.

After the Second World War, Szilard became an active campaigner for the peaceful use of atomic energy and for international control of nuclear weapons, and in 1959 he received the Atoms for Peace award for this work.

tachyon Hypothetical variety of particle that always travels faster than the *speed of light*. The special theory of relativity tells us that no particle which is moving more slowly than the speed of light can ever be accelerated to a speed faster than that of light. But there is a symmetry in the equations which allows for the existence of particles that always travel faster than light, and can never be slowed down below the speed of light.

Such superluminal particles would mirror many of the properties of everyday particles – for example, they would move faster as they *lost* energy, and they would travel backwards in time. There is no direct evidence that tachyons exist, and most physicists believe that the equations which describe the possibility of their existence have no more significance than the 'negative roots' which appear in some quadratic equations and are without physical relevance – for example, if I chose to specify the length of a piece of string as 'the square root of 9' centimetres, the equations would tell you that the string is either 3 cm long or –3 cm long; but you would have no difficulty deciding which 'answer' applied to the real world. This particular symmetry does not seem to be physically meaningful.

Tamm, Igor Yevgenyevich (1895–1971) Russian physicist who shared the Nobel Prize in 1958 with Pavel Cerenkov and Ilya Frank for their investigation of *Cerenkov radiation*. Tamm was born in Vladivostock on 8 July 1895 (26 June, Old Style); he died in Moscow on 12 April 1971.

The award that he shared with Cerenkov and Frank marked the first time that Russians had received the Nobel Prize for Physics, a landmark event in recognizing the importance of scientific work being done in what was then the Soviet Union. Ironically, though, Tamm himself did not regard the work for which he received the prize as his most important contribution to physics (a view shared by many physicists).

After completing his secondary education, Tamm studied at the University of Edinburgh in 1913 and 1914, then returned to Russia where he completed his PhD (awarded by the University of Moscow in 1918) amid the turmoil of the Russian revolution. He taught physics at several universities, becoming head of the physics department at Moscow University in 1930 and head of the Lebedev Institute in 1934. His prestige as a physicist and his international reputation helped him to maintain a relatively independent position even under the Stalinist regime, and he was a strong supporter of relativity theory, which was regarded by many in the Soviet Union at that time as anti-Marxist. Tamm worked with Andrei Sakharov on the Soviet nuclear fusion programme and was active in the Pugwash movement, an international group of scientists opposed to the use of nuclear weapons. He spoke out in support of nuclear disarmament on American TV in 1963.

Among Tamm's many contributions to physics, he showed that although the neutron is electrically neutral it has a *magnetic moment*. The work for which he should have received the Nobel Prize was his theory of *beta decay*, developed in 1934, which suggested that it is a result of a weak force operating between nucleons. Hideki Yukawa took up the idea and described the weak nuclear force in terms of the exchange of *mesons*.

tau   A member of the *lepton* family, a form of heavy *electron*. The tau particle has the same electric charge as the electron, and the same spin (½), but it has a mass of 1.784 GeV (making it the heaviest lepton) and is unstable, with a lifetime of 3 x 10⁻¹³ sec. Although already predicted by theory, the tau was identified only in 1975, in experiments carried out at SLAC. See *generation*.

tau neutrino   See *neutrino*.

Taylor, Richard Edward (1929–)   Canadian physicist who received a share of the Nobel Prize in 1990 for his contribution to the experimental work which confirmed the existence of particles within neutrons and protons – the *quarks*.

Taylor was born in Medicine Hat, Alberta, on 2 November 1929, and studied at the University of Alberta, where he graduated in 1950 then took a master's degree (1952), and at Stanford University, where he received his PhD in 1956. He worked at the Linear Accelerator Laboratory at Orsay, at the Lawrence Berkeley Laboratory, and (from 1962 to 1968) at SLAC, before becoming a professor at Stanford. At the end of the 1960s, he worked at SLAC (with Jerome Friedman and Henry Kendall) on experiments which probed the structure of nucleons (protons and neutrons) using beams of high-energy electrons. Together with the theoretical interpretation of these experiments provided by James Bjorken and Richard Feynman, they showed that the structure within the nucleons is caused by the presence of point-like particles. In the 1970s, further experiments confirmed that the particles could be identified with quarks. See *partons*.

technicolour   An unproven (indeed, untested) theory that attempts to avoid the need for regarding the *Higgs particle* as a truly fundamental entity (which causes problems in some versions of particle theory) by suggesting that it is made up of other entities in a way reminiscent of (and deliberately modelled on) the way *protons* and *neutrons* are made up of quarks. The theory borrows from the *colour* theory of *quantum chromodynamics*, and invokes the existence of 'techni-fermions', spin ½ particles which are held together by a technicolour force to make up techni-mesons that are absorbed by detectable particles and give them mass. The specific idea of technicolour may be incorrect, but it indicates one approach to the problem of how the particles get their mass which differs from (or at least modifies) the Higgs mechanism.

Teller, Edward (1908–)   Hungarian-born American physicist famous as 'the father of the hydrogen bomb', but this owes more to his political activities than to his scientific work. Born in Budapest on 15 January 1908, Teller studied at the local Institute of Technology in Budapest, then at several German universities, culminating in the award of a PhD from Leipzig University in 1930. He left Germany in 1933, when the Nazis came to power, and worked briefly in Copenhagen and London before becoming professor of physics at George Washington University, in Washington, DC, in 1935.

In Washington, Teller worked with George Gamow on the rules which govern *beta decay*. In 1941 he became a US citizen and moved to Columbia University in New York; the next year he moved on to the University of Chicago, where he worked on nuclear fission with Enrico Fermi. Although Teller was a member of the *Manhattan Project*, after initially working on the fission ('atomic') bomb, before that was completed he started work on the design of the fusion ('hydrogen') bomb. This research was not pursued intensively immediately after the end of the Second World War, when Teller returned to Chicago; but after the Soviet Union exploded its first fission bomb in 1949 Teller was instrumental in persuading the US government to go ahead with a crash programme to develop a fusion bomb, which was achieved by 1951 using the designs of the Polish-born mathematician Stanislaw Ulam.

Teller also played a major part in getting the Lawrence Livermore Laboratory, near Berkeley, California, established as a second American nuclear weapons research centre (the first being at Los Alamos, where the atomic bomb was developed). Teller's testimony at the investigation into Robert Oppenheimer's activities that 'I would personally feel more secure if public matters could rest in other hands' was instrumental in Oppenheimer losing his security clearance and being slurred as an unpatriotic communist sympathizer.

Teller was associate director at Livermore from 1954 until he retired in 1975; he was also a professor at the University of California from 1953 to 1975. Although officially retired, he retained close links with Livermore in the late 1970s and 1980s, and in the 1980s he was a moving force behind the establishment of the 'Star Wars' Strategic Defense Initiative.

He is thought to be the inspiration for the movie character Dr Strangelove.

temperature A measure of the amount of heat there is in a substance – in other words, how fast the *atoms* and *molecules* that the substance is made of are moving. *Electromagnetic radiation* also has a temperature, which is a measure of how energetic the radiation is, and is related to the number of *photons* with different *wavelengths* present in the radiation (see *black body radiation*).

The temperature scale used by scientists is the *Kelvin scale*, which is based on measurements of the rate at which heat flows between objects with different temperatures. If there is no heat flow, there is no temperature difference; the bigger the heat flow (always from the hotter object to the cooler one), the bigger the temperature difference.

A fundamental feature of the Kelvin scale, which is a natural scale based on the thermodynamic properties of objects, not an artificial 'man-made' scale, is that there is an *absolute zero* of temperature, at which a body cannot give up any more heat. This occurs at $0\,K$ (roughly $-273\,°C$), where the substance still contains energy, but cannot give it up. This is related to the property of quantum *uncertainty*. The actual energy, or temperature, of an object is always blurred a little by quantum uncertainty – but if the substance had precisely zero energy, there would be no uncertainty about its energy, which is not allowed by the quantum rules.

It is a fundamental law of nature that heat cannot, of its own accord, flow from a cooler object to a hotter object. See *second law of thermodynamics*.

tensor Originally developed in the 19th century as an abstract mathematical concept, a tensor can be regarded as a generalization of a *vector*. A tensor is specified in terms of a set of coordinates, so its form is changed if you choose to work in a different set of mathematical coordinates; but they have the property that any equation involving tensors that is true in one set of coordinates is true when it is written in the same form in any other set of coordinates. Prompted by his friend Marcel Grossmann, in the second decade of the 20th century Albert Einstein discovered that this was just what he needed in order to develop the general theory of relativity. By expressing the equations of the general theory in tensor terms, he derived equations which hold true in any system of coordinates, which means for any observer in the Universe, no matter how they are moving. Tensor equations represent a deep truth that is independent of the observer, and this also makes them important in quantum physics.

One simple physical way to think of a tensor is as providing information about the rate at which things are changing (the acceleration) at any point. If a two-dimensional rubber sheet were being pulled in different directions by different forces, it would be expanding at different rates in different directions. The rate at which it is expanding in each direction at each point could be represented by a set of numbers making up a tensor field. It is precisely this image, with a two-dimensional stretched rubber sheet replaced by four-dimensional stretched spacetime, that makes tensors so useful in the general theory of relativity.

In mathematical terms, the best way to put this in perspective is to say that a scalar is represented by a single number, a vector is represented by a row of numbers, and a tensor is represented by a two-dimensional grid of numbers, like a matrix (see *matrices*). The difference between a tensor and a matrix is that a matrix is an array of numbers that represents the phenomenon being discussed in a particular coordinate system, but a tensor equation is the same in all coordinate systems – to a theoretical physicist, the tensor *is* the real, physical object.

tensor field   A *field* defined by the value of a *tensor* at every point. Gravity is an example of a tensor field, and *gravitons* are therefore tensor particles.

tensor particle   Particle associated with a tensor field, and therefore having spin 2. See *vector particle*.

TeV   1,000 billion *electron volts*. Abbreviation for tera electron volt.

tevatron   Particle accelerator at *Fermilab*, which can accelerate protons to energies of 1 TeV. These energies were first achieved in the mid-1980s.

Thales of Miletus (about 625–550 BC)   Greek philosopher, born in Miletus (now in Turkey). He is thought to have been a successful merchant who visited Egypt and learned geometry there. He probably developed some of the geometrical ideas written down in a systematic fashion by Euclid some 250 years later. He also developed what would now be regarded as attempts at scientific explanations of phenomena such as earthquakes, instead of 'explaining' them as the work of the gods, and is regarded as the first scientific thinker.

theory of everything (TOE)   Any theory which attempts to combine *gravity* and some form of *grand unified theory* (GUT) in one package. The most likely candidate to succeed in this aim seems at present to be some form of superstring theory (see *string theory*, *supersymmetry*).

The GUTs describe only the combination of the electroweak and strong forces (see *forces of nature*), but a true TOE must include gravity as well.

thermal neutrons   = *slow neutrons*.

thermionics   See *Richardson, Owen Williams*.

thermodynamics   The branch of science that deals with the relationship between heat and other forms of energy. Thermodynamics was developed in the 19th century when, as well as its intrinsic scientific interest, it was of great practical importance in the age of steam engines. The basics of thermodynamics can be summed up in four laws. The first, so obvious that it was added to the other three later as a kind of after-thought, is known as the 'zeroth' law. It says that if two systems are each in thermal equilibrium with a third system (which in everyday language means that they are at the same temperature) then they are in thermal equilibrium with each other. The first law says that the total energy of a system remains constant, although it may be changed from one form to another – so **kinetic energy**, for example, is converted into heat energy in the brakes of a car when the car slows down. (The law is often remembered in the form 'work is heat, and heat is work', or 'you can't get something for nothing'.) The **second law of thermodynamics** is arguably the most fundamental law in science, and says that heat can never flow spontaneously from a system at a lower temperature to a system at a higher temperature. Work has to be done to make this happen (for example, in a refrigerator), and work is heat (so cooling the inside of a refrigerator warms the world outside even more than the amount that the 'fridge cools down). This is closely related to the idea of **entropy** and the **arrow of time**; it means that the Universe is running down ('you can't even break even'). The third law says that the change in entropy of a system changing from one state to another approaches zero as its temperature approaches zero on the **Kelvin scale**. This helps in defining an absolute scale of entropy, and also means that no real system can ever be cooled all the way down to 0 K ('you can't get out of the game').

thermodynamic temperature scale   See *Kelvin scale, temperature*.

thermonuclear reactions   See *nuclear reactions*.

Thomson, Benjamin (1753–1814)   See *Rumford, Count*.

Thomson, George Paget (1892–1975)   British physicist who received the Nobel Prize for Physics in 1937 for his experimental work showing that electrons behave as waves. The son of J. J. Thomson, who received the Nobel Prize for Physics in 1906 for his experimental work showing that electrons behave as particles. Both were right, and both prizes were deserved; see **wave–particle duality**.

Thomson was born in Cambridge on 3 May 1892. He was educated at Trinity College, Cambridge, graduating in 1913 and becoming a Fellow of Corpus Christi College. After wartime service in France in the infantry, and then as an aeronautical expert with the Royal Flying Corps at Farnborough, Thomson returned to Cambridge in 1919. He became professor of natural philosophy at the University of Aberdeen in 1922; it was there that he carried out the experiments which showed, in 1927, that a beam of electrons passing through a thin metal foil is diffracted, and therefore electrons are waves. The same discovery was made independently (using a different technique) by Clinton Davisson, who shared the Nobel Prize with Thomson, and his assistant Lester Germer, who didn't.

In 1930 Thomson was appointed professor of physics at Imperial College, London, where he became interested in nuclear physics and was one of the first people to alert the British government to the possibility of an atomic bomb. He chaired the Committee on Atomic Energy which established the atomic bomb project in Britain in the Second World War (the 'Maud' Committee). He was sent to Canada in 1942 as a wartime scientific adviser, liaising with the United States on the **Manhattan Project**, and returned to England to serve as a scientific adviser to the Air Ministry; he was knighted in 1943. Thomson returned to Imperial College after the war, having techni-cally been on secondment from his post there during all these activities.

From 1952 to 1962 Thomson was master of Corpus Christi College. He lived in Cambridge during his retirement, and died there on 10 September 1975.

Thomson, J(oseph) J(ohn) (1856–1940)   British physicist (always known by his initials, as 'J.J.') who received the Nobel Prize for Physics in 1906 for his experimental work showing that electrons behave as particles. The father of George Thomson, who received the Nobel Prize for Physics in 1937 for his experimental work showing that electrons behave as waves. Both were right, and both prizes were deserved; see *wave–particle duality*.

Born at Cheetham Hill, near Manchester, on 18 December 1856, Thomson was the son of an antiquarian bookseller. He studied at Owens College, Manchester (later the nucleus of Manchester University), and was originally intending to become an engineer. But the engineering course required the payment of an additional fee, and when Thomson's father died in 1873 the family could no longer afford this premium. Aided by a scholarship, Thomson switched to physics, chemistry and mathematics. In 1876 he won a scholarship to Trinity College, Cambridge, graduating in mathematics in 1880. He became a Fellow of Trinity and remained academically attached to the college and the university for the rest of his life (although he did visit Princeton in 1896 and Yale in 1904).

From 1880 onwards, Thomson worked at the **Cavendish Laboratory**, initially under Lord Rayleigh, whom he succeeded as Cavendish Professor of Experimental Physics in 1884. In his sixties by the end of the First World War, Thomson resigned this post in 1919 (being succeeded by Ernest Rutherford), having been appointed

J.J. Thomson (1856–1940).

master of Trinity College, the first scientist to receive this appointment; he held the post until he died, in Cambridge, on 30 August 1940. He was buried in Westminster Abbey.

There is a certain irony about Thomson's position as a professor of experimental physics, since although he had brilliant insight and devised experiments that shed new light on fundamental physics, he was notoriously clumsy, and the experiments had to be carried out by other people under his direction. His son George once said that although J.J. 'could diagnose the faults of an apparatus with uncanny accuracy it was just as well not to let him handle it' (Barbara Cline, *The Questioners*, p. 13). Nevertheless, Thomson presided over the Cavendish in its glory years at the end of the 19th century and the beginning of the 20th century, and one of the superb experiments that he devised, using electric and magnetic fields to balance the electric and magnetic properties of a moving charged particle, showed in 1897 that **cathode rays** were actually charged particles, which he called 'corpuscles', but which were soon named electrons by the Dutch physicist Hendrik Lorentz.

These particles could only be pieces that were somehow chipped off from atoms. It is hard to appreciate today what a dramatic discovery this was at that time, when atoms were still thought to be indivisible. In a lecture to the Royal Institution in 1897, Thomson said, with masterly understatement, that 'the assumption of a state of matter more finely subdivided than the atom of an element is a somewhat startling one'.

Thomson's Nobel Prize was awarded in 1906 and he was knighted in 1908; seven of the physicists who worked as his assistants at the Cavendish later received Nobel Prizes of their own. This was a measure both of the prestige of the Cavendish, which (in no small measure thanks to Thomson's presence) attracted the best physicists to work there, and also of Thomson's inspired leadership. He was also extremely well liked by his colleagues.

Thomson's later work, after the identification of the electron, included studies of what are now called positively charged *ions* (formerly 'canal rays'), pointing the way for Francis Aston's work on *isotopes*.

Thomson, William   See *Kelvin, Lord.*

Thomson model   See *atomic physics.*

t'Hooft   See *Hooft, Gerardus 't.*

thought experiment (*gedanken experiment*)   An experiment which is not intended to be carried out as a practical reality, but is 'all in the mind'. The idea is that by using our understanding of the laws of physics we can construct imaginary experiments and predict their outcomes, thereby highlighting features of those laws which may not be obvious at first sight. The classic example in quantum physics is the experiment involving *Schrödinger's cat.*

It is hard to see how that experiment could ever be carried out (at least, not with a real cat). Sometimes, though, the thought experiment is adapted and made practicable by later generations of scientists. Albert Einstein was particularly fond of inventing thought experiments to demonstrate what he regarded as the absurdity of quantum physics (see *EPR experiment, 'clock in the box' experiment*). One of these, the EPR experiment, evolved into the *Aspect experiment*, which provided conclusive experimental proof of the weirdness of the quantum world.

three-jet event   See *gluons*.

'three-man paper'   See *matrix mechanics*.

time   Everybody knows what time is, but nobody can *explain* what it is. In physics, the important thing about time is that it provides a reference system (a set of coordinates) in which events can be ordered. One event comes before or after another in this system. But it is important that, although this defines an *arrow of time*, there is no suggestion anywhere in the laws of physics that time actually flows from the past through the present and into the future. All times have equal status.

This shows up most forcefully in *relativity theory*, where time is regarded as the fourth dimension, on an equal footing with the familiar three dimensions of *space*. You can imagine all of space and time represented as a four-dimensional spacetime map, on which all of history, the present and the future of the Universe can be represented. This raises interesting questions about the nature of destiny and free will – is the future 'already there' in some sense, just waiting for our consciousness to move over it? But the uncertainty inherent in *quantum physics* suggests that a better theory of space and time, merging relativity and quantum theory, may restore the vagueness of the future to the description of spacetime.

Scientists are on more secure footing when using time simply as a measure of the interval that has elapsed (in seconds or some other suitable units) between two events, or the time taken for some process to occur. Happily, this is the straightforward usage of the term in most of this book, and the deeper philosophical issues can be ignored except where they are explicitly referred to.

time reversal symmetry   The symmetry, denoted by the letter T, which says that the laws of physics are the same if all the motions of a physical process are reversed. In the simplest example, a film showing a collision between two billiard balls looks equally plausible whether we run the movie forwards or backwards. The same is true of the vast majority of particle interactions, so that if one allowed interaction is described by a *Feynman diagram*, in almost all cases we can reverse the direction of time in the diagram and get an allowed interaction (the exceptions are a very few processes involving the weak interaction; see *CP conservation*, *CPT conservation*).

But although each individual collision seems to be symmetric in time, if we show a film of a break being made at pool, with all the balls scattering from an initially ordered state, it is obvious which way the film should be run. This appearance of an *arrow of time* when going from simple systems to complex systems is still not understood, because time reversal symmetry seems to be built in at a fundamental level in both classical physics and quantum physics.

See *transactional interpretation*.

time travel   To the surprise of most physicists, and the delight of most science fiction writers, research carried out in the late 1980s showed that genuine time travel is not forbidden by the laws of physics as at present understood. This does not mean that it will be easy to build a working time machine; but it does mean that it may not be impossible. More importantly, it means that there may be naturally occurring objects in the Universe that act as time machines.

The kind of time machine recently investigated (theoretically!) in some detail involves wormholes – tunnels through spacetime which may, according to the equations of the general theory of relativity, connect a *black hole* in one part of spacetime with a black hole in another part of spacetime. This is such an obvious idea that it has been freely used by science fiction writers for decades, but until the mid-1980s it was generally accepted by physicists that such objects could not 'really' exist, and that a better understanding of Einstein's equations would prove this. They had to change their tune as a result of a careful investigation of wormholes carried out by Kip Thorne and his colleagues at Caltech – an investigation triggered by a science fiction story.

It happened like this. Carl Sagan, a well-known astronomer, had written a novel in which he used the device of travel through a black hole to allow his characters to travel from a point near the Earth to a point near the star known as Vega. Although he was aware that he was bending the accepted rules of physics, this was, after all, a work of fiction. Nevertheless, as a scientist himself, Sagan wanted the science in his story to be as accurate as possible, so he asked Thorne, an established expert in gravitational theory, to check it out and advise on how it might be tweaked up. At the end of 1985, musing on the problem of how to hold a wormhole open, Thorne realized that this could be done by threading the wormhole with so-called exotic matter. The critical factor is that this exotic matter has an enormous tension, sufficient to hold the wormhole open in spite of the inward tug of gravity. Nobody has ever found anything that could do the job – but, so far from being forbidden by the laws of physics, such exotic matter almost exactly matches the description of **cosmic string**. If anything like cosmic string existed, and could be captured, it would be capable of holding wormholes open.

Sagan gratefully accepted Thorne's modification to his fictional 'star gate', and the wormhole duly featured in the novel, *Contact*, published in 1985. But this was still only presented as a shortcut through space. Neither Sagan nor Thorne realized at first that what they had described would also work as a shortcut through time. Thorne seems never to have given any thought to the time travel possibilities opened up by wormholes until, in December 1986, he went with his student, Mike Morris, to a symposium in Chicago, where one of the other participants casually pointed out to Morris that a wormhole could also be used to travel backwards in time.

The point is that space and time are treated on an essentially equal footing by Einstein's equations. So a wormhole that takes a shortcut through spacetime can just as well link two different times as two different places. Indeed, any naturally occurring wormhole would most probably link two different times.

Other physicists who were interested in the exotic implications of pushing Einstein's equations to extremes (including Igor Novikov, in what was then still the Soviet Union) were encouraged to go public with their own ideas once Thorne was seen to endorse the investigation of time travel, and the Caltech work led to the growth of a cottage industry of time travel investigations at the end of the 1980s and into the 1990s. The bottom line of all this work is that, while it is hard to see how any civilization could build a wormhole time machine from scratch, it is much easier to envisage how a naturally occurring wormhole might be adapted to suit the time travelling needs of a sufficiently advanced civilization. This raises all kinds of interesting paradoxes and possibilities, which are discussed in the books cited below. In the present context, though, the most relevant discovery from this work is that there is an intimate connection between time travel and the quantum-mechanical description of the Universe.

We can see how this provides new insights into the workings of the Universe, by looking at the billiard-ball equivalent of the so-called granny paradox, in which a time traveller accidentally causes the demise of his maternal grandmother before his own mother has been born. We do this by imagining a time tunnel set up with its two mouths close together. If a billiard ball is fired into the appropriate mouth of the time tunnel in just the right way, it will emerge from the other mouth in the past, and just have time to travel across the intervening space to collide with itself before it enters the tunnel, knocking the earlier version of itself out of the way. So it never travels through time, the collision never takes place, and therefore the earlier version of the billiard ball *does* enter the time tunnel … and so on. This is the self-inconsistent solution to the problem, and the conventional wisdom says that it must be rejected – that the Universe cannot possibly operate like that.

The reason why physicists such as Thorne and Novikov are confident that it is acceptable to dismiss the self-inconsistent solution is that they have found that, in any situation of this kind, there is always another solution of the equations that gives a self-consistent picture starting from the same initial circumstances.

An example of a self-consistent solution to this kind of billiard-ball problem is when the ball approaches the time tunnel and is struck a glancing blow by an identical billiard ball that has just emerged from one mouth of the time tunnel, knocking the first ball into the other mouth of the tunnel. As the first ball emerges from the other mouth of the tunnel, it collides with the younger version of itself, knocking itself into the tunnel. Thorne, Novikov and their colleagues have found not only that there are no billiard-ball problems of this kind which do not have at least one self-consistent solution, but that every problem of this kind that they can think of has an infinite number of self-consistent solutions.

Imagine a billiard ball that passes neatly in a straight line between the two mouths of the time tunnel. Or does it? Suppose that when the ball is midway between the two mouths it is struck a violent blow by a fast-moving ball that emerges from one mouth. The 'original' ball is knocked sideways, travels through the tunnel and becomes the 'second' ball – but in the collision it is deflected back on to exactly the same path, or trajectory, that it was following before the collision. As far as any distant observer is concerned, it still looks as if the single ball has passed smoothly in a straight line between the two mouths; and you can imagine similar patterns involving two, three or more circuits by the ball around the time tunnel. There seems to be more than one acceptable way to describe the ball's behaviour.

All of this is reminiscent of the way the Universe operates at the quantum level. There is a choice of realities, just as there is in the example of **Schrödinger's cat**. The billiard ball seems to be perfectly normal before it gets near the time tunnel, then interacts with the tunnel system in many different ways, forming a **superposition of states**, before it emerges on the other side behaving, once again, in a perfectly normal fashion. What Thorne calls the 'plethora' of self-consistent solutions to the same billiard-ball/wormhole problem would be deeply troubling, if it were not for the fact that quantum theorists have already worked out how to handle such multiple realities.

The technique they use was the one first developed by Richard Feynman in the 1940s, the **sum over histories** approach. It is as if the particle is aware of all the possible routes it might take, and decides where it is going on that basis.

Of course, all of this usually applies down at the quantum level, on the scale of atoms and below. Quantum *uncertainty* is very small and has a negligible influence on our everyday world, so that real billiard balls, for example, behave just as if they are following classical trajectories. But the presence of a traversable time tunnel in effect creates a new kind of uncertainty, in the region between the mouths of the tunnel, operating on a much larger scale. Thorne, Novikov and their colleagues have found that the sum over histories approach works perfectly in this new situation, describing solutions to problems involving billiard balls that travel through time tunnels.

If you start out with an initial state of the ball as it approaches the time tunnel from far away, then the sum over histories approach gives you a unique set of probabilities which tell you when and where the ball is likely to emerge on the other side, clear of the region containing closed loops in time. It doesn't tell you how the billiard ball gets from one place to another, any more than quantum mechanics tells you how an electron moves within an atom. But it does tell you, precisely, the probability of finding the billiard ball in a particular place, moving in a particular direction, after its time tunnel encounter. Moreover, the probability that the ball starts out moving along one classical trajectory and ends up moving along a different one turns out to be zero. From a distance an observer will not see the ball to have been deflected at all by its encounters with itself, and unless you look closely you will not notice anything peculiar going on. 'In this sense,' says Thorne, 'the ball "chooses" to follow, in each experiment, just one classical solution; and the probability for following each of the solutions is predicted uniquely.'

And there is a bonus. In the sum over histories approach, strictly speaking we are not ignoring the self-inconsistent solutions after all. They are still there, in the addition of probabilities, but they make such a tiny contribution to the overall sum that they have no real influence over the outcome of the experiment.

There is one more very strange feature of all this. Because the billiard ball is, in some way, 'aware' of all the possible trajectories – all the possible future histories – open to it, its behaviour anywhere along its world line depends to some extent on the paths open to it in the future. Because there are many different paths that such a ball can follow through a time tunnel, but far fewer that it can follow if there is no time tunnel to pass through, this means that it will behave differently, in principle, if it has a time tunnel to go through than if it has not. Although it would be very difficult indeed to measure such an influence, according to Thorne this means that it ought to be possible, in principle, to carry out a set of measurements on the behaviour of billiard balls before any attempt to construct such a time machine has been made, and to work out from the results whether or not a successful attempt to construct a time tunnel will be made in the future. This, he says, is 'a quite general feature of quantum mechanics with time machines'.

Further reading: John Gribbin, *In Search of the Edge of Time*; Kip Thorne, *Black Holes and Time Warps*. Further viewing: *Contact* (the movie).

Ting, Samuel Chao Chung (1936–) American particle physicist who received the Nobel Prize for Physics in 1976 (jointly with Burton Richter) for the discovery of the *J/psi particle*.

Ting was born on 27 January 1936, in Ann Arbor, Michigan, where his father (his parents were Chinese nationals) was studying at the University of Michigan. He was raised in mainland China and later (after the Chinese revolution) in Taiwan, where his father was a professor at the National Taiwan University.

In 1956 Ting emigrated to the United States, where he studied at the University of Michigan, graduating in 1959 and receiving his PhD in 1962. He spent a year at CERN, and joined the faculty of Columbia University in 1964. He spent a year (1966–7) at DESY, then moved in 1967 to MIT, which has remained his academic base. The experiments which led to the award of his Nobel Prize were carried out by a team led by Ting, at the Brookhaven National Laboratory at Upton, Long Island. He later worked on collaborative experiments using PETRA and LEP, studying the properties of *gluons*.

TOE See *theory of everything*.

Tokamak Any kind of *fusion* reactor in which the hot plasma (ionized gas) in which nuclear reactions are taking place is confined within a toroidal chamber by magnetic fields. The name is an acronym derived from the Russian for 'toroidal magnetic chamber'. It is widely regarded as the most promising prospect for generating usable amounts of electricity by fusion.

Tomonaga, Sin-itiro (1906–1979) Japanese physicist who received the Nobel Prize for Physics in 1965 (jointly with Richard Feynman and Julian Schwinger) for the theory of *quantum electrodynamics*.

Tomonaga was born in Tokyo on 31 March 1906. His family moved to Kyoto in 1913, when his father was appointed professor of philosophy at Kyoto University,

Sin-itiro Tomonaga (1906–1979).

and in 1929 Tomonaga received his first degree from Kyoto University, where he was a contemporary and friend of Hideki Yukawa. He spent three more years at Kyoto, working as an unpaid research assistant, then moved to the Institute of Physical and Chemical Research, in Tokyo, where he worked with Yoshio Nishina on aspects of quantum electrodynamics, including the production of electron–positron pairs. He was deeply influenced by the papers of Paul Dirac and would later pass on this influence to a new generation of Japanese physicists.

From 1937 to 1939 Tomonaga worked in Leipzig, where he was influenced by Werner Heisenberg. The work he carried out there earned him a doctorate from Tokyo University in 1939. In 1940 he became a lecturer at Tokyo Bunrika University (which became part of the Tokyo University of Education in 1949), and in 1941 was promoted to professor of physics there. It was at about this time that he began work on what would become his theory of QED, but during the Second World War this took a back seat as he worked on microwave systems, important in radar.

At the end of the war, he completed his version of the theory of QED, working under appalling conditions in the devastated remains of Tokyo. (Apart from the lack of food or housing, Tomonaga was cut off from contact with Western scientists; with no money to pay for subscriptions to learned journals, he learned of the discovery of the *Lamb shift* from an article in *Newsweek*.) His work appeared in print roughly at the same time as the theories of Feynman and Schwinger, but had actually been worked out first. This led to a year (1949) at the Institute for Advanced Study, in Princeton.

In the mid-1950s, Tomonaga was instrumental in establishing the Institute for Nuclear Study at the University of Tokyo. From 1956 to 1962 he was President of the Tokyo University of Education, and from 1963 to 1969 he was President of the Science Council of Japan. He also wrote a highly successful textbook of quantum mechanics, and two books at a more popular level. He died of cancer, in Tokyo on 8 July 1979.

Further reading: Silvan Schweber, *QED and the Men Who Made It*.

top One of the six *flavours* that distinguish different types of *quark*. Previously referred to as truth.

Townes, Charles Hard (1915–) American physicist who received the Nobel Prize for Physics in 1964 for his work on the physics of *masers* and *lasers*.

Born at Greenville, South Carolina, on 28 July 1915, Townes graduated from Furman University in Greenville in 1935, then moved to Duke University (MA 1937) and Caltech (PhD 1939). From 1939 to 1947 he worked at the Bell Laboratories, then joined the faculty of Columbia University, becoming a full professor in 1950. It was there that he conceived and constructed (in 1953) the first maser, and went on to suggest the possibility of lasers. He moved to MIT in 1961, and on to the University of California, Berkeley, in 1967. Townes had turned his attention to astronomy, where he pioneered the techniques of infrared and microwave astronomy. In 1968, with his colleagues at Berkeley, Townes discovered the first polyatomic molecules (water and ammonia) identified in space. He retired in 1986.

transactional interpretation An interpretation of *quantum physics* developed by John Cramer of the University of Washington, Seattle. It makes exactly the same predictions about the outcomes of experiments as other *quantum interpretations*, but offers a different perspective on what is going on, which many people find easier to live with than, say, the *Copenhagen interpretation* or the *many worlds interpretation*.

The inspiration for Cramer's transactional interpretation is the **Wheeler–Feynman absorber theory** of radiation. Thanks to the existence of both **retarded waves** and **advanced waves** in the solutions to **Maxwell's equations**, this says that if you poke an electron in a laboratory here on Earth, in principle every charged particle in, say, the Andromeda galaxy, more than 2 million light years away, *immediately* knows what has happened, even though any retarded wave produced by poking the electron here on Earth will take more than 2 million years to reach the Andromeda galaxy.

Even supporters of the Wheeler–Feynman absorber theory usually stop short of expressing it that way. The conventional version (if anything about the theory can be said to be conventional) says that our electron here on Earth 'knows where it is' in relation to the charged particles everywhere else, including those in the Andromeda galaxy. But it is at the very heart of the nature of feedback that it works both ways. If *our* electron knows where the Andromeda galaxy is, then for sure the Andromeda galaxy knows where our electron is. The result of the feedback – the result of the fact that our electron has to be considered not in isolation but as part of an holistic electromagnetic web filling the Universe – is that the electron resists our attempts to push it around, because of the influence of all those charged particles in distant galaxies, even though no information-carrying signal can travel between the galaxies faster than light (see **radiation resistance**).

The original version of the Wheeler–Feynman theory was, strictly speaking, a classical theory because it did not take account of quantum processes. In order to apply the absorber theory ideas to quantum mechanics, you need a quantum equation, which, like Maxwell's equations, yields two solutions, one equivalent to a positive energy wave flowing into the future, and the other describing a negative energy wave flowing into the past. At first sight, Schrödinger's famous **wave equation** doesn't fit the bill because it describes a flow in only one direction, which (of course) we interpret as from past to future. But as all physicists learn at university (and most promptly forget), the most widely used version of this equation is incomplete. As the quantum pioneers themselves realized, it does not take account of the requirements of relativity theory. In most cases, this doesn't matter, which is why physics students, and even most practising quantum mechanics, happily use the simple version of the equation. But the full version of the wave equation, making proper allowance for relativistic effects, is much more like Maxwell's equations. In particular, it has two sets of solutions – one corresponding to the familiar simple Schrödinger equation, and the other to a kind of mirror image Schrödinger equation describing the flow of negative energy into the past.

This duality shows up most clearly in the calculation of probabilities in the context of quantum mechanics. The properties of a quantum system are described by a mathematical expression, the **state vector**. In general, this is a **complex number**. The probability calculations needed to work out the chance of finding an electron (say) in a particular place at a particular time actually depend on calculating the square of the state vector corresponding to that particular state of the electron.

But calculating the square of a complex variable does not simply mean multiplying it by itself. Instead, you have to make another variable, a mirror image version called the complex conjugate, by changing the sign in front of the imaginary part –

if it was + it becomes –, and vice versa. The two complex numbers are then multiplied together to give the probability. But for equations that describe how a system changes as time passes, this process of changing the sign of the imaginary part and finding the complex conjugate is equivalent to reversing the direction of time! The basic probability equation, developed by Max Born back in 1926, itself contains an explicit reference to the nature of time, and to the possibility of two kinds of Schrödinger equation, one describing advanced waves and the other representing retarded waves. The remarkable implication is that, ever since 1926, every time a physicist has taken the complex conjugate of the simple Schrödinger equation and combined it with this equation to calculate a quantum probability, he or she has actually been taking account of the advanced wave solution to the equations, and the influence of waves that travel backwards in time, without knowing it. There is no problem at all with the mathematics of Cramer's interpretation of quantum mechanics because the mathematics, right down to Schrödinger's equation, is exactly the same as in the standard Copenhagen interpretation. The difference is, literally, only in the interpretation.

The way Cramer describes a typical quantum 'transaction' is in terms of a particle 'shaking hands' with another particle somewhere else in space and time. You can think of this in terms of an electron emitting electromagnetic radiation which is absorbed by another electron, although the description works just as well for the state vector of a quantum entity which starts out in one state and ends up in another state as a result of an interaction – for example, the state vector of a particle emitted from a source on one side of the experiment with two holes and absorbed by a detector on the other side of the experiment.

One of the difficulties with any such description in ordinary language is how to treat interactions that are going both ways in time simultaneously, and are therefore occurring instantaneously as far as clocks in the everyday world are concerned. Cramer does this by effectively standing outside of time, and using the semantic device of a description in terms of some kind of pseudotime. This is no more than a semantic device – but it certainly helps to get the picture straight.

It works like this. When an electron vibrates, on this picture, it attempts to radiate by producing a field which is a time-symmetric mixture of a retarded wave propagating into the future and an advanced wave propagating into the past. As a first step in getting a picture of what happens, ignore the advanced wave and follow the story of the retarded wave. This heads off into the future until it encounters an electron which can absorb the energy being carried by the field. The process of absorption involves making the electron that is doing the absorbing vibrate, and this vibration produces a new retarded field which exactly cancels out the first retarded field. So in the future of the absorber, the net effect is that there is no retarded field.

But the absorber also produces a negative-energy advanced wave travelling backwards in time to the emitter, down the track of the original retarded wave. At the emitter, this advanced wave is absorbed, making the original electron recoil in such a way that it radiates a second advanced wave back into the past. This 'new' advanced wave exactly cancels out the 'original' advanced wave, so that there is no effective radiation going back in the past before the moment when the original emission occurred. All that is left is a double wave linking the emitter and the absorber, made

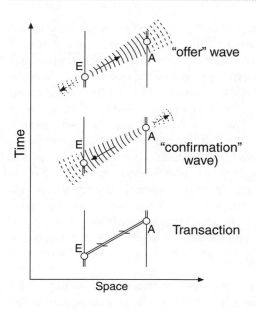

Transactional Interpretation. The way two quantum entities establish an a-temporal 'handshake' in the Transactional Interpretation of quantum mechanics.

up of half of a retarded wave carrying positive energy into the future and half of an advanced wave carrying negative energy into the past (in the direction of negative time); see *Wheeler–Feynman absorber theory*.

Because two negatives make a positive, this advanced wave *adds* to the original retarded wave as if it too were a retarded wave travelling from the emitter to the absorber. In Cramer's words: 'The emitter can be considered to produce an "offer" wave which travels to the absorber. The absorber then returns a "confirmation" wave to the emitter, and the transaction is completed with a "handshake" across spacetime.' But this is only the sequence of events from the point of view of pseudotime. In reality, the process is atemporal; it happens all at once.

The situation is more complicated in three dimensions, but the conclusions are exactly the same.

Cramer is at pains to stress that his interpretation makes no predictions that are different from those of conventional quantum mechanics, and that it is offered as a conceptual model which may help people to think clearly about what is going on in the quantum world; a tool which is likely to be particularly useful in teaching, and which has considerable value in developing intuitions and insights into otherwise mysterious quantum phenomena. But there is no need to feel that the transactional interpretation suffers in comparison with other interpretations in this regard, because none of them is anything other than a conceptual model designed to help our under-standing of quantum phenomena, and all of them make the same predictions.

If there is one particular link in [the] event chain that is special [says Cramer], it is not the one that ends the chain. It is the link at the beginning of the chain when the emitter, having received various confirmation waves from its offer wave, reinforces one of them in such a way that it brings that particular confirmation wave into reality as a completed transaction. The atemporal transaction does not have a 'when' at the end.

This dramatic success in resolving all of the puzzles of quantum physics has been achieved at the cost of accepting just one idea that seems to run counter to common sense – the idea that part of the quantum wave really can travel backwards through time. At first sight, this is in stark disagreement with the commonsense intuition that causes must always precede the events that they cause. But on closer inspection it turns out that the kind of time travel required by the transactional interpretation does not violate the everyday notion of causality after all – and nor does all of this atemporal handshaking across the Universe necessarily remove that most prized of our human attributes, our freedom of will.

In the quantum world, the probability 'offer' wave radiates to all possible partner particles, and many provisional 'acceptances' may be returned, but only one is accepted by the original particle (chosen by the familiar rules of probability) and made real in this way.

It should be no surprise that the way the transactional interpretation deals with time differs from common sense, because the transactional interpretation explicitly includes the effects of **relativity theory**. The Copenhagen interpretation, by contrast, treats time in the classical, 'Newtonian' way, and this is at the heart of the inconsistencies in any attempt to explain the results of experiments like the **Aspect experiment** in terms of the Copenhagen interpretation. If the velocity of light were infinite, the problems would disappear; there would be no difference between the local and non-local descriptions of processes involving **Bell's inequality**, and the ordinary Schrödinger equation would be an accurate description of what is going on – the ordinary Schrödinger equation is, in effect, the correct 'relativistic' equation when the speed of light is infinite. Cramer has actually found a rather subtle link between relativity and quantum mechanics, and this is at the heart of his interpretation.

How does the atemporal handshaking affect the possibility of free will? At first sight, it might seem as if everything is fixed by these communications between the past and the future. Every photon that is emitted already 'knows' when and where it is going to be absorbed; every quantum probability wave, slipping at the speed of light through the slits in the experiment with two holes, already 'knows' what kind of detector is waiting for it on the other side. We are faced with the image of a frozen Universe, in which neither time nor space has any meaning, and everything that ever was or ever will be just *is*.

But in my time frame, decisions are made with genuine free will and no certain knowledge of their outcomes. It takes time (in the macroscopic world) to make the decisions (both human decisions and quantum 'choices' like those involved in the decay of an atom) which make the atemporal reality of the microscopic world.

Further reading: John Gribbin, *Schrödinger's Kittens*.

transistor   See **semiconductors, solid-state devices**.

transition   See **quantum leap**.

transmutation of elements   The alchemists' dream of converting one element into another is now carried out routinely by nuclear physicists. For example, plutonium can be made by bombarding uranium with neutrons. Natural radioactivity also involves transmutation, which was first demonstrated by Ernest Rutherford and Frederick Soddy at the beginning of the 20th century. In a pair of papers published in 1903, Rutherford and Soddy demonstrated that radioactivity involves the change of atoms (strictly speaking, nuclei) of one chemical element into those of another element, by emitting a charged particle – either an electron (beta ray) or an alpha particle (a helium nucleus).

triple alpha process   See *carbon resonance*.

TRISTAN   An electron–positron storage ring facility at Tsukuba, in Japan. It became operational in 1987 and reached energies of around 64 GeV.

tritium   See *hydrogen*.

truth   See *top*.

tunnel effect   See *tunneling*.

tunnelling   A result of the *uncertainty principle* of *quantum physics*, which allows particles to penetrate barriers, such as the barriers around *nuclei*. When two *protons*, for example, approach one another, they are repelled by the positive electric charge that they each carry, which stops them touching. But if they have enough momentum, they can come so close that their quantum wave functions overlap, enabling them to interact (this happens naturally inside stars, as part of the process of nuclear fusion by which stars generate energy and hold themselves up against the inward tug of gravity). It is as if one proton 'tunnelled' through the electric barrier between them. Similarly, the tunnel effect explains how *alpha particles* can escape from nuclei during *alpha decay*, even though they are restrained by the strong inter-action (see *forces of nature*), and the ones that do escape seem not to have enough energy to have overcome this restraint. This explanation of alpha decay was, in fact, the first indication that quantum tunnelling operates in this way, and was proposed by George Gamow at the end of the 1920s.

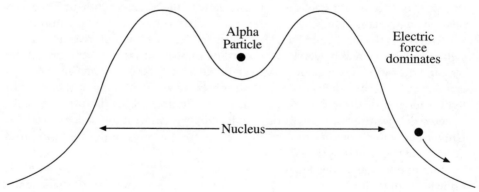

Tunnelling. An alpha particle in a nucleus does not have enough energy to climb out of its potential well. But it can escape because quantum uncertainty means that there is a finite probability that the alpha particle is actually outside the central region of the nucleus, on the slope where the electric force dominates. It only has to be outside for an instant for it to roll away down the hill. In effect, it 'tunnels through' the barrier. This only happens in nuclei where the barrier is low enough and narrow enough for tunnelling to occur.

Tyndall, John (1820–1893)   Irish physicist who was a great popularizer of science (he succeeded Michael Faraday at the Royal Institution in London), and who discovered the *Tyndall effect*, which helps to explain why the sky is blue.

Tyndall was born at Leighlinbridge, Carlow, on 2 August 1820. After attending school in Carlow, he worked as a surveyor and engineer in Ireland and England, working his way up from being a chain carrier to the head of a surveying team on the railways. He was always interested in science, and although largely self-taught, he became a teacher of mathematics and science at Queenswood College, a Quaker school in Hampshire, in 1847.

In 1848, using money he had saved while working on the railways, Tyndall left to study physics at the University of Marburg, in Germany, where he was taught by Robert Bunsen. He was awarded a PhD in 1850, studied briefly in Berlin, returned to Queenswood College in 1851, and became professor of natural philosophy at the Royal Institution in 1853, working with Faraday. When Faraday retired in 1867, Tyndall succeeded him as director (the actual title in those days was superintendent) of the RI.

Tyndall did much to popularize science, not just in Britain but in the United States, where he toured in 1872 and 1873, giving well-attended popular lectures (a kind of scientific Charles Dickens). He also wrote popular books, some based on his lectures.

Tyndall discovered the effect which now bears his name in 1869. He carried out a great deal of interesting but not spectacularly important work, and helped to found the science journal *Nature*. He became seriously ill in 1886, and retired from the Royal Institution the following year. He died at Hindhead, in Surrey, on 4 December 1893, as a result of an accidental overdose of drugs.

Tyndall effect   The scattering of light by tiny particles suspended in a liquid (a 'colloidal solution'). A beam of light will pass right through an absolutely pure liquid (such as a glass of water) without being scattered sideways, but will be scattered if even invisibly small particles are suspended in the liquid. The effect was first studied scientifically by John Tyndall in 1869; it is similar to the way that a 'beam of sunlight' becomes visible when it illuminates dust particles floating in the air (see p.353).

Because shorter wavelengths of light are scattered more easily than longer wavelengths, Tyndall suggested that the blue colour of the sky might be caused by particles suspended in the air, scattering light right around the sky. He was partly right. This does explain why tobacco smoke looks blue, and dust and other material in the air does indeed explain the red of sunset and sunrise in this way, by scattering blue light sideways and leaving red behind (see *spectrum*). But as Albert Einstein showed in the first decade of the 20th century, the blueness of the entire sky is actually caused by scattering not from dust particles, but from the molecules of the atmosphere itself.

UA1   Short for Underground Area 1, the name of an experiment at CERN which monitored proton–antiproton collisions from 1981 onwards, discovered the W and Z particles (1983), and found evidence for the top quark (1984). See *UA2*.

UA2   Short for Underground Area 2, the name of an experiment at CERN (see also *UA1*) which discovered the W and Z particles in 1983.

Uhlenbeck, George Eugene (1900–1988)   Dutch-American physicist who was a member of the first team to introduce the concept of electron spin.

Uhlenbeck was born on 6 December 1900, in Batavia (now Djakarta) in what was then the Dutch East Indies. He took both his first degree and his PhD (awarded in

UA1. The main detector of the UA1 experiment at CERN, which discovered the W and Z particles and the top quark. The people in the photograph give an idea of the scale of the detector. See p.415.

1927) at the University of Leiden. While still a student, in 1925 he and Samuel Goudsmit explained an experiment involving the splitting of a beam of silver atoms into two streams by a magnetic field (the Stern–Gerlach experiment) by suggesting that the silver atoms possess an intrinsic spin ('up' or 'down'). They suggested that this is because electrons have spin, and since there are an odd number of electrons in a silver atom (47), all but one of the electron spins cancel each other out (23 up and 23 down), leaving one unit of spin (up for half the atoms and down for the other half) to provide a 'handle' by which the magnetic field could get a grip on the silver atoms.

This key insight explains, among other things, how two electrons can occupy the same energy level in an atom, even though they are *fermions* – they are distinguished by their spin. Spin was later put on a secure theoretical basis by the work of Paul Dirac. Many physicists were baffled that this important discovery was never rewarded with the Nobel Prize.

After receiving his PhD, Uhlenbeck went to the United States, where he worked at the University of Michigan from 1927 to 1935. He then returned to Holland and became professor of physics at the State University of Utrecht, where he worked from 1935 to 1939, before finally emigrating to the United States, where he worked once again at the University of Michigan, becoming professor of physics in 1939. He moved on in 1960 to the Rockefeller Medical Research Center at the State University of New York, where he stayed until he retired in 1974, when he was appointed professor emeritus. He died in Boulder, Colorado, on 31 October 1988.

UKAEA Acronym for United Kingdom Atomic Energy Authority, set up in 1954. It is responsible for research into and development of nuclear power in the UK.

Ulam, Stanislaw Marcin (1909–1985) Polish-born American mathematician who designed the first hydrogen bomb. Ulam was born in Lwow (then in the Polish part of the Austrian Empire, now Lvov in the Ukraine) on 13 April 1909. He was educated locally and received his doctorate from the Polytechnic Institute in Lwow in 1933.

He emigrated to the United States in 1936, joining the staff at the Institute for Advanced Study in Princeton. He moved to Harvard University in 1939, and was on the staff of the University of Wisconsin, Madison, from 1941 to 1943. He then became an American citizen and joined the **Manhattan Project**, staying on at Los Alamos until 1965. He worked with Edward Teller on the hydrogen bomb project, providing the scientific input while Teller concentrated on the politics.

After 1965 Ulam held teaching posts at several American universities. He died in Santa Fe on 13 May 1985.

ultraviolet catastrophe   See *black body radiation*.

ultraviolet radiation   *Electromagnetic radiation* with wavelengths shorter than those of visible light, beyond the blue/violet end of the visible *spectrum*, but longer than those of *X-rays*. Covers the band of wavelengths from 10 to 320 nanometres.

uncertainty   In quantum physics, uncertainty is a precise and definite thing. There are pairs of parameters, known as conjugate variables, for which it is impossible to have a precisely determined value of each member of the pair at the same time. The most important of these uncertain pairs are position/momentum and energy/time.

The position/momentum uncertainty is the archetypal example, first described by Werner Heisenberg in 1927. It means that no entity can have both a precisely determined *momentum* (which essentially means velocity) and a precisely determined position at the same time. This is not the result of the deficiencies of our measuring apparatus – it is not just that we cannot measure both the position and momentum of, say, an electron at the same time, but that an electron *does not have* both a precise position and a precise momentum at the same time. At any instant, the electron itself cannot know both where it is and where it is going. (Some reference books still tell you that quantum uncertainty is solely a result of the difficulty of measuring position and momentum at the same time; do not believe them!)

The uncertainty in position multiplied by the uncertainty in momentum is always greater than the parameter $\hbar$, *Planck's constant* divided by 2π. So although you can (in principle) get as near to this limit as you like, the more precisely one parameter is determined, the less accurately the other one is constrained. This is related to the basic *wave–particle duality* of the quantum world. A particle (in the everyday sense of the word) is capable of being precisely located at a point, but a wave is not. Alternatively, you can think of it as another manifestation of the probabilistic nature of the quantum world – where everything is governed by the rules of probability, nothing is certain. What is special about quantum uncertainty is the way you can trade off one kind of uncertainty against another (provided you choose the right pairs to trade).

For example, when an electron is passing through an experimental apparatus, such as the *double-slit experiment*, there is considerable uncertainty about its position, but less uncertainty about the direction it is going in; when it is detected after passing through the experiment, there is very little uncertainty about its position, but considerable uncertainty about where it has come from (which slit it has passed through).

The momentum/position uncertainty is important in quantum *tunnelling*; the energy/time uncertainty is important in allowing the existence of *virtual particles*. But quantum uncertainty does not noticeably affect large objects (objects more massive than molecules) because Planck's constant is so small – 6.6 x 10^{-34} joule seconds. Uncertainty is very important for electrons, which have a mass of 10^{-26} g. The amount of uncertainty in the position of an object caused by its wave nature is inversely proportional to its mass, so for everyday objects the uncertainty is absolutely tiny, although in principle it is still there.

See also *wave packet*.

uncertainty principle   Werner Heisenberg's formal statement that the amount of quantum *uncertainty* in the simultaneous determination of both members of a pair of conjugate variables is never zero. Nobody ever specified a form of words that spells out a formal definition of the principle, but we like the way Heisenberg himself put it at the end of the paper he published in 1927 setting out the nature of quantum uncertainty: 'We *cannot* know, as a matter of principle, the present in all its details.'

undivided whole   See *Bohm, David Joseph*, and *hidden variables*.

unified field theory   Another name for *theory of everything*. A tantalizing will o' the wisp idea that drives physicists insane.

up   One of the six *flavours* that distinguish different types of *quark*.

upsilon   Very massive (9.46 GeV) short-lived (10^{-20} sec) *meson*, with zero charge and spin 1, made up of a *bottom*–antibottom *quark* pair. It was discovered in 1977 at Fermilab.

vacuum   In quantum physics, the vacuum is not nothing at all, but seethes with activity. Thanks to quantum *uncertainty*, *virtual particles* are constantly being produced and disappearing. A good way to think of the vacuum is as a *superposition of states* for many different kinds of *field*. To keep things simple, think first of just one kind of field, the electromagnetic field. The different states of the field are like the different notes that can be obtained from a single plucked guitar string. Like the *energy levels* of an electron in an *atom*, they form a kind of energy staircase, with the steps spaced out at distances corresponding to the energy of a single photon. When an atom emits a photon, the energy of the corresponding frequency of the vacuum field is increased by one unit, matching the decrease in energy of the atom. The temporary appearance of a virtual photon corresponds to the field energy moving up a step all by itself, and then falling back again, like a guitar that plays random notes all by itself.

In addition to the electromagnetic field, in all its states, you have to imagine also a field for electrons, one for protons, and so on for every kind of particle and interaction there is. And each of these vacuum fields exists in a superposition of many states. All of these fields fill the Universe and together form the vacuum, making it a very many-stringed instrument. The vacuum is the lowest-energy state of spacetime filled with these fields.

See also *Casimir effect*.

vacuum fluctuation   The temporary appearance of *virtual particles* out of the *vacuum*.

vacuum polarization   Change in the structure of the *vacuum* caused by the presence of electric charges or conductors. The vacuum is a sea of virtual particle–antiparticle pairs, but for simplicity think only of electron–positron pairs popping in and out of existence as *virtual particles*. Near a real electron, there will be a tendency, even during the brief lifetime of each electron–positron pair, for the virtual positron to move towards the electron and the virtual electron to be repelled from it. This results in the shielding that gives the electron its measured charge, rather than the infinitely large *bare charge* predicted without quantum fluctuations of the vacuum (see *renormalization*). Similar effects involving other interactions can also alter the structure of the vacuum.

It is important to emphasize that these effects, bizarre though they seem in terms of the everyday world, have indeed been observed. An extension of the experiments which demonstrate the *Casimir effect* provides more evidence that the structure of 'empty space' is affected by what is in the space.

When two conducting plates are placed parallel and extremely close together in a vacuum, they are pulled towards one another by the Casimir effect. In experiments carried out at Yale University in the early 1990s, a similar experiment was set up, but with the conducting plates (tiny glass plates coated with gold) wedged together to make a 'V'. The V was just a few millionths of a metre wide at the top, and atoms of sodium were sent through the gap of the V at different heights. The actual separation of the plates was measured at each height to an accuracy of 5 billionths of a metre, using *interference* fringes produced by pure monochromatic light. So the experimenters knew exactly how closely the atoms were passing the plates, and could calculate how much they should be influenced by the structure imposed on the vacuum by the presence of the conductors. As the atoms emerged on the other side of the V, they were monitored by having laser beams bounced off them.

The observed behaviour of the atoms exactly matched the predictions of quantum theory, after allowing for the way the structure of the vacuum was altered by the presence of the V-shaped conducting plates, and did not match the predictions of classical theory, with a structureless vacuum. 'Nothing at all' could be seen to have a measurable effect on the individual sodium atoms.

valence quarks   The constituent 'real' *quarks* that make up a *baryon* (three quarks each) or a *meson* (two quarks each). This excludes all of the *virtual particles* that must also be present inside the baryons and mesons to explain the way they behave when they are probed in scattering experiments. The *parton model* explains the observed results of these scattering experiments by saying that each *nucleon* (for example) is like a box filled with partons, including not only the three (in this case) valence quarks, but also a sea of virtual quark–antiquark pairs and *gluons*.

valency   A measure of the ability of *atoms* of a particular *element* to link up with other atoms. It is measured in terms of the number of hydrogen atoms that can combine with a single atom of the element in question. So oxygen, for example has a valency of 2 because it combines with hydrogen to make H_2O, while carbon has a valency of 4 because it combines with hydrogen to make CH_4. This is consistent with the way carbon and oxygen combine to form CO_2, although things are not always that simple and compounds which do not make full use of the valency (such as CO) can also be formed in some circumstances. The valency is related to the number of electrons available in the outermost *shell* of an atom to form *chemical bonds* with other atoms. Sometimes referred to as valence.

van de Graaf generator   A simple, rugged machine for producing large voltage differences, invented by the American physicist Robert van de Graaf (1901–67) at the end of the 1920s. Small versions of these machines, with a characteristic dome-shaped top, are often found in schools; much larger versions are still used in research

Van de Graaf generator. The build-up of electric charge on a metal sphere, produced by a van de Graaf generator, provides a hair-raising experience. If you stand on a well-insulated platform and touch the charged ball, the 'like charges' that accumulate on your body repel the hairs from one another.

laboratories (as well as in medical and industrial applications) to provide the initial potential difference used to accelerate charged particles, such as electrons, into a beam before they are injected into the giant machines where they are raised to very high energies.

van der Meer, Simon (1925–) Dutch physicist and engineer who worked as senior engineer at CERN and developed a technique known as stochastic cooling, which made it possible to produce more intense beams of antiprotons in particle accelerators. This was a key contribution to the experiments at CERN which led to the discovery of the W and Z particles; in 1984 van der Meer shared the Nobel Prize for Physics with Carlo Rubbia for this work.

Born in The Hague, on 24 November 1925, van der Meer graduated in engineering from the Technische Hogeschool, in Delft, in 1952, and worked with Philips, in Eindhoven, from 1952 to 1955. He then joined CERN, where he stayed. He was awarded honorary doctorates by the University of Geneva (in 1983) and the University of Amsterdam (in 1984).

van der Waals, Johannes Diderik (1837–1923) Dutch physicist who derived a widely used equation describing the behaviour of gases – an equation of state.

Born at Leiden on 23 November 1837, van der Waals worked as a primary school teacher before gaining admission to the University of Leiden in 1862 and graduating in 1865. He then worked as a secondary school teacher, becoming headmaster of a school in The Hague in 1866, but continued his studies of physics part time, and submitted a thesis for his doctorate to the University of Leiden in 1873, when he was 35. This work, on the nature of gases and liquids, made his name as a scientist, and in 1877 he became a professor of physics at the University of Amsterdam. Most of his subsequent work built from the ideas in his thesis, including the equation for which he is best remembered. One of the key ingredients in this was an allowance for *van der Waals forces*, an idea which later provided a useful analogy for the way the *colour* force operating between *quarks* gives rise to the strong interaction (see *forces of nature*) operating between *nucleons*.

Van der Waals retired in 1907 (being succeeded as professor of physics at Amsterdam by his son) and received the Nobel Prize for his work in 1910. After 1912 he did very little work because of ill health; he died in Amsterdam on 8 March 1923.

van der Waals forces   Forces of attraction that operate between atoms and molecules because of the way electric charge in the atoms and molecules is distributed over a finite volume, instead of being concentrated at a point. They are named after the Dutch physicist Johannes van der Waals.

The best way to picture this is by thinking of the electron cloud around an atom or molecule being attracted by the positively charged nucleus of another atom, and vice versa. This happens because half of the electron cloud belonging to the other atom is on the other side of the atom, so the positively charged nucleus is only partially shielded from view. Similarly, the two electron clouds will repel each other slightly, as do the two positively charged nuclear regions, but this is not sufficient to overcome the attractive force until the molecules or atoms get very close together. The result is that atoms and molecules bouncing off one another in a liquid or gas (or off the walls of the container) are slightly more 'sticky' than they would otherwise be, which affects the way the fluids behave (see diagram on p.422). The effect was explained in quantum-mechanical terms by Fritz London, in the 1930s.

The point of closest approach at which the repulsive and attractive van der Waals forces exactly cancel out is, in effect, the radius of an atom – its van der Waals radius. In an analogous way, the strong force (see *forces of nature*) operating between *nucleons* can be regarded as a leakage of the colour force operating between *quarks* inside the nucleons, and the radius of a nucleon can be thought of as the equivalent of a van der Waals radius for the bag of quarks making up a nucleon.

van der Waals radius   See *van der Waals forces*.

van Vleck, John Hasbrouck (1899–1980)   American physicist who received a share of the 1977 Nobel Prize for Physics for his work on magnetic theory.

Born in Middletown, Connecticut, on 13 March 1899, van Vleck graduated from the University of Wisconsin in 1920, and received his master's degree (1921) and PhD (1922) from Harvard. In 1923 van Vleck moved to the University of Minnesota, where he became a full professor in 1927. He was professor of physics at the University of Wisconsin from 1928 to 1234, then returned to Harvard, where he spent the rest of his career (apart from visits to Leiden and Oxford), retiring in 1969. He worked on radar during the Second World War. The work for which he received the Nobel Prize was largely carried out in the 1930s, and used quantum theory to explain the magnetic, electrical and optical properties of many elements in terms of the way electrons in one atom are influenced by other nearby atoms (this work later

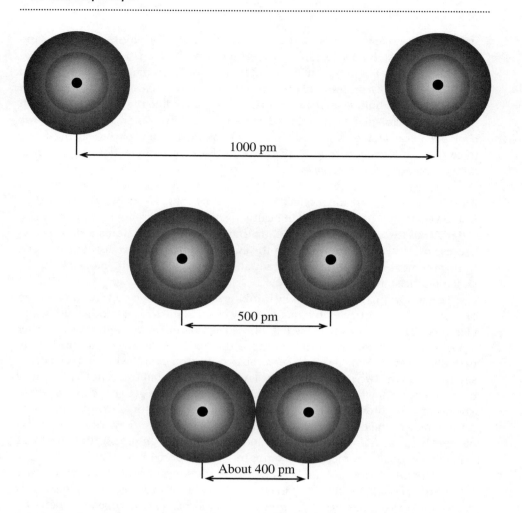

Van der Waals forces. When two atoms are relatively far apart, the positive charge on each nucleus is only partly screened by the part of the electron cloud on the side nearest the other atom, and is attracted to the other atom's electron cloud. At about the point where the electron clouds touch, the two positively charged nuclei become aware of each other and are repelled from one another. This defines the van der Waals' radius of the atom. The electron clouds also contribute to this repulsion. All things are made of atoms, little entities that move around constantly, attracting each other in this way when they are a little distance apart, but repelling each other if they are squeezed together.

turned out to be important in the development of the *laser*). When asked if he could explain his theory of magnetism to a non-scientist, van Vleck replied, 'No.'

Van Vleck stayed in Cambridge, Massachusetts, after he retired from Harvard and died there on 27 October 1980.

variational principle Any mathematical principle which says that some particular integral is a minimum or a maximum. The most relevant example in quantum physics is the principle of least *action*; a specific example of this principle at work is that light travels in straight lines, minimizing the total (that is, integral) amount of time it takes for a journey between two points (this is known as the principle of least time). The basics of variational analysis go back to the ancient Greek interest in geometry. Pierre Fermat stated the principle of least time for light in the 17th century, and a little later Pierre de Maupertuis proposed the principle of least action in the context of mechanics. In the mid-18th century Leonard Euler developed the first general rules for this kind of analysis, and it was developed further by Joseph Lagrange and, notably, William Hamilton. Variational principles are important in the formulation of the general theory of relativity, and were used by Louis de Broglie and Erwin Schrödinger in the development of *wave mechanics*.

vector   A quantity which has both magnitude and direction. The speed of, say, a car tells you how fast it is going; the velocity tells you both how fast it is going and which direction it is heading in – for example, 60 miles per hour heading north.

vector boson   A particle associated with a *vector field*. Because of the structure of a vector field, all such vector particles have spin 1 and are therefore *bosons*. They are the particles that carry the *forces of nature* – the *photon*, the *W* and *Z particles*, and the *gluons*. The *graviton* is probably an even more complicated tensor particle, which has spin 2.

vector field   A *field* which has a built-in sense of direction. The electric field is the archetypal example of a vector field. A tiny charged particle placed in an electric field will move in a certain direction (if the particle happens to have negative charge, it will move away from other negative charges and towards positive charge), accelerating at a rate which depends on the strength of the field. Even if the field is perfectly uniform, the tiny charged particle will still move along a *line of force*. See also *scalar field*.

vector particle   See *vector boson*.

velocity   See *vector*.

Veltman, Martin   See *Hooft, Gerard 't*.

virtual pairs   Particle/antiparticle pairs, such as an *electron* and a *positron*, produced out of the *vacuum* by *quantum fluctuations*.

virtual particles   There are two kinds of virtual particle.

One kind are particles produced out of the *vacuum* by *quantum fluctuations*. Many such particles can be produced only in *virtual pairs* in order to preserve the existing balance of properties such as electric charge in the Universe. But particles which are not constrained by these *conservation laws*, notably *photons*, can be produced without any mirror image counterpart.

The particles which join the vertices in a *Feynman diagram* are also virtual particles and can never be detected directly, even though they are of key importance in determining the way 'real' particles interact. This kind of virtual particle does not have to worry about the conservation laws, which are obeyed overall during the interaction. For example, an *electron* can interact with a *neutrino* by emitting a virtual W⁻ particle which is absorbed by the neutrino. The effect is that the original electron becomes a neutrino, and the original neutrino becomes an electron, so charge and other properties are conserved overall; but the virtual W⁻ does not have to be accompanied by an equivalent antiparticle on its journey (indeed, if it were, the interaction would leave the original particles unchanged).

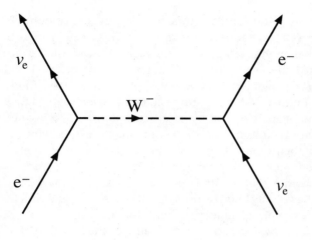

Virtual particles. A quantum interaction can be thought of as occurring when one of the virtual particles that 'belongs' with one particle is transferred to another particle. In this case, an electron and a neutrino have come close enough for a W⁻ in the cloud of virtual particles around the electron to be transferred to the cloud around the neutrino. From a distance, it looks as if the electron has turned into a neutrino and the neutrino has turned into an electron. At a more subtle level, the interaction can be seen as involving a W⁻ from the electron annihilating with a W⁺ from the neutrino.

The photons which are exchanged between charged particles during electromagnetic interactions (for example, if two electrons approach each other and are repelled because they each carry negative charge) are virtual photons.

All quantum particles can be thought of as being surrounded by a cloud of virtual particles (and pairs) of various kinds, which are being emitted and (usually) reabsorbed by the parent particle. The lifetime of each of these virtual particles (and therefore the distance it can travel from its parent particle) depends on its mass-energy and the leeway allowed by *uncertainty*; the more massive and energetic a particle is, the shorter its lifetime and the shorter its range. Interactions occur when another particle comes close enough for one or more of the virtual particles in the cloud to be absorbed by the other particle instead.

This central idea of quantum interactions occuring by the *exchange* of particles can be pictured in another way (as can most quantum processes; see *models*), this time by imagining two particles, A and B, each surrounded by a cloud of virtual particles which includes pairs of W⁻ and W⁺ particles. If the two clouds overlap, a W⁻ from the cloud around particle A may annihilate with a W⁺ from the cloud around particle B. This leaves each of the clouds with a spare particle – an unpaired W⁺ with particle A and an unpaired W⁻ with particle B. These left-over particles have to be absorbed by their parents, with the same effect as if particle A has gained a W⁺ and particle B has lost a W⁺, or as if particle A has lost a W⁻ and particle B has gained a W⁻ (gaining a W⁺ is exactly the same as losing a W⁻, and vice versa). But there is no need to think of an actual particle being emitted from either particle A or particle B and travelling across to its counterpart; this is simply a useful device which makes the Feynman diagrams easy to work with.

virtual photon   See *virtual particle*.

virtual processes   Quantum processes that go on below the threshold of detectability, but which influence the observed properties of quantum entities. The archetypal example comes from *quantum electrodynamics*, where a charged particle (such as an *electron*) is envisaged as surrounded by a cloud of *virtual photons* which it is constantly emitting and reabsorbing. It is only by allowing for the influence of these virtual processes that it is possible to calculate a value for the *magnetic moment* of the electron which precisely matches the experimentally determined value. When the virtual processes are allowed for, theory and experiment agree to an accuracy better than 1 part in 10 billion.

virtual quantum = *virtual particle*.

von Fraunhofer See *Fraunhofer, Josef von*.

von Klitzing See *Klitzing, Klaus von*.

von Laue See *Laue, Max Theodor Felix von*.

von Neumann See *Neumann, John von*.

Walton, Ernest Thomas Sinton (1903–1995) Irish physicist who shared the Nobel Prize for Physics with John Cockcroft in 1951 for their joint development of the first 'atom smasher' (see *accelerator*, *Cockcroft–Walton machine*) in 1932.

Born in Dungarvan, County Waterford, on 6 October 1903, Walton studied at Trinity College, Dublin, where he was awarded his first degree in 1926 (with double first-class honours in experimental science and mathematics). Although he moved to Cambridge in 1927, obtaining his PhD in 1931 and working at the Cavendish Laboratory until 1934, Walton also received an MSc (in 1928) and MA (in 1934) from Trinity. He became a Fellow of Trinity College, Dublin, in 1934, and professor of

Ernest Walton (1903–1995).

natural and experimental philosophy there in 1946. His work with Cockcroft at the Cavendish was partly inspired by George Gamow, who was in Cambridge in 1929 and calculated the probability of 'splitting the atom' by bombarding *nuclei* with *protons.*

Walton retired in 1974, and died, in Belfast, on 25 June 1995. His son Alan also became a physicist.

wave   An oscillating disturbance that moves through a medium, or through space. The familiar picture of waves on the sea or ripples on a pond is essentially the image that physicists have of the kind of waves that occur in quantum physics.

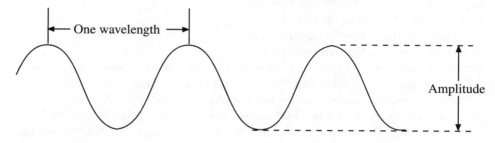

Wave. The basic properties of a wave.

wave equation   Any equation that describes mathematically the way that a wave propagates. In quantum mechanics, the term is used specifically to refer to the wave equation discovered by Erwin Schrödinger, which describes the wave nature of a quantum entity such as an electron. The Schrödinger wave equation is also known as the wave function.

wave function   The mathematical description of a quantum system in terms of waves. This can be very simple, as in the case of a Schrödinger equation that describes a single electron, or very complicated (for the ultimate complication, see *wave function of the Universe*). See also *wave equation.*

wave function of the Universe   In principle, it would be possible to describe the entire Universe by a mathematical expression based on the Schrödinger wave equation; this would be the wave function of the Universe. Because there is nothing 'outside' the Universe which can 'observe' the Universe and trigger a collapse of the wave function (see *Copenhagen interpretation*), this can be understood only in terms of a *sum over histories* or *many worlds interpretation*. The complicated wave equation that describes such a situation is known as the Wheeler–De Witt equation. Every solution to this equation describes a different possible universe, and it is impossible to solve the equation to provide a description of our particular Universe. But cosmologists have studied the general nature of the Wheeler–De Witt equation in order to work out what the general properties of universes must be, and to try to use these properties to tell us something about the nature of our Universe. Stephen Hawking and James Hartle, in particular, have claimed that the natural boundary condition that must be applied to any universe described by the Wheeler–De Witt equation (including our Universe) is that it should have no boundary – that is, that spacetime must be closed in the four-dimensional equivalent of the two-dimensional surface of a sphere, being finite in size but having no edges. This is still speculation – albeit informed speculation – but represents the most remarkable marriage of quantum physics with cosmology.

wavelength   The distance from peak to peak (or trough to trough) in a *wave*.

wave mechanics   The version of quantum physics that was developed initially by Erwin Schrödinger in 1926, and that was almost immediately established as the preferred framework in which to solve problems involving quantum interactions, largely because of the fact that physicists were already familiar with the language of *wave equations* in which wave mechanics was couched. For the same historical reasons (plus the weight of tradition that has grown up since the 1920s), this is still the way most students are introduced to quantum physics, which is unfortunate because the *sum over histories* approach provides much better insight and leads more naturally to more advanced work in the subject.

The idea behind wave mechanics came from the work of Louis de Broglie, via Albert Einstein. De Broglie's concept of electron waves had appeared on the scene in 1925, but this was still before the crucial experiments confirming the wave nature of the electron had been carried out (see *Davisson, Clinton Joseph* and *Thomson, George Paget*). It was in one of Einstein's papers, published in February 1925, that Schrödinger read Einstein's comment on de Broglie's work: 'I believe that it involves more than merely an analogy.' This encouraged him to take up the idea and develop it further.

De Broglie pointed the way to wave mechanics with his idea that electron waves 'in orbit' around an atomic nucleus had to fit a whole number of wavelengths into each orbit, so that the wave neatly bit its own tail, like the worm Ouroboros. This could happen only for whole numbers of wavelengths, but for different whole numbers for different wavelengths, which could be thought of as being related to one another like the different harmonics that can be played on a single violin string.

Schrödinger used the mathematics of waves to calculate the atomic energy levels allowed in such a situation. At first, he was disappointed because his results did not match the energy levels calculated from *spectroscopy*. In fact, there was nothing wrong with his calculations as such, but he had not taken account of the quantum *spin* of the electron – which is not surprising, since at that time, early in 1925, the idea of electron spin had not yet been developed, although Wolfgang Pauli was about to come up with it. So Schrödinger put the work to one side for several months. He came back to this work when he was asked to give a talk explaining de Broglie's ideas, and then he found that if he left out relativistic effects from his equations he could get a good agreement with spectroscopy in situations where relativistic effects were known to be unimportant. Paul Dirac later showed that electron spin is essentially a relativistic property, and once the idea of spin was established the full version of the Schrödinger equation including spin was found to match up extremely well with spectroscopic studies of atoms. So Schrödinger's great work was actually published in 1926, hot on the heels of the key papers on *matrix mechanics*, although he had begun work on wave mechanics before Werner Heisenberg had his insight that led to the development of matrix mechanics.

The equations in Schrödinger's variation on the quantum theme are wave equations, members of the same family of equations that describe waves in the everyday world – ripples on a pond, or sound waves moving through the air, or whatever. This is why the world of physics greeted them with such enthusiasm. But the two approaches to quantum mechanics discovered in the mid-1920s could not have been more different philosophically. Heisenberg deliberately discarded any physical picture of the atom and dealt only in quantities that could be measured by experiment; at the heart of his theory, though, was the idea that electrons are particles. Schrödinger started out from the idea that the atom is a 'real' entity; at the heart of his theory was the idea that electrons are waves. But both approaches produced sets of equations that exactly described the behaviour of things that could be measured in the quantum world.

Schrödinger himself, the American Carl Eckart, and Dirac all soon independently showed that the different sets of equations were, in fact, mathematically equivalent to one another, different views of the same mathematical world, like the two sides of a coin. But because of the cosy familiarity of wave equations and their appearance in so many aspects of physics (from optics to hydrodynamics, **Maxwell's equations** of electromagnetism, and much more), the impression still persists among many people who ought to know better that there is something more fundamental about wave mechanics than any of the other formulations of quantum mechanics.

This is particularly ironic, since the one person who was deeply troubled by the discovery that wave mechanics is exactly equivalent to the other formulations was Schrödinger himself. He had deliberately set out to restore sanity (as he saw it) to quantum physics, and to get rid of such puzzles as instantaneous **quantum leaps**. But all the puzzles he sought to avoid were still there, even in the wave mechanics version of quantum theory. You may, for example, have noticed that wave mechanics works properly only if allowance is made for electron spin, at which point you should have asked, 'But how can a wave spin?'

wave packet   A train of *waves* that is confined within a small region of space. Most waves that we encounter in the everyday world are spread-out things – like ripples on a pond, for example, which spread out over a long distance so that it is hard to see where the string of ripples (the wave train) begins and ends. But in order to reconcile the idea that a quantum entity such as an electron can behave either as a wave or as a particle (see **wave–particle duality**), in 1927 Werner Heisenberg came up with the idea of treating quantum entities as wave packets, and linked this with the idea of quantum *uncertainty*.

The appropriate image is of a little package of waves, a short wave train which extends over only a small distance, roughly equivalent to the size of the equivalent particle. There is no difficulty about constructing such wave packets, and the mathematics and physics involved are very well known. The essential requirement to

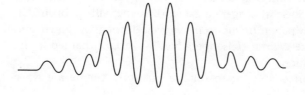

Wave packet. By adding many waves with different wavelengths together, you can make a localized wave packet as small as you like.

make a wave packet that is located in a small space is to use many waves of different wavelengths, chosen so that they interfere with one another in such a way that they cancel out everywhere outside the wave packet, and reinforce one another inside the wave packet. The smaller the wave packet is, the more variety of waves you need to do the job. Conversely, if you use a few waves with similar wavelengths, you get a very large, spread-out wave packet. This has nothing to do with quantum effects as yet; it happens in the everyday world, and can be used to make wave packets in the form of ripples on a pond and the like.

But the spread in wavelengths required to localize a wave packet does have implications for quantum physics because the wavelength is related to momentum. A pure wave with a single wavelength has a definite momentum, but a mixture of waves with different wavelengths has an uncertain momentum. The more accurately you want to locate a wave packet (the smaller you want to make it), the more different waves you need, and the more uncertainty there is in the momentum. The more precisely you want to specify the momentum, the fewer waves you must use, and the more spread-out the wave packet becomes, increasing the uncertainty in its position.

This was, in fact, the route which led Heisenberg to the uncertainty principle which now bears his name. But don't run away with the idea that an electron *is* a wave packet. How can a wave packet have electric charge, or spin? As in all descriptions of the quantum world, it is merely an analogy, a crutch for our inadequate imaginations.

wave–particle duality   The idea that quantum entities may behave either as waves or as particles, depending on the circumstances. This does not mean that the entities are waves, or that they are particles; we have no way of knowing what they are, and can build up a picture of what is going on in a particular experiment only by making analogies with things in the everyday world (see *models*). We can say what quantum entities are *like*, not what they *are*.

The two classic examples, which forced physicists to develop this concept in the context of quantum theory, are light and electrons. At the beginning of the 20th century, it was firmly established from experiment that light is a form of electromagnetic wave, and that electrons are tiny particles. Within 30 years, there was equally convincing experimental evidence that light is made up of particles (called photons) and that electrons are waves. It became appreciated that this wave–particle duality is related to the inherent *uncertainty* in the quantum world – a particle is usually regarded as having a precise location in space, but a wave is intrinsically a spread-out thing, and the uncertainty in the position of, say, an electron can be linked to its wave nature. This is a particularly useful analogy, since the ideas of uncertainty and wave–particle duality can be related mathematically (see *wave packet*).

There is nothing that can be done to make you feel happier if you find the idea of wave–particle duality distasteful. That is just the way the world is, and the only comfort is that because *Planck's constant* is so small the effect is important only on a small scale, for atoms and smaller entities. But it is *very* important on the quantum scale, and this has recently been underlined by an experiment that would have impressed (and perhaps baffled) the quantum pioneers. The one crumb of comfort that Niels Bohr and the Copenhagen school of thought clung to in the face of wave–particle duality was that a quantum entity could never be seen behaving both as a particle and as a wave at the same time. But even that security blanket has now been removed.

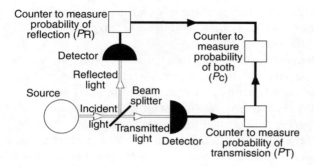

Wave-particle duality 1. If light comes in the form of particles, each photon must go either one way or the other around this experiment. Half of them are reflected at the beam splitter, and half of them are transmitted, but always as a whole photon. The final counter should never record the arrival of two half-photons, one by each route, simultaneously. This is exactly what researchers find; it shows that light behaves as a stream of particles.

At the end of the 1980s, three Indian scientists came up with a new suggestion for an experiment which could show single photons behaving both as particles and as waves at the same time. Dipankar Home, Partha Ghose and Girish Agarwal suggested sending single photons one at a time through an experiment in which the photons had a choice of paths to take. The 'crossroads' was provided by a kind of beam splitter, made of two right-angle triangular prisms almost touching one another.

The prisms are simple triangular blocks of transparent material, with one of the corners of the triangle making a right angle. With a single prism of this kind, when a beam of light comes in perpendicular to one of the 'square' faces, so that it hits the 'hypotenuse' face at 45 degrees on the inside, it is totally reflected, through a right angle, and comes out of the other 'square' face. If the two prisms are touching one

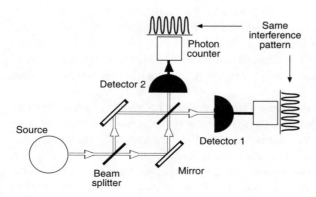

Wave-particle duality 2. If you change the experiment to allow the light travelling by each path to meet up and recombine, you get an interference pattern. This shows that light behaves as a wave.

Wave-particle duality 3. The latest twist on the theme uses a different kind of beam splitter. Only waves can tunnel across the gap between the prisms. But still you find that the final coincidence counter never records the simultaneous arrival of a half photon from each route. The same photons are caught behaving as particle and wave at the same time.

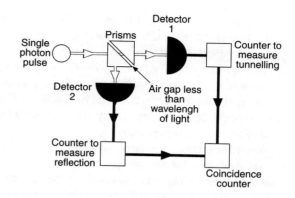

another, hypotenuse to hypotenuse, to make a square block, light coming in perpendicular to one of the faces goes straight through the block and out the other side without being reflected at all. But if there is a tiny gap between the two hypotenuse faces, some of the light is reflected and some of it tunnels across the gap and carries on in a straight line (see *tunnel effect*).

The gap has to be really small for the trick to work – smaller than the wavelength of the light involved. In effect, if the gap is smaller than one wavelength, some of the light can reach out across the gap without noticing it is there, and proceed on its way. As usual in the quantum world, probabilities and statistics come into the story. The smaller the gap, the bigger the proportion of the light that tunnels; so for a precisely set gap and a particular wavelength of light, exactly half the light will be transmitted and half will be reflected. The essential point, though, is that only waves can tunnel in this way. Classical particles cannot tunnel.

This experiment was actually carried out, using pure single photons, by Yutaka Mizobuchi and Yoshiyuki Ohtaké, of Hamamatsu Photonics. Some idea of the subtlety of their experiment can be gleaned from the fact that the size of the gap had to be controlled to within a few ten-billionths of a metre, about one-tenth of the wavelength of the light involved. Detectors were placed in the positions where the two beams of light, corresponding to reflection or transmission, should emerge from the prisms. These detectors record the precise time of arrival of individual photons. Single photons cannot be split in half, so there is a precise fifty-fifty chance that an individual photon arriving at the air gap will be reflected or transmitted. So if the two counters 'click' only in anticoincidence (that is, never at the same time as each other), that will be proof that light is travelling through the experiment in the form of photons.

But any photons that follow the straight-through path in the experiment can have done so only by tunnelling – in other words, by behaving as waves. When the experiment was carried out, the researchers did indeed find half the photons in each channel, confirming that they were behaving like waves at the gap and tunnelling. They also found that the detectors clicked in perfect anticoincidence, confirming that the photons were behaving as particles at the gap and not splitting in half. The experiment observes the same photons acting as both wave and particle at the same time (when confronted by the air gap), contradicting Niels Bohr's basic tenet of *com-*

431

plementarity. 'Three centuries after Newton,' says Home, 'we have to admit that we still cannot answer the question "what is light?".'

In 1951 Albert Einstein remarked, in a letter to his old friend Michelangelo Besso: 'All these fifty years of conscious brooding have brought me no nearer to the answer to the question "what are light quanta?" Nowadays every Tom, Dick and Harry thinks he knows it, but he is mistaken.' That still holds true as we approach the end of the 20th century, nearly 100 years after Einstein started brooding about light.

wave train   See *wave packet*.

weak interaction   See *forces of nature*.

weak isospin   See *isotopic spin*.

weakly interacting massive particle   See *WIMP*.

Weber, Wilhelm Eduard (1804–1891)   German physicist who made pioneering investigations of electrodynamics.

Born at Wittenberg on 24 October 1804, Weber studied at Halle, where he obtained his PhD in 1826 for work on the acoustics of organ pipes, and became a lecturer there. He was promoted to assistant professor in 1828, moving on to become professor of physics in Göttingen in 1831. It was there that Weber's interest in electricity and magnetism was fired by Karl Gauss, but he was among a group of academics sacked in 1837 for objecting to interference by the King of Hanover in the State Constitution. After several years in the academic wilderness, but still based in the city of Göttingen and carrying out research unofficially, he became professor of physics in Leipzig in 1843, and was reinstated as a professor in Göttingen in 1849, where he worked again with Gauss. His most important work was undertaken there in the 1850s.

Weber carried out experiments which determined the ratio of two key quantities, one a measure of electrostatic behaviour and the other a measure of electromagnetic behaviour. The ratio, now known as Weber's constant, turned out to be the speed of light; the importance of this link between electromagnetism and light soon became clear from the work of James Clerk Maxwell.

Weber was good at applied physics, and in 1833 he built an electric telegraph 3 km long to link his laboratory with the observatory where Gauss was based. This was the first practical telegraph ever built. During his time in academic semi-purdah, he organized (with Gauss) a network of observing stations to study the magnetic field of the Earth, and in 1871 he was one of the first people to put forward the idea that atoms are made up of a central positive charge surrounded by negatively charged particles. This prescient idea was largely ignored at the time.

Weber died in Göttingen on 23 June 1891.

weber   A unit used to measure magnetic fields (symbol Wb).

Weber's constant   See *Weber, Wilhelm Eduard*.

Weinberg, Steven (1933–)   American physicist who shared the Nobel Prize for Physics with Sheldon Glashow and Abdus Salam in 1979, for his contribution to the development of the theory of the electroweak interaction (the unification of electromagnetism and the weak force; see *forces of nature*).

Born in New York City on 3 May 1933, Weinberg attended Bronx High School, where he was an exact contemporary of Glashow, and then studied at Cornell University. He obtained his first degree in 1954, then (after a brief spell in

Copenhagen) moved to Princeton (PhD 1957). He worked at Columbia University for two years, and in 1959 moved to the Berkeley campus of the University of California, where he stayed until 1969. He then moved to MIT, and in 1973 became Higgins Professor of Physics at Harvard and also a senior scientist at the Smithsonian Astrophysical Observatory, reflecting his interest in the physics of the very early Universe. This led to the publication of a best-selling book, *The First Three Minutes*, in 1977. In 1986 Weinberg became professor of physics and astronomy at the University of Texas, Austin.

Weinberg's most important work, for which he received the Nobel Prize, was carried out during his time at Berkeley. Glashow, Salam and Weinberg made their contributions to electroweak theory independently of one another; one of the specific contributions made by Weinberg was the prediction of **neutral currents**.

Weinberg angle   A **mixing angle** which is used specifically in the context of the electroweak interaction (see **forces of nature**). In the language of **non-Abelian gauge theory**, there is 'room' in the description of the weak interaction for three particles which interact with equal strength – the W^+, W^- and an equivalent neutral particle, the W^0. The appropriate mathematical description is in terms of a **group** known as special unitary group 2, or SU(2); this is a simpler version of the **SU(3)** group which underpins the theory of **quantum chromodynamics**. But the actual neutral particle associated with the weak interaction, the Z^0, does not interact in quite the same way that this hypothetical W^0 particle would – most noticeably, the mass of the Z^0 is slightly greater than the mass of each of the W particles.

The explanation is that there is another kind of neutral particle allowed by group theory, a singlet boson known as B^0 which belongs all on its own in a U(1) group. The two neutral particles, the W^0 and the B^0, 'mix' with one another in different proportions, as quantum superpositions, to make up the two neutral particles that we actually see, the photon (which carries the electromagnetic interaction) and the Z^0 (which carries the **neutral current** of the weak interaction). The amount of this mixing is measured in terms of the Weinberg angle, and can be determined from experiments involving electroweak interactions.

Curiously, this way of describing the mixing was first proposed by Sheldon Glashow, in 1961, and not by Steven Weinberg.

Weinberg–Salam model   See **forces of nature**.

Wheeler, John Archibald (1911–)   American physicist who became one of the leading experts on the general theory of relativity, and who, earlier in his career, worked with Richard Feynman on a version of the idea of **action at a distance** that helped provide the inspiration for the **transactional interpretation** of quantum mechanics (see **Wheeler–Feynman absorber theory**). A theme running through much of his work during his long career has been the search for a **unified field theory**.

Born in Jacksonville, Florida, on 9 July 1911, Wheeler studied at Johns Hopkins University, receiving his PhD in 1933. He then spent two years in Copenhagen, working in Niels Bohr's institute, where he learned the details of quantum theory. On his return to the United States, he became an assistant professor at the University of North Carolina, and in 1938 moved to join the faculty at the University of Princeton, where he was the PhD supervisor for Feynman. Wheeler worked with Bohr on the theory of **nuclear fission**, using George Gamow's **liquid drop model**, and in 1939 they

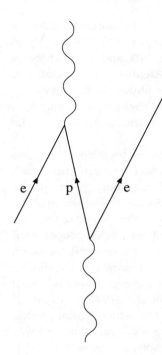

Wheeler 1. A gamma ray gives up its energy to make an electron-positron pair. A little later, the positron meets another electron and annihilates, producing another gamma ray.

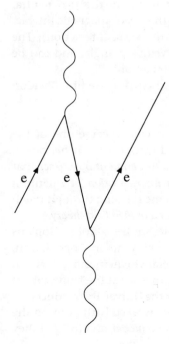

Wheeler 2. Wheeler realized that the interaction between electrons and photons could be described in another way. An electron moving forwards in time radiates a gamma ray (or meets a gamma ray coming back from the future), and is bounced backwards in time itself. It then (if that is the right term) radiates a gamma ray back further into the past, and recoils forwards in time once again. A positron is an electron going backwards in time, and photons do not distinguish between past and future.

wrote a joint paper pointing out the feasibility of using uranium-235 in making an atomic bomb; but he was not a member of the team based in Los Alamos that worked intensively on the *Manhattan Project*. In 1947 he became professor of mathematics at Princeton, and he was a member of the team working on the development of the hydrogen bomb from 1949 to 1951 (see *nuclear fusion*). He remained in the Princeton post until 1976 (but made extended working visits to overseas universities and to other research centres in the USA), then moved to the University of Texas, Austin.

Wheeler was also the research supervisor, in the 1950s, of Hugh Everett, and at that time espoused the *many worlds interpretation* of quantum mechanics, but later changed his mind and rejected it because 'it carries too great a load of metaphysical baggage'.

He is the originator of the idea that at the *Planck scale* the fabric of spacetime dissolves into a quantum foam, and he has pointed out that by taking the *Copenhagen interpretation* at face value and applying it to the whole Universe, we are led to conclude that the *wave function* of the Universe itself (right back to the Big Bang) might have been collapsed by our observations of it. If this is correct, by observing the Universe we have brought both it and ourselves into existence. This makes the entire Universe a kind of *delayed choice experiment*. Wheeler has also carried out fundamental work on the theory of *black holes*.

For someone whose career has been filled with exotic ideas and what some might regard as wild speculations, Wheeler presents a sober and conservative appearance, dressing neatly in traditional business suits and looking more like the stereotype of an old-fashioned banker than that of a crazy professor. He is possibly the cleverest physicist of his generation not to win the Nobel Prize.

Wheeler delayed choice experiment   See *delayed choice experiment*.

Wheeler–De Witt equation   See *wave function of the Universe*.

Wheeler–Feynman absorber theory   A version of the idea of *action at a distance*, developed by Richard Feynman and John Wheeler in the early 1940s.

Feynman and Wheeler were led to this idea by the puzzle of *radiation resistance*. All objects resist being pushed about – this is the property known as *inertia*. But if you try to accelerate a charged particle, perhaps by shaking it to and fro with a magnetic field, you discover that it has an extra inertia, over and above the inertia you would find for a particle with the same mass but no electric charge. This extra inertia makes it harder to move the charged particle.

The most common reason for shaking electrons to and fro is to make them radiate electromagnetic energy, in line with *Maxwell's equations*. This is what goes on in the broadcast antennae of TV and radio stations. It takes energy to make the electrons in the antenna oscillate and radiate the signal you want to broadcast, and it takes more energy (requiring a more powerful transmitter) than it would to shake equivalent uncharged particles, which do not radiate, by the same amount, hence the name 'radiation resistance'.

One curious feature of the classical description of electrons (and all other charged particles) and electromagnetic *fields* is that the interaction between each electron and the field (the *self-interaction*) actually has two components. The first component looks as if it ought to represent ordinary inertia, but is infinite for a point charge. But the second term exactly gives the force of radiation resistance.

Feynman needed some interaction to act back on the electron and give it radiation resistance when it was accelerated, and he wondered whether this back-reaction might come from other electrons (strictly speaking, any other charged particles) rather than from the 'field'. As physicists do when trying to get to grips with such problems, he considered the simplest possible example – in this case, a universe in which there were only two electrons. When the first charge shakes, it produces an effect on the second charge, which shakes in response. But now, because the second charge is shaking, there must be a back-reaction which shakes up the first charge. Perhaps this could account for radiation resistance.

Feynman took the idea to Wheeler, his thesis supervisor, who, as it happened, had already noticed an interesting feature of Maxwell's equations. He pointed out that these equations have two sets of solutions. One corresponds to a wave moving outward from its source and forward in time at the speed of light; the other (usually ignored) corresponds to a wave converging on its 'source' and moving backwards in time at the speed of light (or, if you like, moving forwards at *minus* the speed of light, –*c*). The waves corresponding to the usual solution of the equations are called retarded waves, because they arrive somewhere at a later time than they set out on their journey (the journey time is 'retarded' by the speed of light); the other solution corresponds to so-called advanced waves, which arrive before they set out on their journey (the journey time is 'advanced' by the speed of light). If the back-reaction from the second electron involved only advanced waves, Wheeler realized, its influence on the first electron would arrive at exactly the right time to cause radiation resistance, because it would have travelled the same distance at the same speed, but backwards in time.

Wheeler set Feynman the task of calculating what mixture of advanced and retarded waves would be required to produce the correct form of radiation resistance. Between them, Wheeler and Feynman also proved that in the real Universe, full of charged particles, all the interactions would cancel out in the right way to produce the same radiation resistance that they had calculated for the simple case.

A key ingredient of the model is that a wave has both an **amplitude** and a **phase**; it was through this work on the absorber theory of radiation (as he made clear in his Nobel address) that Feynman began to develop the tools to create his version of **quantum mechanics** and the **sum over histories** approach to quantum physics. But that still lay in the future. In the autumn of 1940, he found that you need a mixture of exactly half advanced waves and half retarded waves generated by each charge every time it shakes, using the solution of Maxwell's equations that is completely symmetrical in time.

Wheeler discovered that the Dutch physicist Adriaan Fokker had reached a similar conclusion in a series of papers published between 1929 and 1932; but Feynman's version was much more straightforward and easier to understand, while Fokker had never developed his ideas further. The half retarded wave goes out from the first electron forwards in time, while the half advanced wave goes out backwards in time. When the second electron shakes in response, it produces another half retarded wave, which is exactly out of step with the first wave and so precisely cancels out the remaining half retarded wave for all later times; and a half advanced wave, which goes back down the track of the first wave to the original electron, in step with

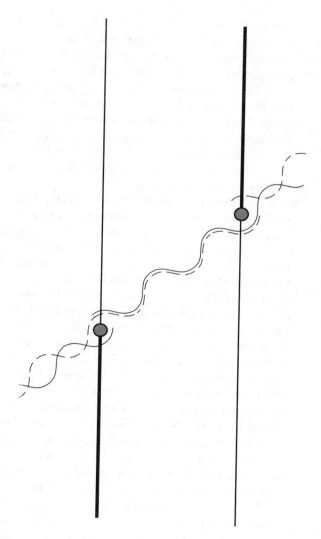

Wheeler-Feynman absorber theory.
When a charged particle jiggles (left),
it radiates both into the past and into
the future. This causes another
charged particle to jiggle in the future
(right). The jiggling of that particle
sends another wave both into the
future and into the past. The two sets
of waves cancel out everywhere
except in the region between the two
charged particles. But because one
wave goes forwards in time and the
other travels backwards in time, the
connection is made instantaneously.
See also *transactional interpretation*.

that wave, reinforcing the original half-wave to make a full wave, matching the usual solution to Maxwell's equations.

This half advanced wave arrives at the first electron, of course, at the moment it started to shake, and causes the radiation resistance. Then it continues back into the past, cancelling out the original half advanced wave from the first electron. The result is that between the two electrons there is a single wave exactly matching the conventional solution to Maxwell's equations, but everywhere else the wave cancels out, and radiation resistance emerges automatically from the equations, while the infinite self-energy never appears.

It is because of this essential role of the absorber in determining the way radiation is emitted that it is called the 'absorber theory' of radiation.

Wheeler decided that Feynman's next task, in the spring of 1941, should be to give a talk describing the work on direct action at a distance and time-symmetric electrodynamics. Even though this was purely an informal, 'in-house' talk, his audience would include Eugene Wigner, Henry Norris Russell, one of the greatest astronomers of the time, John von Neumann, Wolfgang Pauli and Albert Einstein. After the presentation, Pauli spoke up, saying that he didn't think the theory could possibly be right, and turned to Einstein to ask if he agreed. 'No,' replied Einstein, softly. 'I find only that it would be very difficult to make a corresponding theory for gravitational interaction.' In fact, some progress has even been made recently in applying these kinds of ideas to gravity (see *Mach's Principle*).

Further reading: John and Mary Gribbin, *Richard Feynman: A life in science*.

white dwarf   See *neutron star*.

Wien, Wilhelm Carl Werner Otto Fritz Franz (1864–1928)   German physicist who received the Nobel Prize for Physics in 1911 for his work on the laws governing the radiation of heat, which provided insight into *black body radiation* (see *Wien's Law).*

Wien was born on 13 January 1864, in Gaffken, in what was then a rural part of East Prussia. Wien's father was a successful farmer, and the family moved to another farm at Drachenstein when the boy (their only son) was two. After being taught privately at home, Wien entered the University of Göttingen in 1882, but stayed only for one term before dropping out to travel. In 1884, as he later put it, he 'really came into contact with physics for the first time,' at the laboratory of Hermann Helmholtz at the University of Berlin. He stayed there long enough to receive a PhD in 1886 (for work on the diffraction of light), then returned home, feeling obliged to help his parents run the farm. But in 1890 a drought ruined the crops and the family had to sell up. Wien went to work with Helmholtz (now at the new Institute for Science and Technology in the Charlottenburg district of Berlin) and his parents moved to Berlin. Wien's work on black body radiation was mainly carried out between 1892 and 1896, when he was a lecturer at the University of Berlin.

After teaching at the technical college in Aachen for three years (1896–9), and then spending a year as professor of physics at the University of Giessen, Wien succeeded Wilhelm Röntgen as professor of physics at the University of Würzburg in 1900. He stayed there until 1920, and was then once again Röntgen's successor, this time at the University of Munich, a post he held until he died, in Munich, on 30 August 1928.

As well as the work on black body radiation for which he is primarily remembered, Wien studied the way electricity is conducted by gases, made the first measurements (in 1905) of the energy of X-rays, and estimated their wavelengths. He also made a major impact as an excellent teacher and lecturer who inspired generations of students of physics.

Wien's Law   A relationship which gives the temperature of a *black body* in terms of the wavelength at which it radiates the maximum amount of energy in its spectrum. It is named after the German physicist Wilhelm Wien.

A graph of the amount of energy radiated by a black body at different wavelengths rises smoothly from lower energies at shorter wavelengths to a peak at some intermediate wavelength, then slides down smoothly again towards lower energies at longer wavelengths. The position of the peak moves towards shorter wavelengths at higher temperatures, and the temperature of the black body (in *kelvin*) is given simply by dividing the wavelength of the peak emission (in micrometres) into the number 2,900; this is Wien's Law. So if the peak in the black body curve is at, say, 4 micrometres (0.004 mm), the temperature of the object is 725 K.

This is a useful way of measuring the temperature of an object (provided that it is radiating roughly like a black body) from a few measurements of the intensity of its emission at different wavelengths around the peak in the spectrum.

Wigner, Eugene Paul (1902–1995) Hungarian-born American physicist who received a share of the Nobel Prize for Physics in 1963 for his work on the theory of *quantum physics*, especially the law of conservation of *parity*.

Born in Budapest on 17 November 1902, he was originally named Jeno Pal Wigner. He studied chemical engineering at the Berlin Technische Hochschule (Technical Academy), graduating in 1924, but after gaining his PhD in 1925, he immediately turned his attention to quantum physics and carried out postdoctoral work in this subject at the University of Göttingen. Wigner was then a lecturer at the Technical Academy in Berlin, from 1928 to 1930. In 1930 he moved to Princeton University in the United States, where he spent the rest of his career, apart from two years (1936–8) as professor of physics at the University of Wisconsin and wartime work at the University of Chicago and Oak Ridge, Tennessee. He became a US citizen in 1937.

Together with John von Neumann (who attended the same high school as Wigner in Budapest), Wigner introduced many of the key mathematical concepts used in modern *field theory* into quantum mechanics, including *group theory* and concepts involving *symmetry* in space and time. He also worked on the theory of neutron absorption, which is important in the design and construction of *nuclear reactors*. Wigner was one of the people (with Edward Teller and Leo Szilard) who persuaded Albert Einstein to write the letter to President Franklin Roosevelt which led to the setting up of the *Manhattan Project*. During the Second World War, he worked with Enrico Fermi in Chicago on the construction of the first 'atomic pile', as the nuclear reactor was then called.

Wigner's sister Margit was the wife of Paul Dirac, whom Wigner used to refer to as 'my famous brother-in-law'. He retired in 1971, and died, in Princeton, on 1 January 1995.

Wigner's friend An extension of the *Schrödinger's cat* thought experiment, devised by Eugene Wigner, who was intrigued by the central role of consciousness in determining the nature of reality in the *Copenhagen interpretation* of quantum mechanics. He pointed out that the 'cat in a box' experiment could be carried out inside a windowless laboratory with locked doors, where only one person ('Wigner's friend') looked into the box to see if the cat were dead or alive. Does the *superposition of states* representing the cat collapse into single state when the friend looks into the box? Or does the friend now become part of the superposition of states, until somebody opens the door (or rings up on the telephone) to find out the outcome of the experiment?

It looks very much as if we are trapped in an infinite regress. Perhaps Wigner himself becomes part of the superposition once he knows the answer, and his wave function can collapse only when somebody else asks him about the outcome of the experiment – and so on. Taken to its extreme, the implication is that the *wave function of the Universe* cannot collapse, because there is no observer outside the Universe to observe it (but see *participatory universe*).

On the other hand, if wave functions collapse as a result of observations by individual observers (Wigner's friends), how come we all agree on so much that happens in the world? If you and I each take a peek into the box, why do we invariably both see the same thing, either a dead cat or a live cat?

These puzzles do not arise in other *quantum interpretations*.

Wilson, Charles Thomson Rees (1869–1959) British physicist who invented the *cloud chamber* (which he actually invented in order to study clouds, but soon realized could be used as a particle detector), and who shared the Nobel Prize for Physics with Arthur Compton in 1927.

Wilson was born on 14 February 1869, on a farm near Glencorse, outside Edinburgh. He was the youngest of eight children; his father, a sheep farmer, died when the boy was four, and his widow took the younger members of the family to Manchester. Wilson studied at Owens College (which later became the University of Manchester), where he received his first degree in 1887, then moved on to the University of Cambridge, where he graduated in 1892. Wilson then taught for four years at Bradford Grammar School, before returning to Cambridge as a research student in 1896. In 1900 he was appointed university lecturer in physics (the university did not award PhDs then), in 1918 became reader in meteorology, and in 1925 became Jacksonian Professor of Natural Philosophy in the University of Cambridge, a post he held until he retired in 1934 and moved to Scotland.

Even in retirement, Wilson continued to carry out research, and he published his last paper, on thunderstorm electricity, at the age of 87; his investigations of the phenomenon had included an aeroplane flight to observe thunderclouds over the outer isles of Scotland in 1955, when he was 86. He died at Carlops, near Edinburgh, on 15 November 1959.

WIMP Acronym for weakly interacting massive particle. WIMPs have not yet been detected, but are required by cosmology if (as many cosmologists suspect) the Universe contains enough *dark matter* to ensure that gravity will one day reverse the present expansion, pulling everything together in a Big Crunch. Present understanding of the Big Bang, in which the Universe as we know it was born, says that most of this matter cannot be in the form of *baryons*. WIMPs would not interact with baryonic matter except through gravity, or in direct collisions, such as those between a WIMP and the nucleus of an atom. The name does *not* mean that they take part in the weak interaction; it was chosen by a cosmologist who was more interested in puns than in particle physics.

The idea of WIMPs is more than a cosmological speculation, because particles with exactly the properties required to match the cosmological calculations are predicted independently by the *grand unified theories* and by *supersymmetry*; two of the best candidates for WIMPs are the *axion* and the *lightest supersymmetric partner*.

wire chamber An improved version of the *spark chamber*, developed by Frank Kiernan at CERN in the 1960s. The parallel metal plates of a spark chamber are replaced in this kind of particle detector by sheets of parallel wires, each wire separated from its neighbour by a gap of about a millimetre. Georges Charpak and his colleagues later developed the *drift chamber* and the *multiwire proportional chamber*, also using planes of parallel wires, but in different ways.

Wollaston, William Hyde (1766–1828) English physicist and chemist who discovered the elements rhodium and palladium, and developed techniques for processing metals such as platinum into a malleable form. Wollaston was one of the first supporters of John Dalton's atomic theory, and in 1808 he pointed out that an understanding of the three-dimensional arrangement of atoms in a substance would be useful in explaining its properties.

Born in East Dereham, Norfolk, on 6 August 1766, Wollaston was one of seventeen children of a clergyman (one of his brothers, Francis, became a professor of chemistry at the University of Cambridge). He studied at the University of Cambridge, where he graduated in medicine in 1793. Wollaston set up a private research laboratory, where he worked until 1797, then practised as a doctor in London until 1800. Having partially lost his sight, he then went into business and made a fortune using a technique he developed for preparing platinum (a metal very resistant to heat and acids, and useful in industry to make containers used in the preparation and storage of sulphuric acid). The success of this process (which he kept secret until shortly before he died, and which is now known as powder metallurgy) made him financially independent and able to carry out whatever scientific investigations he liked. He also donated large sums of money to scientific societies to support their activities.

Among many other things, Wollaston was a friend of Thomas Young and a supporter of the wave theory of light. He was the first person to notice the dark lines in the *spectrum* of the Sun, but failed to follow this up. Wollaston died (of a brain tumour) in London, on 22 December 1828, leaving substantial bequests to both the Royal Society and the Geological Society.

world line   See *path integral.*

wormhole   See *black hole.*

W particles   Two of the three *bosons* which (together with the *Z particle*) carry the weak interaction (see *forces of nature*). The W⁺ has one unit of positive charge, the W⁻ has one unit of negative charge. They each have a mass of 83 GeV, closely matching the predictions of the electroweak theory, a lifetime of 10^{-25} sec and one unit of spin. They were first detected at CERN in 1983.

See *charged current.*

Wu, Chien-Shiung (1912–1997)   Chinese-born American physicist who carried out the experiments that confirmed the violation of *parity* in some processes involving the weak interaction (see *forces of nature*). She died in Manhattan on 16 February 1997.

Born in Shanghai on 31 May 1912, Wu studied at the National Central University of China, where she graduated in 1934. In 1936 she moved to the United States, where she worked under the supervision of Ernest Lawrence at the Berkeley campus of the University of California, and was awarded her PhD in 1940. After two years of postdoctoral work at Berkeley, she taught briefly at Smith College (in Northampton, Massachusetts) and then at Princeton University, before becoming (in 1944) a member of the staff at Columbia University, where she was appointed as a full professor in 1957.

It was while she was at Columbia, in 1956, that Wu carried out the experiments which confirmed the prediction of parity non-conservation in beta decay, made only a few months earlier by T. D. Lee and Chen Ning Yang; this led almost equally swiftly to the award of the Nobel Prize to Lee and Yang (but not to Wu).

Partly building from these discoveries, Richard Feynman and Murray Gell-Mann independently developed a new theory of the weak interaction, which they then published together in 1958. Wu and her colleagues carried out experiments confirming the accuracy of this theory in 1963. After carrying out many other important experiments in particle physics, Wu became interested in biological problems and investigated the structure of haemoglobin. She retired in 1981.

X boson   See *grand unified theories.*

xi particles   Two *baryons* which each contain a couple of strange *quarks.* The xi-minus, discovered in 1952, is made up of a down quark and two strange quarks. It has a mass of 1.321 GeV, −1 units of charge, and spin ½. Its lifetime is 1.6×10^{-10} sec. The xi-zero, discovered in 1959, is made up of an up quark and two strange quarks. It has a mass of 1.315 GeV, no charge, and spin ½. Its lifetime is 3×10^{-10} sec.

X particle   = *X boson.*

X-rays   *Electromagnetic radiation* with *wavelengths* shorter than those of *ultraviolet radiation*, but longer than those of *gamma rays*, in the range from 12 billionths of a metre (12 nanometres, or 12×10^{-9} m) to about 12 trillionths of a metre (12×10^{-12} m). At these very short wavelengths, corresponding to high-energy *radiation*, X-rays are more usually described in terms of the *energy* carried by individual *photons*, which ranges from about 100 *electron volts* for longer-wavelength X-rays up to about 100,000 eV for shorter-wavelength X-rays. The distinction between X-rays and gamma rays is arbitrary and not clear cut; what some people call very high-energy X-rays others regard as low-energy gamma rays. There is a similar overlap between low energy X-rays and high-energy ultraviolet radiation. Lower-energy X-rays are sometimes called soft X-rays, while higher-energy X-rays are correspondingly referred to as hard X-rays.

X-rays were discovered by the German physicist Wilhelm Röntgen in 1895; they used to be known as Röntgen rays. X-rays are produced when a stream of energetic *electrons* bombards the surface of a material object. Like many physicists in the 1890s, Röntgen was experimenting with the radiation which emanates from a wire that carries an electric current through an evacuated tube (cathode rays). He was using a *Crookes tube* to produce these rays (now known to be a stream of electrons) and to study their effect on a fluorescent screen, which produces flashes of light when struck by energetic radiation. Röntgen happened to have another fluorescent screen lying on a bench near his cathode ray experiment, and spotted tell-tale flashes coming from it while his experiment was running. Working intensively over a period of six weeks, he identified the cause as this secondary radiation, which he called X-radiation, because *x* is traditionally the unknown quantity in a mathematical equation.

By 28 December 1895 he was able to announce not only his discovery of the phenomenon, but a description of the basic properties of the rays – that they travel in straight lines and cause shadows, pass through all bodies to some extent, cause fluorescence and fog photographic plates, but are not affected by magnetic fields.

X-rays are produced when an electron in a low *energy level* in an *atom* is knocked out of the atom, leaving a hole. An electron from a much higher energy level can fall all the way into this hole, releasing an X-ray photon in the process.

Yang, Chen Ning (1922–)   Chinese-born American physicist, known to his friends as Frank (he adopted the name 'Franklin' after reading a biography of Benjamin Franklin), who shared the Nobel Prize for Physics with T. D. Lee in 1957, for their joint prediction of the violation of *parity* conservation in *beta decay* (see *CP violation*). The prediction had been confirmed by experiments carried out by Chien-Shiung Wu in 1956, soon after it had been made; this makes the time from the work being carried out to the award of the Nobel Prize one of the quickest ever.

Born in Hofei, China, on 22 September 1922, Yang was the son of a professor of mathematics. He studied at the National Southwestern Associated University of Kunming (BSc 1942), where he met Lee, and received his MSc in 1944 from Tsinghua University, which had actually moved to Kunming during the Sino-Japanese War. At the end of the Second World War, Yang went to the University of Chicago to work for his PhD (awarded in 1948), where he was again a contemporary of Lee. After a post-doctoral year at Chicago, Yang moved to the Institute for Advanced Study, in Princeton, where he became a professor in 1955. He became a US citizen in 1964, and in 1966 he moved to the State University of New York, at Stony Brook, as Einstein Professor of Physics and director of the Institute of Theoretical Physics. In 1954 Yang also worked with Robert Mills (born 1924) on a version of *non-Abelian gauge theory* important in *quantum field theory*; this is sometimes known as Yang–Mills theory.

His collaboration with Lee continued even when they were at different research centres, and together they were among the theorists who argued in 1960 that there was more than one kind of neutrino (see *Lederman, Leon Max*); they also predicted the existence of the charged W boson and the existence of weak neutral currents, carried by what is now known as the Z particle.

Yang–Mills field See *isotopic spin*.

Yang–Mills theory The version of *non-Abelian gauge theory* developed by Robert Mills and Chen Ning Yang in 1954, which underpins modern *quantum field theory*.

Young, Thomas (1773–1829) English physicist who made a major contribution to the establishment of the wave theory of light at the beginning of the 19th century, in particular using the *double-slit experiment*.

Young was born at Milverton, in Somerset, on 13 June 1773. He was a child prodigy who could read at the age of two, mastered Latin and Greek as a child, learned several Middle Eastern languages before he was fourteen, and had not only read but understood both of Isaac Newton's great books, the *Principia* and *Opticks*, by the time he was seventeen. He studied medicine, initially at St Bartholomew's Hospital in London, where during his first year as a student (1793) he wrote a paper on the way the eye focuses which was so important that the following year he was elected as a Fellow of the Royal Society. In 1794 he studied in Edinburgh and in 1795 he moved to Göttingen, where he received his MD in 1796. In 1797 he moved to Cambridge to complete his studies, which took another two years. This period he largely spent indulging in his scientific interests, earning the nickname 'Phenomenon Young', and carrying out a series of experiments involving sound and light that he described in a book published in 1800. One of his main conclusions was that the wave theory proposed by Christopher Huygens was a better description of the behaviour of light than the corpuscular theory proposed by Newton.

The year before the book appeared, Young had completed his medical studies and started practising medicine in London; since his great uncle (also a doctor) had left him £10,000 and a house in London in 1797, however, Young had no need to earn his living from medicine, and in 1801 he became professor of natural philosophy at the Royal Institution. In the same year, he explained how colour vision works. Unfortunately, he proved to be a very poor lecturer and gave up the post in 1803.

After four more years of scientific experiments, in 1807 he turned once again to medicine, which was his main occupation until 1817 (in 1811 he became a physician at St George's Hospital), although in 1816 he did write a letter to François Arago (see **Fresnel, Augustin Jean**) suggesting that **polarization** might be explained if light propagated as a transverse wave (see **Maxwell's equations**). After 1817 Young became increasingly occupied by his posts as secretary of the Commission on Weights and Measures, and foreign secretary of the Royal Society. He also became keenly interested in Egyptology, and was instrumental in the deciphering of the Rosetta Stone.

Young's contributions to the theory of light were chiefly made during the early years of the 19th century (the key work on interference was carried out in 1801). He died in London on 10 May 1829.

Young's slit experiment   See *double-slit experiment.*

Yukawa, Hideki (1907–1981)   Japanese physicist who received the Nobel Prize for Physics in 1949 (the first Japanese to be honoured in this way) for his prediction of the existence of *pions*.

Born in Kyoto, on 23 January 1907 (as Hideki Ogawa; he married Sumiko Yukawa in 1932 and took her family name). He had three brothers who became university professors, and two sisters who married professors. Yukawa's father (Takuji Ogawa) was professor of geology at Kyoto University, where Yukawa himself was educated, graduating with a master's degree in 1929. He then moved to Osaka University where he worked as a lecturer while studying for his PhD, which was awarded in 1938. In 1939 he returned to Kyoto as professor of theoretical physics, and stayed there for the rest of his career, apart from visits to several other universities, notably including an extended visit to the United States taking in Princeton in 1948–9, and Columbia University from 1949 to 1953.

It was in 1935 that Yukawa proposed (in the first paper he published!) that there must be a 'new' kind of force which holds the *nucleus* of an atom together in spite of the tendency of the positive charge on all the protons in the nucleus to blow it apart. He predicted that this force must be carried by a previously undetected kind of particle with a very short range (only about 10^{-13} cm). Because the lifetime of a *virtual particle* is related inversely to its mass, and the distance it can travel depends on its lifetime, this led Yukawa to predict that the mass of his hypothetical particle would be about 200 times that of the electron (about one-ninth of the mass of a nucleon).

In 1936 Carl Anderson discovered the *muon*, which had about the right mass and was at first identified as Yukawa's proposed particle; it soon became clear, though, that this was not the carrier of what is now called the strong interaction (see *forces of nature*). But in 1947 Cecil Powell discovered the pion, which proved to be exactly what the theory of the strong interaction required.

Yukawa also made an important contribution to the theory of the weak interaction, taking up the suggestion (made by Igor Tamm in 1934) that *beta decay* is a result of a force operating between nucleons, and suggesting that the force could be described in terms of the exchange of *mesons*. He retired in 1970, and died in Kyoto on 8 September 1981.

Zeeman, Pieter (1865–1943)   Dutch physicist who investigated the effect of magnetic fields on light, and who discovered the *Zeeman effect*.

Born at Zonnemaire, in Zeeland, on 25 May 1865; Zeeman studied at the University of Leiden, where he was a pupil of Hendrik Lorentz. After completing his PhD (in 1893), Zeeman stayed on in Leiden as a *Privatdozent*. In 1897 he became a lecturer at the Free University of Amsterdam, where he was promoted to professor of physics in 1900.

It was shortly before he left Leiden that Zeeman began his investigation of the effect of a magnetic field on light, following the earlier, largely unsuccessful, work by Michael Faraday. This led to the discovery of the first hint of the Zeeman effect (in the form of a broadening of the spectral lines of sodium) in 1896 and its detailed investigation (involving complete splitting of a line in the cadmium spectrum into three components) in 1897. After Zeeman moved to Amsterdam, he had access only to second-rate experimental facilities, so his discovery was largely developed by other researchers. (Among other things, the laboratory was on a busy road, and was shaken so severely by traffic that the delicate apparatus needed for accurate spectroscopic studies became unusable.) But he was awarded a share of the second ever Nobel Prize for Physics, in 1902 (with Lorentz, who had predicted the Zeeman effect). He was also a first-class teacher, and among his other research Zeeman measured the speed of light in dense and moving media.

In 1923 a new laboratory (later named after Zeeman) was established in the Free University of Amsterdam, with Zeeman as director; but by then it was too late for him to benefit much personally from the improved facilities. He retired in 1935, and died, in Amsterdam, on 9 October 1943.

Zeeman effect   Splitting of spectral lines in the light from an *atom* when it is placed in a strong magnetic field. The effect is caused by the distortion in the shape of the electron *orbitals* caused by the magnetic field, which changes the spacing of *energy levels* in the atom (the Zeeman shift). See also *Stark effect*.

Part of the splitting can be understood in classical terms, and it was on this basis that Hendrik Lorentz predicted the Zeeman effect, and Pieter Zeeman discovered it. But closer investigation in the wake of Zeeman's discovery revealed initially puzzling details (the anomalous Zeeman effect) that were in due course explained as a result of quantum effects, notably electron *spin*.

Observations of the Zeeman effect in the light from the Sun and stars make it possible to measure their magnetic fields.

Zeeman shift   See *Zeeman effect*.

zero-point energy   The energy which is still associated with a particle or system (over and above its mass-energy) at the absolute zero of temperature, 0 K. This minimum energy cannot be precisely zero because of quantum *uncertainty*. In *quantum field theory*, the lowest energy state of a field (its ground state) is also non-zero, for the same reason, giving the quantum vacuum a complex structure, which can be probed experimentally (see, for example, *Casimir effect*).

Z particle   One of the three *bosons* that (together with the *W particles*) carry the weak interaction (see *forces of nature*). The Z particle has no charge and is sometimes referred to as the Z^0. It has a mass of 93 GeV, closely matching the predictions of the electroweak theory, a lifetime of 10^{-25} sec, and a spin of 1. First detected at CERN in 1983.

See *neutral current*.

Zweig, George (1937–)   Russian-born American physicist who was one of the two independent inventors of the concept of what are now known as *quarks*.

Zweig was born in Moscow on 20 May 1937. His parents had been born in what is now Poland, but was then part of the Austro-Hungarian Empire, so they were Austrian citizens. Zweig's father grew up in Vienna, but he and his wife were living in Germany when Hitler came to power in 1933, and left for Russia because they were afraid of being persecuted as Jews. When George was born, they had a choice of giving him Austrian or Russian citizenship, and chose the latter because of the Austrian attitude towards Jews, although as Zweig now comments ruefully, 'the Russians weren't perfect either'. In 1937, shortly after George Zweig was born, his parents left Moscow and went back, with the baby, to Vienna to try to persuade his father's parents to flee from the war that was by then obviously imminent. The older Zweigs refused, and after the Anschluss by which the Nazis took over Austria, in 1938 George Zweig's parents became increasingly desperate to escape from the coming conflict themselves. They succeeded only because his mother's brother had left Poland some time before to go to America, and had enough influence to persuade Senator Vandenberg to attach an amendment to a Senate bill, listing 50 people who would be allowed to enter the United States as refugees. The Zweig family were on the list and were among the last refugees from Hitler's empire to reach the United States before the Second World War began in Europe. Zweig's paternal grandparents stayed in Austria, and died in Auschwitz in 1943.

Zweig became a citizen of the United States in 1942, when his parents were naturalized, but this was never recognized by the Soviet Union, which always claimed him as a Soviet citizen, making it inadvisable for Zweig to travel behind what used to be the Iron Curtain. He studied at the University of Michigan (BSc 1959) and then moved to Caltech, where he completed his PhD in 1963. He had initially started research in experimental physics, working on a high-energy experiment at the Bevatron, but became frustrated by the difficulty of getting any meaningful results, and switched to theory, under the guidance of Richard Feynman. Feynman 'exerted his influence', says Zweig, 'both through his work and outlook. Solutions to problems were invariably based on simple ideas. Physical insight balanced calculational skill. And work was to be published only when it was correct, important, and fully understood. This was a stern conscience who practised what he preached.'

During his time as a frustrated experimenter, Zweig had occasionally discussed his work with Murray Gell-Mann, and it was Gell-Mann who suggested that he should seek guidance from Feynman. But in the autumn of 1962, when Zweig was switching from experiment to theory, Gell-Mann departed on an extended visit to MIT (as a visiting lecturer), and Zweig and Gell-Mann did not meet again, or have any communication, until Zweig returned from a visit to CERN almost two years later.

After completing the work for his PhD in 1963, Zweig spent a year working at

George Zweig (1937–).

CERN, in Geneva, where he developed his model of structure within the proton and neutron, which he described in terms of three sub-baryonic particles that he called aces. The same concept was being developed at the same time by Gell-Mann, although neither of them knew of the other's work; it was Gell-Mann who gave the entities the name 'quarks' and managed to make this name stick, even though his proposal was initially much more tentative than Zweig's.

From a perspective more than 30 years on from this work, it is hard to appreciate just how audacious this idea was. In the early 1960s, the nucleons were regarded as fundamental and indivisible building blocks of nature (much as atoms had been regarded before the 1890s); the really outrageous requirement of the ace/quark model was that the hypothetical sub-baryonic particles would each have a fractional electric charge, either ⅓ or ⅔ of the magnitude of the charge on an electron.

Some idea of just how outrageous this idea seemed at the time can be gleaned from the extremely cautious way in which Gell-Mann put forward the idea. In a paper published in 1964, he wrote:

> It is fun to speculate about the way quarks would behave if they were physical particles of finite mass (instead of purely mathematical entities as they would be in the limit of infinite mass) … a search for stable quarks of charge –⅓ or +⅔ and/or

stable diquarks of charge $-\frac{2}{3}$ or $+\frac{1}{3}$ or $+\frac{4}{3}$ at the highest energy accelerators would help to reassure us of the non-existence of real quarks!

Even Gell-Mann, to judge from this passage, did not believe that quarks were real. He regarded them as a mathematical device to aid calculations, and urged the experimenters to comfort the theorists by proving that quarks were not real, physical particles!

Zweig, with the confidence of youth, had no such inhibitions, and wrote up his ideas in the form of two papers which were circulated as CERN 'preprints'. In what is clearly a style strongly influenced by Feynman, Zweig's papers use graphic visual imagery to put his ideas across, as well as the mathematics. He used geometrical shapes (triangles, circles and squares) to represent his aces, linking them with lines to make the pairs and triplets corresponding to known particles (the way they are now regarded as being held together by the exchange of **gluons**). With this powerful imagery, you can see the way aces/quarks combine as easily as a small child can see how to fit a triangular block into a triangular hole, and it is a great pity that the idea was never taken up and used to teach the quark model.

But Zweig soon found that he had made a mistake – not scientifically, but politically. The papers were never formally published because of the opposition of other scientists to them. In 1981 Zweig recalled that:

> The reaction of the theoretical physics community to the ace model was not benign. Getting the CERN report published in the form that I wanted was so difficult that I finally gave up trying. When the physics department of a leading university was considering an appointment for me, their senior theorist, one of the most respected spokesmen for all of theoretical physics, blocked the appointment at a faculty meeting by passionately arguing that the ace model was the work of a 'charlatan'.

By proposing the ace/quark model, which is now regarded as a jewel in the crown of particle physics, Zweig actually damaged his career prospects.

Zweig returned to Caltech in 1964 and became a junior professor there in 1967. He later (in 1983) moved to the Los Alamos National Laboratory, in New Mexico, but remained a visiting associate at Caltech. In the late 1960s and early 1970s, Zweig worked on defence projects and much of this work is still classified. He then took up neurobiology and, through investigating the way the ear transforms sound into a form that is interpreted by the nervous system, he discovered a new way to extract information from any kind of signal. This led to the construction of a device called SigniScope that emulates the mechanical response of the inner ear to sound, and an understanding of how this represents music led to the design of a music synthesizer that was used to create part of the sound track for the first *Star Trek* film.

In 1985 Zweig founded a company, Signition, Inc., which developed an improved version of SigniScope to analyse the structure of speech and its relationship to hearing. A third version of the device has now been developed as a software package to analyse many kinds of signal and image.

To somebody who is not privy to the inner deliberations of the Nobel Committee, it is totally baffling that Zweig's fruitful theory of fundamental particles, which has now been amply confirmed by experiment and is a cornerstone of the *standard model* of particle physics, has not been marked by the award of a Nobel Prize. Perhaps, though, it is some comfort to him to know that he is the last word in particle physics.

Further reading: Andrew Pickering, *Constructing Quarks*.

BIBLIOGRAPHY

Books referred to in the text, together with other books of interest, are listed here. Those pitched at a significantly more technical level than the present volume are marked with an asterisk.

John Barrow, *The World within the World* (Oxford University Press, Oxford, 1988)

John Barrow, *Theories of Everything* (Oxford University Press, Oxford, 1991)

John Barrow and Frank Tipler, *The Anthropic Cosmological Principle* (Oxford University Press, Oxford, 1986)

*John Bell, *Speakable and Unspeakable in Quantum Mechanics* (Cambridge University Press, Cambridge, 1987)

David Cassidy, *Uncertainty: The life and science of Werner Heisenberg* (Freeman, New York, 1992)

Barbara Cline, *The Questioners* (Crowell, New York, 1965)

Barbara Cline, *Men Who Made a New Physics* (University of Chicago Press, Chicago, IL 1987)

Frank Close, *The Cosmic Onion* (Heinemann, London, 1983)

Frank Close, Michael Marten and Christine Sutton, *The Particle Explosion* (Oxford University Press, Oxford, 1987)

*G. D. Coughlan and J. E. Dodd, *The ideas of Particle Physics* (Cambridge University Press, Cambridge, 2nd edn, 1991)

Paul Davies, *Other Worlds* (Dent, London, 1980)

Paul Davies and Julian Brown (eds), *Superstrings* (Cambridge University Press, Cambridge, 1988)

*Paul Davies (ed.), *The New Physics* (Cambridge University Press, Cambridge, 1989)

Paul Davies and John Gribbin, *The Matter Myth* (Viking, London, 1991)

David Deutsch, *The Fabric of Reality* (Viking, London, 1997)

John Emsley, *The Elements* (Oxford University Press, Oxford, 2nd edn, 1991)

C. W. F. Everitt, *James Clerk Maxwell* (Scribner's, New York, 1975)

Richard Feynman, Robert Leighton and Matthew Sands, *The Feynman Lectures on Physics,* Vol. III (Addison-Wesley, Boston, MA, 1965)

Richard Feynman, *The Character of Physical Law* (Penguin, London, 1992; originally published in 1965)

Richard Feynman, *QED: The strange theory of light and matter* (Princeton University Press, Princeton, 1985)

Richard Feynman, *Six Easy Pieces* (Addison-Wesley, Boston, MA, 1995)

A. P. French (ed.), *Einstein: A centenary volume* (Harvard University Press, Cambridge, MA, 1979)

A. P. French and P. J. Kennedy (eds), *Niels Bohr: A centenary volume* (Harvard University Press, Cambridge, MA, 1985)

Harald Fritzsch, *Quarks* (Penguin, London, 1992)

George Gamow, *The Great Physicists from Galileo to Einstein* (Dover, New York, 1961)

George Gamow, *Thirty Years that Shook Physics* (Dover, New York, 1966)

Ted and Ben Goertzel, *Linus Pauling* (Basic Books, New York, 1995)

John Gribbin, *In Search of Schrödinger's Cat* (Bantam, London, 1984)

John Gribbin, *In Search of the Big Bang* (Black Swan, London, and Bantam, New York, 1986; revised edition, Penguin, London, 1998)

John Gribbin, *In the Beginning* (Penguin, London, and Little, Brown, New York, 1993)

John Gribbin, *In Search of the Double Helix* (Penguin, London, 1995)

John Gribbin, *In Search of the Edge of Time* (Penguin, London, 1995)

John Gribbin, *Schrödinger's Kittens* (Weidenfeld and Nicolson, London, 1995)

John Gribbin, *Companion to the Cosmos* (Weidenfeld and Nicolson, London, 1996)

John and Mary Gribbin, *Richard Feynman: A life in science* (Viking, London, 1997)

John Gribbin and Martin Rees, *The Stuff of the Universe* (Penguin, 1995)

Brian Hatfield (ed.), *Feynman Lectures on Gravitation* (Addison-Wesley, Boston, MA, 1995)

Nick Herbert, *Quantum Reality* (Anchor Press/Doubleday, New York, 1985)

Tony Hey and Patrick Walters, *The Quantum Universe* (Cambridge University Press, Cambridge, 1987)

Gerard 't Hooft, *In Search of the Ultimate Building Blocks* (Cambridge University Press, Cambridge, 1996)

Fred Hoyle, *Home is Where the Wind Blows* (University Science Books, Mill Valley, CA, 1994)

*Vincent Icke, *The Force of Symmetry* (Cambridge University Press, Cambridge, 1995)

*Max Jammer, *The Conceptual Development of Quantum Mechanics* (McGraw-Hill, New York, 1966)

Martin Krieger, *Doing Physics* (Indiana University Press, Bloomington, 1992)

Leon Lederman and David Schramm, *From Quarks to the Cosmos* (Scientific American/Freeman, New York, updated edn, 1995)

*Jagdish Mehra, *The Beat of a Different Drum* (Oxford University Press, Oxford 1994)

Walter Moore, *Schrödinger: Life and thought* (Cambridge University Press, Cambridge, 1989)

Yuval Ne'eman and Yoram Kirsch, *The Particle Hunters* (Cambridge University Press, Cambridge, 1986)

Heinz Pagels, *The Cosmic Code* (Simon and Schuster, New York, 1982)

Abraham Pais, *Subtle is the Lord* (Oxford University Press, Oxford, 1982)

*Abraham Pais, *Inward Bound* (Oxford University Press, Oxford, 1986)

F. David Peat, *Superstrings* (Scribner's, New York, 1988)

*Andrew Pickering, *Constructing Quarks* (Edinburgh University Press, Edinburgh, 1984)

Susan Quinn, *Marie Curie* (Heinemann, London, 1995)

Alastair Rae, *Quantum Physics: Illusion or reality?* (Cambridge University Press, Cambridge, 1986)

Ed Regis, *NANO!* (Bantam Press, New York, 1995)

*Silvan Schweber, *QED and the Men Who Made It* (Princeton University Press, Princeton, NJ, 1994)

Emilio Segre, *A Mind Always in Motion* (University of California Press, Berkeley, CA, 1993)

Bibliography

....

Ruth Sime, *Lise Meitner: A life in physics* (University of California Press, Berkeley, CA, 1996)

Christine Sutton (ed), *Building the Universe* (Blackwell/New Scientist, London, 1985)

Kip Thorne, *Black Holes and Time Warps* (Norton, New York, 1994)

Robert Weber, *Pioneers of Science* (Adam Hilger, Bristol, 2nd edn, 1988)

Steven Weinberg, *The Discovery of Subatomic Particles* (Freeman, New York, revised edn, 1990)

*Bruce Wheaton, *The Tiger and the Shark* (Cambridge University Press, Cambridge, 1983)

*John Wheeler and Wojciech Zurek, *Quantum Theory and Measurement* (Princeton University Press, Princeton, NJ, 1983)

*Andrew Whitaker, *Einstein, Bohr and the Quantum Dilemma* (Cambridge University Press, Cambridge, 1996)

Michael White and John Gribbin, *Stephen Hawking: A Life in Science* (Penguin, London, and Plume, New York, 1992)

Michael White and John Gribbin, *Einstein: A life in science* (Simon and Schuster, London, and Plume, New York, 1993)

David Wilson, *Rutherford* (Hodder and Stoughton, London, 1983)

Harry Woolf (ed.) *Some Strangeness in the Proportion* (Addison-Wesley), Boston, MA, 1980).

PICTURE ACKNOWLEDGEMENTS

Additional credits where necessary are given in brackets after the page number.

The Cavendish Laboratory, Cambridge: 402.

CERN: 39, 65, 66, 72, 114, 154, 326.

Hulton Getty: 49.

Image Select/Ann Ronan Picture Library: 12 (CERN), 82, 95, 104, 132, 145, 148, 161 (CERN), 192 (CERN), 199, 251, 285, 328, 334 (CERN), 353, 355, 393, 396 (CERN).

Los Alamos National Laboratory Information Office: 447.

Mary Evans Picture Library: 41, 52, 56, 59, 60, 62, 83, 84, 97, 107, 117, 129, 152, 181, 229, 274, 290, 347, 350, 358, 408, 425.

Science Photo Library: 91 (C. Powell, P. Fowler & D. Perkins), 105 (Erich Schrempp), 109 (David Parker), 120 (David Scharf), 134 (CERN), 142 (Bruce Iverson), 191 (Library of Congress), 232, 253 (NASA), 266, 323 (Jerry Mason), 364 (Stanford Linear Accelerator Center), 366 (Professor Peter Fowler), 368 (Physics Dept., Imperial College, London), 416 (CERN), 420 (Peter Menzel).

Science & Society Picture Library: 209, 218.

Starland Picture Library: 261 (NOAO).

TIMELINE 1	TIMELINE 2	TIMELINE 3
Birth dates of scientists who made significant contributions to our understanding of the quantum world.	*Key dates in the development of physical science and (especially) of our understanding of the quantum world.*	*Key dates in history, with emphasis on events important in a scientific context.*

TIMELINE 3

About **15000–10000** BC: The world warms out of the latest Ice Age. Cave paintings.

About **12000** BC: Dog domesticated.

10000 BC: World human population 3 million.

About **8000** BC: Invention of trade tokens (early 'money'), agriculture. Pottery.

About **6000** BC: Development of irrigation. Weaving.

About **4500** BC: Copper smelting.

About **4000** BC: Sumerian city of Ur founded.

About **3500** BC: Invention of writing and the wheel.

About **3000** BC: Egyptians use sailing ships. Bronze in widespread use. Pyramid of Giza built. Stonehenge first stage built. Population 100 million.

About **2500** BC: Beaker people spread across Europe.

About **2000** BC: Rise of Babylon. Stonehenge stone circles added. Spoked wheels developed in Asia Minor. Minoan civilization flowers in Crete.

Birth dates of scientists	Key dates in science	Key dates in history
		About **1800** BC: The first alphabet. Babylonians use multiplication tables.
		1700 BC: Judaism founded by Abraham.
		About **1650** BC: Volcanic island of Santorini (Thera) explodes; probable source of Atlantis legend. Bronze ploughs in use in Vietnam.
		About **1500** BC: Chinese writing.
		About **1350** BC: Exodus of Jews from Egypt.
		About **1200** BC: Iron working developed. Trojan Wars.
		878 BC: Carthage founded.
		About **800** BC: Olmec pyramids built in Mexico. Homer writes *The Iliad*.
		776 BC: First Olympiad.
		710 BC: Egypt conquered by Ethiopian invaders.
About **625** BC: Thales of Miletus. The first scientist; a Greek philosopher who proposed that the world is made of water, and that the Earth is a disc which floats on water. Also developed attempts at scientific explanations of phenomena such as earthquakes, rather than regarding them as the work of the gods.		About **600** BC: Greek science begins. Aesop writes his fables. Chinese invent fumigation of houses to destroy pests.
		605–562 BC: Nebuchadnezzar creates Hanging Gardens of Babylon.
611 BC: Anaximander of Miletus. The first philosopher to suggest that the surface of the Earth is curved.	**585** BC: Thales of Miletus correctly predicts a solar eclipse.	**565** BC: Taoism founded.
		538 BC: Persians conquer Babylon.
580 BC: Pythagoras. Taught that the Earth is a sphere and that the planets move in circles – not for any truly scientific reason, but because of a mystic belief that circles were the 'perfect' form. He		**528** BC: Beginning of Buddhism in India.
		525 BC: Persians conquer Egypt.

Birth dates of scientists	Key dates in science	Key dates in history
is also credited with discovering the famous property of right-angled triangles which bears his name, and which turns out to be an immensely useful tool in the investigation of relationships in spacetime.		About **520 BC**: Anaximander devises a cylindrical model of the Earth.
About **500 BC**: Leucippus. A disciple of Zeno and the teacher of Democritus, Leucippus is credited with inventing the atomic theory, although very little is known about his life and work.	**500 BC**: Pythagorians teach that the Earth is a sphere. **5th century BC**: The idea that matter is made up of fundamental, indivisible particles arises in Greek thought, giving us the word 'atom', from the Greek *atomos*.	About **500 BC**: Pythagoreans argue that the Earth is a sphere. Steel manufactured in India.
About **494 BC**: Empedocles. One of the earliest proponents of the idea that everything is made up of four 'elements' – fire, air, water and earth.		**490 BC**: Greeks defeat Persians at Marathon. **478 BC**: Athenian empire established. **470 BC**: Socrates born.
About **460 BC**: Democritus of Abdera. Suggested that the world is made up of only vacuum and atoms – an infinite number of tiny, hard, indestructible particles which combine with one another in different ways to produce the variety of everything in the world, both living and non-living.	**450 BC**: Empedocles proposes the 'four element' idea. **430 BC**: Democritus of Abdera develops the idea that everything is made of atoms.	**430 BC**: Plato born.
About **427 BC**: Plato. Greek philosopher who based much of his thinking on the idea of perfection, arguing that the Earth must be a perfect sphere and that all other objects in the Universe moved in perfect circles around the Earth.		
4th century BC: Hui Shih. Introduced the idea of the 'small unit', the smallest possible entity that could exist in nature, with nothing inside itself.	**4th century BC**: The Chinese philosopher Hui Shih introduces the idea of the 'small unit', the smallest possible entity that can exist in nature, with nothing inside itself.	**400 BC**: First settlement at site of London. **390 BC**: Aristotle born.
388 BC: Heraklides of Pontus. Greek philosopher and astronomer who taught that the Earth turns on its axis once every 24 hours – an idea that did not become widely accepted for 1,800 years.		

Birth dates of scientists

384 BC: Aristotle. Among his many interests, Aristotle wrote about cosmology; building on the ideas of his predecessors, he came up with the model of the Universe as a series of concentric spheres, centred on the Earth and rotating about it. Also established the idea that everything in the material world is composed of four 'elements' – fire, earth, air and water.

342 BC: Epicurus. Taught that the Universe is made up of innumerable indestructible atoms, which differ from one another only in size, shape and position, moving in an infinite void.

306 BC: Euclid born.

Early 3rd century BC: Aristarchus of Samos. The first person to attempt to work out the relative distances to the Sun and Moon after it was realized that the Earth is round. Aristarchus realized that the apparent movement of the stars across the sky is caused by the Earth's rotation.

About **273 BC:** Eratosthenes of Cyrene. Carried out the first reasonably accurate calculation of the size of the spherical Earth.

Key dates in science

Third Century BC: Euclid gathers together the knowledge of his time, and writes it down; this knowledge includes the geometric ideas now known as Euclidean geometry, which formed the basis of mathematical teaching in many places until well into the 20th century.

250 BC: Archimedes establishes the basics of mechanics and hydrostatics.

Key dates in history

348 BC: Plato writes *The Republic.*

340 BC: Philip II of Macedonia rules in Greece.

336 BC: Alexander the Great succeeds Philip.

323 BC: Alexander dies after conquering most of the known world.

312 BC: First aqueduct built to bring water to Rome.

300 BC: Alexandria Museum built.

280 BC: Colossus of Rhodes completed.

265 BC: Archimedes discovers the law of specific gravity (an object displaces its own weight when floating in water).

264 BC: First Punic War (between Rome and Carthage) begins.

About **260 BC:** Construction of Great Wall of China begins.

240 BC: Chinese astronomers observe Halley's Comet – the earliest recorded visit of the comet.

Birth dates of scientists

2nd century BC: Hipparchus of Nicaea. Measured the length of the year to an accuracy of 6.5 minutes, and made the first realistic estimates of the distances to the Sun and Moon.

About **95** BC: Lucretius (Titus Lucretius Carus). Wrote a poem, *De rerum natura*, which propounded the ideas of Epicurus, including the notion of atoms moving in an infinite void.

About **100** AD: Ptolemy. Greek astronomer who wrote the *Almagest*, a thirteen-volume description of the Universe as understood at that time. This Ptolemaic system held sway for more than 1,000 years.

Key dates in science

165 BC: Chinese astronomers record sunspots – first accurately dated observations.

46 BC: Julian calendar brought into use in the Roman Empire by Julius Caesar. In order to bring the calendar back in line with the seasons, the year 46 BC has 445 days in it!

44 BC: Major eruptions of Mount Etna. Volcanic pollution blocks sunlight and cools the Earth, causing crop failures in China.

79 AD: Pliny the Younger provides the first detailed written description of a major volcanic eruption, the Vesuvian outburst that destroyed Pompeii.

132 AD: In China, Zhang Heng invents the first seismograph. It indicates the direction of an earthquake by dropping balls shaken from the mouths of

Key dates in history

About **235** BC: Eratosthenes correctly calculates the size of the Earth.

218 BC: Hannibal crosses the Alps.

About **200** BC: Growth of the Roman Empire.

146 BC: Rome destroys Carthage after taking over Greece.

About **140** BC: Invention of paper in China. *Venus de Milo* sculpted; artist unknown.

About **100** BC: Great Wall of China completed. Chinese ships visit east coast of India, navigating with the aid of the magnetic compass (lodestone).

54 BC: Julius Caesar invades Britain.

46 BC: Julian Calendar introduced in Rome.

44 BC: Julius Caesar killed. Roman conquest of Britain begins.

40–4 BC: Reign of Herod the Great.

5 BC: Christ born. World population 250 million.

28 AD: Jesus crucified.

66 AD: Mark's Gospel written.

79 AD: Pompeii and Herculaneum destroyed by eruption of Vesuvius.

About **100** AD: Paper first used for writing.

122 AD: Romans construct Hadrian's Wall.

Birth dates of scientists	Key dates in science	Key dates in history
	bronze dragons by the earthquake wave.	**286 AD**: Roman Empire divided into Eastern and Western halves, initially for administration.
		330 AD: Constantinople founded.
		389 AD: Library at Alexandria destroyed.
	400 AD: The term 'chemistry' used for the first time by scholars in Alexandria.	**395 AD**: Roman Empire completely split, with two separate emperors.
		410 AD: Visigoths sack Rome.
		432 AD: St Patrick sets out to convert the Irish.
		433 AD: Attila becomes leader of the Huns.
		455 AD: Vandals sack Rome.
		476 AD: Last remnant of Roman Empire in Italy destroyed by German invaders.
		About **500 AD**: The abacus.
		517 AD: Buddhism introduced into China.
		525 AD: Introduction of the Christian calendar.
		570 AD: Mohammed born.
		About **600 AD**: Invention of printing press – Chinese woodblocks. Windmills built in Persia.
		613 AD: Mohammed begins to teach openly.
		616 AD: Visigoths conquer Spain.
		620 AD: Flight of Mohammed from Mecca (the hegira); this event marks the start of the Islamic calendar.
		632 AD: Spread of Arab Empire begins following the death of Mohammed.

Birth dates of scientists	Key dates in science	Key dates in history
	635 AD: Unknown Chinese scholar writes down the rule that the tail of a comet always points away from the Sun.	**670** AD: Venerable Bede born.
		697 AD: Carthage destroyed by Arabs.
		699 AD: *Beowulf* completed.
		711 AD: Arab invasion of Spain.
		731 AD: Mayan Empire at the start of its greatest flowering.
		732 AD: Defeat of Arabs at Poitiers marks the limit of the western expansion of Islam.
		748 AD: First printed newspaper, in Beijing.
		768 AD: Charlemagne becomes King of the Franks.
		About **790** AD: Irish monks settle in Iceland.
		About **800** AD: Spread of Vikings begins.
	827 AD: Ptolemy's *Megale syntaxis* is translated into Arabic as the *Almagest*.	
	840 AD: First Arab record of observations of sunspots.	**849** AD: Alfred the Great born.
		863 AD: Invention of Cyrillic alphabet.
	880 AD: Arab chemists distil alcohol from wine.	**900** AD: Greenland discovered by Norse.
		924 AD: Death of Alfred the Great of England.
		929 AD: Good King Wenceslas murdered.
About **965** AD: Alhazen (Abi Ali al-Hassan ibn al-Haytham). Born in Basra, now part of Iraq. The greatest scientist of the Middle Ages. His greatest scientific contribution was his work on optics. This work was translated into Latin at the end of the 12th century, and influenced the thinking of, among others, Roger Bacon. Key insights included an		**975** AD: Modern arithmetical notation introduced into Europe by the Arabs.
		982 AD: First Viking settlement in Greenland.
	Around **1000**: Alhazen publishes his seven books on optics.	About **1000**: Islamic science flourishes. Chinese use coal as

Birth dates of scientists

argument that sight is caused not by the eye sending out rays to scan the outside world, but by the eye receiving light. Alhazen also measured both the reflection and refraction of light, tried to explain the occurrence of rainbows, and studied the Sun during an eclipse using a camera obscura.

Key dates in science

1054: Chinese astronomers witness the supernova explosion that created the Crab nebula. This explosion is temporarily brighter than Venus, being visible in daylight for 23 days.

1066: Large comet (now known to be a visit of Halley's Comet) seen.

Key dates in history

fuel. In India, the mathematician Sridhara realizes the importance of the zero.

1005: Science library founded in Cairo.

1036: Modern musical notation introduced by Guido d'Arrezzo.

1050: First use of moveable type to print books in China.

1066: William of Normandy defeats Harold at the Battle of Hastings.

1086: Domesday Book.

1099: Jerusalem falls to the Crusaders.

About **1100**: Magnetic compass in use in China. Sinchi Roca becomes the first King of the Incas.

1150: First rockets used in China. University of Paris founded.

1155: Oldest known printed map produced in China.

1157: Richard the Lionheart born.

1167 or **1168**: Formal foundation of the University of Oxford.

1174: Construction of the Tower of Pisa begins.

1204: Crusaders sack Constantinople.

1206–27: Genghis Khan conquers vast area of the Eurasia.

1213: Beginnings of the University of Cambridge.

Birth dates of scientists	Key dates in science	Key dates in history
		1215: Magna Carta.
		About **1250**: Invention of the quill pen.
		1260: Genghis Khan's grandson, Kublai Khan, becomes Emperor of China.
		1271: Marco Polo sets out on his great journey to the east.
1285: William of Ockham. Developer of the idea of 'Ockham's razor'. This idea states that if there are two possible explanations for something, and one explanation is simpler than the other, then the simpler explanation should be preferred.		**1294**: Kublai Khan dies.
		1295: Marco Polo returns to Italy.
		About **1300**: Beginning of the Ottoman Empire which will rule a large part of the Mediterranean and Middle East until 1923.
		1305: Giovanni Pisano completes his sculpture, the *Madonna and Child*.
		1307: Dante starts work on his *Divine Comedy*.
		About **1310**: First mechanical clocks in Europe.
		1311: Notre Dame Cathedral completed in Paris.
		1314: Battle of Bannockburn.
		1331: Black Death emerges in China and eventually spreads to Europe.
		1338: Beginning of the 'Hundred Years' War' between England and France.
	1350s: Jean Buridan develops the idea of 'impetus', a forerunner of the modern concept of inertia. He rejects the idea that planets and other 'heavenly bodies' are pushed along by angels, and says that impetus is all that is needed to do the job.	**1350**: Black Death (plague) in Europe.
		1356: First use of cannon in warfare, in China.
		1366: Petrarch writes *Canzoniere*.
		1386: Heidelberg University founded.

Birth dates of scientists	Key dates in science	Key dates in history
		1387: Chaucer starts work on *The Canterbury Tales*.
		1388: Chaucer writes *Troilus and Criseyde*.
		1398: Delhi destroyed by Tamburlaine.
		1409: Donatello completes his sculpture, *David*.
		1419: Publication of Baccaccio's *Decameron*.
		1426: Masaccio's painting, the *Virgin Enthroned*.
		1431: Joan of Arc burned at the stake.
		1440s: Printing press using moveable type developed in Europe by Gutenberg.
		1452: Completion of the Medici Palace in Florence. Richard III of England born. Leonardo da Vinci born.
		1453: Constantinople falls to the Turks and becomes part of the Ottoman Empire.
		1454–5: Gutenberg Bible printed.
		1455–85: Wars of the Roses.
1473: Copernicus, Nicolaus (Mikolaj Kopernigk). Polish astronomer (and doctor) who set out the idea that the Sun, and not the Earth, is the centre of the Solar System.		**1473**: Michelangelo paints the ceiling of the Sistine Chapel.
		1474: Caxton prints the first book in English.
About **1480**: Magellan, Ferdinand. First European to describe the Magellanic Clouds in detail.		**1478**: Boticelli paints *Primavera*. Establishment of the Spanish Inquisition.
		1489: First use of + and – signs in mathematics.
	1490: Leonardo da Vinci studies capillary action of liquids in narrow tubes.	**1491**: Henry VIII born.

Birth dates of scientists	Key dates in science	Key dates in history
		1492: Columbus discovers the islands off the east coast of Central America. First globe made by geographer Martin Behaim.
		1498: Vasco da Gama voyages to India round the Cape of Good Hope.
		1513: Machiavelli publishes *The Prince*.
	16th century: The first reflecting telescope, a telescope that gathers light and magnifies an image primarily through the use of a curved mirror, is developed by Leonard Digges.	**1516:** Thomas More publishes *Utopia*.
		1517: Martin Luther begins the Protestant movement.
	1519: Ferdinand Magellan describes the Magellanic Clouds in detail.	**1519:** Ferdinand Magellan commences the voyage that will end with one of his ships (but not Magellan, who is killed en route, in the Philippines) completing the first circumnavigation of the globe.
		1532: Pizarro conquers Peru.
		1533: Elizabeth I born.
	1540: The German astronomer Peter Apian records the first European discovery of the fact that comet tails always point away from the Sun.	**1543:** Publication of Nicolaus Copernicus's book *On the Revolutions of Heavenly Bodies*.
1548: Bruno, Giordano. Italian monk who was an early supporter of the proposal made by Copernicus that the Earth moves round the Sun. Burned at the stake for heresy in Rome.	Second half of the 16th century: Leonard Digges probably makes the first refracting telescope.	**1550:** Tobacco growing begins in Spain.
		1551: Leonard Digges invents the theodolite. Titian paints Prince Felipe of Spain.
		1558: Elizabeth I becomes Queen of England.
1564: Galileo (Galilei, Galileo). First person to make systematic observations of the heavens using a telescope. Disproved commonly held theory that heavy objects fell faster than lighter objects. Realized the importance of actually carrying out experiments to test theories.	**1572:** Supernova observed by Tycho Brahe.	**1564:** Shakespeare born. Horse-drawn carriage introduced into England from the continent.
		1568: Mercator introduces his eponymous map projection.

Birth dates of scientists

1571: Kepler, Johannes. Discovered the three laws of planetary motion that helped lead Newton to his universal law of gravity.

Key dates in science

1576: Thomas Digges discards the Ptolemaic idea of the stars being attached to a single crystal sphere surrounding the Earth, and suggests that the stars are distributed into an endless infinity of space.

1581: Galileo investigates the behaviour of pendulums.

1582: Gregorian calendar, the calendar widely used on Earth today, is introduced by Pope Gregory XIII. It replaces the less accurate Julian calendar. The Gregorian calendar is introduced first to Catholic countries.

Probably **1583:** Galileo makes the observation that, provided the length of a pendulum is fixed, the time it takes for the pendulum to complete one swing is always the same, whether it swings through a large arc or a small one.

Key dates in history

1582: Pope Gregory XIII reforms the calendar. As a result, this is the shortest year on record.

1586: Russian expansion east of the Urals begins. Walter Raleigh introduces tobacco smoking to England.

1588: Spanish Armada defeated.

1599: First performance of Shakespeare's *Much Ado about Nothing*.

1592: Gassendi, Pierre. Carried out the experiment of dropping a ball from the top of the mast of a moving ship and showing that it landed at the foot of the mast, demonstrating that motion is relative. He also espoused the atomic theory, believed that light was a stream of particles, and gave the aurora borealis its name.

1596: Brahe, Tycho. Made accurate measurements of the positions of stars and the movements of planets, paving the way for Kepler to discover the laws of planetary motion.

1596: Descartes, René du Perron. Invented the techniques of coordinate geometry (Cartesian geometry).

1604: Supernova observed by Kepler.

1600: First performance of Shakespeare's *Hamlet*.

1605: In *Advancement of Learning*, Francis Bacon encourages the scientific investigation of the world. Gunpowder Plot fails to blow up the English Parliament. Cervantes writes *Don Quixote*.

Birth dates of scientists	Key dates in science	Key dates in history
		1607: English begin to settle in Virginia.
	1609: Learning of the invention of the telescope, Galileo becomes the first person to use a refracting telescope for astronomical observations.	**1609**: Galileo uses a telescope to study the Moon and planets. Kepler's laws published.
	1610: Johannes Kepler becomes the first person to realize that the darkness of the night sky directly conflicts with the idea of an infinite Universe filled with bright stars. He concluded that the Universe must therefore be finite – that, in effect, when we look through the gaps between the stars we see the dark end of the Universe.	**1610**: French colony established in Quebec. **1611**: King James Bible published. **1615**: Completion of St Peter's in Rome. **1618**: The Thirty Years' War begins.
1627: Boyle, Robert. English pioneer of chemistry, best remembered for 'Boyle's Law', which says that the volume of a given mass of gas at constant temperature is inversely proportional to the pressure on it.	**1610**: Galileo's book *The Starry Messenger* is published, recording his observations of thousands of stars invisible to the naked eye.	**1620**: Pilgrims land at Plymouth Rock.
1629: Huygens, Christiaan. Invented the first successful pendulum clock; designed and built improved astronomical telescopes; developed a complete wave theory of light; discovered Titan; and recognized the nature of Saturn's rings.	**1631**: Pierre Gassendi is the first person to observe a transit of Mercury across the face of the Sun.	**1632**: Christopher Wren born. **1633**: Galileo's trial.
1635: Hooke, Robert. Hooke's many contributions included ideas about gravity, coining the world 'cell' in its biological context, and an early attempt to make a watch powered by a spring. His theories of light led him to bitter argument with Isaac Newton.		**1636**: Harvard College founded. **1637**: Descarte's *Discourse on the Method of Rightly Conducting Reason and Seeking Truth in the Sciences* published. **1638**: Birth of Louis XIV. **1641**: First pendulum clock built by Galileo's son.
1642: Newton, Isaac. Discovered the law of gravity and the laws of motion that bear his name;	**1643**: Torricelli makes the first barometer.	**1642**: English Civil War begins.

Birth dates of scientists

invented the mathematical technique of calculus; carried out important studies in optics, including the design of a new kind of reflecting telescope. Spelled out what quickly became the scientific method of formulating hypotheses and testing them with experiments.

1644: Rømer, Øle. Measured the speed of light in 1675, using observations of eclipses of the moons of Jupiter to reveal how long it takes light to cross the orbit of the Earth.

1656: Halley, Edmond. Realized that the comet which now bears his name is in a regular orbit around the Sun under the influence of gravity, in accordance with the law of gravity established by Newton. Key evidence that scientific laws apply to the entire Universe. Also compiled a star catalogue, detected the proper motion of some stars using historical records, and initiated a research programme that led to a good estimate of the distance from the Earth to the Sun.

Key dates in science

1647: First map of the Moon made by Johannes Hevelius.

1654: Grand Duke Ferdinand II of Tuscany invents the thermometer.

1655: Titan, the largest moon of Saturn, is discovered by Christiaan Huygens.

1656: Huygens identifies the true nature of the rings of Saturn.

1659: Huygens observes surface markings on Mars.

1660: Boyle publishes his law relating gas pressure and volume.

1663 John Gregory proposes the design of reflecting telescope that becomes known as the Gregorian telescope.

1664: Robert Hooke discovers the Great Red Spot of Jupiter.

1665: Plague closes Cambridge University and sends Isaac Newton back home to Woolsthorpe, where he makes many of his great discoveries.

1668: Isaac Newton reinvents the reflecting telescope first invented by Leonard Digges in the 16th century, and becomes the first person to put the invention to practical use.

Key dates in history

1649: Charles I beheaded.

1650: Bishop Ussher sets the date of the Creation at 4004 BC. Cyrano de Bergerac suggests seven ways of flying to the Moon.

1651: In *Leviathan*, Thomas Hobbes says that man's life is 'solitary, poor, nasty, brutish, and short'.

1656: Huygens develops an accurate pendulum clock.

1660: Restoration of the monarchy in England.

1661: Construction of Versailles Palace begins.

1663: Royal Society receives its Charter.

1664: Descartes' *Treatise on Man* says that animals and people are 'mechanical' objects with no 'vital force', or soul.

1665: Royal Society starts publication of its *Philosophical Transactions*. Rembrandt van Rijn paints *Juno*. Plague strikes England.

1666: Great Fire of London. French Royal Academy of Sciences founded.

1670: Molière writes *Le Bourgeois Gentilhomme*.

Birth dates of scientists	Key dates in science	Key dates in history
	1671: The distance to Mars is first measured reasonably accurately, by a team of French astronomers observing the position of the planet on the sky from Cayenne, in French Guiana, while a team in Paris note its position at the same time.	
	1672: Cassegrain publishes the design of the telescope which becomes known as a Cassegrain telescope. The design was not put into practice until the 18th century.	
	1675: Charles II founds the Royal Greenwich Observatory. Cassini discovers the gap in the rings of Saturn now known as the Cassini division.	
	1676: The finite speed of light is determined by Øle Rømer.	**1676**: British Museum founded.
		1680: The first clocks with minute hands (previously, they only showed the hours).
		1683: Turks besiege Vienna.
1685: Berkeley, (Bishop) George. Argued that all motion is relative and must be measured against something.		**1685**: Birth of Johann Sebastian Bach.
	1687: Newton's great work, *Philosophiae Naturalis Principia Mathematica* is published at the urging of Edmond Halley. It gives the three fundamental laws describing the dynamical behaviour of objects, proving that the orbits of the planets around the Sun can be explained by an inverse square law of gravity.	**1688**: In the 'Glorious Revolution' in England, the Catholic James II is replaced by the Protestant William and Mary, of the Dutch House of Orange.
	1690: Huygens publishes *Treatise on Light*, which fully develops his wave theory of light.	**1690**: Locke publishes his *Essay Concerning Human Understanding*.
		1692: Salem witchcraft trials.
1693: Bradley, James. Discovered the aberration of starlight and used this to determine the speed of light, arriving at a figure equivalent to 308,300 km/sec, close to the modern value of 299,792 km/sec.		**1697**: Birth of Canaletto.

Birth dates of scientists

1698: de Maupertuis, Pierre Louis Moreau. The first person to formulate the principle of least action.

1701: Celsius, Anders. Suggested a temperature scale based on two fixed points: 0 degrees for the boiling point of water, and 100 degrees for the melting point of ice. Soon after his death this scale was reversed, so that the boiling point of water was 100 degrees, and the melting point of ice 0 degrees.

1707: Euler, Leonhard. Developed the idea of the principle of least action and the calculus of variations, pointing the way for the later work of Joseph Lagrange, which was developed into a key tool in the path integral approach to quantum mechanics, by Richard Feynman. Among other achievements he also introduced mathematical notations, such as π, e and i, which have become standard.

1724: Kant, Immanuel. German philosopher who was also interested in cosmology and who, in an essay published in

Key dates in science

1700s: Immanuel Kant suggests that distant nebulae might be complete star systems beyond the Milky Way.
Johannes Kepler discovers the three laws of planetary motion, using data gathered by Tycho Brahe. These laws become known as Kepler's Laws.

1704: Isaac Newton publishes his *Opticks*, which sees light as a stream of tiny particles, or corpuscles, like miniature cannon balls.

1705: Edmond Halley publishes his prediction of the return of the comet that now bears his name.

1712: First volume of John Flamsteed's star catalogue published.

1714: Gabriel Fahrenheit devises a mercury thermometer and uses the temperature scale that will later be named after him.

1718: Halley becomes the first person to realize that stars move across the sky and are not really 'fixed'.

Key dates in history

1698: Steam-powered pump to remove water from mines patented. Place Vendôme completed in Paris.

1701: Yale University founded.

1702: The first daily newspaper, the *Daily Courant*, published in London. Jethro Tull invents the machine drill, for planting seeds.

1704: The *Boston Newsletter* is the first American newspaper.

1706: Benjamin Franklin born.

1707: Union of England, Wales and Scotland to form Great Britain.

1710: Completion of St Paul's in London.

1711: David Hume born. Alexander Pope writes his *Essay on Criticism*.

1712: Newcomen develops an improved steam engine.

1714: British Government offers a prize of £20,000 for a technique to find longitude at sea.

1715: First Jacobite rebellion.

1719: Daniel Defoe writes *Robinson Crusoe*.

1720: Collapse of the 'South Sea Bubble', a speculative venture in England.

1723: Adam Smith born.

1725: Birth of Robert Clive ('Clive of India'). Vivaldi writes *The Four Seasons*.

Birth dates of scientists

1755, made one of the first suggestions that the planets formed from a cloud of material around the Sun.

1724: Michell, John. First person to come up with the notion of black holes.

1731: Cavendish, Henry. Made pioneering investigations in chemistry and used a torsion balance experiment, devised by John Michell, to make the first accurate measurements of the mean density of the Earth and the strength of the gravitational constant. Much of his work was unpublished during his lifetime, including an anticipation of Ohm's Law and of much of the work of Michael Faraday and Charles Coulomb. He also showed that gases could be weighed, and that air is a mixture of gases.

1736: Coulomb, Charles Augustin de. French physicist who discovered Coulomb's Law, which says that the force between two small charged spheres is proportional to the product of the two charges divided by the square of the distance between them.

1736: Lagrange, Joseph Louis. Developed techniques that proved invaluable in the formulation of group theory, and whose Lagrangian function provides a simple way to describe the behaviour of entities as diverse as light rays, planets in their orbits and subatomic particles scattering from one another in particle accelerators.

1743: Lavoisier, Antoine Laurent. The father of modern chemistry. His single most important contribution was to disprove the phlogiston theory, realizing that burning involves a substance

Key dates in science

1729: James Bradley discovers aberration, an apparent shift in the position of a star caused by the finite speed of light and the motion of the Earth in its orbit around the Sun. Through this he is able to determine the speed of light, arriving at a figure equivalent to 308,300 km/sec, close to the modern value of 299,792 km/sec.

1735: John Harrison builds his first marine chronometer in an attempt to win the prize offered by the British Board of Longitude for a way of keeping time accurately at sea.

1738: Daniel Bernoulli describes the behaviour of a gas in terms of the motion of many tiny particles, which bounce around, colliding with one another and with the walls of their container.

1742: Anders Celsius invents the temperature scale which now bears his name (formerly the Centigrade scale).

1744: Jean-Phillipe Loÿs de Chésaux estimates that there would be a star visible in every direction we look into space,

Key dates in history

1726: Jonathan Swift writes *Gulliver's Travels*.

1732: George Washington born.

1737: Göttingen University founded. William Hogarth paints *The Good Samaritan*.

1739: Royal Society of Edinburgh founded. Hume writes his *Treatise on Human Nature*.

1745: Second Jacobite rebellion.

Birth dates of scientists

combining with oxygen (which he named) from the air, not losing 'phlogiston'.

1749: Laplace, Pierre Simon, Marquis de. Updated Newton's investigations of planetary movements, to account for perturbations in the strict elliptical orbits propounded by Kepler. Was also one of the first people to consider the possibility of black holes.

1753: Rumford, Count (Benjamin Thompson). Proved that heat is a form of energy, not a fluid (caloric) as others had thought.

Key dates in science

provided that the Universe were (in modern terms) 10^{15} light years or more across. To explain why this is not so, he suggests that empty space simply absorbs the energy in the light from distant stars, so that the light gets fainter and fainter as it travels through the Universe.

Pierre de Maupertuis states the principle of least action, a formal version of the idea that objects (including light rays) follow the quickest possible paths.

1746: Leonhard Euler uses Christiaan Huygens' wave theory of light to explain refraction.

1749: Benjamin Franklin invents the lightning rod.

Second half of 18th century: Boyle takes the first steps towards the modern understanding of elements, and the way in which different elements combine to form compound substances.

1750: Pierre de Maupertuis publishes his *Essai de Cosmologie*, the first work to propound the principle of least action.

1752: The Gregorian Calendar is introduced to Britain, and some other parts of the world. To catch up with the Gregorian Calendar, eleven days were omitted, so that the day after 3 September became 14 September.

Benjamin Franklin carries out his famous kite experiment.

1753: George Richmann is killed by lightning while copying Franklin's famous kite experiment.

1755: Immanuel Kant suggests, in his book *Universal Natural History and Theory of the Heavens*, that the planets condensed out of a cloud of primordial matter.

Key dates in history

1746: Princeton University founded.

1747: Johnson's dictionary published.

1748: First blast furnace constructed at Bliston in England.

About **1750**: Population of China reaches 225 million.

1751: Calendar reform in Britain makes 1 January the official first day of the year.

1755: University of Moscow founded.

1756: Birth of Wolfgang Amadeus Mozart.

Birth dates of scientists

1758: Olbers, Heinrich Wilhelm Matthäus. Publicized the puzzle of why the sky is dark at night, developed an improved method of calculating the orbits of comets, and discovered two of the minor planets.

1766: Dalton, John. English chemist who pioneered the use of atomic theory to explain chemical reactions, arguing that the atoms of different chemical elements can be distinguished by differences in their weights.

1766: Wollaston, William. Physician and physicist who first noticed dark lines in the solar spectrum, although he did not appreciate their significance at the time. Also discovered the elements rhodium and palladium, and developed techniques for processing metals such as platinum into a malleable form.

1773: Brown, Robert. Scottish botanist who noticed, in 1827, that pollen grains suspended in water can be seen under the microscope to be in continuous erratic movement. He had no explanation for this Brownian motion, as it became known.

1773: Young, Thomas. Made a major contribution to the

Key dates in science

1758: Halley's Comet reappears, as predicted.

1760: Daniel Bernoulli discovers that electricity obeys an inverse square law similar to the law of gravity.

1767: John Michell suggests that stars which appear close together on the sky are really physically associated in space, and not the result of a chance juxtaposition at quite different distances along the line of sight – binary stars hypothesized.

1769: Transit of Venus across the face of the Sun observed by (among others) Captain James Cook, in Tahiti.

Key dates in history

1758: The 'Imperial' system of weights and measures formally established in Britain.

1761: Rousseau publishes *La nouvelle Héloïse*. John Harrison's 'Number Four' chronometer taken to the West Indies under test.

1763: Boston Massacre.

1764: Richard Arkwright patents his spinning jenny.

1765: James Watt develops an improved steam engine. John Harrison receives the first half of his prize.

1768: Publication of the *Encyclopaedia Britannica* starts, initially in weekly instalments.

1769: First meeting of the American Philosophical Society.

1770: Ludwig van Beethoven born.

1771: Discovery of oxygen. Tobias Smollett writes *The Expedition of Henry Clinker.*

1773: John Harrison receives the second half of his prize at the age of 80, after the British Board of Longitude is told by King George III to stop delaying the award. Boston Tea Party.

Birth dates of scientists	Key dates in science	Key dates in history

Birth dates of scientists

establishment of the wave theory of light at the beginning of the 19th century, chiefly using the double-slit experiment.

1775: Malus, Etienne Louis. Discovered the polarization of light.

1776: Avogadro, Amedeo. Showed, using the discoveries of Joseph Gay-Lussac, that the chemical formula for water is H_2O, not HO, and in 1811 published the paper in which he set out the idea that equal volumes of gas at the same temperature contain equal numbers of molecules (Avagadro's Law).

1777: Gauss, Karl Friedrich. German mathematician and astronomer who pioneered the development of non-Euclidean geometry (important in the general theory of relativity) and did important work in calculating orbits of planets, as well as extensive work in mathematics and physics.

1777: Oersted, Hans Christian. Discovered that an electric current produces a magnetic field.

Key dates in science

1774: Nevil Maskelyne determines the mass of the Earth by measuring the amount by which a mountain deflects a plumb line from the vertical.
    Priestley discovers oxygen.

1776: Pierre Simon de Laplace claims that, if all the forces acting on all objects at any one time were known, then the future would be completely predictable.
    Charles Messier compiles a catalogue of more than 100 star clusters and other objects that might be mistaken for comets.

1783: John Michell becomes the first person to suggest that there might exist 'dark stars' whose gravitation is so strong that light cannot escape from them, presenting his ideas to the Royal Society. Basing his calculations on Newton's theory of gravity, and on the corpuscular theory of light, he assumed that the particles of light would be affected by gravity in the same way as any other objects.

1785: Coulomb publishes his law, which says that the force between two small charged spheres is proportional to the product of the two charges, divided by the square of the distance between them.

Key dates in history

1775: James Watt's steam engine patented.

1776: American Declaration of Independence. Adam Smith's *Wealth of Nations*.

1777: First performance of Mozart's Concerto no. 9.

1778: James Cook discovers Hawaii.

1779: World's first iron bridge built in Coalbrookdale, England.

1781: James Watt patents a system for developing rotary motion from a steam engine.

1783: Montgolfier brothers build and fly hot-air balloons. On 21 November, Jean de Rozier and François Laurent, in a Montgolfier balloon, become the first humans to fly.

1784: Benjamin Franklin invents bifocal spectacles.

1785: First balloon crossing of the English Channel. Seismograph invented.

1786: First experiments with gas lighting.

Birth dates of scientists

Key dates in science

Key dates in history

1787: Fraunhofer, Joseph von. The first person to study the rainbow pattern produced by passing light through a prism in detail, under intense magnification. To his surprise, he discovered that there are many dark lines in the spectrum of white light from the Sun. Fraunhofer counted 574 lines in the solar spectrum, and found many of the same lines in light from Venus and from many stars. The dark lines in the spectrum of the Sun which he studied now bear his name.

1788: Fresnel, Augustin Jean. French physicist who played a major part in establishing the wave nature of light. The key ingredient to his theory was that he envisaged light as transverse waves, not as longitudinal waves.

1788: Gay-Lussac, Joseph Louis. Made an important contribution to the understanding of the behaviour of gases. His most important scientific work was his discovery of the law that gases combine with one another in simple whole-number proportions by volume, published in 1808 and known as Gay-Lussac's law.

1791: Faraday, Michael. English chemist and physicist whose greatest achievements concerned his studies of electricity and magnetism, and who was largely responsible for introducing the concepts of *fields* and *lines of force* into physics.

1788: Lagrange publishes his great book, *Analytical Mechanics*.
   Antoine Lavoisier publishes the first list of elements based on Boyle's definition of compound substances, although he includes substances now known to be compounds and includes caloric (heat) as an element.

1790s: The metre is defined, by the National Assembly in revolutionary France, as 1 ten-millionth of the distance from the North Pole to the equator.

1789: The storming of the Bastille on 14 July triggers the French Revolution.

1791: Metric system introduced in France; it is officially adopted in 1795.

1792–1815: Napoleonic Wars.

1796: Pierre Laplace suggests, independently of John Michell, that there might exist 'dark stars' whose gravitation is so strong that light cannot escape from them. He also proposes the 'nebular hypothesis' for the origin of the Solar System.

1798: Henry Cavendish determines the mass of the Earth, establishing that it has an average density 5.5 times that of water.

1796: Edward Jenner develops vaccination for smallpox.

1797: First use of iron railways, for horse-drawn waggons.

1798: Thomas Malthus publishes (initially anonymously) his *Essay on the Principle of Population*.

Birth dates of scientists

Key dates in science

Key dates in history

1799: Discovery of the Rosetta Stone.

1800: Herschel discovers infrared radiation.

1800: Richard Trevithick builds a high-pressure steam engine. The following year, he builds a steam-powered vehicle. World population now some 870 million.

1801: First known asteroid, Ceres, discovered.

1801: Jean Lamarck publishes early ideas on evolution, unaware of similar work by Erasmus Darwin.

1802: Thomas Young publishes the first of his papers on the wave theory of light, utilizing the double-slit experiment.
   William Wollaston becomes the first person to notice dark lines in the solar spectrum, although he does not appreciate their significance at this time.

1803: Doppler, Christian Johann. Predicted what is now known as the Doppler effect, in 1842.

1803: John Dalton publishes the first table of atomic weights.

1803: Successful trials of Robert Fulton's steam-powered boat on the Seine.

1804: Jacobi, Carl Gustav Jacob. One of the first people to appreciate the importance of symmetry and invariance in physics. His Jacobian determinants, which developed further some of the ideas of Hamilton, became an important feature of quantum mechanics.

1804: Napoleon becomes Emperor of France.

1804: Weber, Wilhelm Eduard. Carried out experiments which determined the ratio of two key quantities, one a measure of electrostatic behaviour and the other a measure of electromagnetic behaviour. This ratio is now known as Weber's constant, and turned out to be the speed of light.

1805: Hamilton, William Rowan. A prodigy, Hamilton discovered the principle of least action in the context of light paths while during his second year as an undergraduate. His greatest contribution to a later generation

1805: Battle of Trafalgar.

1806: Beaufort invents his wind scale.

1807: Young introduces the concept, and the word, 'energy'.

1807: First use of gas to light London streets. Improved

Birth dates of scientists

of quantum physicists was his discovery of a way to restate the equations of motion devised by Joseph Lagrange; the Hamiltonian interpretation became widely used in wave mechanics. He also studied pure mathematics, developing an algebra of four dimensions, known as quaternion theory, which splits into a one-dimensional component and a three-dimensional component, and coined the term 'vector' in its modern mathematical context.

1811: Bunsen, Robert. Worked with Gustav Kirchoff in formulating the basic principles of spectroscopy. One of Bunsen's assistants modified a device invented by Michael Faraday, coming up with what is now known as the Bunsen burner. This provides a clean, hot flame in which different substances can be heated until they glow, radiating light at their own characteristic spectral wavelengths, which can be studied and analysed.

1814: Ångström, Anders Jonas. Swedish pioneer of spectroscopy. After his death, his name was

Key dates in science

1807 and 1808: Etienne Malus discovers the polarization of light. By his hypothesis light consists of a stream of particles, each with a specific orientation. In double refraction, there are two routes that a beam of light can follow – one permits light of one polarization to pass, the other permits light of the other polarization to pass. Therefore the beam splits.

1808: John Dalton publishes the first volume of his book *New System of Chemical Philosophy*, which puts forward for the first time the idea that atoms of different elements have different weights, and concludes that atoms can be neither created nor destroyed – chemical reactions represent simply a rearrangement of the atoms.

Joseph Gay-Lussac publishes his discovery of the law that gases combine with one another in simple whole-number proportions by volume, which is to become known as Gay-Lussac's Law.

William Wollaston observes that an understanding of the three-dimensional arrangement of atoms in a substance would be useful in explaining its properties.

1811: Amedeo Avogadro realizes that equal volumes of gas at the same temperature and pressure contain equal numbers of molecules (Avogadro's Law).

1812: Hans Oersted predicts that an electric current should produce a magnetic field.

1814: Joseph von Fraunhofer becomes the first person to study in detail the rainbow pattern

Key dates in history

steamboat tested in the East River off New York by Robert Fulton.

1808: Humphrey Davy invents the electric arc light. Richard Trevithick builds a passenger railway in London.

1809: Charles Darwin born.

1810: Foundation of the University of Berlin.

1812: Napoleon invades Russia.

1814: George Stephenson's first steam locomotive starts work.

Birth dates of scientists

given to a unit of wavelength used in spectroscopy which is now obsolete.

1814: Foucault, (Jean Bernard) Leon. Made the first laboratory measurements of the speed of light accurate to within 1 per cent, invented the gyroscope, and devised the pendulum method which bears his name for demonstrating the rotation of the Earth.

1820: Tyndall, John. A great popularizer of science. Discovered the Tyndall effect, which helps to explain why the sky is blue, in 1869.

1821: Loschmidt, Johann Joseph. Made the first reasonably accurate calculation of the size of the molecules present in air, obtaining a figure about 30 times smaller than modern estimates.

Key dates in science

produced by passing light through a prism, using intense magnification. To his surprise he discovers (independently of Wollaston) that there are many dark lines in the spectrum of white light from the Sun. Fraunhofer soon counts 574 lines in the solar spectrum, and finds many of the same lines in light from Venus and from many stars.

1816: Augustin Fresnel develops his version of the wave theory of light.
   Thomas Young estimates that 'particles of water' must have a size in the range from 5 to 25 billionths of a centimetre, only about ten times bigger than the modern estimate.

1817: Fresnel submits his explanation of diffraction to the French Academy of Sciences. The theory suggests a wave theory of light which runs contrary to the thought of most of the judges. One of the judges, Simeon Poisson, uses the theory to predict that if light is a wave there should be a bright spot in the centre of the shadow cast by a circular object, caused by diffraction. The experiment is carried out, the bright spot (now known as the Poisson spot) found, and Fresnel duly awarded the prize.

1820: Fresnel develops a type of lens, now known as a Fresnel lens, which will find widespread use in concentrating the beams of light from lighthouses.
   Hans Oersted for the first time measures the magnetic field produced by an electric current.

1821: Catholic Church lifts its ban on teaching the Copernican theory.
   Fresnel explains polarization in terms of transverse waves, rather than the particles suggested by Malus.

Key dates in history

1815: Battle of Waterloo. First steam-powered warship built in America. Humphrey Davy invents the coal miners' safety lamp.

1816: Rossini writes *The Barber of Seville*. Stethoscope invented.

1819: Paddle steamer *Savannah* crosses the Atlantic. Birth of Queen Victoria.

1821–32: Greek wars of independence (from Ottoman Empire).

1823: Charles Macintosh invents a waterproof coat.

Birth dates of scientists

1824: Kelvin (Baron Kelvin of Largs, William Thomson). Made many contributions to thermodynamics and the theory of electromagnetic radiation; was responsible for the success of the first transatlantic cable communications system. Kelvin's main contribution to astronomy dealt with the ages of the Earth and the Sun. Also devised (in 1848) the 'absolute' scale of temperature now known as the 'Kelvin' scale.

1824: Kirchoff, Gustav Robert. Explained the Fraunhofer lines as due to absorption of light by different elements present in the Sun's atmosphere, and went on to formulate the basic principles of spectroscopy, working with Robert Bunsen. His law of emission (Kirchoff's Law) was to lead to the concept of a black body, and played an important part in the early development of quantum theory.

1825: Balmer, Johann Jakob. Studied spectroscopy and found a simple formula which relates the frequencies of lines in the visible spectrum of hydrogen to one another.

1826: Cannizzaro, Stanislao. The first person to appreciate the difference between molecular weight and atomic weight, drawing on the work of Amedeo Avogadro to show that common gases such as hydrogen exist as molecules. Also the first person to draw up a table of atomic and molecular weights based on the atomic weight of hydrogen as the fundamental unit of mass; this was an important step towards the periodic table.

1826: Riemann, (Georg Friedrich) Bernhard. First person to develop a comprehensive mathematical description of curved space (non-Euclidean

Key dates in science

1826: Heinrich Olbers publishes a landmark paper outlining the problem with having an infinitely large Universe and a dark night sky. He reaches a conclusion similar to that reached by de Chésaux – that the light from distant stars is absorbed in a thin gruel of material between the stars.

1827: Robert Brown discovers that tiny particles of pollen suspended in water can be seen under the microscope to be in continuous erratic movement.

Around **1827:** William Hamilton finds that he can modify the equation of motion derived by

Key dates in history

1825: First passenger steam train.

1826: Delacroix paints Greece in the *Ruins of Missolonhi*.

Birth dates of scientists

geometry), providing the mathematical framework later used by Einstein in his general theory of relativity.

1830: Meyer, (Julius) Lothar. Discovered the periodic pattern of the chemical elements, independently of Dmitri Mendeleyev, and for this shared the award of the Davy Medal of the Royal Society with Mendeleyev in 1882.

1831: Maxwell, James Clerk. His many achievements included developing the three-colour theory of colour vision, and the equations which describe the behaviour of electromagnetic radiation.

1832: Crookes, William. British physicist and chemist best known for his experiments with high-voltage discharge tubes.

1834: Mendeleyev, Dmitri Ivanovich. Chemist who came up with the idea of the periodic table of the elements, and correctly predicted the existence of unknown elements, on the basis of gaps in that table.

1835: Stefan, Josef. The first person to identify the relationship between the rate at which an object radiated energy and its temperature, a relationship which became known as the Stefan-Boltzmann Law.

Key dates in science

Lagrange into a set of equations which is in some ways simpler than Lagrange's equations. This version of the equations will be widely used in the early days of quantum mechanics.

1830: Joseph Henry discovers the principle of the dynamo, but does not publish his discovery. He publishes after the same discovery is announced by Michael Faraday.
  First volume of *Principles of Geology* published by Charles Lyell.

1831: Michael Faraday and Joseph Henry independently discover electromagnetic induction.
  Earliest version of the second law of thermodynamics developed by Benoit Clapeyron.

1835: Halley's Comet makes its second return since Halley's death.
  August Comte says that mankind will never know the chemical composition of the stars.
  Gustave Coriolis describes the force which now bears his name.

Key dates in history

1830: Spread of English-speaking Americans west begins.

1831: Darwin begins his voyage on the *Beagle*.

1832: Reform Bill extends the franchise in Britain.

1833: Karl Gauss and Wilhelm Weber develop an electric telegraph which sends signals between two stations 2 km apart. Babbage builds his 'difference engine'; Ada Lovelace writes the first computer program, to run on it.

1834: Louis Braille perfects system of reading for the blind.

1835: Samuel Colt patents his revolver. John Constable paints *The Valley Farm*.

Birth dates of scientists	Key dates in science	Key dates in history

1836: Lockyer, (Joseph) Norman. Discovered and named helium, 30 years before it was found on Earth. In 1868 Lockyer and French astronomer Jules Janssen independently of one another, noticed a feature in the spectrum of the sun that they had never seen before. Janssen sent his data to Lockyer, who confirmed the feature and decided it was caused by the presence in the Sun of an unknown element. This he called helium, from the Greek word for the Sun, *helios*.

1837: Newlands, John Alexander Reina. Developed the idea of a periodicity in the properties of the chemical elements independently of, and earlier than, Dmitri Mendeleyev. Unfortunately, however, his ideas were not accepted at the time.

1837: Samuel Morse patents his electric telegraph.

1837: van der Waals, Johannes Diderik. Derived a widely used equation describing the behaviour of gases – an equation of state.

1838: Mach, Ernst. Most familiar today because of the Mach number, which gives the speed of an object relative to the speed of sound in the medium through which the object is travelling. Mach also made profound contributions to the way scientists think about the Universe and the models they use to describe it, both on the large scale and the small. These ideas influenced both Einstein and some of the pioneers of quantum mechanics.

1838: The first stellar distance determined by parallax, the distance to 61 Cygni, is published. Parallax is the apparent movement of an object across the sky when it is seen from two different points, and can be used to calculate the distance to the object by triangulation.

1838: Morley, Edward Williams. Carried out an experiment with Albert Michelson which famously failed to find any evidence for an 'ether' through which light was transmitted.

Birth dates of scientists

1842: Rayleigh, Lord (John William Strutt, Third Baron Rayleigh of Terling Place). Received the Nobel Prize for Physics in 1904 for his discovery of the element argon, made jointly with William Ramsay. Rayleigh also wrote an epic book on *The Theory of Sound*, published in two volumes in 1877 and 1878. However, to physicists his most important contribution was the Rayleigh-Jeans Law, an equation which described the distribution of energy at different wavelengths in black body radiation. This equation accurately describes longer wavelengths of radiation.

1844: Boltzmann, Ludwig. Suggested that the Universe might be a gigantic statistical freak. In a uniform Universe, according to Boltzmann's interpretation of Poincaré's work, it will very occasionally happen that all the particles in one part of the universe will be moving in just the right way to create stars, galaxies or a Big Bang. A region of order would, by chance, form out of some of the more permanent state of disorder. Also made important contributions to the understanding of thermodynamics, and extended James Clerk Maxwell's theory of the distribution of velocities among gas molecules.

1845: Röntgen, Wilhelm Conrad. Discovered X-rays and, as a result, was awarded the first ever Nobel Prize for Physics in 1901.

Key dates in science

1842: The Doppler effect first predicted by Christian Doppler. This effect shows how a change may be made in the frequency of light, or in the pitch of a sound, by the motion of the object emitting the light or making the sound.

Julius Mayer is the first person to state the law of conservation of energy.

1845: A 73-inch reflecting telescope is built at Birr Castle, in central Ireland, by the Earl of Rosse. This telescope is the most powerful in the world for many years.

Key dates in history

1839: Louis Daguerre describes a technique for making photographs. Charles Goodyear develops a technique for 'vulcanizing' rubber.

1840: Penny post introduced in UK.

1843: First tunnel under the Thames opened. Screw-driven steamer *Great Britain* crosses the Atlantic.

1844: Friedrich Engels writes *The Condition of the Working Class in England.*

1845: First publication of *Scientific American*.

Birth dates of scientists	Key dates in science	Key dates in history
	The first galaxy (then known as nebula) in which spiral structure is seen is discovered by Lord Rosse. This is the whirlpool galaxy, a particularly striking and photogenic example of a disc galaxy with well-defined spiral arms, and is also known by its catalogue numbers as M51 or NGC 5194.	
	Faraday suggests that light is a form of electromagnetic wave.	
	1846: Faraday suggests that light is 'a high species of vibration in the lines of force which are known to connect particles', attempting to dismiss the notion of the ether.	**1846**: Foundation of the Smithsonian Institution in Washington, DC. **1846–51**: Potato famine in Ireland.
	1847: James Joule independently discovers the law of conservation of energy.	
	William Thomson proposes the idea that electric and magnetic fields are distributed throughout space, a step towards the comprehensive field theory of electromagnetism developed by James Clerk Maxwell a little later.	
	1848: Julius Mayer calculates that the Sun would cool in only 5,000 years if it had no source of energy.	**1848**: Karl Marx and Friedrich Engels produce the *Communist Manifesto*.
	Hippolyte Fizeau suggests that the Doppler effect should also apply to light, and predicts a redshift in light from an object moving away from the observer.	
	Edgar Allen Poe speculates that the sky might be dark at night, even in a Universe that is infinite in extent, if the distances involved are so immense that the light from many stars has yet to reach us.	
	William Thomson, later Baron Kelvin of Largs, defines the absolute scale of temperature that is to become known as the Kelvin scale.	
	1849: Fizeau measures the speed of light to within 5 per cent of the accepted modern value.	**1849**: California gold rush.

Birth dates of scientists	Key dates in science	Key dates in history

Key dates in science

William Thomson (later Lord Kelvin) coins the term 'thermodynamics'.

1850: Leon Foucault shows that the speed of light is greater in air than it is in water.

Late 19th century: Henri Poincaré shows that an 'ideal' gas, trapped in a box, must eventually pass through every possible arrangement of particles that is allowed by the laws of thermodynamics. However, even a small box of gas might contain 10^{22} atoms, and it would take that many atoms a time much longer than the age of the Universe to pass through every possible arrangement of atoms.

Ludwig Boltzmann suggests that the Universe might be a gigantic statistical freak. It is possible, he suggests, that we are simply a chance collection of atoms – a state of temporary order within a more permanent disorder.

Birth dates of scientists

1851: Fitzgerald, George Francis. Best known for his suggestion that moving objects are shortened in the direction of their motion (the Fitzgerald contraction).

Key dates in science

1851: Leon Foucault makes use of a long pendulum to demonstrate the rotation of the Earth – such a pendulum becomes known as Foucault's pendulum. As the Earth rotates, the pendulum remains swinging in the same plane. This means that the pendulum appears to drift slowly, relative to the ground.

William Thomson develops the concept of absolute zero of temperature and shows that it corresponds to –273°C; he also formulates the second law of thermodynamics.

Key dates in history

1851: Crystal Palace opened by Queen Victoria for the Great Exhibition.

Birth dates of scientists

1852: Becquerel, (Antoine) Henri. Discovered radioactivity, finding that no outside agency was needed to make uranium salts produce radiation that could penetrate sheets of paper and leave its trace on a photographic plate.

Key dates in science

1852: Leon Foucalt invents the gyroscope.

Birth dates of scientists	*Key dates in science*	*Key dates in history*

1852: Michelson, Albert Abraham. Made many determinations of the speed of light and carried out an experiment with Edward Morley which famously failed to find any evidence of an 'ether' through which light was transmitted.

1852: Poynting, John Henry. Determined the density of the Earth and made an accurate measurement of the gravitational constant. In 1903 he suggested that radiation from the Sun can have the seemingly paradoxical effect of making small particles in orbit around the Sun spiral inwards until they are consumed by the Sun's heat. This idea was taken up and developed in the context of relativity theory by Howard Robertson, in the 1930s. He also worked on electromagnetism and showed that the flow of electromagnetic energy across a surface can be expressed in terms of a vector quantity, now known as the Poynting vector.

1852: Ramsay, William. Discovered the inert gases helium, neon, krypton, xenon and radon – the only person to discover an entire group of the elements in the periodic table. Received the Nobel Prize for Chemistry, for doing so, in 1904.

1853: Lorentz, Hendrik Antoon. Received the Nobel Prize for Physics in 1902, for his work on the theory of electromagnetism, in which he developed James Clerk Maxwell's theory of electromagnetism, providing the bridge between Maxwell's work and Albert Einstein's special theory of relativity. He also coined the name 'electron', in 1899, and developed, independently of George Fitzgerald, the Lorentz transformation equations, which describe the way space and time are distorted for objects travelling

Birth dates of scientists	Key dates in science	Key dates in history

at a sizeable fraction of the speed of light.

1853: Onnes, Heike Kamerlingh. Carried out pioneering investigations of low-temperature physics.

1854: Poincaré, Henri. Showed that an 'ideal' gas, trapped in a box, must eventually pass through every possible arrangement of particles that is allowed by the laws of thermodynamics. However, even a small box of gas might contain 10^{22} atoms, and it would take that many atoms a time much longer than the age of the Universe to pass through every possible arrangement. Also made, in 1906, an independent derivation of many of the results of Albert Einstein's special theory of relativity, and showed that the transformation equations discovered by Hendrik Lorentz form a group.

1854: Rydberg, Johannes Robert. Discovered a mathematical formula to give the frequencies of the lines in the spectrum from an element; a formula which includes a number named the Rydberg constant, in his honour.

1856: J.J. Thomson. Received the Nobel Prize in 1906 for his experimental work showing that electrons behave as particles; in 1937 his son was also to receive the Nobel Prize for Physics for showing that electrons behave as waves. Both were right.

1857: Hertz, Heinrich Rudolph. German physicist who was the first to produce long-wavelength radiation – radio waves – artificially, in 1888. In doing so he proved that electromagnetic waves are generated exactly as the equations discovered by James Clerk Maxwell had predicted.

1854: Hermann von Helmholtz suggests that the Sun is kept hot by gravitational energy, released as it shrinks slowly under its own weight.

1857: Leon Foucalt begins to manufacture glass mirrors coated with a thin film of silver to use in astronomical telescopes.

1854: First electric telegraph link between London and Paris.

1854–6: Crimean War.

1855: Louvre opens in Paris.

1856: Henry Bessmer invents a cheap process for manufacturing steel.

Birth dates of scientists

1858: Planck, Max Ernst Karl Ludwig. Proposed, in 1900, that electromagnetic radiation can be emitted or absorbed only in definite units, which he called quanta. This explained the nature of radiation from a black body, and laid the foundations for quantum theory. Planck did not suggest that light actually existed as quanta, but thought of this process as something to do with the way the oscillating charges within atoms worked. For this work, he was awarded the Nobel Prize in 1918.

1859: Arrhenius, Svante August. Suggested the Earth might have been 'seeded' with life from space, in the form of micro-organisms riding on dust particles. First person to explain that, when a chemical compound dissolves in water, it dissociates into electrically charged ions.

1859: Curie, Pierre and **1867:** Curie, Marie. French husband-and-wife team who shared the Nobel Prize for Physics in 1903, along with Henry Becquerel, for their pioneering investigations of radioactivity. Marie Curie also received the Nobel Prize for Chemistry in 1911 for her discovery of radium and polonium.

1862: Bragg, Sir William Henry. Studied crystal structure using X-rays, in partnership with son William, usually known as Lawrence. For this they won the Nobel Prize, the only father-and-son team to do this.

Key dates in science

1859: Gustav Kirchoff and Robert Bunsen develop the use of spectroscopy for chemical analysis.
   Gustav Kirchoff states Kirchoff's Law for the first time. This law demands that, at a given temperature, the rate of emission of electromagnetic energy given by an object is equal to the rate at which the object absorbs electromagnetic energy of the same wavelength (frequency).
   Kirchoff detects sodium in the Sun.

Early 1860s: James Clerk Maxwell develops a statistical treatment of the behaviour of gases, based on the idea that a gas is made up of a large number of particles in rapid, random motion.

1861: Anders Ångstrom begins a study of the Solar System that will show hydrogen is present in the Sun.
   Crooke discovers the element thallium.
   Kirchoff's Law is proven, by him, for the first time.

1862: The companion star to Sirius, predicted by Friedrich Bessel, is found by American telescope maker Alvan Clark. This is the first white dwarf to be discovered and is known as Sirius B. A white dwarf is a star with

Key dates in history

1858: First Atlantic telegraph cable.

1859: Construction of the Suez Canal begins; it will be completed in 1869. Publication of the *Origin of Species*. First internal combustion engine (using gas as its fuel) developed by Jean Lenoir. First oil well drilled in Titusville, PA.

1860: First vehicle driven by an internal combustion engine developed by Jean Lenoir.

1861: Unification of Italy. First telegraph linking San Francisco to New York. American Civil War begins.

1862: Richard Gatling invents the machine gun.

1862: Lenard, Philipp Eduard Anton. Studied the photoelectric effect, providing the experimental evidence which Albert Einstein explained in terms of photons.

about the same mass as the Sun, but occupying a volume about the same as that of the Earth.

Foucault obtains a measurement of the speed of light, 298,005 km/sec, within 1 per cent of the best modern determination.

The proof of his law leads Gustav Kirchoff to develop the idea of a black body, and black body radiation, which will lead Max Planck to introduce the idea of quanta into physics.

Maxwell finds that the electromagnetic waves he invokes to explain the behaviour of electric and magnetic fields must travel at the speed of light, a finding he publishes in the paper *On Physical Lines of Force*.

1863: William Huggins uses spectroscopy to show that the same chemical elements exist in the atmospheres of stars as exist on Earth.

Ernst Mach publishes *Die Mechanik*, in which he spells out the idea now known as Mach's Principle, which says that an object possesses inertial mass only by virtue of the presence of all the other masses in the Universe.

John Tyndall publishes a discussion of the atmospheric greenhouse effect.

1863: National Academy of Sciences founded in USA.

1864: Minkowski, Hermann. One of Einstein's teachers, Minkowski came up with the idea of four-dimensional spacetime as the stage on which physical interactions take place.

1864: Wien, Wilhelm. Awarded the Nobel Prize, in 1911, for his work on the laws governing the radiation of heat, which provided insight into black body radiation (see **Wien's Law**).

1864: James Clerk Maxwell publishes his work *A Dynamical Theory of the Electromagnetic Field*. This paper solves all classical problems of electric and magnetic phenomena in four equations; in doing this, it also predicts the existence of radio waves.

Birth dates of scientists

1865: Zeeman, Pieter. Investigated the effect of magnetic fields on light, and discovered the Zeeman effect, the splitting of spectral lines in the light from an atom when it is placed in a strong magnetic field. Zeeman also measured the speed of light in dense and moving media.

1866: Lebedev, Pyotr Nikolayevich. The first person to measure the pressure produced by light, an effect predicted by James Clerk Maxwell. Using very light apparatus in an evacuated chamber, he actually measured the pressure of light, and suggested that this pressure explains why comet tails always point away from the Sun.

1868: Millikan, Robert Andrews. Received the Nobel Prize for Physics in 1923 for measuring the charge on the electron. He also carried out experiments which proved that Albert Einstein's theory of the photoelectric effect was correct, and later studied cosmic rays.

1868: Sommerfield, Arnold Johannes Wilhelm. Made major contributions to the quantum theory of the structure of the atom, and was an influential teacher. Sommerfield encouraged Niels Bohr to develop the 'Bohr model' of the atom, and then developed this model further by introducing the idea of elliptical orbits.

1869: Wilson, Charles Thomson Rees. Invented the cloud chamber, and shared the Nobel Prize for Physics with Arthur Compton in 1927.

Key dates in science

1865: Johann Loschmidt calculates the number of molecules of gas in a cubic centimetre of air, obtaining a figure about 30 times smaller than modern estimates.
John Newlands produces the first periodic table of the elements. However, this work is ridiculed by establishment critics.

1866: A transatlantic telegraph cable link is established, using principles proposed by William Thomson, later Baron Kelvin.

1868: Lockyer and Janssen, independently of one another, discover helium, present in the Sun.

1869: Dmitri Mendeleyev publishes the first periodic table, in which the elements are grouped according to their chemical properties.
John Tyndall discovers the Tyndall effect – the scattering of light by tiny particles suspended in a liquid – and suggests that the reason the sky looks blue might

Key dates in history

1865: Lewis Carroll (Charles Lutwidge Dodgson) writes *Alice's Adventures in Wonderland*. American Civil War ends.

1866: First successful transatlantic telegraph cable.

1867: Publication of *Das Kapital*.

1869: First publication of the science journal *Nature*. First railway line linking the west and east coasts of the US completed. Suez Canal opens.

Birth dates of scientists	**Key dates in science**	**Key dates in history**

be because of particles
suspended in the air.

1870: Perrin, Jean Baptiste.
Demonstrated the reality of
atoms by studying **Brownian
motion,** for which he was
awarded the Nobel Prize for
Physics in 1926.

1870s: William Crookes, studying
the effects of electric currents
passed through traces of gas in
discharge tubes, finds a stream of
positively charged particles are
repelled from the anode. Crookes
called these 'molecular rays'; they
would now be called ions.

1870: Vladimir Ilyich Lenin
(originally Ulyanov) born. Franco-
Prussian War.

1871: Rutherford, Ernest. Made
many pioneering investigations
of the structure of the atom and
radioactivity, for which he was
awarded the Nobel Prize for
Chemistry in 1908. Rutherford
was responsible for the discovery
of the atomic nucleus, and the
trasmutation of the elements; he
also determined three types of
radiation, which he termed
alpha, beta and gamma, and
invented the process of
radioactive dating.

1871: The Cavendish Laboratory
is established in Cambridge,
England.

1872: De Sitter, Willem. One of
the first people to apply Einstein's
equations of the general theory
of relativity to provide a
mathematical model of the
Universe.

1872: Sarah Bernhardt begins to
work with the Comédie-Français
in Paris.

1873: Shwarzschild, Karl.
Discovered the solution to Albert
Einstein's equations of the
general theory of relativity that
describes what are now known as
black holes.

1874: Lyman, Theodore. Lyman's
scientific work concentrated
entirely on investigations of the
ultraviolet part of the spectrum,
where he famously discovered
the **Lyman series** of lines, in the
spectrum of hydrogen.

1874: Cavendish Laboratory
completed.

1874: First exhibition of
Impressionist paintings in Paris.
Levi Strauss invents blue jeans
with rivets.

1874: Stark, Johannes. Stark's
best-known piece of work came
in 1913 when he showed that,
when positively charged ions
pass through a strong electric
field, there is a splitting of the

Birth dates of scientists	*Key dates in science*	*Key dates in history*

lines in the spectrum of light from the ions – the 'Stark effect'. In his personal life, Stark was almost unique among the prominent scientists of the time, in being a vigorous anti-semite and a keen supporter of the Nazi Party.

1875: Lewis, Gilbert Newton. Invented the idea of the covalent bond, in which atoms are envisaged as sharing pairs of electrons.

1876: Adams, Walter. American astronomer who obtained a spectrum showing that Sirius B, although far fainter, is actually as hot as its companion star Sirius. In order to be both hot and faint, the star must be small – little bigger than the Earth. It was the first white dwarf to be discovered.

1877: Aston, Francis William. Invented the technique of mass spectroscopy, using it to show that many common elements come in different varieties (called isotopes), which have the same atomic number but different atomic masses.

1877: Jeans, James Hapwood. Wrote many popular works and made radio broadcasts about the Universe, and was largely responsible for publicizing the idea of the heat death of the Universe. Earlier in his career he contributed to the development of the kinetic theory of gases, and made early applications of the idea of quanta to the investigation of specific heats and the nature of black body radiation.

1875: Crookes invents his famous 'radiometer', a small four-bladed paddle wheel mounted in a glass vessel in a vacuum. Each blade of the panel is black on one side and shiny on the other. When light shines on it, it moves rapidly because the black side of the panel is hotter than the shiny side, so that the molecules of air left in the vessel bounce off the dark side more vigorously and push it.
　Lorentz publishes his PhD thesis, titled *The Theory of Reflection and Refraction of Light,* which shows how to solve Maxwell's equations at a boundary between two materials.
　Maxwell says that atoms have a more complex structure than rigid bodies.

1877: The two moons of Mars are found by Asaph Hall.
　Giovanni Schiaparelli reports that he has seen channels (mistranslated as 'canals') on Mars.

1875: Georges Bizet writes *Carmen.*

Birth dates of scientists	Key dates in science	Key dates in history
1877: Soddy, Frederick. Received the Nobel Prize for Chemistry in 1921 for his investigations of radioactive decay and the study of isotopes. Identified, with Rutherford, the two radioactive series that begin with uranium and thorium, respectively, and end with lead.		
1878: Meitner, Lise. One of the first people to investigate nuclear fission, Meitner discovered protactinium with her colleague Otto Hahn, and, with her nephew Otto Frisch, published the first explanation of uranium experiments indicating the occurrence of nuclear fission, which she named.	**1878:** Rayleigh publishes the first volume of his work *The Theory of Sound*; the second volume follows the year after.	**1878:** Telephone invented.
1879: Einstein, Albert. Developed both the special theory of relativity and the general theory of relativity, as well as making major contributions to the development of quantum theory.	**1879:** Josef Stefan becomes the first person, following experiments by John Tyndall, to identify the relationship between the rate at which an object radiates energy and its temperature – a relationship which becomes known as the Stefan–Boltzmann Law.	**1879:** Forerunner of the modern electric light bulb invented independently by Joseph Swan in England and Thomas Edison in America. Lev Davidovich Bronstein (Leon Trotsky) born. First performance of *Eugen Onegin* by Tchaikovsky. Mary Cassatt paints *The Cup of Tea*. First electric railway, in Berlin.
1879: Hahn, Otto. Awarded the Nobel Prize for Chemistry in 1944, for his role in the discovery of nuclear fission.		
1879: Laue, Max Theodor Felix von. Predicted the diffraction of X-rays by crystals, thereby proving that they were short-wavelength electromagnetic waves, for which he received the Nobel Prize for Physics in 1914.		
1879: Richardson, Owen Williams. Awarded the Nobel Prize for Physics in 1929, for his investigation into the emission of electrons from hot surfaces, a process he named thermionics. The discovery of this process was a key step in the development of electron tubes, used in many modern devices including television sets.	**1880s:** Bishop Berkeley's suggestion that all motion is relative, and must be measured against something, is taken up by Ernst Mach, who suggests that if we want to explain the equatorial bulge of the Earth as due to	**1880:** Ned Kelly captured.

Birth dates of scientists

1881: Davisson, Clinton Joseph. Shared the Nobel Prize for Physics in 1937 with George Thomson for their independent confirmation of the wave nature of the electron.

1882: Born, Max. German physicist who made major contributions to the development of quantum theory, and introduced the idea that the outcome of experiments or interactions involving quantum entities is not directly deterministic, but involves probability in an intimate way. Worked with Heisenberg and Jordan on a scientific paper which produced the first complete, self-consistent version of quantum mechanics.

1882: Eddington, Sir Arthur Stanley. The first astrophysicist, Eddington carried out the crucial test of Einstein's general theory of relativity, developed the application of physics to an understanding of the structure of stars, and was a great popularizer of science in the 1920s and 1930s.

1882: Geiger, Hans Wilhelm. German physicist who invented the eponymous counter, for measuring radioactivity.

1882: Noether, (Amalie) Emmy. Made important contributions to the development of the concepts of symmetry groups and non-commutative (non-Abelian) fields.

1883: Hess, Victor Francis. Discovered cosmic rays, through a series of balloon flights in 1911 and 1912 which showed that penetrating radiation comes through the atmosphere, much being absorbed before reaching the ground. For these discoveries he shared the Nobel Prize for

Key dates in science

centrifugal forces, 'it does not matter if we think of the Earth as turning round on its axis, or at rest while the fixed stars revolve around it' – it is the relative motion that is responsible for the bulge. This principle is later to be dubbed Mach's Principle by Einstein.

Key dates in history

1881: First practical electric generation and distribution system.

1882: Electric light introduced in New York. Franklin Delano Roosevelt born.

1883: System of four time zones officially adopted in the USA. Gottlieb Daimler invents the first version of the modern internal combustion engine, and tests it in a boat.

Birth dates of scientists	Key dates in science	Key dates in history

Physics with Carl Anderson in 1936.

1884: Debye, Peter Joseph Willem. Awarded the Nobel Prize for Chemistry in 1936 for his work on the structure of molecules, working on specific heat, X-ray diffraction and the idea that molecules have permanent electric dipoles.

1884: An international committee establishes the meridian through the Royal Greenwich Observatory as the 'prime meridian' from which longitude is to be measured.

Svante Arrhenius submits a doctoral thesis containing the essence of his work on solutions, which for the first time explains that when a chemical compound dissolves in water it dissociates into electrically charged ions.

1885: Bohr, Niels Hendrik David. Danish physicist who was awarded the Nobel Prize in 1922 for his theoretical model of the structure of the atom, based on spectroscopy and the principles of quantum physics.

Balmer gives an explanation for the way lines in a spectrum are distributed. He succeeds in finding an equation that not only reproduces the relationship between the four bright lines known in the spectrum of hydrogen at that time, but also predicts the existence of a fifth line, at the edge of visibility, that was soon detected.

1885: Automobile and motorbike invented.

1886: Siegbahn, (Karl) Manne Georg. Received the Nobel Prize for Physics in 1924 for his pioneering work in X-ray spectroscopy. This provided important confirmation of the accuracy of the Bohr model of the atom.

John Newlands finally publishes the paper concerning the periodic relationship between the properties of the chemical elements which had been ridiculed by the establishment in 1866.

1887: Hertz, Gustav. His main work concerned techniques to separate different isotopes of an element.

Ludwig Boltzmann refines the Stefan–Boltzmann Law, first derived by Josef Stefan in 1879.

1887: Mosely, Henry Gwyn Jeffreys. A pioneer of X-ray spectroscopy, and the first person to realize that the atomic number of an element is a measure of the charge on the nucleus of an atom of the element. This led to a better understanding of the periodic table, and to the prediction of several elements that were soon found.

1887: Michelson–Morley experiment carried out, in an attempt to detect the motion of Earth through the ether, by measuring differences in the speed of light determined along the line of the Earth's motion and at right angles to that line. These experiments show, through demonstrating that no effects attributable to the motion of the Earth can be seen in any measurements of the speed of light, that the ether does not exist.

1887: Alternating-current electric motor invented by Nikola Tesla.

1887: Schrödinger, Erwin. Developed the wave mechanics formulation of quantum physics, and in 1933 received the Nobel Prize for Physics for this work.

Joseph Lockyer's spectroscopic study of the Sun leads to the discovery of helium.

Birth dates of scientists

1888: Friedmann, Aleksandr Aleksandrovich. Worked in practical applications such as hydromechanics and meteorology, but is best remembered for his solutions to Albert Einstein's equations of the general theory of relativity.

1888: Raman, Chandrasekhara Venkata. Discovered the Raman effect in 1928, and was awarded the Nobel Prize as a result, in 1930. The Raman effect is caused by the scattering of monochromatic light as it passes through a transparent medium. The way the light is scattered provides information about the molecules within the medium.

1888: Stern, Otto. A member of the team which carried out the experiment which revealed the existence of electron spin, the Stern–Gerlach experiment. Stern received the Nobel Prize in 1943 for his development of the technique used in this experiment and for his measurement of the magnetic moment of the proton.

1889: Hubble, Edwin Powell. Proved that many objects classified as nebulae are other galaxies; discovered the relationship between redshift and distance, and inferred that the Universe is expanding.

1890: Bragg, (William) Lawrence. Studied crystal structure using X-rays, in partnership with his father William; in 1915 they became the only father and son to be jointly awarded the Nobel Prize, in recognition for this work.

1891: Bothe, Walther Wilhelm Georg Franz. Won a half-share in the Nobel Prize for Physics in 1954 for developing an improved particle detector technique – the

Key dates in science

1888: The long-wavelength radiation predicted by Maxwell is first produced artificially by the German physicist Heinrich Hertz.

First measurements of the velocities of stars using the Doppler effect.

1890s: Edward Pickering establishes the first standard system of measuring the temperature of a star by taking measurements at two separate wavelengths of light from that star, the comparison of which enable him to work out the star's temperature.

John Poynting determines the density of the Earth, and makes an accurate measurement of the gravitational constant.

Key dates in history

1888: William Burroughs patents his adding machine. Oscar Wilde writes *The Happy Prince*.

1889: Eiffel Tower completed.

1890: London Underground opens.

Birth dates of scientists	Key dates in science	Key dates in history

coincidence counter, derived from the Geiger counter – which he used in his work on cosmic rays.

1891: Chadwick, James. Discovered the neutron, a particle with no electric charge and almost the same mass as the proton.

1891: Humason, Milton Lasell. Worked with Edwin Hubble on the redshift survey which led to the discovery that the Universe is expanding.

1892: Broglie, Louis-Victor Pierre Raymond, Prince de. Received the Nobel Prize for Physics in 1929 for the proposal that all material 'particles', such as electrons, could also be described in terms of waves. This wave–particle duality lies at the heart of quantum physics.

1892: Compton, Arthur Holly. American physicist who received the Nobel Prize for Physics in 1927 for his discovery of the Compton effect, an increase in the wavelength of X-rays (or gamma rays) when they are scattered by electrons.

1892: Thomson, George Paget. Received the Nobel Prize in 1937 for his experimental work showing that electrons behave as waves; the son of J.J. Thomson, who received the Nobel Prize in 1906 for showing that electrons behave as particles. Both were right.

1893: Wilhelm Wien discovers the relationship between the wavelength of maximum intensity of 'black body' radiation and its temperature; this enables astronomers to measure the temperatures of stars.

1893: Mormon Temple completed in Salt Lake City. Zip fastener invented. Mao Tse-tung born.

1894: Bose, Satyendra Nath. Indian physicist who made one outstanding contribution to quantum theory, the development of what became known as Bose–Einstein statistics.

1894: Philipp Lenard shows that cathode rays can pass right through a piece of metal foil, so they cannot (as Crookes had guessed) be electrically charged molecules.

William Ramsay, Lord Rayleigh and William Crookes are the first to discover a 'noble' gas, argon.

1894: Manchester Ship Canal completed. Guglielmo Marconi develops a radio transmitter that will ring a bell at a distance of 10 m.

1894: Kapitza, Piotr Leonidovich. Made major contributions to the understanding of low-

Birth dates of scientists	*Key dates in science*	*Key dates in history*

temperature physics, and was awarded the Nobel Prize for Physics as a result, in 1978. Discovered superfluidity, which he named in 1938.

1894: Klein, Oskar. A pioneer of the idea that spacetime may be made up of more than four dimensions. See *Kaluza–Klein theory.*

1894: Kramers, Hendrik Anthony. Predicted (with Werner Heisenberg) Raman scattering. Also developed the idea of 'renormalization', which he discussed in 1947. By making use of this idea it is possible to cancel out the intrusive infinities of an equation by balancing them against one another.

1894: Lemaître, Georges Édouard. Developed the first version of what became the Big Bang description of the Universe.

1895: Tamm, Igor Yevgenyevich. Shared the Nobel Prize, in 1958, for the investigation of the Cerenkov radiation. Also showed that, although the neutron is electrically neutral, it has a magnetic moment. Tamm's most important work is believed by many to have been his theory of beta decay, developed in 1934, which suggested that it is a result of a weak force operating between nucleons.

1895: X-rays are discovered by Wilhelm Röntgen; initially called Röntgen rays.
Helium is identified on Earth, by William Ramsay and William Crookes.
Jean Baptiste Perrin produces his study of cathode rays, which provides a rough indication of the ratio of the charge on an electron to its mass, and paves the way for the definitive work by J. J. Thomson which establishes that the electron behaves as a particle.

1895: Lumière brothers present moving pictures to the public.

1896: Germer, Lester Halbert. Worked with Clinton Davisson on the experiment that proved electrons are waves, for which Davisson received a share of the Nobel Prize.

1896: Milne, Edward Arthur. Made pioneering theoretical investigations into the structure of stars; in the early 1930s showed, with William McCrea, that the expanding Universe

1896: First photographic atlas of the Moon published by Lick Observatory.
Henri Becquerel, stimulated by the discovery of X-rays, discovers another form of radiation, produced spontaneously by atoms of uranium.
Charles Wilson surmises, through the use of the cloud chamber, that the droplets are condensed around electrically charged particles (ions), and

1896: Discovery of radioactivity.

predicted by the general theory of relativity could actually be described very well in the context of classical mechanics and Isaac Newton's theory of gravity.

1897: Blackett, Patrick Maynard Stuart. British physicist who developed the *cloud chamber* and received the Nobel Prize as a result in 1948.

1897: Cockcroft, John Douglas. English physicist who, with Ernest Walton, carried out the first experiment in which one element was transmuted into another artificially (popularly referred to as the first 'splitting of the atom').

1897: Joliot-Curie, Irène and **1900:** Joliot-Curie, (Jean) Frédéric. Shared the Nobel Prize for Chemistry in 1935, for their work on artificial radioactivity. This work involved bombarding aluminium with alpha radiation. After the source of alpha rays was removed, the aluminium emitted positrons for several minutes. Some of the aluminium nuclei had each absorbed an alpha particle and been transformed into nuclei of a radioactive form of phosphorus, which decays into silicon with a half life of about 3.5 minutes. This was the first recognized production of artificial radioactivity, and clear evidence that the transmutation of elements was in some cases a practical possibility.

1898: Atkinson, Robert D'escourt. Atkinson's chief contribution to science was to show, with Fritz Houtermans, how energy could in principle be produced inside stars by fusion.

1898: Rabi, Isidor Isaac. Won the Nobel Prize for Physics in 1944, for the development of the resonance method which led to the measurement of the magnetic moment of the

proves this by operating the cloud chamber beside a source of X-rays, and seeing it fill with condensation as the X-rays ionize the atoms within the chamber.

1897: J. J. Thomson discovers that the so-called cathode rays are in fact a stream of tiny charged particles broken off from atoms. These particles he calls 'corpuscles', but they soon become known as electrons, a name suggested by Dutch physicist Hendrik Lorentz.

1898: Radioactivity named by Marie Curie.
   Ludwig Boltzmann publishes details of his calculations concerning atoms.

1897: Camille Pisarro paints *Boulevard des Italians*.

Birth dates of scientists	Key dates in science	Key dates in history

electron. This was a key step in the development of quantum electrodynamics. His work on the behaviour of atoms and nuclei forms part of the basis of practical applications such as lasers and atomic clocks. The measurement of the Lamb shift also used a development of Rabi's techniques.

1898: Szilard, Leo. Played a major part in the development of the atomic bomb, because of concern that a similar weapon might be developed by Nazi Germany, but opposed the idea of using it on the Japanese. Szilard also solved the problem of Maxwell's demon.

1898: Zwicky, Fritz. Made many contributions to astrophysics, studied supernovae, compact galaxies and clusters of galaxies, and was one of the few people to take the idea of neutron stars seriously before the discovery of pulsars.

1899: Auger, Pierre. Discovered the Auger effect, a process in which an atom with an excess of energy can return to a lower energy state by emitting an electron, instead of emitting electromagnetic radiation. The emitted electron is called an Auger electron.

1899: Ernest Rutherford discovers that there are two forms of radioactivity, which he calls alpha and beta radiation; a third form, gamma radiation, is identified later.

Philipp Lenard commences his investigations of the photoelectric effect; he also studies the nature of cathode rays, and is awarded the Nobel Prize for Physics for this work in 1905.

Hendrik Lorentz coins the name 'electron'.

1899: Aspirin marketed.

1900: Joliot-Curie, Frédéric. Shared the Nobel Prize for Chemistry in 1935, with wife Irène, for their work on artificial radioactivity. Also calculated the possibility of a runaway chain reaction of the kind later used in the atomic bomb, but with the outbreak of war decided not to publish the results, placing them

1900: Max Planck announces the discovery that the nature of black body radiation can be explained only if light is emitted or absorbed by atoms in discrete quanta (photons). The essential feature of Planck's discovery is that there is a limit to how small a change in energy an atom can experience – this corresponds, in

1900: First offshore oil wells. Ferdinand von Zeppelin builds his first dirigible airship. Invention of the thermionic valve.

Birth dates of scientists

in a sealed letter deposited with the French Academy of Sciences.

1900: London, Fritz Wolfgang. Best known for his collaboration with Walter Heitler, which provided the first quantum-mechanical description of the covalent bond in the hydrogen molecule. This was the birth of quantum chemistry. Also calculated the strength of what are now known as *van der Waals forces*, and investigated superconductivity.

1900: Pauli, Wolfgang. Made many contributions to the development of quantum theory, including the Pauli exclusion principle, which is crucial to an understanding of white dwarf and neutron stars, and for which he received the Nobel Prize for Physics in 1945. He was also the first to propose the existence of the neutrino.

1900: Uhlenbeck, George Eugene. A member of the first team to introduce the concept of electron spin. This key insight explains, among other things, how two electrons can occupy the same level in an atom, even though they are fermions.

1901: Fermi, Enrico. Took up Wolfgang Pauli's idea of a 'new' particle, used to explain anomalous behaviour observed during beta decay, and coined the name 'neutrino' for this particle. Is best remembered as the leader of the group that built the first nuclear reactor; known in astronomy for his claim that the fact that the Earth has not been colonized by aliens 'proves' that we are the only civilization in the Galaxy.

1901: Heisenberg, Werner. Developed the uncertainty principle during the 1920s. This shows that there is an intrinsic

Key dates in science

modern terminology, to the emission or absorption of a single photon.

The Rayleigh–Jeans Law is first developed by Lord Rayleigh. This law accurately describes the long-wavelength end of the spectrum of black body radiation. However, as it is based on the wave theory of light, it is unsuccessful when dealing with shorter wavelengths.

Early 20th century: Theodore Lyman studies the nature of the spectral features in hydrogen now known as the Lyman lines, or Lyman series.

Key dates in history

1901: First electric typewriter. First transatlantic radio communication. Paul Gauguin paints *The Gold in their Bodies*.

| Birth dates of scientists | Key dates in science | Key dates in history |

Birth dates of scientists

uncertainty associated with knowledge of position and momentum in the quantum world. Awarded the Nobel Prize for Physics, in 1932, for a subtle explanation of features in the spectrum of molecular hydrogen. Formulated matrix mechanics, the first complete, self-consistent theory of quantum physics.

1901: Lawrence, Ernest Orlando. Invented the cyclotron, an accelerator which can be used for many purposes, including the production of radioactive isotopes for use in medicine, and the synthesis of plutonium and neptunium. Received the Nobel Prize for Physics in 1939, and after the war worked with his brother John on the application of physics to medicine, and with him invented the use of neutron beams to treat cancer.

1901: Pauling, Linus Carl. Applied quantum physics to establish the foundation of quantum chemistry in the late 1920s and early 1930s, for which he won the Nobel Prize for Chemistry in 1954. He was also awarded a second Nobel Prize, for Peace, in 1962, for work in the campaign to end nuclear weapons testing. He determined the molecular structure of amino acids and proteins, and later became convinced that massive doses of vitamin C could not only protect against the common cold and other ailments, but also help the body fight cancer.

1902: Brattain, Walter Houser. Shared the Nobel Prize for Physics in 1956 with William Shockley and John Bardeen for their discovery of the transistor effect.

1902: Dirac, Paul Adrien Maurice. Developed his own version of quantum theory, known as operator theory or quantum

Key dates in science

1902: Lord Kelvin puts forward the first modern model of the atom. This becomes known as the Thomson model.

Key dates in history

1902: New York's 'Flatiron' building completed.

algebra, and showed that both Heisenberg's matrix mechanics and Schrödinger's wave mechanics were special cases of his own operator theory, and therefore exactly equivalent to one another. Dirac also introduced the idea of second quantization to quantum physics, and predicted the existence of antimatter.

1902: Goudsmit, Samuel Abraham. A member of the first team to introduce the concept of electron spin.

1902: Jordan, (Ernst) Pascual. Worked with Max Born and Werner Heisenberg in Göttingen on the formulation of the matrix mechanics version of quantum theory. He was one of the pioneers of the theoretical work in the 1920s and 1930s which paved the way for the theory of quantum electrodynamics. He also carried out important work on the theory of gravity.

1902: Strassman, Fritz. Collaborated with Otto Hahn on the first experiments that clearly revealed the phenomenon of fission of atomic nuclei, and were interpreted in those terms by Lise Meitner and Otto Frisch.

1902: Wigner, Eugene Paul. Received a share of the Nobel Prize for Physics in 1963 for his work on the theory of quantum physics, especially the law of conservation of parity. Wigner introduced, together with John von Neumann, many key mathematical concepts used in modern field theory into quantum mechanics, also working on the theory of absorption.

1903: von Neumann, John. A brilliant mathematician whose most influential contribution to quantum theory was nevertheless

Rutherford and Frederick Soddy show that, when an atom emits either alpha radiation or beta radiation, it is turned into an atom of a different element.

1903: John Poynting suggests that radiation from the Sun can have the seemingly paradoxical effect of making small particles in

1903: Wright brothers' first powered flight. First use of the term 'atomic energy'.

Birth dates of scientists	*Key dates in science*	*Key dates in history*

a major mistake. In a book published in 1932 von Neumann gave a 'proof' that no 'hidden variables' theory could ever properly describe the workings of the quantum world. This proof rested on a basic mistake involving commutation, and held back the investigation of the 'hidden variables' theory for decades.

1903: Powell, Cecil Frank. Developed the technique of studying elementary particles from the tracks they leave in photographic emulsion. Awarded the Nobel Prize for Physics, in 1950, for his discovery of the pion.

1903: Walton, Ernest Thomas Sinton. Shared the Nobel Prize for Physics with John Cockcroft in 1951 for their joint development of the first 'atom smasher' in 1932.

1904: Cerenkov, Pavel Alekseyevic. Received the Nobel Prize for Physics in 1958 for the discovery and interpretation of the phenomenon now known as Cerenkov radiation.

1904: Frisch, Otto Robert. Worked, with Lise Meitner, on the theory of fission. Also worked with Rudolf Peierls; together they were probably the first people to appreciate the possibility of triggering an explosive chain reaction using uranium-235, and alerted the British government to the possibility of a nucelar bomb.

1904: Gamow, George ('Joe'). Made the first calculations of conditions in the Big Bang, predicted the existence of the background radiation, and played a part in cracking the genetic code of DNA, the molecule of life.

orbit around the Sun spiral inwards until they are consumed by the Sun's heat.

1904: Hendrik Lorentz develops the equations now known as Lorentz transformations, to describe how electromagnetic fields would look to observers moving at different velocities.

1904: Construction of the Panama Canal begins; it will be completed in 1914.

Birth dates of scientists

Key dates in science

Key dates in history

1904: Heitler, Walter. Worked with Fritz London on the quantum-mechanical description of the covalent bond.

1904: Kronig, Ralph de Laer. Came up with the idea of electron spin in 1925, but was dissuaded from publishing it after discussions with Wolfgang Pauli, Werner Heisenberg and Hendrik Kramers, all of whom were sceptical about the idea.

1904: Oppenheimer, (Julius) Robert. Worked on quantum theory in the 1920s and 1930s, and is best known for the part he played in the production of the atomic bomb – he was the first Director of the Los Alamos Scientific Laboratories, where the bomb was developed.

1905: Anderson, Carl David. Received the Nobel Prize in 1936 for the discovery of the positron – the first proof of the existence of antimatter.

1905: Mott, Nevill Francis. Awarded the Nobel Prize for Physics in 1977 for his work on semiconductors. Also worked on the quantum theory of atomic collisions and wrote a series of important textbooks.

1905: Segre, Emilio. Received the Nobel Prize for Physics in 1959 for the discovery of the antiproton. He was also the discoverer of technetium – the first entirely man-made new element to be identified – and was involved with the discovery of two other elements, astatine and plutonium.

1905: Einstein's special theory of relativity unites space and time in one mathematical description, dealing with the dynamical relations between objects moving at constant speeds in straight lines.

Einstein suggests that light (previously regarded purely as an electromagnetic wave) could be described in terms of particle-like quanta; what are now called photons. This is a key step in the development of quantum theory.

Einstein estimates the size of molecules as a few ångströms (a few hundred-millionths of a centimetre), in line with modern estimates.

Einstein sends for publication his classic paper on Brownian motion, the curious zigzag dance of pollen grains suspended in water, which is explained in terms of the buffeting that the grains receive from the molecules themselves.

The Rayleigh-Jeans Law, first propounded by Lord Rayleigh in 1900, is refined by James Jeans.

1905: Special theory of relativity.

Birth dates of scientists

1906: Bethe, Hans Albrecht. Worked out the mechanisms by which stars derive their energy from nuclear fission reactions. Also made major contributions to the development of quantum physics in the 1940s.

1906: Gödel, Kurt. Austrian-born American mathematician who found a solution to Einstein's equations that permits time travel.

1906: Goeppert-Mayer, Marya. Physicist who received a share of the Nobel Prize for Physics in 1963, for her work in developing the shell model of the nucleus.

1906: Ruska, Ernst August Friedrich. Received the Nobel Prize for Physics in 1986 for work on the development of the electron microscope.

1906: Tomonaga, Sin-itiro. Shared the Nobel Prize for Physics in 1965, for the theory of quantum electrodynamics.

1907: Jensen, (Johannes) Hans (Daniel). Awarded the Nobel prize for Physics in 1963, jointly with Marya Goeppert-Mayer, for their suggestion, made independently of each other at the end of the 1940s, that the structure of the nucleus could be described in terms of a series of shells.

1907: London, Heinz. Studied superconductivity, superfluidity and techniques for separating isotopes.

1907: Peierls, Rudolf Ernst. Investigated, among other things, quantum theory and the theory of nuclear reactions. In 1940 Peierls wrote, with Otto Frisch, a paper *On the Construction of a 'Superbomb'*, which showed there was a realistic possibility of developing

Key dates in science

1906: Poincaré produces an independent derivation of many of the results of Albert Einstein's special theory of relativity.

1907: Einstein's equivalence principle is formulated, stating that acceleration and gravity are equivalent.

Key dates in history

1906: Great San Francisco Earthquake. Kellogg's cornflakes go on sale for the first time.

1907: Invention of colour photography. Pablo Picasso displays *Les Demoiselles d'Avignon*.

Birth dates of scientists	Key dates in science	Key dates in history

nuclear weapons. Peierls became the leader of the British group on the Manhattan Project, and was knighted in 1968.

1907: Yukawa, Hideki. Became the first Japanese to receive the Nobel Prize for Physics, in 1949, for his prediction of the existence of pions.

1908: Bardeen, John. The first person to be awarded the Nobel Prize for physics on two separate occasions. His work on semiconductors led to the development of the transistor and, in 1956, to his first Nobel Prize. With Robert Schrieffer and Leon Cooper, he developed a theory of superconductivity, for which he received a second Nobel Prize in 1972.

1908: Heike Onnes becomes the first person to liquefy helium.

Hermann Minkowski shows that the uniting of space and time in one mathematical description is mathematically equivalent to a description in terms of four-dimensional Euclidean geometry, so that the combination of space and time in spacetime could be regarded as the four-dimensional equivalent of a flat sheet of paper.

1908: First Model T Ford.

1908: Frank, Ilya Mikhailovitch. Received the Nobel Prize in 1958 (shared with Pavel Cerenkov and Igo Tamm) for his investigations of the Cerenkov effect.

1908: Landau, Lev Davidovich. Predicted the existence of neutron stars, developed a theory of superfluidity and was awarded the Nobel Prize for Physics in 1962, chiefly for his theoretical studies of the behaviour of liquid helium.

1908: Teller, Edward. Famous as 'the father of the hydrogen bomb', although this owes more to his political than scientific work. Worked with George Gamow on the rules which govern beta decay. Later worked on nuclear fission with Enrico Fermi. Teller both worked on the Manhattan Project and was later intrinsic in the development of the fusion bomb. Thought to be the inspiration for Dr Strangelove.

Birth dates of scientists

1909: Casimir, Hendrik Brugt Gerhardt. Dutch physicist who, in a wide-ranging career, made particularly important contributions to the theory of superconductivity, and predicted the Casimir effect, a quantum force which pulls two parallel metal plates, placed a short distance apart, towards one another.

1909: Ulam, Stanislaw Marcin. Designed the first hydrogen bomb, working with Edward Teller.

1910: Chandrasekhar, Subrahmanyan. Best known for his theoretical investigations of stars at the end point of their lives, especially white dwarf stars, and the mathematical theory of black holes.

1910: Shockley, William Bradford. Shared the Nobel Prize for Physics in 1956 with John Bardeen and Walter Brattain for their invention of the transistor.

1911: Alvarez, Luis Walter. American physicist who received the Nobel Prize in 1968 for his work in high-energy particle physics, including the development of the hydrogen bubble chamber technique.

Key dates in science

1909: Karl Bohlin proposes that the Sun is not at the centre of the Milky Way.

Hans Geiger and Ernest Marsden, working under the direction of Rutherford, discover that when a beam of alpha particles is fired at a thin sheet of metal most of the particles go straight through, but occasionally one bounces back from the foil.

Jean Baptiste Perrin, through painstaking observations of Brownian motion, establishes that this process occurs exactly in the way required if the particles involved are being buffeted by atoms, providing the definitive experimental support for the theory proposed in 1905 by Albert Einstein.

1909–15: Robert Millikan is carrying out the work which will lead him to measure the charge on the electron, for which he will receive the Nobel Prize for Physics in 1923.

1910: Protons are identified by the British physicist J. J. Thomson, who also discovered the electron.

Einstein proves that the sky is blue because light is scattered by the molecules in the atmosphere.

Charles Wilson fires alpha and beta radiation through a cloud chamber and, for the first time, sees tracks of individual particles as thread-like clouds.

Theodor Wulff uses an early detector to measure the natural background radiation on top of the Eiffel Tower, and finds more than at ground level. This is the first indication of cosmic radiation.

1911: Ernest Rutherford discovers that an atom consists of a tiny central nucleus surrounded, somehow, by a cloud of electrons.

Heike Onnes discovers superconductivity.

Key dates in history

1909: Robert Peary and Matthew Henson are the first modern explorers to reach the North Pole. First electric toaster on sale. Louis Blériot flies from Calais to Dover.

1911: Roald Amundsen's party are the first people to reach the South Pole; they also return home safely. First escalators introduced, at Earl's Court station in London. Henri Matisse paints *The Red Studio*.

Birth dates of scientists

1911: Fowler, William Alfred. Received the Nobel Prize in 1983 for his work with Fred Hoyle on the theory of how elements are manufactured inside stars – stellar nucleosynthesis. After this work, he worked with Hoyle and Robert Wagoner on the theory of how light elements were made in the very early Universe – Big Bang nucleosynthesis.

1911: Wheeler, John Archibald. Became one of the leading experts on the general theory of relativity. Worked with Richard Feynman on a version of *action at a distance* that helped provide inspiration for the *transactional interpretation* of quantum mechanics. Much of his work has been dedicated to the search for a *unified field theory*.

1911: Wu, Chien-Shiung. Carried out the experiments that confirmed the violation of parity in some processes involving the weak interaction.

1913: Lamb, Willis Eugene, Jr. Received the Nobel Prize for Physics in 1955 for his precision measurements of details of the spectrum of hydrogen. These included studies of fine structure and measurements of the Lamb shift, a key observational step on the road to the development of the theory of quantum electrodynamics.

Key dates in science

1912: Victor Hess makes manned balloon flights which show that the intensity of radiation is five times greater at an altitude of 5,000 m than it is at sea level, conclusively proving the existence of cosmic radiation.

The measure of energy known as the electron volt is introduced. This is equal to the energy gained by a single electron when it is accelerated across an electric potential difference of 1 volt.

1913: Niels Bohr completes his model of the atom, the first model taking on board ideas from quantum theory.

Frederick Soddy shows that lead comes in different varieties, explaining this, and the existence of different forms of other elements, in terms of isotopes, which he names.

Johannes Stark shows that when positively charged ions,

Key dates in history

1912: *Titanic* sinks.

1913: Henry Ford uses an assembly line to speed up production of his cars. Geiger counter invented. First performance of Stravinsky's *Rite of Spring*.

Birth dates of scientists   *Key dates in science*   *Key dates in history*

known as 'canal rays', pass through a strong electric field there is a splitting of the lines in the spectrum of light from the rays – the 'Stark effect'.

1914: Davis, Ray, Jr. Devised and built the first experiment to measure the flux of neutrinos coming from the sun.

1914: Zel'dovich, Yaakov Borisovitch. One of the pioneers of the theoretical investigation of the kind of particle interactions that went on in the Big Bang and of calculations of the amount of helium they should have produced.

1914: Vesto Slipher, working at Lowell Observatory, discovers that eleven out of fifteen of the objects then known as spiral nebulae (now called galaxies) showed redshifts in their light.
  Arthur Eddington suggests that the spiral nebulae are galaxies.
  Theodore Lyman discovers the **Lyman series** of lines in the ultraviolet part of the spectrum of hydrogen.

1914–18: First World War.

1914: Brassière patented. Panama Canal opens.

1915: Hofstadter, Robert. Won a share of the Nobel Prize for Physics in 1961 for his work on the internal structure of the nucleus. Hofstadter found that the proton and the neutron have similar sizes and shapes, and developed an overall picture of the structure of the nucleus.

1915: Hoyle, Sir Fred. Explained how elements are manufactured inside stars by nucleosynthesis, but is best known for his promotion of the idea that the Universe is eternal, known as the Steady State model of the Universe.

1915: Shull, Clifford Glenwood. Received the Nobel Prize for Physics in 1994 for his contribution to the development of neutron scattering techniques for studying the structure of matter.

1915: Townes, Charles Hard. Received the Nobel Prize for Physics in 1964 for his work on masers and lasers. Also discovered the first polyatomic molecules (water and ammonia) identified in space.

1915: Albert Einstein's general theory of relativity is presented to the Prussian Academy of Science. The theory describes what happens when the combination of space and time is distorted by the presence of matter.
  Robert Millikan proves, contrary to his expectations, that Einstein's notion of light quanta is correct.
  Alfred Wegener publishes his theory of continental drift.

1915: First transcontinental telephone conversation in the USA. Marc Chagall paints *Birthday*.

Birth dates of scientists

1916: Dicke, Robert Henry. Best know for his investigations of the cosmic background radiation.

1916: Prokhorov, Alexander Mikhailovich. Shared the Nobel Prize for Physics in 1964 with Nikolai Basov and Charles Townes, for his contribution to the development of masers and lasers.

Key dates in science

1916: Solutions to the equations of the general theory of relativity that describe wormholes are found shortly after the theory is developed, although they are not interpreted in this way at the time.
   Schwarzschild sends his calculations of the exact mathematical description of the geometry of spacetime around a spherical mass to Einstein, who presents them to the Prussian Academy of Sciences.
   Heinrich Reissner becomes the first to develop the Reissner–Nordstrøm solution to Albert Einstein's equations of the general theory of relativity, which describes a black hole which is not rotating but has an electric charge. This is not likely to occur in the real Universe because such an object would attract opposite charge from its surroundings and soon become an electrically neutral Schwarzschild black hole. The solution was developed by Reissner independently of Nordstrøm, who later arrived at the same conclusions.
   Einstein presents Karl Schwarzschild's analysis of the implications of the general theory of relativity to the Prussian Academy of Sciences. Schwarzschild calculated the exact mathematical description of the geometry of spacetime around a spherical mass, showing that for any mass there is a critical radius, now called the Schwarzschild radius, which corresponds to such an extreme distortion of spacetime that, if the mass were to be squeezed inside the critical radius, space would close around the object and pinch it off from the rest of the Universe, becoming, in effect, a self-contained Universe in its own right, from which nothing (not even light) can escape. This is the simplest kind of black hole – a Schwarzschild black hole – although the metric actually

Key dates in history

1916: Frank Lloyd Wright completes work on the Imperial Hotel in Tokyo.

Birth dates of scientists	*Key dates in science*	*Key dates in history*

describes spacetime in the vicinity of any spherical concentration of mass.

Gilbert Lewis publishes the idea of the covalent bond, in which atoms in a molecule are envisaged as sharing pairs of electrons.

Einstein introduces the idea of momentum associated with the quantum of light.

Einstein, investigating *spontaneous emission*, is the first person to appreciate that an excited atom might be triggered into releasing a quantum of energy (a photon) and falling back into its ground state, by receiving a nudge from another photon with the same wavelength as the one the atom is primed to release.

1917: Bohm, David Joseph. Made major contributions to the interpretation of quantum mechanics. Worked on the Manhattan Project. Developed an alternative version of quantum theory, variously referred to as the pilot wave, undivided whole or hidden variables interpretation.

1917: Rainwater, (Leo) James. Received the Nobel Prize in 1975 for his work on the theory of the structure of the nucleus. Apart from this work, his research interests included pion scattering studies and measurements of neutron cross-sections.

1917: Applying his general theory of relativity to the world, Einstein discovers that the equations say that the Universe must be either expanding or contracting, but not stable. To stabilize the theory, Einstein introduces his 'cosmological constant' as a parameter which makes the model of the Universe, described in terms of the general theory of relativity, static. He later calls this 'cosmological constant' his greatest mistake.

Willem de Sitter also finds an 'expanding Universe' solution to Einstein's equations.

1917: Russian Revolution. Clarence Birdseye invents frozen food. John Singer Sargent paints John D. Rockefeller. John F. Kennedy born.

1918: Feynman, Richard Phillips. Nobel Prize-winning quantum physicist and great teacher. Reformulated quantum mechanics to put it on a secure logical foundation in which classical mechanics is naturally incorporated; also made major contributions to the theory of superfluidity in liquid helium, and to studies of both the strong and weak forces, and provided insights into the way to approach

1918: Gunnar Nordstrøm develops the Reissner–Nordstrøm solution to Albert Einstein's equations of the general theory of relativity, which describes a black hole which is not rotating but has an electric charge. This is not likely to occur in the real Universe because such an object would attract opposite charge from its surroundings and soon become an electrically neutral Schwarzschild black hole. The

1918: First radio link between Britain and Australia.

Birth dates of scientists

a quantum theory of gravity. Developed the path integral approach to quantum physics, from which he derived the clearest and most complete version of quantum electro-dynamics.

1918: Reines, Frederick. One of the first to detect neutrinos (strictly, antineutrinos) experimentally. These were detected in experiments carried out with Clyde Cowan, which studied a stream of radiation produced by the Savannah River reactor. For this work he received the Nobel Prize for Physics, in 1995. Reines has also developed improved techniques for detecting elementary particles using scintillation counters, and studied cosmic rays.

1918: Schwinger, Julian Seymour. Received the Nobel Prize for Physics in 1965 for his part in the development of the theory of quantum electrodynamics.

1918: Siegbahn, Kai Manne Borje. Received the Nobel Prize for Physics in 1981 for his pioneering work in electron spectroscopy.

1919: Bondi, Sir Hermann. One of the three original proponents of the Steady State model of the Universe, along with Tommy Gold and Fred Hoyle.

1919: Cowan, Clyde Lorrain, Jr. Worked with Frederick Reines on the experiment that first detected neutrinos, in 1956.

Key dates in science

solution was developed by Nordstrøm independently of Reissner.

Harlow Shapley provides the first accurate estimate of the size of the Milky Way, and correctly locates the Sun in the outer part of the Galaxy.

Lise Meitner and Otto Hahn discover the radioactive element protactinium.

1919: The bending of starlight caused by the gravity of the Sun is measured, during a solar eclipse, and is found to match Einstein's prediction rather than Newton's.

Rutherford discovers that sometimes when a fast-moving alpha particle strikes a nucleus of nitrogen, the nucleus is changed into one of oxygen, and a hydrogen nucleus is emitted. He has achieved the alchemists' dream of transmutation of the elements.

Theodor Kaluza proposes the unification of gravity and electromagnetism in one five-dimensional model, which is later going to be refined to form the

Key dates in history

Birth dates of scientists

Key dates in science

Key dates in history

Kaluza–Klein model.
   Francis Aston invents the mass spectrometer.

1920: Chamberlain, Owen. Shared the Nobel Prize with Emilio Segre in 1959 for their discovery of the antiproton.

1920: Pippard, Alfred Brian. Used microwave radiation to study the behaviour of the surface of superconductors. He introduced the idea of 'coherence', in which many electrons in a superconductor behave like a single quantum entity.

1920: Sakharov, Andrei Dimitrievich. The leading scientist involved in the development of the Soviet hydrogen bomb, Sakharov later became known as a dissident under the Soviet regime, and was awarded the Nobel Peace Prize in 1975 for his efforts to obtain a nuclear test-ban treaty. He also proposed, in the 1960s, a mechanism for the formation of matter, as opposed to antimatter, in the Big Bang.

1920: Steinberger, Jack. Shared the Nobel Prize for Physics in 1988 with Leon Lederman and Max Schwartz, for their joint work which revealed the existence of the muon neutrino.

Early **1920s:** The Compton effect, an increase in the wavelength of X-rays or gamma rays when scattered by electrons, is discovered by Arthur Compton. This effect establishes once and for all the reality of photons, proving Einstein right.

1920s: The work of Edwin Hubble, and other observers, shows not only that our Milky Way is just one galaxy among many in the Cosmos, but that galaxies are moving apart from one another as the Universe expands. This dispels the need for Einstein's cosmological constant.
   Linus Pauling makes the first full description of chemical bonds as spread-out clouds of charge, whose shape and orientation are determined by quantum probability.

1920: William Draper proposes the existence of a neutral counterpart to the proton – the neutron.

1920s: Radio broadcasting takes off.

1920–33: Prohibition in the USA.

1920: John Thompson patents the submachine gun.

1921: Mao Tse-tung involved in forming the Chinese Communist Party. First motorway in the world opens in Berlin.

1922: Basov, Nikolai Gennadiyevich. Shared the Nobel Prize for Physics in 1964 for his work on the development of masers and lasers. Suggested the idea of using semiconductors to make lasers in 1958.

1922: Bohr, Aage Niels. Received the Nobel Prize for Physics in 1975 for his work on the theory of the structure of the nucleus.

1922: Lederman, Leon Max. Investigated muon decay and

1922: Aleksander Friedmann finds a family of solutions to Einstein's equations, describing different model universes.
   Francis Aston is awarded the Nobel Prize for Chemistry, for developing a technique to measure the masses of ions.

1922: Benito Mussolini comes to power in Italy. Tomb of Tutankhamun discovered.

Birth dates of scientists	Key dates in science	Key dates in history

confirmed, in 1962, the existence of the muon neutrino. Lederman was also involved in the discovery of evidence for the bottom quark, and shared the 1988 Nobel Prize for Physics with his colleagues Melvin Schwartz and Jack Steinberger.

1922: Yang, Chen Ning. Shared the Nobel Prize for Physics with T.D. Lee, in 1957, for the prediction of the violation of parity conservation in beta decay. Also, with Lee, predicted the existence of the charged W boson and of weak neutral currents, carried by what is now known as the Z particle.

1923: Anderson, Philip Warren. Awarded the Nobel Prize in 1977 for his work on the behaviour of electrons in disordered (non-crystalline) solids. His work on semiconductors paved the way for the devices now used in computer memories.

1923: Dyson, Freeman John. Best known for his clarification of the theory of quantum electrodynamics, which showed that the theories of Richard Feynman, Julian Schwinger and Sin-itiro Tomonaga were equivalent to one another.

1923: Fitch, Val Lodgson. Shared the Nobel Prize for Physics in 1980 with James Cronin for their discovery that the decay of neutral kaons violates *CP conservation*.

1923: Charpak, Georges. Awarded the Nobel Prize in 1992 for work on developing the kind of detector known as a multiwire proportional chamber. Much of his academic life has been devoted to seeking ways to pick out very rare particle events from a mass of data.

1923: Arthur Compton's investigations of the Compton effect, the inelastic scattering of X-rays, show electromagnetic quanta behaving exactly like particles, exchanging both energy and momentum in collisions with electrons.

1923: Great earthquake in Tokyo; more than 130,000 killed.

Birth dates of scientists

1925: Esaki, Leo. Shared the Nobel Prize for Physics in 1973 with Ivar Giaever and Brian Josephson for the discovery of tunnelling in semiconductors.

1925: van der Meer, Simon. Worked as senior engineer at CERN and developed a technique known as stochastic cooling, which made it possible to produce more intense beams of antiprotons in particle accelerators, and in 1984 shared the Nobel Prize for Physics with Carlo Rubbia for this work.

1925: Ne'eman, Yuval. Developed the eightfold way idea independently of Murray Gell-Mann and played a significant part in establishing the theoretical basis for the quark model.

Key dates in science

1924: Louis de Broglie suggests that all material 'particles', such as electrons, can also be described in terms of waves. This wave–particle duality lies at the heart of quantum physics.

Eddington suggests that white dwarf stars are made of degenerate matter.

Satyendra Bose finds a way to derive Planck's equation for black body radiation using a statistical approach based entirely on the idea that light is made up of tiny particles (photons).

Statistical rules outlining the behaviour of bosons ('particles' similar in nature to the photons of light) are developed by Satyendra Bose and Albert Einstein.

Wolfgang Pauli comes up with the idea that the fourth quantum number, previously a mystery, describes the electron's 'spin', which can only point up or down, giving the required double-valued quantum number.

1925: Wolfgang Pauli formulates the exclusion principle, an expression of a law of nature which prevents any two electrons (or other members of the particle family now known as fermions) from existing in exactly the same quantum state.

George Uhlenbeck and Samuel Goudsmit suggest that silver atoms possess an intrinsic spin – a consequence of electron spin.

Robert Millikan names cosmic rays.

Pierre Auger discovers the Auger effect, a process in which an atom with an excess of energy can return to a lower energy state by emitting an electron, instead of emitting electromagnetic radiation. The emitted electron is called an Auger electron.

Heisenberg formulates what is to become known as matrix mechanics, the first complete, self-consistent theory of quantum physics. Max Born and Pascual

Key dates in history

1924: First use of insecticides.

1925: Publication of Franz Kafka's *The Trial*.

Birth dates of scientists	*Key dates in science*	*Key dates in history*
	Jordan collaborate with Heisenberg in developing it into a complete theory. Paul Dirac propounds his own version of quantum theory, known as operator theory or quantum algebra. Manne Siegbahn and his colleagues show that X-rays are affected by refraction in the same way that light is.	
1926: Glaser, Donald Arthur. Invented the bubble chamber. **1926:** Kendall, Henry Way. Received a share of the Nobel Prize in 1990 for his contribution to the experimental work which confirmed the existence of particles within neutrons and protons – the quarks. **1926:** Lee, Tsung Dao. Shared the Nobel Prize for Physics in 1957 with Chen Ning Yang for their joint theoretical work which predicted the non-conservation of parity under certain conditions. Together with Yang he was among the theorists who argued in 1960 that there was more than one kind of neutrino, and they also predicted the existence of the charged W boson and the existence of weak neutral currents, carried by what is now known as the Z particle. **1926:** Mottelson, Ben Roy. Shared the Nobel Prize for Physics in 1975 with Aage Bohr and James Rainwater, for their work on the theory of the structure of the atomic nucleus. **1926:** Nishijima, Kazuhiko. Arrived at the idea of 'strangeness' independently of Murray Gell-Mann. Also predicted that there has to be a second kind of neutrino, to accompany the muon in the same way that the electron neutrino accompanies the electron.	**1926:** The idea that probability plays a key role in the quantum world is developed by Max Born. This means that the outcome of a quantum experiment depends on chance, in the same way that the number that comes up on a roulette wheel depends on chance. The rules by which fermions behave are developed by Enrico Fermi and Paul Dirac, requiring that the total number of fermions can never be increased or decreased in isolation in a particle interaction; if a fermion is produced, it must always be accompanied by the appropriate antiparticle. So the total number of fermions in the Universe is always the same, and was determined by conditions that operated in the Big Bang. Erwin Schrödinger publishes his development of the wave mechanics formulation of quantum physics. This Hamiltonian version of quantum mechanics is to become the standard methodology, and is still used as an introduction to the subject today. Paul Dirac shows that both Heisenberg's matrix mechanics and Schrödinger's wave mechanics are special cases of his own operator theory and are therefore exactly equivalent to one another. Gilbert Lewis introduces the term 'photon' for the particle of light. Robert Goddard develops the first liquid-fuelled rocket.	**1926:** Talking pictures ushered in with *The Jazz Singer*. John Logie Baird invents a television system, based on a mechanical scanning system. Robert Goddard launches a liquid-fuelled rocket to a height of 56 m.

Birth dates of scientists

1926: Salam, Abdus. Received the Nobel Prize for Physics in 1979 for his contribution to the development of the theory of the electroweak interaction.

1927: Maiman, Theodore Harald. Developed the laser, by building on the principles of the maser, while working at the Hughes Laboratories in Miami, in 1960.

1927: Muller, (Karl) Alex. Shared the Nobel Prize for Physics with Georg Bednorz in 1987, for their discovery of high-temperature superconductivity.

Key dates in science

The five-dimensional unification of gravity and electromagnetism is extended by Oskar Klein to include quantum effects, and becomes known as Kaluza–Klein theory, after the two researchers who pioneer the work. This unification is largely ignored for many years because it requires not just one but several 'extra' dimensions to include the effects of the more complicated weak and strong interactions, which were discovered just after the initial triumph of the Kaluza–Klein theory.

Heisenberg discovers the central role of uncertainty in the quantum world, and suggests that wave–particle duality could be explained if entities such as electrons were real little particles guided to their destinations by 'pilot waves'.

1927: Walter Heitler and Fritz London use the mathematical formalism of Schrödinger's quantum-mechanical wave equation to calculate the change in energy when two hydrogen atoms, each with its own single electron, combine to form one molecule with a pair of shared electrons.

Werner Heisenberg develops the uncertainty principle, which shows that there is an intrinsic uncertainty associated with position and momentum in the quantum world. You can never know both the position and the momentum of an entity such as an electron at the same time – if you try to measure the momentum, this has the effect of enhancing the 'waviness' of the entity and making it spread out so that its position is uncertain; if you try to measure its position precisely, this creates uncertainty in its momentum.

Edwin Hubble establishes that the Milky Way is just one galaxy among many.

Niels Bohr presents the (then

Key dates in history

1927: Charles Lindbergh flies the Atlantic.

Birth dates of scientists	*Key dates in science*	*Key dates in history*
	untitled) Copenhagen interpretation of quantum physics to a conference in Tomo, Italy.	
	Paul Dirac introduces the idea of second quantization to quantum physics, pointing the way for the development of quantum field theory.	
	Experiments show that a beam of electrons passing through a thin metal foil is diffracted, and that electrons are therefore waves.	
	Georges Lemaitre publishes his version of what becomes known as the Big Bang model of the Universe.	
1928: Bell, John Stuart. Developed a practicable way to test some of the stranger predictions of quantum theory, using 'Bell's inequality'.	**1928**: The tunnel effect, which enables particles to penetrate barriers, such as the barriers around nuclei, is proposed as an explanation of alpha decay by George Gamow. The theory argues that when two protons (for example) approach one another inside a star, although they are repelled by the positive electric charge that they each carry, which stops them touching, they can nevertheless come so close that their quantum wave functions overlap, enabling them to interact. This may also be applied to alpha particles escaping from a nucleus.	**1928**: First Mickey Mouse cartoon. Penicillin discovered.
1928: Friedman, Jerome Isaac. Received a share of the Nobel Prize in 1990 for his contribution to the experimental work which confirmed the existence of particles within neutrons and protons – the quarks.		
	George Gamow invents the ***liquid drop model*** of the nucleus. This model treats the nucleus as a drop of liquid, with the individual nucleons regarded as behaving like molecules in a liquid, and the whole being held together by a kind of nuclear surface tension. At about the same time, Gamow also works with Robert Atkinson and Fritz Houtermans on calculations of the rate at which nuclear reactions take place inside stars.	
	The Raman effect is first observed by Chandrasekhara Raman. An example of this effect would be the scattering of a beam of light, of a single colour, as it passes through a transparent	

Birth dates of scientists	Key dates in science	Key dates in history

medium. This 'inelastic scattering' is caused by an interaction with the molecules within the medium, and provides information about the medium.

Paul Dirac becomes the first to create a form of quantum mechanics which also takes account of the requirements of the special theory of relativity – 'relativistic quantum mechanics'. Among other things, this development leads to the prediction of the existence of antimatter.

Eugene Wigner introduces the concept of parity.

1929: Gell-Mann, Murray. Won the Nobel Prize for Physics in 1969 for his work on the classification of fundamental particles, and was one of the people who introduced the idea of quarks, which he named.

1929: Giaever, Ivar. Shared the Nobel Prize for Physics in 1973 with Leo Esaki and Brian Josephson for his work on the tunnel effect.

1929: Higgs, Peter Ware. Developed the first field theory of particle interactions, in which the forces of nature are explained in terms of the exchange of gauge bosons. This pointed the way to the development of the electroweak theory, which utilizes the notion of the Higgs particle, an as yet hypothetical particle, to explain why the carriers of the electroweak force have mass.

1929: Mossbauer, Rudolf Ludwig. Awarded the Nobel Prize for Physics in 1961 for his discovery of the transitions of electrons between energy levels.

1929: Taylor, Richard Edward. Received a share of the Nobel Prize in 1990 for his contribution to the experimental work which confirmed the existence of quarks.

1929: Through the discovery of the redshift in light from galaxies beyond the Local Group, Hubble infers that the Universe is expanding, removing the need for Einstein's cosmological constant in its original form.

Leo Szilard solves the puzzle of Maxwell's demon. James Clerk Maxwell had proposed a thought experiment in which a miniature demon inhabited a box with two halves. Every time this demon saw a fast molecule, he would open a trap-door in the middle, and allow that molecule into one of the halves. Every time the demon saw a slow molecule, he would allow it into the other side. In this way, it seems, we encounter a theoretical violation of the second law of thermodynamics. Szilard states that this is not the case, because the energy used by the demon in identifying the different molecules is always greater than the heat gained.

Heisenberg and Pauli introduce the Lagrangian version of quantum theory.

Late **1920s**: The Copenhagen interpretation – for many years the standard interpretation of quantum mechanics – is developed by Niels Bohr and others. This interpretation will be

1929: FM radio broadcasts begin. Foam rubber put on the market.

seriously challenged only in the 1980s and 1990s.

Pascual Jordan develops the operator theory of quantum mechanics, in which the fundamental wave equation of Erwin Schrödinger's version of the theory is interpreted in terms of a mathematical expression known as an operator.

Robert van de Graaf develops the van de Graaf generator.

1930: Cooper, Leon Niels. Won a share of the Nobel Prize for Physics in 1972 for his work with John Bardeen and Robert Schrieffer on the theory of superconductivity.

1930: Everett, Hugh. Developed the many worlds interpretation of quantum mechanics while a graduate student at Princeton in the 1950s.

Early **1930s:** Frédéric and Irène Joliot-Curie bombard aluminium with alpha radiation. After the source of alpha rays is removed, the aluminium emits positrons for several minutes. Some of the aluminium nuclei have each absorbed an alpha particle, transforming into nuclei of a radioactive form of phosphorus, which decays into silicon with a half life of about 3.5 minutes. This is the first recognized production of artificial radioactivity, and shows that transmutation of elements is, in some cases, a practical possibility.

1930s: Heisenberg espouses the idea that the nucleus is made up of protons and neutrons, suggesting that nucleons could be regarded as different states of the same basic entity, distinguished by a property which Wigner called *isotopic spin*.

Albert Einstein and Nathan Rosen discover that the Schwarzschild solution actually represents a black hole as what they called a bridge (now known as an Einstein–Rosen bridge) between two regions of flat spacetime. Although this was considered a mathematical curiosity at the time, it was generally accepted by scientists that there had to be some law of nature preventing the existence of wormholes.

Einstein predicts the 'Einstein ring', a gravitational lens effect in

1930s: Worldwide population boom begins.

1930: Foundation of the Institute for Advanced Study, in Princeton. Frank Whittle patents the jet engine.

Birth dates of scientists	*Key dates in science*	*Key dates in history*

which light or other electromagnetic radiation from a distant point source (such as a quasar) is spread into a ring on the sky by the gravity of an intervening object (such as a galaxy) along the line of sight.

Nevill Mott works on the quantum theory of atomic collisions.

1930: Fritz London calculates the strength of what are now known as *van der Waals forces*.

Wolfgang Pauli proposes the existence of the particle later called the neutrino, to explain where the 'missing' energy in beta decay goes.

Paul Dirac publishes *The Principles of Quantum Mechanics*, the first systematic treatment of the subject.

Oppenheimer uses Paul Dirac's equation of the electron to show that there should be a positively charged counterpart to the electron, with the same mass as the electron.

Peter Debye probes the structure of molecules with X-rays.

1931: Cronin, James Watson. American physicist who shared the Nobel Prize for Physics in 1980 with Val Fitch for their discovery that the decay of neutral kaons violates CP conservation.

1931: Penrose, Roger. Made major contributions to the development of the theory of black holes.

1931: Richter, Burton. Received the Nobel Prize for Physics in 1976 for the discovery of the J/psi particle.

1931: Schrieffer, John Albert. Shared the Nobel Prize for Physics in 1972 with John Bardeen and Leon Cooper for their theory of superconductivity.

1931: Subrahmanyan Chandrasekhar discovers that there is no way for a white dwarf star with more than about 1.4 times the mass of the Sun to hold itself up against collapse once its nuclear fuel is exhausted. The implication is that any star left with more mass than this Chandrasekhar limit at the end of its lifetime would collapse indefinitely, forming what is now called a black hole.

Pauling publishes his classic paper on *The Nature of the Chemical Bond*, which uses quantum mechanics to explain how two electrons, one from each of two atoms, can be shared to make a bond between the two atoms.

The first cyclotron is built, by Ernest Lawrence at Berkeley.

1931: First clinical use of penicillin.

Birth dates of scientists

1932: Glashow, Sheldon Lee. Shared the Nobel Prize for Physics in 1979 with Abdus Salam and Steven Weinberg for their contributions to the development of the electroweak theory, pointing the way towards the *standard model* of particle physics. This was a unified theory of the electromagnetic and weak forces.

1932: Schwartz, Melvin. Shared the Nobel Prize for Physics in 1988 with Leon Lederman and Jack Steinberger for their joint discovery of the muon neutrino.

1933: Penzias, Arno Allan. Together with Robert Wilson, Penzias was the accidental discoverer of cosmic background radiation.

1933: Weinberg, Steven. Shared the Nobel Prize for Physics in 1979, for his contribution to the development of the theory of the electroweak interaction; one of the specific contributions made by Weinberg was the prediction of neutral currents.

1934: Bjorken, James David. Played a major part in the development of the *parton* model of nucleons, which led to the establishment of the idea of *quarks* as a fundamental building block of matter.

1934: Cramer, John. Developed the transactional interpretation of quantum mechanics in the 1980s.

Key dates in science

1932: James Chadwick discovers the neutron, a particle with no electric charge and almost the same mass as the proton. The nucleus is then explained as a collection of protons and neutrons held together by the strong nuclear interaction, or strong force.

Carl Anderson discovers the positron.

John Cockcroft and Ernest Walton use the first particle accelerator to 'split the atom'.

John von Neumann publishes his fallacious proof that no 'hidden variables' theory could every properly describe the workings of the quantum world.

Hans Bethe and Enrico Fermi suggest that interactions between charged particles could be described as being mediated by photons.

1933: Fermi takes up the suggestion, made by Wolfgang Pauli, that a 'new' particle is needed to explain details of beta decay, and Fermi is responsible for giving that particle its name – the neutrino.

Grete Hermann points out the flaw in von Neumann's proof that no 'hidden variables' theory could every properly describe the workings of the quantum world; however, her work is ignored because of the strength of von Neumann's reputation.

Otto Stern measures the magnetic moment of the proton.

1934: Louis de Broglie introduces the term 'antiparticle'.

Pavel Cerenkov discovers the radiation effect now named after him. Cerenkov radiation takes the form of bluish light and is produced by charged particles moving through a transparent medium at a speed greater than the speed of light in the medium.

Walter Baade and Fritz Zwicky suggest that 'a supernova represents the transition of an

Key dates in history

1932: Oil found in Arabia. First production of Noel Coward's *Design for Living*. Imperial Airways begin a regular service from London to Cape Town. RCA demonstrates an electronic version of television.

1933: Adolf Hitler becomes Chancellor of Germany.

1934: 'Cat's eyes' invented.

Birth dates of scientists	Key dates in science	Key dates in history

Birth dates of scientists

1934: Kerr, Roy Patrick. Has made important contributions to the understanding of black holes. He developed the Kerr solution of the equations of the general theory of relativity.

1934: Rubbia, Carlo. Received the Nobel Prize for Physics in 1984 in recognition of the efforts of a research team at CERN, which he headed, which had discovered the W and Z particles in 1983.

1935: Peebles, (Phillip) James Edwin. Has been especially interested in the physics of the Big Bang, the large-scale structure of the Universe, and the background radiation.

1936: Ting, Samuel Chao Chung. Shared the Nobel Prize for Physics in 1976 for the discovery of the J/psi particle.

1936: Wilson, Robert Woodrow. Together with Arno Penzias, Wilson was the accidental discoverer of the cosmic background radiation.

Key dates in science

ordinary star into a neutron star'.
Irène and Frédéric Joliot-Curie discover artificial radioactivity.
Leo Szilard realizes that a chain reaction might be triggered in a suitable radioactive substance; patenting this discovery, he assigns the patent to the Admiralty in Britain as a security measure.
Igor Tamm develops his theory of beta decay, which suggests that it is a result of a weak force operating between nucleons.

1935: Hideki Yukawa proposes that there must be a force which holds the nucleus together in spite of the tendency of the positive charge of the protons to blow it apart. He predicts the existence of a particle, which carries the strong force between protons, neutrons and other hadrons, binding an atomic nucleus together. This particle will in 1947 be discovered by Cecil Powell, and named the pion.
Einstein and two of his colleagues publish a paper drawing attention to one of the seemingly paradoxical features of quantum mechanics – the EPR paradox.

1936: Carl Anderson and his student Seth Neddermeyer announce the discovery of cosmic ray particles with mass larger than that of the electron, but smaller than that of the proton – these particles become known as muons.
George Gamow and Edward Teller work on an early version of the theory of beta decay.
Emilio Segre discovers a previously unknown element in the heavily irradiated parts of an old accelerator at Berkeley. The detection of this element, which Segre terms technetium, is the first identification of a previously unknown man-made chemical element.

Key dates in history

1935: Development of the beer can. Richter invents his earthquake scale.

1936: First paperback books published by Penguin. Edward VIII abdicates. Buddy Holly born.

1936–9: Spanish Civil War.

Birth dates of scientists

1937: Zweig, George. One of the two independent inventors of the concept of what are now called quarks, which he described as aces. At the time, he was criticized as a 'charlatan' for the proposal of this model, now regarded as intrinsic to modern particle physics.

1940: Josephson, Brian David. Awarded the Nobel Prize for Physics in 1973, for work that he carried out while still a student, when, in 1962, he discovered (or invented) the Josephson junction.

Key dates in science

1938: The CNO cycle, the process of nuclear fusion reactions that provides the energy source inside hot, massive stars, is worked out by Hans Bethe and, independently, Carl von Weizsäcker.

The proton–proton chain is first proposed as the source of solar energy by Hans Bethe and Charles Critchfield.

Otto Hahn, working with Strassman, splits the uranium nucleus.

Piotr Kapitza, investigating the behaviour of liquid helium, discoves superfluidity.

1939: George Volkoff and Robert Oppenheimer publish a paper showing that there is an upper mass limit (the Oppenheimer–Volkoff limit) above which no stable neutron star can exist.

Robert Oppenheimer and Hartland Snyder publish what is now regarded as the first clear description of the astrophysics of black holes.

First cyclotron built in California.

1940s: George Gamow and Ralph Alpher become the first people to attempt to describe the conditions in the Big Bang quantitatively. Applying the developing understanding of quantum physics, they investigate the kind of nuclear interactions that would have occurred at the birth of Universe, and find that primordial hydrogen would have been partly converted into helium. This discovery becomes known as the 'alpha, beta, gamma' theory.

Steady State hypothesis put forward by Hermann Bondi, Tommy Gold and Fred Hoyle; this has since been refuted by

Key dates in history

1937: Japanese invasion of China. Whittle develops a working jet engine. Hindenburg disaster.

1938: Lazlo Biro patents the ballpoint pen. Nylon commercially available in the USA. Jean-Paul Sartre writes *La Nausée*. First xerox copies made.

1939–45: Second World War.

1939: First flight of a jet airplane, the He 178, in Germany.

1940: First antibiotics. First use of freeze drying to preserve foods. John Lennon born. First electron microscope demonstrated.

Birth dates of scientists	Key dates in science	Key dates in history
	evidence that the Universe is changing as time passes and almost certainly originated in a Big Bang a finite time ago. The Dutch physicist Hendrik Casimir suggests the existence of a quantum force which pulls two parallel metal plates, placed a short distance apart, towards one another. The force is very small, but it has been measured, proving Casimir right. Richard Feynman and John Wheeler develop the Wheeler–Feynman absorber theory.	
	1940: Otto Frisch and Rudolf Peierls write a paper *On the Construction of a 'Superbomb'* which demonstrates that there is a realistic possibility of developing nuclear weapons. Emilio Segre and his colleagues discover astatine, a previously unknown element.	
1941: Schwartz, John. Pioneered the development of the theory of 'string' in its modern particle physics context.	**1941**: First use of the term 'nucleon'.	**1941**: Japanese attack Pearl Harbor.
	1942: First controlled nuclear chain reaction in a uranium 'pile' at the University of Chicago.	**1942**: Nuclear energy released in the first controlled nuclear chain reaction.
1943: Bell Burnell, (Susan) Jocelyn. Discovered pulsars while still a student, in 1967. **1943**: Klitzing, Klaus von. Awarded the Nobel Prize for Physics in 1985 for his work on the quantum Hall effect. This deals with the behaviour of electrons moving in a very thin layer of material – for example, at the semiconductor surface of a so-called field effect transistor. This has important applications in electronics.	Feynman presents his thesis, in which he propounds for the first time the notion of path integrals, a way of adding up quantum probabilities which is based upon the use of Lagrangian functions, instead of the Hamiltonian approach previously used in quantum mechanics. This theory of quantum mechanics is more complete and satisfying than the earlier model of Schrödinger. Just postwar: Sin-itiro Tomonaga completes his theory of QED.	**1943**: First electronic computer. First practical nuclear reactor becomes operational at Oak Ridge, Tennessee. Los Alamos laboratory established. **1945**: Atomic bomb.
1946: 't Hooft, Gerard. Dutch physicist who found a way to renormalize the electroweak theory of Abdus Salan and Steven Weinberg, removing the infinities which plagued it and thereby making it respectable.	Mid to late **1940s**: QED, quantum electrodynamics, is developed independently by three great physicists, Julian Schwinger, Sin-itiro Tomonaga and Richard Feynman.	**1946**: First meeting of the United Nations. Carbon dating technique developed by Willard Libby. First general-purpose computer, ENIAC.

Birth dates of scientists	Key dates in science	Key dates in history

Towards the end of the **1940s**: Marya Goeppert-Mayer and Hans Jensen develop, independently, the idea that the structure of the nucleus could be described in terms of a series of shells.

Late **1940s**: Strange particles are identified for the first time, in cosmic ray showers.

1946: First synchrocyclotron.

1947: Binnig, Gerd Karl. German physicist who shared the Nobel Prize in Physics in 1986 for the development of the scanning tunnelling microscope.

1947: The Lamb shift, a small difference in the energy levels of two possible quantum states of the hydrogen atom, revealed by the splitting of a line in the spectrum of hydrogen into two components, is measured for the first time by Willis Lamb. Before the end of the 1940s it will be explained in the context of quantum electrodynamics, as a result of the way charged particles such as electrons interact with electromagnetic fields.

1947: First supersonic flight. India freed from British rule.

1947: Guth, Alan Harvey. American physicist who came up with the idea of inflation and gave the theory its name.

    First practical linear accelerator accelerates protons to energies of 32 MeV.
    Research into cosmic rays discovers the pion (pi-meson) particle, the existence of which had been predicted by Hideki Yukawa in 1935.
    Hendrik Kramers completes his work on renormalization.

1948: Linde, Andrei Dmitrivitch. Played a major role in the development of the theory of the very early Universe known as inflation; also made important contributions to the understanding of fermions in the context of gauge theories.

1948: Ralph Alpher and Robert Herman extend the alpha beta gamma theory to predict that the Universe today must be filled with background radiation at a temperature of about 5 K.
    The first atomic clock is developed by the US National Bureau of Standards, based on measurements of the vibrations of atoms of nitrogen oscillating back and forth in ammonia molecules, at the rate of 23,870 vibrations per second.
    Julian Schwinger completes his work on quantum electrodynamics, or QED.

1948: George Orwell writes *1984*.

Birth dates of scientists	Key dates in science	Key dates in history
	Richard Feynman publishes his theory of path integrals, developed for his thesis in 1942 but not formally published for six years because of his involvement with the Manhattan Project. Feynman also demonstrates that renormalization works in a fully relativistic version of QED (quantum electrodynamics). Without renormalization, QED would not work; however, Feynman himself described this as 'a dippy process'.	
	1949: Freeman Dyson shows that the theories of quantum electrodynamics propounded independently by Feynman and Schwinger (and Tomonaga, whose approach is similar to Schwinger's) are mathematically equivalent to one another.	**1949**: First LP records. First flight of the Comet, the world's first commercial passenger-carrying jet aircraft.
1950: Bednorz, George. Received the Nobel Prize, jointly with Alex Muller, in 1987 for his work on high-temperature superconductivity.	Early **1950s**: The bubble chamber is invented by Donald Glaser. Hugh Everett, puzzled by the Copenhagen interpretation, rejects the notion of the collapse of the wave function and argues that we should treat each outcome of every possible quantum event as existing in a real world – the many worlds theory, in which ours is just one in an infinity of alternative realities which exist.	**1950s**: Television broadcasting takes off. **1950**: Korean War begins. First charge card introduced by Diner's Club.
	1950s: Fred Hoyle realizes that three alpha particles can get together to make carbon-12 if the carbon nucleus itself has an excited energy state corresponding to the combined energy of three alpha particles. David Bohm points out that it is possible to explain the photoelectric effect in terms of light as a continuous wave which carries energy that can be accepted by the atoms in a metal surface only in definite lumps.	
	1951: Willis Lamb moves to Stanford University, where he is to study the hyperfine structure of the spectrum of helium.	**1951**: First commercially available electronic computer sold by Remington Rand.

Birth dates of scientists	Key dates in science	Key dates in history
	Charles Townes invents the maser, a device which produces an intense beam of microwaves from the stimulated emission of radiation by excited atoms.	
	The fusion bomb is developed, with the aid of Edward Teller, using the designs of Stanislaw Ulam.	
	1952: Walter Baade's revision of the Cepheid distance scale doubles the size and age of the known Universe.	**1952**: Invention of the transistor radio. Mount Everest climbed.
	The xi-minus particle is discovered.	
	David Bohm suggests, as a thought experiment, a variation on the EPR theme, involving the behaviour of photons.	
	A particle accelerator known as the Cosmotron begins working at the Brookhaven National Laboratory.	
	1953–6: The family of three particles known as 'sigma', one positively charged, one negative and one neutral, are discovered. These are early examples of strange particles.	**1953**: Structure of DNA deciphered; hydrogen bomb. Invention of the maser. Edmund Hillary and Norgay Tensing climb Mount Everest. Francis Crick and James Watson publish their discovery of the structure of DNA.
	1953: Iosif Samuilovich Shklovskii becomes the first to suggest that radio waves and X-rays from the Crab nebula are produced by synchrotron radiation. Synchrotron emission occurs when electrons, freed from their atoms, move in spirals in a strong magnetic field. This often produces radio emission, but there is so much energy available in the Crab that the electrons also radiate visible light.	
	Charles Townes constructs the first maser, at Columbia University.	
	1954: The Bevatron, a particle accelerator which can accelerate protons to an energy of 6 GeV, begins operating at the Lawrence Berkeley Laboratory in California.	**1954**: TV dinners go on sale in the USA. First sub-four-minute mile. Maiden flight of Boeing 707.
	CERN, a European particle	

Birth dates of scientists	*Key dates in science*	*Key dates in history*

physics research centre which will become home to the Large Electron–Positron Collider (LEP), is founded.

UKAEA, United Kingdom Atomic Energy Authority, is set up. This authority will become responsible for research into the development of nuclear power in the UK.

First use of the term 'baryon'.

1955: Smithsonian Institution moves from Washington to Harvard.

Niels Bohr organizes the first Atoms for Peace conference in Geneva.

Owen Chamberlain and Emilio Segre discover the antiproton, using the Bevatron particle accelerator.

Mid-**1950s**: The assumption that parity must be conserved in all particle interactions is shown to be flawed by some puzzling features of the decay of particles then known as theta and tau.

1956: Neutrinos are proven to exist, in experiments carried out by Clyde Cowan and Frederick Reines, which monitor the flood of neutrinos emitted by a nuclear reactor at Savannah River in the United States.

Richard Feynman makes the proposal that the theta and tau are different states of the same particle, which itself has no definite parity, and that parity is not always conserved; this theory will be taken up by Chen Ning Yang and Tsung Dao Lee, and proved by Chien-Shiung Wu, in the same year.

1956: Hungarian uprising crushed by Soviet Union. First transatlantic telephone cable. Suez crisis. Development of video tape recorder.

1957: Geoffrey and Margaret Burbidge, Willy Fowler and Fred Hoyle publish 'B^{2}FH' paper describing how all the naturally occurring varieties of nuclei except primordial hydrogen and helium are built up inside stars by nucleosynthesis.

1957: First artificial Earth satellite, Sputnik 1.

Birth dates of scientists	Key dates in science	Key dates in history
	The first artificial satellite, Sputnik 1, is launched from the Soviet Union on 4 October.	
	Robert Dicke publishes a paper pointing out that the size of the Universe is 'not random but conditioned by biological factors'.	
	Hugh Everett publishes the many worlds interpretation of quantum mechanics.	
	Kazuhiko Nishijima predicts that there must be a second kind of neutrino, to accompany the muon in the same way that the electron neutrino accompanies the electron.	
	Julian Schwinger also predicts that there must be two different kinds of neutrino, associated with the electron and the muon respectively – the tau particle has yet to be discovered.	
	John Bardeen, Leon Cooper and Robert Schrieffer explain superconductivity as a quantum phenomenon. This becomes known as the BCS theory.	
	1958: The International Atomic Time Scale, based on atomic clock data and the standard time signal used today, commences on 1 January at 0 min 0 sec GMT.	**1958**: Fidel Castro leads revolution in Cuba. Microchip invented in USA.
	NASA is created to handle all non-military aspects of the US space programme.	
	Nikolai Gennadiyevich Basov suggests using semiconductors to make lasers.	
	Richard Feynman and Murray Gell-Mann independently develop a new theory of the weak interaction, which they publish together.	
	1959: Freeman Dyson suggests that an intelligent species with an expanding population, confined to a single planetary system like our own Solar System, will eventually rearrange the raw materials provided by those planets to build a hollow sphere around the parent star.	**1959**: First commercial photocopier. First successful hovercraft crosses the English Channel.

Birth dates of scientists	Key dates in science	Key dates in history
	Lunik III obtains the first views of the far side of the Moon.	
	Richard Feynman suggests nanotechnology as a real possibility, in a talk at Caltech.	
	The xi-zero particle is discovered.	
	Early **1960s**: Leon Lederman, Melvin Schwartz and Jack Steinberger discover the muon neutrino, the type of neutrino associated with the muon.	**1960**: First lasers. First weather satellite.
	1960s: Frank Kiernan develops the wire chamber, an improved version of the spark chamber, at CERN.	
	1960: T. D. Lee, Chen Ning Yang and Max Lederman develop the idea that there is more than one kind of neutrino.	
	Theodore Maiman develops the laser.	
	1961: Murray Gell-Mann and Yuval Ne'eman propose, independently, the classification scheme for elementary particles which is to become known as the eightfold way. Ne'eman predicts the existence of the omega-minus particle.	**1961**: First manned spaceflight, by Yuri Gagarin. Nature of the genetic code in DNA discovered.
	The Weinberg angle, a mixing angle used specifically in the context of the electroweak interaction, is first proposed by Sheldon Glashow.	
	About **1962**: Richard Feynman develops the concept of quantum gravity, using the example of quantum electrodynamics as a template. The major idea that he introduces to the concept of gravity is the perturbation technique. He shows that the entire classical theory of gravity, including the equations of the general theory of relativity, could be derived from a quantum description of interactions between particles that have mass, involving the exchange of gravitons, which are	**1962**: Cuban missile crisis. Telstar is launched and relays the first live TV pictures between the USA and Europe.

531

massless particles that each have two units of quantum spin.

1962: Brian Josephson discovers (or invents) the Josephson junction, the work for which he will receive the Nobel Prize. This shows that with superconductors, quantum effects allow a (low) supercurrent to flow through a thin layer of insulating material, encountering no resistance at all. The behaviour of Josephson junctions is very sensitive to the magnetic fields or voltages applied to them, so they can be used as very effective measuring devices.

Feynman makes the single loop correction, or perturbation calculation, for quantum gravity.

Second neutrino (the mu-neutrino) is discovered.

1963: Roy Kerr discovers the Kerr solution, the mathematical description of a rotating, uncharged black hole, which is the most likely to exist in our Universe.

Hydroxyl becomes the first interstellar molecule to be detected.

Chien-Shiung Wu and her colleagues carry out experiments confirming the accuracy of Feynman and Gell-Mann's theory of the weak interaction.

1963: John F. Kennedy assassinated.

1964: The Davis detector, designed to detect solar neutrinos, is installed 1,500 m below the ground, in the Homestake Gold Mine at Lead, South Dakota. Here, Ray Davis and colleagues will monitor the flux of solar neutrinos from the Sun. The results of this experiment will conclude that there are only about one-third of the expected number of solar neutrinos.

Peter Higgs, and the partnership of Robert Brout and François Englert, independently come up with the notion of a

1964: Tonkin incident leads to the start of the US–Vietnam War.

Birth dates of scientists	*Key dates in science*	*Key dates in history*

previously unknown kind of boson, the Higgs particle, which has mass. The existence of such a boson would allow any photon-like particle to obtain mass by swallowing up a Higgs boson, and might serve to explain why the carriers of the electroweak force have mass.

John Bell shows how Bohm's variation on the EPR theme, suggested in 1952, might, in principle, form the basis of a real experiment.

It is discovered that the decay of a particle known as the neutral kaon (K), to produce particles known as pions, does not conserve CP.

The omega-minus particle predicted by Yuval Ne'eman is discovered.

The idea of quarks is introduced, in a paper by Gell-Mann, and independently by George Zweig (as 'aces'), although at this time only three different kinds of quark are needed to explain all known hadrons. The strange quark is therefore seemingly without a partner. Walter Greenberg quickly realizes that what he calls 'parastatistics', a hitherto pure mathematics concept, can be applied to the problem of distinguishing the members of triplets of otherwise identical quarks from one another.

1965: Roger Penrose shows that according to the general theory of relativity any object which contracts within its event horizon must collapse all the way to a singularity, a point of infinite density and zero volume where the laws of physics break down and literally anything can happen.

The Kerr–Newman solution, a solution to Einstein's equations of the general theory of relativity that describes a rotating, electrically charged black hole, is developed by Ezra Newman and

1965: First communications satellite.

colleagues, starting from the Kerr solution.

Arno Penzias and Robert Wilson, at the Bell Research Laboratories, accidentally find a persistent source of interference that proves to be cosmic background radiation. This is accepted as proof that the Big Bang really happened.

Yoichiro Nambu and M. Y. Han collaborate in developing a version of Greenberg's highly mathematical theory of quarks which can be understood by physicists. The idea of labelling the property that distinguishes otherwise identical quarks from one another with the names of colours is introduced.

Late **1960s**: Georges Charpak, at CERN, devises the multiwire proportional chamber, a type of particle detector which is to become an integral component of most particle physics experiments.

Neuron stars are discovered (accidentally), in the form of pulsars.

Scattering experiments are carried out at SLAC, in which electrons are fired into protons at high energies; these scatter in a way reminiscent of the scattering of alpha particles from gold atoms, and suggest the existence of smaller particles.

Richard Feynman introduces the idea of partons, specifically to avoid prejudging the question of whether quarks exist.

1966: John Wheeler and Franco Pacini speculate that the source of energy emitted by the Crab nebula might be a spinning neutron star.

Fred Hoyle, Bob Wagoner and Willy Fowler develop the understanding of Big Bang nucleosynthesis.

John Bell establishes that the von Neumann proof that no 'hidden variables' theory could

1966–8: Cultural Revolution in China.

ever properly describe the workings of the quantum world is fatally flawed. This opens the way for further investigation into forms of the 'hidden variables' theory, which by the 1990s will become one of the most exciting areas of the development of models of the quantum world.

SLAC begins operating.

1967: The first pulsars are discovered by Jocelyn Bell, a radio astronomer working in Cambridge under the supervision of Antony Hewish.

The Davis detector, developed through the efforts of theorist John Bahcall and experimenter Ray Davis, and making use of chlorine (in the form of perchlorethylene) to monitor the arrival of neutrinos from the Sun, commences operations. This detector finds only about one-third of the expected number of solar neutrinos, leading to what is now known as the 'solar neutrino problem'.

The term 'black hole' is used for the first time in its modern astronomical context, by John Wheeler.

Abdus Salam and Steven Weinberg independently make the discovery of a way to describe the weak interaction and electromagnetic interaction between bosons in one mathematical formalism, as a single force, the electroweak interaction.

Andrei Sakharov publishes a paper showing that there is a tiny asymmetry in the laws of physics. This means that under the conditions operating in the Big Bang roughly a billion and one baryons were produced to every billion baryons, and the Universe is composed of those particles that were not annihilated – the rest being transformed into cosmic microwave background radiation.

Key dates in science

Key dates in history

1968: Soviet troops crush the 'Prague spring' in Czechoslovakia.

1969: First manned Moon landing.

1969: First humans on the Moon.

End of **1960s** and beginning of **1970s**: John Friedman, Henry Kendall and Richard Taylor, working at SLAC on experiments which probe the structure of nucleons using beams of high-energy electrons, show (together with the theoretical interpretation of these experiments provided by James Bjorken and Richard Feynman) that the structure within the nucleons is caused by the presence of point-like particles, which further experiments showed could be identified as quarks.

1970: First Boeing 747s introduced on transatlantic routes. Floppy disks used for storing computer data.

Early **1970s**: John Clauser, Michael Horne, Abner Shimony and Richard Holt create the first good test of Bell's theorem. Their experiment is motivated by Clauser's desire to prove that the world is local – that there is no 'spooky action at a distance'. In fact they find that Bell's inequality, and common sense, is violated.

Sheldon Glashow and two colleagues at Harvard, John Iliopoulos and Luciano Maiani, revive the idea of a fourth quark, charm. This is taken up by Murray Gell-Mann and Harald Fritzsch, who develop a field theory approach to describe the various interactions of these particles.

1970s: The drift chamber, the electronic equivalent of a bubble chamber, is developed by Georges Charpak at CERN.

The jet effect, that of a shower of particles moving in the same direction, generated from a quark, antiquark or gluon that has been either knocked out of its 'parent' particle or created out of energy in a particle accelerator experiment, is observed for the first time at SPEAR, in electron–positron collisions.

Eric Drexler, at MIT, reinvents the idea of nanotechnology first suggested by Feynman, and makes it a practical possibility.

1970: Sheldon Glashow extends and generalizes the notion of electroweak interaction of bosons propounded by Abdus Salam and Steven Weinberg in 1967.

Charge coupled devices (CCDs) are invented at the Bell Laboratories in the United States.

John Schwarz and André Neveu find a way to describe fermions using string theory.

1971: Gerard 't Hooft shows that it is possible – though not proven – that all massive particles get their mass from taking on Higgs particles.

1971: Direct dialling introduced for telephone calls between Europe and the USA. First pocket electronic calculator put on sale by Texas Instruments.

1972: One of the most important pioneering gamma ray observatories, SAS II, is launched by NASA.

Fermilab, the US National Accelerator Laboratory at Batavia, near Chicago, opens. This is the machine that will find the evidence for the top quark, in the mid-1990s.

International Atomic Time, the standard of international timekeeping, is established by the Bureau Internationale de l'Heure in Paris. Time is measured by 80 atomic clocks in 24 countries, which each send their information to Paris where it is co-ordinated to 1 millisecond.

1973: First Skylab missions.

Neutral current, a weak interaction in which no charge is carried by the boson that mediates the interaction, is observed for the first time. These reactions are mediated by the Z particle, and provide important confirmation of the accuracy of

1973: Genetic engineering techniques developed; first oil crisis increases cost of energy.

Birth dates of scientists	Key dates in science	Key dates in history

the gauge theory of the weak interaction.

1974: Frank Tipler suggests (in a paper published in *Physical Review D*) that it might be possible to build a time machine. The trick involves making a naked singularity that rotates extremely rapidly. The effect of the rotation is to twist spacetime, tipping it over so that one of the space dimensions is replaced by the time dimension, enabling a spaceship to make a journey through an area which seems to be ordinary space, but is in fact time.

1974: Richard Nixon resigns in the wake of the Watergate scandal.

Evidence for the existence of the J/psi particle – a massive meson, with a mass of 3,097 MeV (more than three times as heavy as a proton), composed of a charmed quark and an anticharmed quark, and therefore possessed of no charm at all – is found by two groups in the United States, working independently of each other. This discovery, made in November, is jokingly called the 'November revolution'.

John Schwarz and Joel Scherk discover the link between string theory and gravity. The mathematics of string theory provide a description of a particle with zero mass and a spin of 2. This is precisely the description of the graviton, the messenger particle of the gravitational interaction required by any theory of quantum gravity. The first real hint that string theory may provide the 'theory of everything'.

1975: The four-quark theory first becomes known as the 'standard model'.

1975: Last US troops leave Vietnam. First personal computer available in the USA; it has a memory of 256 bytes (a quarter of a kilobyte). E. O. Wilson publishes *Sociobiology*.

Birth dates of scientists	Key dates in science	Key dates in history

1976: Concorde enters service. Unmanned spacecraft land on Mars.

1977: Two spaceprobes known as Voyager are launched by NASA, to send back pictures and other data from the outer planets of the Solar System.
Discovery of the bottom quark.

1977: Apple II personal computer launched.

1978: Motohiko Yoshimura independently develops a model of the Big Bang similar to that developed by Andrei Sakharov in 1967.

1979: Despite having yet to reach full power, PETRA, the Positron–Electron Tandem Ring Accelerator, produces evidence for *gluons*, in the form of *three-jet events*.

End of **1970s**: Tony Klein and colleagues at the University of Melbourne carry out a series of experiments which clearly demonstrate neutron diffraction, confirming the dual wave–particle nature of the neutron.

1980s: Relativists attempting to prove that there is a law of nature preventing the existence of wormholes find they cannot. There is nothing in the general theory of relativity that forbids the existence of wormholes.
John Cramer uses a similar formalism to that found in the Wheeler–Feynman absorber theory as the basis for a new interpretation of quantum theory, the transactional interpretation.
The Einstein ring, a gravitational lens effect predicted by Albert Einstein in the 1930s, is observed for the first time.
The Kaluza–Klein theory is revived, as investigations into the possibility of a grand unified theory increasingly suggest the existence of many extra

1980s: Development of smaller, faster, cheaper computers.

1980: Development of optical fibre communications links.

dimensions, and string theory is increasingly considered seriously. String theory only 'works' in many dimensions, but it produces an enormous bonus – gravity.

Inflation becomes established as the standard model of the very early Universe.

Danny Hillis and colleagues at Thinking Machines Corporation in Cambridge, Massachusetts, develop parallel processing.

Quantum cosmology, the application of quantum physics (including the ideas of the grand unified theories) to describe the very early stages in the life of the Universe and, in some models, the origin of the Universe out of nothing at all, brings within the bounds of respectable physics questions and ideas which previously seemed to belong in the realms of philosophy, metaphysics and religion.

1980: First suggestions that neutrinos may have mass.

PETRA, the Positron–Electron Tandem Ring Accelerator, is completed and reaches its design energy of 19 GeV per beam.

PEP, the Positron–Electron Project, a collider at SLAC designed to reach energies of 36 GeV, is completed; together with PETRA, PEP will provide direct evidence for gluons.

1981: Underground Area 1 (UA1), at CERN, begins monitoring proton–antiproton collisions.

1981: Introduction of the IBM PC.

1982: The first known millisecond pulsar, PSR 1937+211, is identified by astronomers from Princeton University using the Arecibo radio telescope. This pulsar has a period of just 1.6 milliseconds, which means that a star roughly as massive as our Sun is squeezed into a ball of material with the density of an atomic

1982: Space Shuttle begins operations. CD players available.

Birth dates of scientists	Key dates in science	Key dates in history

nucleus, spinning more than 600 times every second.

The results of the Aspect experiment are published, establishing that non-locality rules the quantum world. It is possible to stimulate an atom in such a way that it produces two photons simultaneously, heading off in different directions. Common sense demands each photon starts with a definite, opposing polarization; however, within the Copenhagen interpretation of quantum theory, any quantum entity with a choice of possibilities exists in a state of uncertainty until measured. Only then is there a 'collapse of the wave function', as it settles into one possibility. The counterpart to the photon observed must also have existed in a state of uncertainty, and must also collapse, into the opposing state, when its counterpart is observed. Therefore we encounter what Einstein rejected as 'spooky action at a distance'.

Richard Feynman points out the possibility of building a computer that operates on quantum principles. He further conjectures that a quantum computer could act as a 'universal quantum simulator' which would simulate the behaviour of any quantum system, following the rules of the quantum world rather than the logic of everyday life.

1983: The metre is defined as the distance travelled by light in 1/299,792,458 sec.

Underground Areas 1 and 2 (UA1 and UA2), at CERN, discover the W and Z particles.

1984: UA1 finds evidence for the top quark.

1984: Apple Macintosh introduced.

1985: Kip Thorne realizes that it would be possible to hold a

1985: Mexico City devastated by earthquake.

Birth dates of scientists	*Key dates in science*	*Key dates in history*

Key dates in science

wormhole open using so-called exotic matter. This renders time travel (theoretically) possible.

Mid-**1980s**: Two independent enactments of the delayed choice experiment, one at the University of Maryland and the other at the University of Munich, suggest that photons have some kind of precognition.

Late **1980s**: Research shows that genuine time travel is not forbidden by the laws of physics, as at present understood.

1986: Andrei Linde suggests that the Universe might be part of a self-reproducing system of baby universes.
   Physicists are surprised by the discovery of superconductivity in some materials at relatively high temperatures, above 100 K. This is significant because such temperatures can be reached relatively easily, with the aid of liquid nitrogen. This raises the prospect of practical applications – for example, in computing.

1987: Supernova 1987A becomes the first supernova visible from Earth to the naked eye since the one Kepler observed in 1604. This supernova is the explosion of a star about 160,000 light years away from us, in the Large Magellanic Cloud; the precursor star was later identified on old photographic plates as a supergiant known as Sanduleak – 69° 202.
   First direct evidence of planet-sized objects orbiting other stars.
   TRISTAN, an electron–positron storage ring facility at Tsukuba, becomes operational.

1988: Morris, Thorne and Yurtsever publish their conclusions about the possibility of wormholes linking different

Key dates in history

1986: Chernobyl nuclear disaster. Spaceprobes visit Halley's Comet.

Birth dates of scientists	*Key dates in science*	*Key dates in history*

times, in the journal *Physical Review Letters*.

1989: LEP, the Large Electron–Positron collider, a circular particle accelerator 27 km in diameter which can reach energies of more than 100 GeV, starts working at CERN.

End of 1980s: Japanese researchers carry out a version of the double-slit experiment in which an electron gun fires electrons off one at a time, on to the electron detector. This experiment proves the wave–particle duality of electrons.

Dipankar Home, Partha Ghose and Girish Agarwal present a new suggestion for an experiment which can show single photons behaving as both particles and waves at the same time.

1990s: An experiment at Yale University upon atoms in a vacuum, utilizing the Casimir effect, observes that the behaviour of the atoms exactly matches the predictions of quantum theory, and does not match the predictions of classical theory. A 'pure' vacuum is seen to have a measurable effect upon individual sodium atoms.

1990: The suggestion is made that gravity and the other forces of nature arise naturally from the behaviour of 'string' with dimensions comparable to the Planck length. String theory automatically produces massless particles with two units of quantum spin – gravitons.

1991: JET, Joint European Torus, an experimental reactor used for research into nuclear fusion, generates 1.7 megawatts of power in a pulse lasting 2 seconds – the first sustained burst of power on a large scale from a

1991: Collapse of the Soviet Union.

controlled fusion reaction on Earth.

1992: The satellite COBE (COsmic Background Explorer) finds ripples in the background radiation that are exactly the right size to conform with the standard Big Bang model. This is considered the ultimate triumph of the Big Bang theory.

The sentence finding Galileo guilty of heresy for supporting the Copernican idea that the world revolves around the Sun is formally revoked by Pope John Paul II.

1993: Flaws with the Hubble Space Telescope are largely corrected by astronauts.

1994: Physicists attempting to recreate the conditions that existed in the quark–gluon plasma by colliding massive nuclei together in particle accelerators on Earth produce 'little bangs' which unambiguously confirm that the photons produced in these events are coming from a quark–gluon plasma, confirming the theory of the Big Bang fireball.

The last of the six members of the quark family, known as the top quark, is identified by researchers at Fermilab, in Chicago.

Peter Shor, curious to find whether it is possible, in principle, to use a quantum parallel computer to crack certain kinds of near-uncrackable code, shows that although such codes require (when enormously powerful conventional computers are used) a period of several months, or even years, to crack, a quantum computer no more powerful than a mid-1990s PC could solve such problems in a few seconds.

Birth dates of scientists	Key dates in science	Key dates in history

1995: Researchers at Los Alamos claim to have measured the mass of the electron neutrino, about 5 electron volts.

Researchers at CERN succeed for the first time in making complete antiatoms (of hydrogen) in which a negatively charged antiproton is associated with a positively charged antielectron.

Calculations suggest that tiny wormholes might by topologically equivalent to the string invoked by superstring theories to explain the structure of matter on the smallest scales. If so, black holes may be the missing link required to complete the sought-for theory of everything.

Construction begins on the LHC, or Large Hadron Collider, a particle accelerator being built in the tunnel of the LEP machine at CERN. The machine should begin operating, at energies as high as 14 TeV, in 2003.

1996: An experiment at Los Alamos that has been running for three years has produced evidence for 22 events caused by electron neutrinos in a beam of supposedly pure muon neutrinos, which suggests that muon neutrinos can oscillate to become electron neutrinos, and that one of the two has a mass of a few tenths of an electron volt.

Seth Lloyd, of MIT, proves that Richard Feynman's conjecture of a quantum computer as a 'universal quantum simulator' is correct, and that 'a variety of quantum systems, including quantum computers, can be "programmed" to simulate the behaviour of arbitrary quantum systems whose dynamics are determined by local interactions' (*Science*, vol. 273, p. 1073, 1996).